Biotechnology for Pulp and Paper Processing

Pratima Bajpai

Biotechnology for Pulp and Paper Processing

Springer

Pratima Bajpai
Thapar Research and Development Center Colony
Patiala, India
pratima.rcdp@gmail.com

ISBN 978-1-4899-9147-8 ISBN 978-1-4614-1409-4 (eBook)
DOI 10.1007/978-1-4614-1409-4
Springer New York Dordrecht Heidelberg London

Printed on acid-free paper

Springer is part of Springer Science+Business Media (www.springer.com)

Preface

The pulp and paper (P&P) industry is traditionally known to be a large contributor to environmental pollution due to its large consumption of energy and chemicals. Biotechnological methods, however, offer potential opportunities for changing the industry toward more environmentally friendly and efficient operations compared to the conventional methods. The importance of biotechnology lies in its potential for more specific reactions, less environmentally deleterious processes, energy savings, and capacity to be used in place of nonbiological processes. Increased pulp yield, improved fiber properties, enhanced paper recycling, reduced processing and environmental problems, and energy efficiency are all consequences of biotechnological processes in the pulp and paper industry. The number of possible applications of biotechnology in pulp and paper manufacture has grown steadily during the past 3 decades. Many applications have approached or are approaching commercial reality. Applications that have been successfully transferred to commercial use include xylanases for bleach boosting; cellulases for improved drainage; lipases for pitch removal; cellulase–hemicellulase mixture for deinking and fiber modification; esterases for stickies control; and levan hydrolase, proteases, cellulases, amylases, etc. for slime removal. "Biotechnology for Pulp and Paper Processing" gives updated information on various biotechnological processes useful in the pulp and paper industry; these processes could help in reducing environmental pollution problems, in addition to other benefits. Various chapters deal with latest developments in the areas like Tree improvement, Raw material preparation, Pulping, Bleaching, Deinking, Fiber modification, Slime control, Stickies control, Production of dissolving grade pulp, Shive removal, Vessel picking, Degradation of pollutants, Retting of flax, Treatment of exhaust gasses for removal of odorous emissions, and Biosolids management. *Biotechnology for Pulp and Paper Processing* also includes a chapter on Forest Products Biorefinery. Biorefineries actually can help pulp mills use by-products and residual products of the papermaking process to create additional high-value revenue streams. The major benefits, limitations, and future prospects of these processes have also been discussed.

Patiala, India　　　　　　　　　　　　　　　　　　　　Pratima Bajpai

Contents

List of Figures

List of Tables

Chapter 1
Introduction

1.1 Introduction

The global pulp and paper industry is in physical terms one of the largest industries in the world. It is dominated by North American (United States and Canada), northern European (Finland, Sweden), and East Asian countries such as Japan. Australasia and Latin America also have significant pulp and paper industries. Both India and China are expected to be key in the industry's growth over the next few years.

World production of paper and paper board totals some 380 million tons. Growth is most rapid in Asia, thanks mainly to the quick expansion of industry in China. Asia already accounts for well over a third of total world paper and paperboard production. In North America, by contrast, production is contracting. Consumption of paper and paperboard is increasing ever more rapidly in Asia, in China especially. Asia already accounts for almost 40% of global consumption, while EU and North America account for about one quarter each. Per capita consumption of paper and paperboard varies significantly from country to country and regionally. On average, one person uses about 60 kg of paper a year; the extremes are 300 kg for each US resident and some 7 kg for each African. Only around 35 kg of paper per person is consumed in the populous area of Asia. This means that Asian consumption will continue to grow strongly in the coming years if developments there follow the precedent of the West. In Finland, per capita consumption of paper and paperboard is about 200 kg.

The pulp and paper industry plays an important role in the country's economic growth. It is highly capital intensive and has been periodically affected by overcapacity. It is traditionally known to be a large contributor to environmental pollution due to its large consumptions of energy and chemicals.

This is a difficult time for the pulp and paper industry. Consumer standards are high, and manufacturing is competitive. Cost reduction pressures are causing consolidation of companies through mergers and acquisitions; many research and development laboratories are being downsized, closed, or directed toward short-term objectives and opportunities; and profitability is being constrained by external

P. Bajpai, *Biotechnology for Pulp and Paper Processing*,
DOI 10.1007/978-1-4614-1409-4_1,

factors including globalization, environmental concerns, and competition. There is a need to find new ways to use forest resources more efficiently and with fewer environmental consequences. Emerging technologies based on sustainable use of renewable resources hold promise for the rejuvenation and growth of the pulp and paper industry.

Biotechnology has the potential to increase the quality and supply of feedstocks for pulp and paper, reduce manufacturing costs, and create novel high-value products (Anon 2004, 2005; Mansfield and Esteghlalian 2003; Ojanpera 2004; Viikari et al. 2006, 2009). Biotechnology is defined as the use of biological organisms/systems and processes for practical or commercial purposes. In this broad sense, biotechnology encompasses a diverse array of activities including fermentation, immobilized cell and enzyme technology, cell and tissue culture and monoclonal antibody techniques, although in recent years, the term has been increasingly identified with techniques for genetic transfer and DNA manipulation i.e., genetic engineering. The attractiveness of biotechnology lies in its potential to provide processes/products where nonbiological processes are impractical, to increase specificity in reactions, to provide less environmentally deleterious process, to save energy and by virtue of foregoing to decrease cost. The raw material in forest-based industries is wood and its components. Thus, possibilities for employing biotechnology in these industries are numerous since one of the nature's most important biological processes is the degradation of lignocellulosic materials to CO_2, water and humic substances. In point of fact, biotechnology has been used in the paper industry for quite some time (Bajpai 2006). Waste water treatment systems for the removal of oxygen-demanding substances and suspended solids, fermenting sulfite liquors, preparing starch for paper sizing have long been part of the industry. Improvement in fiber supply by the selection of superior trees is still being done by forest product companies. Even the control of slime and deposits on paper machines can be considered as aspect of biotechnology. However, within the past several years, biotechnologists have sought specific applications for microorganisms/enzymes in the pulp and paper industry. The growth has been fueled by several factors:

- An improved understanding of the interactions between enzymes and the constituents of pulp and paper processing
- An increased need for the industry to adopt environmentally benign technology
- Development of cost-effective technology for the relevant enzymes

Suitable biological treatments in conjunction with less intensive conventional treatment could help solve many of the problems of currently used processes. In response to environmental concerns and regulations, the industry has greatly reduced chlorinated aromatic by-products that can be formed during pulp bleaching, first by reducing the amount of residual lignin in pulps and second by turning to other bleaching agents. An enzyme technology based on microbial xylanases has helped to achieve this goal by reducing or even eliminating the need for chlorine in the manufacture of elemental chlorine free (ECF) and totally chlorine free (TCF) printing and writing paper grades (Bajpai 2004, 2009; Viikari et al. 2002). Enzymes have helped meet

Table 1.1 Biotechnology for the pulp and paper industry in different stages of development

Process	Status
Bleaching of kraft pulp	Commercial scale
Modification of fiber properties for improving beatability	Commercial scale
Improvement of pulp drainage	Commercial scale
Decreasing vessel picking	Commercial scale
Deinking	Commercial scale
Stickies control	Commercial scale
Starch modification	Commercial scale
Removal of pitch in pulp	Commercial scale
Slime control in paper manufacture	Commercial scale
Production of chemicals or fuels from wastes and waste liquors	Commercial scale
Biomechanical pulping	Pilot scale
Biochemical pulping	Pilot scale
Pulp bleaching with laccase mediator system	Pilot scale
Purification of bleach plant effluents	Pilot scale
Production of dissolving pulps	Pilot scale
Use of enzymes for debarking	Laboratory scale
Use of enzymes for retting of flax fibers	Pilot scale

environmental goals in other ways as well. By reducing costs involved in deinking, enzymes have increased the ability of manufacturers to recycle fiber, thereby placing fewer demands on timber resources. Enzymes have been used commercially to reduce paper manufacturing costs or improve the product. Lipases can control the accumulation of pitch during the production of paper from pulps with high resin content, such as sulfite and mechanical pulps from pine. Enzymes also help remove contaminants in the recycle stream. They can reduce the accumulation of adhesives and pitch residues, called stickies, on machines. They can facilitate the deinking of recycled paper and improve pulp drainage, which is particularly important as the amount of recycled fiber in the feedstock stream increases. With higher drainage rates, paper machines are able to operate faster, which again saves capital costs (Bajpai 1999). Xylanases have saved chemical costs for the industry without interfering with the existing process. This technology has increased the bleaching speed in both TCF and ECF processes and, in the case of chlorine dioxide bleaching, has actually increased the throughput of the plant due to debottlenecking at the chlorine dioxide generator. Developments of this last type are viewed very favorably since they enable the industry to make better use of its existing capital equipment. Many other enzyme applications are also possible. These include eliminating caustic chemicals for cleaning paper machines, enhancing kraft pulping, reducing refining time, decreasing vessel picking, facilitating retting, selectively removing fiber components, modifying fiber properties, increasing fiber flexibility, and covalently linking side chains or functional groups.

Table 1.1 presents the current developmental stages of various biotechnological approaches for use in pulp and paper industry.

Most of the commercialized biotechnological applications are based on industrially produced enzymes (Paice and Zhang 2005). Enzymes are mother nature's catalysts that drive the chemical reactions that are in all living things. Enzymes have the following properties:

- They are effective in very small amounts – a few enzyme molecules will catalyze thousands of reactions per second.
- They are unchanged and are not consumed in the reaction.
- They reduce the activation energy of a reaction and therefore increase the speed of reaction.
- They are very specific to a reaction.
- They have a specific pH and temperature range that they are active in.

The fact that enzymes are specific to a certain reaction allows enzymatic products to be tailored to specific needs. The enzymes available presently are more specific with less side activities, more tolerant with respect to pH and temperature, and economically more competitive than those in the late 1980s. In the recent years, there is an increased availability of a whole range of enzymes at reasonable cost. New enzymes can be made to order, based on genome information for the major wood-degrading microorganisms now available in the public domain. Another factor is a concerted research effort by a number of players to develop a cost-effective portfolio of enzyme-based applications in papermaking. The most important commercialized applications of enzymes in the pulp and paper industry include bleaching, energy saving in refining, removal of stickies and pitch, deinking, improvement of paper machine runnability by hydrolysis of slimes or extractives, and enhanced drainage, as well as fiber modification for speciality products (Bajpai 2006). Bio-based unit operations are usually combined with traditional or new chemical and mechanical unit operations to fully benefit the performance of enzymes.

Recently there has been much discussion about biorefineries, aimed at improving the profitability of kraft mills by diversifying the product mix. One idea is to prehydrolyze chips to provide a hemicellulose-rich stream as a by-product. Recently, Oji Paper claimed that hydrolysis of such hemicelluloses by xylanase to give a mixture of xylooligosaccharides results in a product with therapeutic value (Paice and Zhang 2005). Another by-product is xylitol which is widely used as an artificial food sweetener. One suggested biorefinery product is fuel ethanol. It is produced by enzymatic hydrolysis of cellulose substrates such as sawdust, followed by fermentation of the resulting glucose. Although the economics of this process do not currently compete with fuel ethanol production from starch, there has been a significant decrease in cellulase manufacturing costs as the result of USDOE.

References

Anon (2004) Biotechnology sparks an industrial revolution. Solutions 87:40–41

Anon (2005) Biotechnology for pulp and paper manufacture: from tailor made biocatalysts to mill application, Baiona, Spain, 26–29 April 2005

Bajpai P (1999) Application of enzymes in pulp & paper industry. Biotechnol Prog 15(2): 147–157

Bajpai P (2004) Biological bleaching of chemical pulps. Crit Rev Biotechnol 24(11):1–58, CRC Press

Bajpai P (2006). Potential of biotechnology for energy conservation in pulp and paper; energy management for pulp and papermakers, Budapest, Hungary, 16–18 Oct. 2006, Paper 11, 29p

Bajpai P (2009) Xylanases in "Encyclopedia of Microbiology, Third Edition" Vol. 4 (Moselio Schaechter and Joshua Lederberg ed). Academic, San Diego, pp 600–612

Mansfield SD, Esteghlalian AR (2003) Applications of biotechnology in the forest products industry. Applications of enzymes to lignocellulosics, edited by Mansfield S D, Saddler JN, Chapter 1, pp 2–29 [ACS Symposium Series 855, Washington, DC, USA: American Chemical Society, 2003, 468 pp]

Ojanpera K (2004) Biotechnology breaks through to forest industry. Tek. Talous no. 17, 6 May 2004, p 2

Paice M, Zhang X (2005) Enzymes find their niche. Pulp Paper Can 106(6):17–20

Viikari L, Poutanen K, Tenkanen M, Tolan JS (2002) Hemicellulases. In: Flickinger MC, Drew SW (eds) Encyclopedia of bioprocess technology: fermentation, biocatalysis, and bioseparation. Wiley: Chichester (Update: Electronic release)

Viikari L, Gronqvist S, Suurnakki A (2006). Biotechnology: future key for tailoring fibres, 3rd International symposium on emerging technologies of pulping and papermaking. Research progress in pulping and papermaking, Guangzhou, China, 8–10 Oct. 2006, pp 31–33

Viikari L, Suurna kki A, Gronqvist S, Raaska L, Ragauskas A (2009) Forest products: biotechnology in pulp and paper processing, encyclopedia of microbiology, 3rd edn. Elsevier, pp 80–94

Chapter 2
Brief Description of the Pulp and Paper Making Process

2.1 Introduction

Pulp and paper are manufactured from raw materials containing cellulose fibers, generally wood, recycled paper, and agricultural residues. In developing countries, about 60% of cellulose fibers originate from nonwood raw materials such as bagasse, cereal straw, bamboo, reeds, esparto grass, jute, flax, and sisal (Gullichsen 2000). The main steps in pulp and paper manufacturing are: Raw material preparation and handling, Pulp manufacturing, Pulp Washing and Screening, Chemical recovery, Bleaching, Stock Preparation, and Papermaking.

Pulp mills and paper mills may exist separately or as integrated operations. An integrated mill is one that conducts pulp manufacturing on-site. Nonintegrated mills have no capacity for pulping but must bring pulp to the mill from an outside source. Integrated mills have the advantage of using common auxiliary systems for both pulping and papermaking such as steam, electric generation, and wastewater treatment. Transportation cost is also reduced. Nonintegrated mills require less land, energy, and water than integrated mills. Their location can, therefore, be in a more open setting where they are closer to large work force populations and perhaps to their customers. A paper mill can house a single paper machine or several machines. Each machine can make a single grade of paper or a variety of papers. A dedicated machine usually manufactures a commodity grade paper such as liner board or tissue. Machines designed to make specialty grades typically have more operating flexibility and will manufacture many types of paper. The basic process of papermaking remains the same despite the type of paper manufactured or the size of the machine.

P. Bajpai, *Biotechnology for Pulp and Paper Processing*,
DOI 10.1007/978-1-4614-1409-4_2,

2.2 Pulp and Paper Making Process

2.2.1 Pulp Making Process

Manufacturing of pulp starts with raw material preparation (Smook 1992a; Biermann 1996a). This includes debarking (when wood is used as raw material), chipping, and other processes such as depithing (for example, when bagasse is used as the raw material). Cellulosic pulp is manufactured from the raw materials, using chemical and mechanical means. The manufacture of pulp for paper and cardboard employs mechanical (including thermomechanical), chemimechanical, and chemical methods.

Mechanical pulping separates fibers from each other by mechanical energy applied to the wood matrix causing the gradual break of the bonds between the fibers and the release of fiber bundles, single fibers, and fiber fragments (Smook 1992b; Biermann 1996b). It is the mixture of fibers and fiber fragments that gives mechanical pulp its favorable printing properties. In the mechanical pulping, the objective is to maintain the main part of the lignin in order to achieve high yield with acceptable strength properties and brightness. Mechanical pulps have a low resistance to aging which results in a tendency to discolor. The main processes are Stone Groundwood Pulping (SGW), Pressure Groundwood Pulping (PGW), Thermo-Mechanical Pulping (TMP), or Chemi-Thermo-Mechanical Pulping (CTMP). The groundwood pulping process grinds wood into pulp. Usually this involves taking a log and pressing it against a rotating surface to grind off small pieces. The groundwood pulp is then often cooked to soften it. This pulp is used in newsprint and other low cost book grades where it contributes bulk, opacity, and compressibility. Groundwood pulp is economical since all the wood is used; however, it contains impurities that can cause discoloration and weakening of the paper. Chemimechanical processes involve mechanical abrasion and the use of chemicals. Thermomechanical pulps, which are used for making products such as newsprint, are manufactured from raw materials by the application of heat, in addition to mechanical operations. The process involves high-temperature steaming before refining; this softens the interfiber lignin and causes partial removal of the outer layers of the fibers, thereby baring cellulosic surfaces for interfiber bonding. TMP pulps are generally stronger than groundwood pulps, thus enabling a lower furnish of reinforcing chemical pulp for newsprint and magazine papers. TMP is also used as a furnish in printing papers, paperboard and tissue paper. Softwoods are the main raw materials used for TMP, because hardwoods give rather poor pulp strength properties. This can be explained by the fact that hardwood fibers do not form fibrils during refining but separate into short rigid debris. Thus, hardwood TMP pulps, characterized by a high-cleanness, high-scattering coefficient, are mainly used as filler-grade pulps. Chemimechanical pulping and chemithermomechanical pulping (CTMP) are similar but use less mechanical energy and soften the pulp with sodium sulfite, carbonate, or hydroxide. The CTMP pulps show good strength properties, even when using hardwood as a fiber source, and provided that the reaction conditions are appropriate to result in high degrees of sulfonation. Mechanical pulps are weaker than chemical pulps, but cheaper to produce (about 50% of the costs of chemical pulp) and are generally obtained in the yield range of 85–95%. Currently, mechanical pulps account for 20% of all virgin fiber material.

Chemical pulping is used on most papers produced commercially in the world today (Smook 1992b; Biermann 1996b). Traditionally, this has involved a full chemical treatment in which the objective is to remove noncellulose wood components leaving intact the cellulose fibers. In practice, separation of the components is never completely realized. Yet satisfactory compromises are reached in the processes which yields somewhere between 45 and 55% of the wood mass. Chemical pulps are made by cooking (digesting) the raw materials, using the kraft (sulfate) and sulfite processes. The kraft (sulfate) process is the most dominating chemical pulping process worldwide. The term "sulfate" is derived from the makeup chemical sodium sulfate, which is added in the recovery cycle to compensate for chemical losses. In the kraft pulp process the active cooking chemicals (white liquor) are sodium hydroxide (NaOH) and sodium sulfide (Na_2S). Kraft process is applicable to all types of wood species but its chemistry carries with it an inherent potential problem of malodorous compounds. Kraft pulp possesses superior pulp strength properties in comparison to sulphite pulp. Kraft processes produce a variety of pulps used mainly for packaging and high-strength papers and board.

Chemical recovery is an essential part of the pulp production process (Tran 2007; Vakkilainen 2000; Bajpai 2008; Biermann 1996c). Half of the wood raw material is utilized as chemical pulp fiber. The other half is utilized as fuel for electricity and heat generation. In fact, a pulp mill has two main lines. Wood is turned into pulp on the fiber line. Energy is produced on the chemical recovery line from the wood material cooked in the liquor; the cooking chemicals are recovered for reuse. In the chemical recovery line, the black liquor is evaporated and combusted in a recovery boiler, and the energy content of the dissolved wood material is recovered as steam and electricity. The chemical pulping process generates more energy than it uses. A pulp mill generates energy for its own use and energy to sell.

Sulfite process uses different chemicals to attack and remove lignin. The sulphite process is characterized by its high flexibility compared to the kraft process, which is a very uniform method, which can be carried out only with highly alkaline cooking liquor. In principle, the entire pH range can be used for sulphite pulping by changing the dosage and composition of the chemicals (Smook 1992b; Biermann 1996b). Thus, the use of sulphite pulping permits the production of many different types and qualities of pulps for a broad range of applications. The sulphite process can be distinguished according to the pH adjusted into different types of pulping. The main sulphite pulping processes are Acid (bi)sulphite, Bisulphite (Magnefite), Neutral sulphite (NSSC), and Alkaline sulphite.

Each pulping process has its advantages and disadvantages (Smook 1992b; Biermann 1996b). The major advantage of mechanical pulping is its high yield of fibers up to 90%. Chemical pulping yields approximately 50% but offers higher strength properties and the fibers are more easily breached because the mechanical pulping process does not remove lignin. Even with subsequent bleaching, these fibers are susceptible to yellowing. This is the reason that paper grades containing high quantities of mechanical pulp fiber such as newsprint discolor quickly, especially when exposed to sunlight.

After pulp production, pulp is processed in wide variety of ways to remove impurities, and recycles any residual cooking liquor via the pulp washing process. Some pulp processing steps that remove pulp impurities are screening, defibering, and deknotting. Residual spent cooking liquor from chemical pulping is washed from the pulp using pulp washers, called brown stock washers for Kraft and red stock washers for sulfite. Efficient washing is critical to maximize return of cooking liquor to chemical recovery and to minimize carryover of cooking liquor (known as washing loss) into the bleach plant, because excess cooking liquor increases consumption of bleaching chemicals. Specifically, the dissolved organic compounds contained in the liquor will bind to bleaching chemicals and thus increase bleach chemical consumption.

Mechanical pulp can be used without bleaching to make printing papers for applications in which low brightness is acceptable – primarily, newsprint. However, for most printing, for copying, and for some packaging grades, the pulp has to be bleached (Smook 1992c). For mechanical pulps, most of the original lignin in the raw pulp is retained but is bleached with peroxides and hydrosulfites. In the case of chemical pulps (kraft and sulfite), the objective of bleaching is to remove the small fraction of the lignin remaining after cooking (Smook 1992c; Reeve 1996a, b). Oxygen, hydrogen peroxide, ozone, peracetic acid, sodium hypochlorite, chlorine dioxide, chlorine, and other chemicals are used to transform lignin into an alkali-soluble form (Reeve 1989). An alkali, such as sodium hydroxide, is necessary in the bleaching process to extract the alkali-soluble form of lignin.

Pulp is washed with water in the bleaching process. In modern mills, oxygen is normally used in the first stage of bleaching (Bajpai 2005a). The trend is to avoid the use of any kind of chlorine chemicals and employ "total chlorine-free" (TCF) bleaching. TCF processes allow the bleaching effluents to be fed to the recovery boiler for steam generation; the steam is then used to generate electricity thereby reducing the amount of pollutants discharged. Elemental chlorine-free (ECF) processes, which use chlorine dioxide, are required for bleaching certain grades of pulp. The use of elemental chlorine for bleaching is not recommended. Only ECF processes are acceptable, and, from an environmental perspective, TCF processes are preferred. The soluble organic substances removed from the pulp in bleaching stages that use chlorine or chlorine compounds, as well as the substances removed in the subsequent alkaline stages, are chlorinated. Some of these chlorinated organic substances are toxic; they include dioxins, chlorinated phenols, and many other chemicals. It is generally not practical to recover chlorinated organics in effluents, since the chloride content causes excessive corrosion.

2.2.2 Stock Preparation and Paper Making Process

Before pulp can be made into paper, it must undergo several steps called stock preparation (Smook 1992d; Biermann 1996e) Stock preparation is conducted to convert raw stock into finished stock (furnish) for the paper machine. The pulp is prepared for the paper machine including the blending of different pulps, dilution, and the addition of chemicals. The raw stocks used are the various types of chemical pulp, mechanical pulp, and recovered paper and their mixtures. The quality of the finished stock essentially determines the properties of the paper produced. Raw stock is available in the form of bales, loose material, or in the case of integrated mills, as suspensions. Stock preparation consists of several process steps that are adapted to one another as fiber disintegration, cleaning, fiber modification, and storage and mixing. These systems differ considerably depending on the raw stock used and on the quality of furnish required. For instance, in the case of pulp being pumped directly from the pulp mill, the slushing and deflaking stages are omitted. The operations practiced in the paper mills are: Dispersion, Beating/Refining, Metering, and blending of fiber and additives.

Pulpers are used to disperse dry pulp into water to form a slurry. Refining is one of the most important operations when preparing papermaking fibers (Baker 2000, 2005; Bajpai 2005b; Biermann 1996d; Stevens 1992). The term beating is applied to the batch treatment of stock in a Hollander beater or one of its modifications. The term refining is used when the pulps are passed continuously through one or more refiners, whether in series or in parallel. Refining develops different fiber properties in different ways for specific grades of paper. Usually, it aims to develop the bonding ability of the fibers without reducing their individual strength by damaging them too much, while minimizing the development of drainage resistance. So the refining process is based on the properties required in the final paper. Different types of fiber react differently because of differences in their morphological properties. The refining process must take into account the type of fibers. During beating and refining, fibers randomly and repeatedly undergo tensile, compressive, shear and bending forces (Baker 2000; Bajpai 2005b; Biermann 1996d; Stevens 1992). They respond in three ways:

- Fibers develop new surfaces externally through fibrillation and internally through fiber wall delamination.
- Fibers deform, resulting in changes in their geometric shape and the fibrillar alignment along their length. Overall, the fibers flatten or collapse. Fiber curl changes and kinks are induced or straightened. On the small scale, dislocations, crimps, and microcompressions are induced or diminished.
- Fibers break, resulting in changes in length distribution and a decrease in mean-fiber length. A small amount of fiber wall material also dissolves. All these changes occur simultaneously and are primarily irreversible. The extent of the changes depends on the morphology of the fibers, the temperature, the chemical environment, and the treatment conditions. The conditions depend on the design of the equipment and its operating variables such as the consistency, intensity, and amount of treatment. Each pulp responds differently to a given set of conditions and not all fibers within it receive the same treatment.

The furnish (as it is now referred to) can also be treated with chemical additives. These include resins to improve the wet strength of the paper, dyes and pigments to affect the color of the sheet, fillers such as talc and clay to improve optical qualities, and sizing agents to control penetration of liquids and to improve printing properties (Bajpai 2004; Hodgson 1997). After stock preparation, the next step is to form the slurry into the desired type of paper at the wet end of the paper machine.

The pulp is pumped into the head box of the paper machine at this point (Smook 1992e; Biermann 1996f). The slurry consists of approximately 99.5% water and approximately 0.5% pulp fiber. The exit point for the slurry is the "slice" or head box opening. The fibrous mixture pours onto a traveling wire mesh in the Fourdrinier process, or onto a rotating cylinder in the cylinder machine (Biermann 1996f). The Fourdrinier machine is named after its French inventors, the Fourdrinier brothers, and is essentially a table over which the wire moves. Greater quantities of slurry released from the head box result in thicker paper. As the wire moves along the machine path, water drains through the mesh. Fibers align in the direction of the wire travel and interlace to improve the sheet formation. After the web forms on the wire, the task of the remaining portion of the paper machine is to remove additional water. Vacuum boxes located under the wire aid in this drainage.

One of the characteristics inherent in the performing of the sheet on a Fourdrinier paper machine is that all the water is removed through one side of the sheet. This can lead to differences in the sheet properties on one side as opposed to the other. This two-sided property increases as machine speed increases. In response to this, manufacturers developed twin wire and multiple Fourdrinier machines. Manufacturers of such equipment use different engineering designs that can be vertical or horizontal. After the paper web has completed its short forming distance, it continues along the second wire losing water as it travels.

The next stop for the paper is the pressing and drying section where additional dewatering occurs (Smook 1992e; Biermann 1996f). The newly created web enters the press section and then the dryers. As the paper enters the press section, it undergoes compression between two rotating rolls to squeeze out more water. The extent of water removal from the forming and press sections depends greatly on the design of the machine and the running speed. When the paper leaves the press section, the sheet usually has about 65% moisture content. The paper web continues to thread its way through the steam heated dryers losing moisture each step of the way. The process evaporates many tons of water.

Paper will sometimes undergo a sizing or coating process. The web in these cases continues into a second drying operation before entering the calendaring stacks that are part of the finishing operation. Moisture content should be about 4–6% as predetermined by the mill. If the paper is too dry, it may become too brittle. About 90% of the cost of removing water from the sheet occurs during the pressing and drying operations. Most of the cost is for the energy required for drying.

At the end of the paper machine, paper continues onto a reel for winding to the desired roll diameter. The machine tender cuts the paper at this diameter and immediately starts a new reel with the additional paper falling as an endless web.

For grades of paper used in the manufacture of corrugated paperboard, the process is now complete. For those papers used for other purposes, finishing and converting operations will now occur, typically off line from the paper machine. These operations can include coating, calendaring, or super calendaring and winding.

Coating is the treatment of the paper surface with clay or other pigments and/or adhesives to enhance printing quality, color, smoothness, opacity, or other surface characteristics. There is a great demand for paper with a very smooth printing surface.

Various grades of paper, including paperboard, printing, writing and industrial or packaging grades sometimes have coatings. Most coated paper is ground with paper made from mechanical pulp. The term "coated free sheet" describes paper made from ground wood-free fibers being produced from chemical pulp. Three major coated paper categories exist – glossy, dull, and mat. Many people equate coated paper with the gloss stock of a magazine. Books and other products may use dull coated paper to retain the advantages of coated paper while reducing light glare.

Two popular coating methods are air knife and blade coating. In the air knife process, a jet of air acts like a blade to remove excess coating applied to the paperboard. The blade coating process using a flexible blade set in an adjustable angle to remove excess coating across the web. Following the coating operation, the sheet must again be dried and rewound.

Calendering is an on-machine process where the paper passes through a series of polished steel rolls to smooth the paper surface before rewinding on a reel. Besides imparting smoothness, calendering can reduce variations in the sheet and create a higher density sheet. It can also affect the water absorption properties of the paper.

Winding may appear to be a simple process, but anyone who has ever tried to rewind a roll of bathroom tissue after a small child has played with it will think differently. Maintaining proper tension on the reel so that the sheet lies flat and attains proper alignment for both edges is a difficult task. Further complications occur with the higher speeds (up to 6,000 ft/min) of the paper machine. At this rate, the paper web is moving faster than a car at highway speed and paper the length of 20 football fields would wrap on a roll every minute.

Other operations can also take place including cutting, sorting, counting, and packaging. For some products such as tissue and copy paper, the typical paper mill will conduct all of these operations. In most cases, however, the rolls are wrapped and readied for shipment to their final destination.

The nature of paper and papermaking has changed very little over the past 150 years since the introduction of the Kraft Fourdrinier process. However, the techniques and equipment necessary to make paper have changed dramatically. Because of this, we can rely on a consistent supply of high quality graded papers for almost any need we can imagine.

References

Bajpai P (2004) Emerging technologies in sizing. PIRA International, UK, p 159

Bajpai P (2005a) Environmentally benign approaches for pulp bleaching. Elsevier Science BV, The Netherlands, p 277

Bajpai P (2005b) Technological developments in refining. PIRA International, UK, p 140

Bajpai P (2008) Chemical recovery in pulp and paper making. PIRA International, UK, p 166

Baker CF (2000) Refining technology. In: Baker C (ed) Leatherhead. Pira International, UK, p 197
Baker CF (2005) Advances in the practicalities of refining. In: Scientific and Technical Advances in Refining and Mechanical Pulping, 8th Pira International Refining Conference, Pira International, Barcelona, Spain, 28 February-March 2005
Biermann CJ (1996a) Wood and fiber fundamentals. Handbook of Pulping and Papermaking. Academic, San Diego, p 13
Biermann CJ (1996b) Pulping fundamentals. Handbook of Pulping and Papermaking. Academic, San Diego, p 55
Biermann CJ (1996c) Kraft spent liquor recovery. Handbook of Pulping and Papermaking. Academic, San Diego, p 101
Biermann CJ (1996d) Refining and pulp characterization. In: Handbook of pulping and papermaking, 2nd edn. Academic, New York, p 137 (Chapter 6)
Biermann CJ (1996e) Stock preparation and additives for papermaking. Handbook of Pulping and Papermaking. Academic, San Diego, p 190
Biermann CJ (1996f) Paper manufacture. Handbook of Pulping and Papermaking. Academic, San Diego, p 209
Gullichsen J (2000) Fiber line operations. In: Gullichsen J, Fogelholm C-J (eds) Chemical pulping – papermaking science and technology. Fapet Oy, Helsinki, p A19 (Book 6A)
Hodgson KT (1997) Overview of sizing. In: Tappi sizing short course. Session 1, Nashville
Reeve DW (1989) Bleaching chemicals. In: Kocurek MJ (ed) Pulp and Paper Manufacture, Alkaline Pulping, Joint Textbook Committee of the Paper Industry, vol 5. Tappi, Atlanta Georgia, p 425
Reeve DW (1996a) Introduction to the principles and practice of pulp bleaching. In: Dence CW, Reeve DW (eds) Pulp bleaching: principles and practice. Tappi Press, Atlanta, p 1 (Section 1, Chapter 1)
Reeve DW (1996b) Pulp bleaching: principles and practice. In: Dence CW, Reeve DW (eds) Chlorine dioxide in bleaching stages. Tappi Press, Atlanta, p 379 (Section 4, Chapter 8)
Smook GA (1992a) Wood and chip handling. Handbook for Pulp & Paper Technologists, 2nd edn. Angus Wilde Publications, Vancouver, p 20
Smook GA (1992b) Overview of pulping methodology. Handbook for Pulp & Paper Technologists, 2nd edn. Angus Wilde Publications, Vancouver, p 36
Smook GA (1992c) Bleaching. Handbook for Pulp & Paper Technologists, 2nd edn. Angus Wilde Publications, Vancouver, p 163
Smook GA (1992d) Preparation of papermaking stock. Handbook for Pulp & Paper Technologists, 2nd edn. Angus Wilde Publications, Vancouver, p 194
Smook GA (1992e) Paper manufacture – wet end operations. Handbook for Pulp & Paper Technologists, 2nd edn. Angus Wilde Publications, Vancouver, p 228
Stevens WV (1992) Refining. In: Kocurek MJ (ed) Pulp and paper manufacture, vol 6, 3rd edn. Joint Committee of TAPPI and CPPA, Atlanta
Tran H (2007) Advances in the Kraft chemical recovery process, Source 3rd ICEP International Colloquium on Eucalyptus Pulp, 4–7 March. Belo Horizonte, Brazil, p 7
Vakkilainen EK (2000) Chemical recovery. In: Gullichsen J, Paulapuro H (eds) Papermaking science and technology book 6B. Fapet Oy, Finland, p 7 (Chapter 1)

Chapter 3
Tree Improvement

3.1 Introduction

Rapid growth in world population will put increased pressures on land and wood resources. Trees, the raw material of the industry, are a renewable resource. However, we are faced with meeting the increased demand for forest products at a time of increased restrictions on land use and environmental controls. Environmental issues are intertwined with the whole fabric of society and have many facets – social, political, economic, scientific, and ethical. Biologically based processes can be used in the pulp and paper industry to reduce some negative environmental impacts. Similarly, forest biotechnology can solve some problems faced by forest managers (Sykes et al. 1999). Effective management of forested lands is central to our quality of life and the sustainability and health of the planet. Policies, or a lack of policies, in one part of the world cannot be isolated from their impact on the global community. We need to be concerned with tree improvement as it relates to forest health, biodiversity, sustainability, resiliency, and other conditions linked to the global forest resource.

While new breeding techniques, fertilizers, pesticides, and improved cultural methods are conventional ways to improve productivity, genetic engineering is a more controversial alternative. However, biotechnology has the potential for generating forest tree cultivars that cannot be produced by conventional breeding alone. Biotechnological approaches are being investigated for integrating conventional forest tree breeding with forest resource productivity. In this chapter, some possibilities for improving forest trees through hybridization and genetic engineering are presented and a summary of application of genetically altered trees for ameliorating toxins, phytoremediation, is also given.

P. Bajpai, *Biotechnology for Pulp and Paper Processing*,
DOI 10.1007/978-1-4614-1409-4_3, © Springer Science+Business Media, LLC 2012

3.1.1 *Forest Trees in the Age of Modern Genetics*

National Research Council of The National Academy of Sciences, through its work Forestry Research a Mandate for Change, and the American Forest and Paper Association, through Agenda 2020 have recognized the potential of biotechnology through establishing common research priorities for industry (Sykes et al. 1999): (1) Sustainable forestry (2) Selection and hybridization, and (3) Genetic engineering and tree breeding.

In keeping with these research priorities, following emerging applications of biotechnology for forest trees are discussed.

3.1.1.1 Genetic Altering of Trees

Current developments in gene mapping techniques permit researchers to identify trees with desired characteristics e.g., fast growth, resistance to disease or cold temperatures. These traits can be used to breed improved species by the use of conventional methods. Mapping allows researchers to concentrate on specific genes and their components at the molecular level. Identification of gene function allows gene manipulation and the introduction of new and desirable traits not available in the breeding population. Ultimately, such mapping should permit isolation of desired tree genes that could be engineered directly into target tree species. New techniques for identifying gene "markers" facilitate the location of desired genes useful for tree breeding. Once potentially valuable genes are located, they can be cloned and improved strains of the same or other tree species can be created.

An attempt is made to change the chemical structure of trees by genetic engineering. More precisely, research is aimed at structural modifications of lignin, the environmentally pernicious component of wood. The role of the genetic engineering of trees could develop into providing the pulp industry with tailor-made fibers. Genetically modified trees are an example of an integrated technology: by substituting the current raw material, the overall process becomes less polluting. The introduction of tailor-made fibers as raw material offers the possibility of simplifying the current process, possibly by eliminating certain parts of it.

The expensive, energy-intensive process of turning wood into paper costs the pulp and paper industries more than $6 billion a year. Much of that expense involves separating wood's cellulose from lignin, the glue that binds a tree's fibers, by using an alkali solution and high temperatures and pressures. Although the lignin so removed is reused as fuel, wood with less lignin and more cellulose would save the industry millions of dollars a year in processing and chemical costs. Research in U.S shows promise of achieving this goal. By genetically modifying aspen trees, researchers have reduced the trees' lignin content by 45–50% and accomplished the first successful dual-gene alteration in forestry science. Their results are described in Proceedings of the National Academy of Sciences (PNAS). Research shows not only a decrease in lignin but also an increase in cellulose in the transgenic aspens and faster growth of the trees. This is indeed very good news for the wood, paper and pulp industries, which do multibillion-dollar business worldwide. Fast-growing, low-lignin trees offer both economic and environmental advantages, because separating lignin from cellulose – using harsh alkaline chemicals and high heat – is costly and environmentally unfriendly. Harvesting such trees, using them as "crops" with desirable traits, would also reduce pressure on existing forests. Four-year field trials of such trees in France and the United Kingdom show that lignin-modified transgenic trees do not have detrimental or unusual ecological impacts in the areas tested. In previous work, U.S researchers had successfully reduced lignin in aspens by inhibiting the influence of a gene called 4CL. The current research modifies the expression of both 4CL and a second gene, CAld5H, in the trees. This dual-gene engineering alters the lignin structure, and produces the favorable characteristics of lower and more degradable lignin, higher cellulose, and accelerated maturation of the aspens' xylem cells. The research is described in the paper by Zhou et al. (2003). These results are "very significant" and will have dramatic impacts on the future genetic improvement of forest trees for pulp and paper production. The improved tree growth and high cellulose content will increase pulp-yield production, while the reduced lignin content will reduce the pulping cost and energy consumption in the pulping process. The ability to produce high-yield plantations with these desirable characteristics will enable to produce wood more efficiently on less land, allowing natural forests to be managed less intensively – for habitat conservation, esthetics, and recreational uses.

Brunelli (2008) has also reported that wood products containing less lignin and more cellulose would have a very marked effect on the environment but there are also many other factors to take into consideration before such a living organism could be cloned and put to extensive cultivation. Growth rates of the modified tree and the environmental impact of their pollen distribution upon native species would require a vast number of generations before real results could be established. Many years of research would be required before parasite resistance, disease resistance, capacity to adapt to environment, food resource utilization, and environmental stress resistance could be ascertained. To counter potentially harmful results of uncontrollable pollen distribution or insect activity, the trees may require genetic sterilization. Thus far, trees grown to 10 months have been shown to contain 45% less lignin and 15% more cellulose than their natural siblings.

Chen et al. (2001) have reported the results obtained by altering the expression of genes of the monolignol biosynthesis pathway in trees and the effect of these modifications on the lignin polymer and on pulping. The genetic engineering of lignin involved down regulation of caffeic acid *O*-methyltransferase (COMT), down regulation of caffeoyl-CoA *O*-methyltransferase (CCoAMT), overexpression of ferulic acid 5-hydroxylase (F5H), down regulation of 4-coumarate:CoA ligase (4CL), and down regulation of cinnamyl alcohol dehydrogenase (CAD). The results obtained by altering the expression of several genes in the monolignol biosynthesis pathways via genetic engineering confirm that it is possible to modify lignin amount and structure without associated detrimental effects for the plant. For some of the transgenic trees, field trials have been established to produce sufficient quantities of wood for larger scale pulping evaluations. These trials also aim to evaluate whether the beneficial effect on pulping is stable over successive years of growth and to study whether the transgenic plants show any alteration in growth and development or disease and pest resistance, when grown under natural conditions.

Dimmel et al. (2000) have reported two approaches to developing an improved wood raw material; one using CAD-deficient trees, the other increasing the level of natural pulping catalysts in trees. The absence of the CAD enzyme results in a different pool of precursors for lignin production, which possess fewer sites for polymerization, which can lead to a less branched, lower molecular weight lignin. Wood from a 12-year-old CAD-deficient loblolly pine (Pinus taeda) was much more easily delignified under soda, kraft, and soda/anthraquinone conditions, compared to a normal 12-year-old loblolly pine. Attempts to increase the content of anthraquinone pulping catalysts in hardwoods that already produce low levels of these materials could also lead to trees with less lignin. An Abrabidopsis isochorismate synthase (ICS) protein in *E. coli* has been successfully overexpressed and experiments have been performed to deliver the ICS gene into model cottonwood plants via Agrobacterium infection. This will make it possible to test the hypothesis that this gene is the rate-limiting enzyme in anthraquinone biosynthesis.

For more than 25 years, Aracruz Celulose has been developing an intensive research program on Eucalyptus tree improvement, looking at the introduction, evaluation, selection, and recombination of superior trees (Bertolucci et al. 1999). Modern biotechnological tools are being introduced into classic genetic tree improvement programs and there is increased pressure for the definition of the most important forest attributes for each type of product. In 27th EUCEPA conference, Bertolucci et al. (1999) presented Aracruz's research addressed at obtaining clones and varieties "engineered" to specific objectives, such as productivity and good quality attributes.

Modifications to the lignin content or composition of wood could provide economic benefits (Boudet 1996). The OPLIGE project was established to characterize genes involved in lignification, transform models and target plans, conduct molecular and biochemical analyses of transformants, study digestibility and pulp production from transformed plants, and cultivate on a large scale the transformations in field trials of their agronomical properties. It appears that the lignin component of the cell wall can be altered, despite its complex polymer biosynthetic pathway. Genetic manipulation of the genes involved in monolignol biosynthesis can improve the lignin content and composition. The manipulation of other genes in the pathway also impacts on the monomeric composition of lignins. Downstream genes might be suitable for reducing the lignin content of woody species.

At the University of Wisconsin-Madison and Ohio State University, researchers introduced genes with desirable traits from nontree species to poplars and white spruce to make these wood species resistant to insect pests or herbicides and further improve their qualities by genetic manipulations. The resultant trees were protected against defoliating insects and, in some cases, a high percentage of the insects feeding on their leaves were killed (Kleiner et al. 1995). Finnish researchers have identified gene markers for cold hardiness in Scots pine and are using these markers to identify trees that could thrive near the Arctic Circle (Anne 1996). This approach has also been used by researchers at North Carolina State University, the USDA Forest Service in Athens, Georgia, and the New Zealand Forest Research Institute at Rotorua to locate a gene that imparts resistance to a major fungal pathogen in loblolly pine (Dean et al. 1997; Todd et al. 1995).

Cloning permits replication of genetically engineered trees and enables mass production of embryos of identical trees that contain one or more value-added traits. Embryos are inserted into manufactured seed and the seeds are sown following conventional culture in a nursery. Identical trees are advantageous in ensuring a uniform raw material that is relatively predictable in its requirements for conversion to pulp and paper (Cyr et al. 1997).

Another approach to genetic altering of trees, which utilizes "antisense constructs" (nucleic acids that bind to the genes themselves or messenger RNAS) to inhibit enzyme production needed for lignin synthesis, has been used by Eriksson et al. (1996) in their work to minimize the lignin content of trees. Obviously, successful development of this work would revolutionize delignification as it is now known. Some of the tree species selected for genetic engineering include loblolly pine, eucalyptus, poplar, sweet gum, and spruce. These species are either fast growing and especially valuable in the pulp and paper industry. Most tree genetic research is presently conducted at universities or government agencies, often in cooperation with paper companies.

3.1.1.2 Phytoremediation

One of the most fasicinating possibilities of biotechnology is that of genetically improving trees to remediate soil contaminated by toxic wastes. Trees are already used for wastewater cleanup, for site stabilization, and as barriers to subsurface flow of contaminated groundwater. Trees are ideal remediators because they are fast-growing perennial plants with extensive root systems and high transpirational rates (Pullman et al. 1998). Their large biomass is advantageous because it allows higher tolerance for toxic materials and has the capacity for accumulating contaminants. Because plant remediation is done *in situ,* it has the potential to be substantially less expensive than alternative technologies used for detoxification.

The most important methods of phytoremediation are (1) Decontamination, and (2) Stabilization and containment. In decontamination the amount of toxic pollutants in the soil is significantly reduced or eliminated; in stabilization and containment, the plants and their associated microflora do not remove contaminants but rather alter the soil chemistry and sequester, reduce, or eliminate the environmental risk of the toxin (Stomp et al. 1993). Research is being conducted to screen tree species for their ability to tolerate, take up, translocate, sequester, and degrade organic compounds and heavy metal ions. Clonal propagation and genetic engineering techniques already exist for a number of species, which opens the door to the creation of tree "remediation" cultivars (Cunningham et al. 1995). This *in situ* use of plants to stabilize, remediate, and restore a comaminated site is referred to as phytoremediation (McIntyre and Lewis 1997). All plants have the ability to accumulate metals essential for their growth and development; these metals include iron, manganese, zinc, copper, magnesium, molybdenum, and possibly nickel (Salt et al. 1995). Certain plants accumulate heavy metals that have no known biological function: these metals include cadmium, chromium, lead, cobalt, silver, selenium, and mercury. However, significant accumulation of heavy metals is usually toxic to most plants. For some time, botanists have been aware that certain tree species are endemic to soils containing high metal content (Baker and Brooks 1989).

Identification of heavy metal tolerance by some plants has led to the research that exploits this characteristic for removing metal contaminants by establishing selected vegetation on contaminated soil; these plants are called "accumulating" plants. A specific example of this technology is the development of a transgenic yellow poplar developed for remediating mercury-contaminated soil (Rugh et al. 1998). Hyperaccumulating plants promise effective, inexpensive remediation of soil, sediment, and groundwater. Whereas metal-tolerant plants exclude toxic metal ions from uptake, hyperaccumulating plants take up high amounts of toxic metals and other ions. An exciting possibility of applying biotechnology lies in identifying a tree species with the ability to tolerate or accumulate toxic substances such as heavy metals or organic compounds. Once identified, this tree species could be introduced in contaminated areas. Furthermore, genetic modification could accelerate remediation by making the tree a hyperaccumulator, by adapting its growth to diverse climatic conditions, or by enabling faster growth (Rulkens et al. 1998).

Phytoremediation is based on root uptake of contaminants and storage in the plant or partial/complete degradation to less toxic compounds. This type of remediation could promote degradation of organic pollutants by increasing soil organic carbon content or by releasing enzymes that promote microbial activity through the plant roots. Phytoremediation could be useful in ameliorating heavy metals and organic compounds such as 2,4,6-trinitrotoluene (TNT), trichloroethylene (TCE), benzene, toluene, xylene, and ethylbenzene (Anon 1996). The benefits of phytoremediation include the fact that it is done *in situ* and that it is a passive, solar-driven "green" technology. Roots are exploratory, liquid-phase extractors that can find, alter, and/or translocate elements and compounds against large chemical gradients (Cunningham and Berti 1993). This technology is most effective on sites containing a low level of contamination that are widely dispersed over a large area in the upper surface of the soil. Phytoremediation can work side by side with site restoration with minimum site disruption. Additionally, plant biomass can be harvested to remove contaminants from the site and trees will resprout without disturbing the site. In sites where a valuable heavy metal has accumulated, it may be possible to reclaim the metal from the harvested tree. Phytoremediation techniques are less expensive than *ex situ* methods, but they require a long time to work. Long-term site remediation and stabilization using trees makes remediation and restoration synonymous, which lowers costs and is compatible with public objectives. The most appropriate type of remediation for a specific site depends on the degree of pollution and the type of toxic material. More intensive remedies are required for localized, highly contaminated sites. Conventional soil remediation methods are more suitable for these sites. These methods typically involve excavation of contaminated soil followed by extraction of the toxin. This *ex situ* technique is usually extremely expensive. Toxic metal contamination of soil and groundwater is a major environmental and human health problem for which affordable, effective solutions are urgently needed. In agricultural areas, sites are frequently contaminated by a buildup of residual herbicides Atrazine, a commonly used agricultural herbicide, has been the focus of bioremediation researchers (Burken and Schnoor 1997). Research with hybrid poplars has resulted in somaclonal variants that tolerate lethal dosages of herbicides. Another aspect of engineered tolerance, pesticide and herbicide resistance, is especially interesting. If trees could be engineered to be more tolerant of the ubiquitous chemicals in soils, substantially higher yields of forest trees could be realized. Such an application was reported by Meilan et al. (1997) in their work on an engineered resistance to the herbicide Roundup.

Other possibilities for phytoremediation range from removing concentrations of naturally occurring selenium solubilized in irrigation water and accumulated in surrounding groundwater (Bañuelos et al. 1997) to using genetically altered eucalyptus trees for absorbing and metabolizing air pollutants (Sorge 1995). The possibilities seem to be limited only by the imagination of researchers and the toxic material present.

References

Anne SM (1996) Moving forest trees in to the modern genetics era. Science 271(5250):760

Anon (1996) The Hazardous Waste Consultant. Phytoremediation Gets to the Root of Soil Contamination, May/June, pp 1.22–1.28

Baker AJM, Brooks RR (1989) Terrestrial higher plants which hyperaccumulate metallic elements. Biorecovery 1:81

Bañuelos GS, Ajwa HA, Terry N, Zayed A (1997) Phytoremediation of selenium laden soils: a new technology. J Soil Water Conserv 52(6):426–430

Bertolucci FLG, Penchel RM, Rezende GDSP, Claudio-da-Silva E (1999) Tree engineering at Aracruz Celulose: results, challenges and perspectives. 27th EUCEPA conference – Crossing the millennium frontier, emerging technical and scientific challenges, Grenoble, France, 11–14 Oct. 1999, pp 33–38

Boudet AM (1996). Tree improvement through lignin engineering, Proceedings of the European conference on pulp and paper research: the present and the future, Stockholm, Sweden, 9–11 Oct. 1996, pp 336–349

Brunelli E (2008) Genetically modified and cross-bred trees. Ind Carta 46(4):22–25

Burken JG, Schnoor JL (1997) Uptake and metabolism of atrazine by poplar trees. Environ Sci Technol 31(5):1399

Chen C, Baucher M, Christensen JH, Boerjan W (2001) Biotechnology in trees: towards improved paper pulping by lignin engineering. Euphytica 118(2):185–195

Cunningham SD, Berti WR, Huang JW (1995) Phytoremediation of contaminated soils. Trends Biotechnol 13(9):393–397

Cunningham SD, Berti WR (1993) Remediation of contaminated soils with green plants: an overview. In Vitro Cell Dev Biol 29P:207–212

Cyr DR, Binnie S, Grimes S, Klimaszewska K, Finstad K, Loyola I, Percy R, Quan G, Valentine A (1997) From the cradle to the forest: advances in conifer propagation, 1997 Biological sciences symposium, San Francisco, CA, USA, 19–23 Oct. 1997, pp 199–202

Dean JFD, Lafayette KE, Eriksson KL, Merkle SA (1997) Forest Tree Biotechnology. Advances in Biochemical Engineering, Vol. 57. Springer, Berlin, p 1
Dimmel DR, MacKay JJ, Pullman GS, Althen EM, Sederoff RR (2000) Improving pulp production with raw material changes. 2000 Pulping/process and product quality conference, Boston, MA, USA, 5–8 Nov. 2000, 5pp
Eriksson K-EL, LaFayette PR, Merkle SA and Dean JFD (1996). Laccase as a targt for decreasing lignin content in transgenic trees through antisense genetic engineering. In: Srebotnik E, Messner K (eds) Proc. 6th International Conference on Biotechnology in the Pulp and Paper Industry. Facultas-Universitatsverlag, Vienna, pp 310–314
Kleiner K, Ellis D, McCown BH, Raffa K (1995) Field evaluation of transgenic poplar against tenr caterpiller and gypsy moth. Environ Entomol 24(5):1358
McIntyre T, Lewis GM (1997) The advancement of phytoremediation as an innovative environmental technology for stabilization, remediation, or restoration of contaminated sites in Canada. J Soil Contam 6(3):227
Meilan R, Ma C, Eaton J, Hoien E, Taylor M, Holden L, Han K-H, James RR, Stanton BJ (1997) Development of glyphosate-tolerant hybrid cottonwoods. Proceedings 1997 Biological Sciences Symposium. Tappi, Georgia, pp 195–197
Pullman GS, Cairney J, Peter G (1998) Clonal forestry and genetic engineering: where we stand, future prospects, and potential impacts on mill operations. Tappi J 81(2):57–64
Rugh CL, Senecoff JF, Meagher RB, Merkle SA (1998) Development of transgenic yellow poplar for mercury phytoremediation. Nat Biotechnol 16:925
Rulkens WH, Tichy R, Grotenhuis JTC (1998) Remediation of polluted soil and sediment: perspectives and failures. Water Sci Technol 37(8):27–35
Salt DE, Blaylock M, Nanda Kumar PBA, Dushenkov V, Ensley BD, Chet I, Raskin I (1995) Phytoremediation: a novel strategy for the removal of toxic metals from the environment using plants. Biotechnology 13:468–474
Sorge M (1995) Toyota's pollution solution. Automotive Ind 175(12):40
Stomp AM, Han KH, Wilbert S, Gordon MP (1993) Genetic improvement of tree species for remediation of hazardous wastes. In Vitro Cell Dev Biol 29:227–232
Sykes M, Yang V, Blankenburg J, AbuBakr S (1999) Biotechnology: working with nature to improve forest resources and products, TAPPI international environmental conference, vol 2, Nashville, TN, USA, 18–21 Apr. 1999, pp 631–637
Todd D, Pait J, Hodges J (1995) Impact and value of tree improvement in south. J Forestry 83:162–166
Zhou LL, Cheng Y, Sun X, Marita J, Ralph JM, Chiang VL (2003) Combinatorial modification of multiple lignin traits in trees through multigene co-transformation. PNAS 100(18):4939–4944

Chapter 4
Biodebarking

4.1 Introduction

Bark is the outermost layer of tree trunks and branches (Fig. 4.1). It protects the tree from its environment. It is distinct and separable from wood. Bark refers to all the tissues outside of the vascular cambium. It overlays the wood and consists of the inner bark and the outer bark. The inner bark, which in older stems is living tissue, includes the innermost area of the periderm. The outer bark in older stems includes the dead tissue on the surface of the stems, along with parts of the innermost periderm and all the tissues on the outer side of the periderm. The outer bark on trees is also called the rhytidome.

The border between wood and bark is cambium (Fig. 4.1), which comprises only one layer of cells. This living cell layer produces xylem cells toward the inside of the stem and phloem cells toward the outside. The cambial cells divide continuously and have a lower mechanical strength than that of other wood cells. Cambium characteristics include high pectin and protein content and the absence or low concentration of lignin (Simson and Timell 1978; Thornber and Northcote 1961; Kato 1981; Fu and Timell 1972). The cambial tissue consists of intracellular material and primary cell walls. According to a model presented for the primary cell wall of dicotyledonous plants, the following carbohydrate polymers are present: cellulose, pectin, xyloglucan, arabinogalactan, and hydroxyproline-rich glycoprotein. The pectins in primary cell walls of dicotyledons are heteropolymers. In addition to galacturonic acid units, they also contain rhamnose linked to galacturonic acid units in interior chains, whereas galactose and arabinose are present as side chain structures (Aspinall 1980; Dey and Brinson 1984). The primary cell wall structure in coniferous trees has not been studied as closely as that of dicotyledons. The content of pectin compounds in the cambial cells varies between the wood species studied.

P. Bajpai, *Biotechnology for Pulp and Paper Processing*,
DOI 10.1007/978-1-4614-1409-4_4, © Springer Science+Business Media, LLC 2012

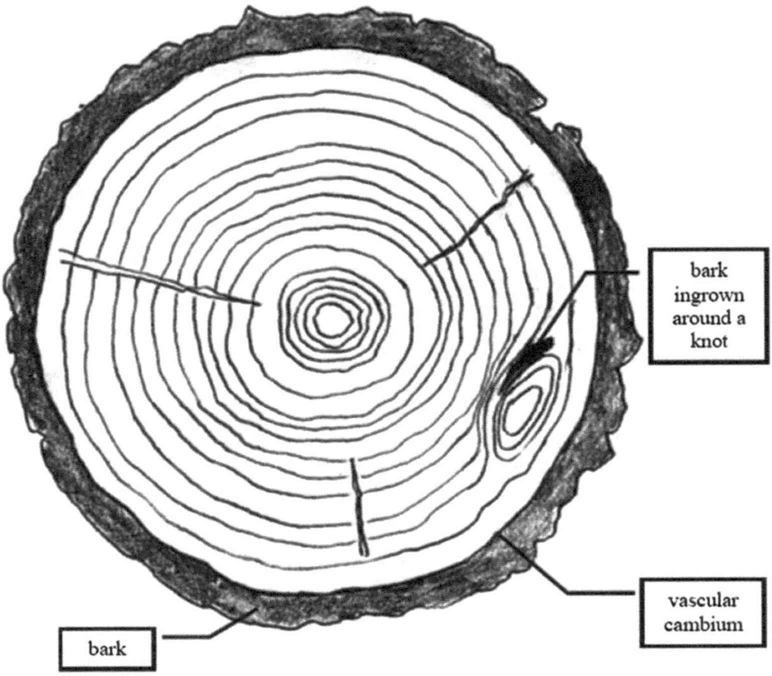

Fig. 4.1 Cross-sectional line drawing of wood

In Betula platyphylla (birch), Fraxinus elatior (ash), Pinusponderosa (pine), and Acer pseudoplatanus (sycamore), the contents of pectic substances in the cambium are 18, 6.6, 8.5, and 15%, debarking respectively (Thornber and Northcote 1961). The cambium of Pinus silvestris (pine) consists mainly of pectic material (partially esterified polygalacturonic acid, arabinan, galactan) (Fu and Timell 1972). In addition, cellulose, glucomannan, and glucurono-araboxylan are present (Meier and Wilkie 1959). In Pinus silvestris, 59% of the galacturonic acid units extracted from cambium were methylated. The cambial tissue of Populus tremuloides (quaking aspen) contains 40% pectins in addition to smaller amounts of arabinogalactan, xyloglucan, xylan, glucomannan, cellulose, and protein (Simson and Timell 1978).

Trees have a cambium layer between the bark and the wood. It is the cambium layer that is the living and continuously growing part of the tree. The cells in this layer divide continuously, which is why they tend to have a lower mechanical strength than cells elsewhere in the tree. In debarking, the aim is to remove the bark together with the cambium layer. Characteristically, the cambium comprises high pectin content. Pectin polymers consist of galacturonic acid, ramnose, arabinose, and galactose. As well, the cambium comprises hemicellulose, cellulose, and protein.

Debarking using conventional commercial procedures usually does not remove all of the barks from logs. It is recognized that up to approximately 3% of bark from coniferous wood and approximately 10% of bark from nonconiferous wood may remain after debarking. Bark has complex anatomy and chemistry. It is a contaminant in the wood supply used for making pulp, decreasing the quality of pulp in proportion to its level. There is very little usable fiber in bark, mostly because bark fibers are very small; and bark consumes chemicals during the pulping and bleaching stages (Smook 1992). Furthermore, it causes dark specks in the final paper product. Some types of bark (e.g., western red cedar and aspen) contain significant quantities of fiber and can be tolerated to an extent in an alkaline pulping system. The relatively high level of nonprocess elements (impurities), such as silica and calcium, interfere with chemical recovery process. For the pulp industry, typical bark tolerances in wood chips are 0.3–0.5%, although the kraft process is more tolerant than the other pulping processes. Bark removed from wood is usually burned as a fuel. Whole-tree chopping in the forest (a practice some argue will become important in the future as it gives a higher yield of wood chips) requires that the chips be cleaned before pulping to remove bark, dust, needles or leaves, twigs, etc.

A significant disadvantage of current mechanical debarking methods and equipment is that in order to achieve a desired degree of debarking it is necessary to continue the debarking process well beyond the time it takes to remove substantially all the bark, in order that pieces which hold steadfastly to the logs can be removed. This results in significant wood loss especially in the trunk areas already completely debarked. Moreover, it leads to increased debarking times and greater energy consumption. Enzymes specific for the hydrolysis of the cambium and phloem layers have been found to facilitate bark removal (Bajpai 1997, 2006, 2009; Viikari et al. 1989, 1991a, b; Wong and Saddler 1992; Ratto et al. 1993; Grant 1992, 1993, 1994; Hakala and Pursula 2007; Ma and Jiang 2002). Enzymes actually weaken the bonds between the bark and wood and break down polymers present in the cells of the cambium layer. The logs may be subjected to enzyme treatment prior to debarking by known methods. If desirable, the enzyme treatment may also be effected after debarking, i.e., part of the bark is first removed, possibly after enzyme treatment, whereupon the logs are subjected to an enzyme treatment designed to weaken the bonds between the wood and the remaining portions of the bark. This allows the remaining bark portions to be removed during a second debarking procedure which may consist of mechanical or some other kind of treatment. The enzyme treatment may also be implemented in other ways in conjunction with the debarking.The enzyme treatment may be implemented by immersing the logs in the treatment solution, or by flushing and/or spraying the logs with the treatment solution. The enzyme treatment has the effect of reducing the detaching resistance of the bark, i.e., it tends to make the bark loosen. This facilitates mechanical debarking and significantly increases the speed thereof. The fact that the bark is more easily removed reduces the amount of energy needed for the debarking. A higher and more constant degree of debarking is achieved. Moreover, enzyme treatment helps reduce wood losses that occur in traditional mechanical debarking as a result of differences in the barking resistance between different trunks or logs. Enzymatic method shows great potential for saving both energy and raw material (Viikari et al. 1989; Ratto et al. 1993).

4.2 Enzymes Used for Debarking

Pectin breaking enzymes, hemicellulases, cellulases and/or proteases, and other enzymes capable of weakening the bonds between wood and bark and/or breaking down polymers present in the cambium have been used. Many commercial preparations of these enzymes are available.

4.3 Application of Enzymes for Debarking

Finnish researchers (Ratto et al. 1993; Viikari et al. 1989, 1991a, b) used debarking enzymes, specific for the hydrolysis of the cambium and phloem layer, from *Aspergillus niger*. A clear dependence was observed between the polygalacturonase activity in the enzyme preparation and reduced energy consumption in debarking. In addition to polygalacturonase, the enzyme mixture produced by *A. niger* also contained other pectolytic and hemicellulolytic activities. The amount of energy needed for the removal of bark was found to decrease to 20% of the reference value (Table 4.1). In this experiment, wood disks were soaked in the enzyme solution and the enzyme was diffused mainly tangentially to the border between wood and bark.

Ratto et al. (1993) studied the effect of enzymatic pretreatment on the energy consumption of wood debarking on the laboratory scale, using enzymes to degrade the cambium layer. Three different pectinases and xylanases were used – a commercial preparation Pectinex Ultra SPL (NOVO) and two preparations produced at VTT biotechnical laboratory: polygalacturonase produced by *A. niger* and a partially purified polygalacturonase obtained from *A. niger* (Bailey and Ojamo 1990; Bailey and Pessa 1990). Xylanase was a commercial preparation, Pentosonase (MKC). The pectinases were dosed (185 nkat/mL) according to their polygalacturonase activity, and the hemicellulase (100 nkat/mL) was dosed according to its xylanase activity. All the enzymes were found to reduce the energy consumption to some extent (Table 4.2). The best result – a 50% decrease in energy consumption, was obtained with Pectinex Ultra SPL. Of the three pectinases, this preparation showed the widest spectrum of the activities of enzymes that hydrolyze the various cambial components. In addition to polygalacturonase, pectin lyase, xylanase, and endoglucanase activities were also detected. The partially purified polygalacturonase with the lowest xylanase and endoglucanase activities was the least efficient

Table 4.1 Effect of pretreatment with polygalacturonase enzyme on energy consumption during debarking of spruce

Polygalacturonase activity (nkat/mL)	Relative energy consumption (%)
0	100
37	75
185	45
195	20

Based on data from Ratto et al. (1993)

Table 4.2 Effect of enzyme treatment on energy consumption during debarking of spruce

Enzyme	Enzyme dose (nkat/mL)				Energy consumption as % of control
	Polygalacturonase	Polymethoxyl-galacturonide lyase	Xylanase	Endoglucanase	
Crude polygalacturonase	185	<0.1	1.1	1.6	77
Partially purified polygalacturonase	185	<0.1	0.2	<0.1	87
Pectin ex ultra SPL	185	0.6	2.0	5.5	50
Pentosanase	0.2	<0.1	100	1.6	82

Based on data from Ratto et al. (1993)

Table 4.3 Effect of enzyme treatment time on energy consumption during debarking of spruce

Treatment time (hour)	Relative energy consumption (%)
0	100
2	98.0
12	62.0
24	50.0

Based on data from Ratto et al. (1993)

Table 4.4 Stability of enzyme in the debarking water

Incubation time (days)	Residual activity (%)
o	100
3	100
8	84.0
10	78.1
15	76.2

Based on data from Ratto et al. (1993)

of the three pectinases. About 18% decrease in energy consumption was obtained with the xylanase preparation.

For Pectinase Ultra SPL, the effects of enzyme dosage and treatment time were studied (Ratto et al. 1993). As much as 80% decrease in energy consumption was obtained with a polygalacturonase dosage of 900 nkat/mL in a 24-h treatment. A moderate effect, a 25% decrease in energy consumption was obtained when only 4% of this activity (40 nkat/mL) was used. For a 50% decrease in energy consumption, a polygalacturonase dosage of 185 nkat/mL was needed. With this dosage, a 40% decrease was obtained in 12 h, whereas only a slight effect was observed after 4 h of treatment (Table 4.3).

Ratto et al. (1993) also studied the stability of the enzyme in process waters containing various components dissolved from wood and bark, to evaluate the possibilities for enzyme recycling. Ultra SPL was incubated in process water from an industrial-scale debarker used for debarking spruce and the residual activity was measured. The polygalacturonase activity was relatively stable, showing more than 70% residual activity after 15 days at 50°C (Table 4.4). Thus, it appears possible to decrease the costs of the enzymatic pretreatment by repeated use of the same enzyme solution.

The efficiencies of the enzymes in the hydrolysis of isolated cambial tissue were also compared. Cambial tissue was isolated from spruce felled during the spring. The polysaccharides in isolated cambium were partially degraded during the isolation, as indicated by the high content of reducing sugars (51% of dry weight) in the reference sample. This effect was probably caused by the endogenous enzymes. However, galacturonic acid was not detected in the reference samples. When the substrate was hydrolyzed with the three pectinases dosed to the same polygalacturonase activity, the most efficient hydrolysis of cambial pectin to galacturonic acid was obtained with Pectinex Ultra SPL (Table 4.5). The amounts of galacturonic acid

Table 4.5 Effects of various pectinases on hydrolysis of isolated cambium

	Hydrolysis products, % of substrate	
Enzyme	Reducing sugars	Galacturonic acid
Pectinex ultra SPL	42	12.3
Crude polygalacturonase	38	3.9
Partially purified polygalacturonase	38	5.4
Pentosonase	35	0.8
Reference		0.4

Based on data from Ratto et al. (1993)

released by crude polygalacturonase and partially purified polygalacturonase were less than half of that released by Pectinex Ultra SPL. In addition to polygalacturonase, Pectinex Ultra SPL contained the highest pectin lyase activity.

Due to chemical complexity at the cambiurn interface and its variation among tree species, research to identify suitable enzymes was conducted. Metra-Serla and Kone Wood have jointly conducted pilot-scale tests (Grant 1992, 1993, 1994).

4.4 Advantages of Biodebarking

Enzymatic method is an attractive approach for debarking. Enzymatic treatments cause significant decreases in energy consumption during debarking. The energy consumed in debarking is decreased as much as 80% after pretreatment with pectinolytic enzymes. The enzymatic treatment also leads to substantial savings in raw material. Enzymes may be able not only to increase existing debarking capacity, thus saving capital investment but also to act as an aid to be used when debarking is difficult. The enzyme treatment may be implemented by immersing the logs in the treatment solution, or by flushing and/or spraying the logs with the treatment solution. The enzyme treatment has the effect of reducing the detaching resistance of the bark, i.e., it tends to make the bark loosen. This facilitates mechanical debarking and significantly increases the speed thereof. The fact that the bark is more easily removed reduces the amount of energy needed for the debarking. A higher and more constant degree of debarking is achieved. Moreover, enzyme treatment helps reduce wood losses that occur in traditional mechanical debarking as a result of differences in the barking resistance between different trunks or logs.

4.5 Limitations and Future Prospects

In logs, the bark forms an effective barrier to the enzyme and may exclude enzyme diffusion. This barrier could be overcome in practice by applying the enzymatic treatment to poorly debarked logs selected after preliminary mechanical debarking. In the case of poorly debarked wood, the major portion of the bark would already be removed and diffusion of the enzyme would therefore be facilitated. In a mill, it might be advantageous to spray poorly debarked logs with enzyme solution before repeating the mechanical debarking. A substantial disadvantage of the current mechanical debarking methods is that the process has to be continued well beyond the time required to remove most of the bark in order to remove few pieces of bark that hold tenaciously to the logs. The enzymatic treatment applied after preliminary debarking could reduce the wood losses that normally occur. Thus, the enzymatic treatment could lead to substantial savings in raw material in addition to the savings in energy. The studies conducted so far have used enzyme preparations with polygalacturonase or xylanase as the main activity. In addition to these, other enzymes that act on the various components of the cambium may also have an effect. There is a need to study the role of each enzyme and the optimal composition of the enzyme mixture for debarking.

References

Aspinall GO (1980) In: Preiss J (ed) The biochemistry of plants, vol. 3. Academic, New York, pp 473–500

Bailey MJ, Ojamo H (1990) Selective concentration of polygalacturonase and β-glucosidase of Aspergillus niger culture filtrate using mineral adsorbents. Bioseparation 1:133–139

Bailey MJ, Pessa E (1990) Strain and process the production of polygalacturonase. Enz Microb Technol 12:266

Bajpai P (1997) Microbial xylanolytic enzyme systems – properties and applications. Adv Appl Microbiol 43:141–194

Bajpai P (2006) Potential of biotechnology for energy conservation in pulp and paper. Energy management for pulp and papermakers, Budapest, Hungary, 16–18 Oct. 2006, Paper 11, 29pp

Bajpai P (2009) Xylanases. In: Schaechter M, Lederberg J (eds) Encyclopedia of microbiology, vol 4, 3rd edn. Academic, San Diego, pp 600–612

Dey PM, Brinson K (1984) Plant cell walls. Adv Carbohydr Chem Biochem 42:226

Fu YL, Timell TE (1972) Polysaccharides in the secondary phloem of Scots pine (Pinus sylvestris L.). Cell Chem Technol 6:517–519

Grant R (1992) Enzymes reveal plenty more potential. Pulp Paper Int 34(9):75–76

Grant R (1993) R&D optimizes enzyme applications. Pulp Paper Int 35(9):56–57

Grant R (1994) Enzymes future looks bright, as range improve and expands. Pulp Paper Int 36(8):20–21

Hakala T, Pursula T (2007) Biotechnology applications in the pulp and paper industry. In: Hermans R, Kulvik M, Nikinmaa H (eds) Biotechnology as a competitive edge for the Finnish forest cluster. ETLA Sarja B 227 Series, Taloustieto Oy, Helsinki, pp 57–63 (Chapter 6)

Kato K (1981) In: Tanner W, Loevus EA (eds) Plant carbohydrates, vol 11. Springer, Berlin, pp 29–46

Ma JH, Jiang C (2002) Prog Pap Recycling 11(3):36–47

Meier H, Wilkie KCB (1959) The distribution of polysaccharides in cell wall of tracheids of pine (Pinus silvestris L.). Holzforschung 13:177–182

Ratto M, Kantelinen A, Bailey M, Viikari L (1993) Potential of enzymes for wood debarking. Tappi J 76(2):125–128

Simson BW, Timell TE (1978) Polysaccharides in cambial tissues of Populus tremuloides and Tilia americana. II. Isolation and structure of a xyloglucan. Cell Chem Technol 12:39

Smook GA (1992) Wood and chip handling. Handbook for Pulp & Paper Technologists, 2nd edn. Angus Wilde Publications, Vancouver, p 20

Thornber JP, Northcote DH (1961) Changes in the chemical composition of a cambial cell during its differentiation into xylem and phloem tissue in trees. Biochem J 1(81):449–455

Viikari L, Rato M, Kantelinen A (1989) Finish Patent Appl 896291

Viikari L, Kantelinen A, Ratto M, Sundquist J (1991a) ACS Symp Ser 460:12

Viikari L, Kantelinen A, Rättö M, Sundquist J (1991b) Enzymes in pulp and paper processing. In: Leatham GF, Himmel ME (eds) Enzymes in Biomass Conversion, ACS Symp Ser 460. American Chemical Society, Washington, pp 426–436

Wong KY, Saddler JN (1992) Trichoderma xylanases, their properties and application. Crit Rev Biotechnol 12(516):413–435

Chapter 5
Biodepitching *

5.1 Introduction

Lipophilic extractives from wood and other lignocellulosic materials, often referred to as wood resin, cause the so-called pitch deposits along the pulp and paper manufacturing processes. Pitch presents a serious problem for the paper industry and has been the subject of many papers published over several decades. Wood resins include alkanes, fatty alcohols, fatty acids, resin acids, sterols, other terpenoids, conjugated sterols, triglycerides (TGs), and waxes. Pitch problem makes the paper machine operation difficult/inefficient and is responsible for reduced production levels, higher equipment maintenance costs, higher operating costs, and an increased incidence of defects in the finished products, which reduces quality and benefits (Back and Allen 2000). Furthermore, process effluents containing wood extractives may be toxic and harmful to the environment (Leach and Thakore 1976; Liss et al. 1997). Pitch problems are greater in mills with a high degree of water circuit closure (Otero et al. 2000).

There are considerable differences in resin content and composition between the different hadwoods and softwoods and even between different parts of the plant (Back 2000). Some differences are also due to the growing conditions, age of the tree, and other genetic and environmental factors. The resin components of softwoods have been extensively studied (Ekman and Holmbom 2000). TGs, resin acids, and fatty acids represent a high percentage of Scots pine extractives, their composition also varying between sapwood and heartwood of the same species, the latter mainly consisting of resin acids. Norway spruce contains similarly high amounts of TGs, resin acids, and steroids. The above lipophilic compounds are often associated to pitch problems (Back and Allen 2000; Ekman and Holmbom 2000). Among hardwoods, silver birch and trembling aspen have been traditionally

*Excerpted from Bajpai et al. (1999), Chap. 2. Wood pretreatment to remove toxic extractives, with kind permission from Springer Science + Business media

P. Bajpai, *Biotechnology for Pulp and Paper Processing*,
DOI 10.1007/978-1-4614-1409-4_5, © Springer Science+Business Media, LLC 2012

used as raw materials for paper pulp production, and therefore, their wood resins have been thoroughly studied (Ekman and Holmbom 2000). A significantly higher proportion of sterols and other unsaponifiable lipids has been reported in birch and aspen compared to softwoods, although TGs predominate in both species. Sterol esters and waxes on one side and triterpenols and saturated fatty acids on the other side (Chen et al. 1995; Bergelin et al. 2005) have been reported to be the main compounds present in pitch deposits in aspen and birch processing, respectively. Enormous information on the composition of lipophilic extractives from another hardwood, eucalyptus has emerged during last decade due to the increasing consumption of this fast-growing hardwood by the pulp and paper industry. Free and conjugated sterols are the main lipophilic components of eucalyptus wood together with other steroids and free fatty acids (Freire et al. 2002; Gutiérrez and del Río 2001; Gutiérrez et al. 1999, 2001a; Rencoret et al. 2007), being also the main responsible for pitch deposition in pulp and paper manufacture (del Río et al. 1998, 1999, 2000; Freire et al. 2005; González-Vila et al. 1997; Gutiérrez et al. 2001b, c; Silvestre et al. 1999). On the contrary, the chemical composition of lipophilic extractives from nonwoody species used by the pulp and paper industry has not been studied to the same extent as that of woody species, although during the last years more knowledge has been produced (del Río and Gutiérrez 2006; Gutiérrez and del Río 2003a, b; Gutiérrez et al. 2004, 2006d, 2008, 2009; Morrison and Akin 2001). Alkanes, fatty alcohols and aldehydes, sterols and waxes are the major compounds identified in nonwoody plants such as flax, hemp, kenaf, sisal, and abaca, which are used in the production of high-quality pulps for specialty papers. In pulping of nonwoody plants, such as hemp, fatty alcohols, alkanes, and sterols are the compounds responsible for pitch deposits (Gutiérrez and del Río 2005).

Pitch problems are also dependent upon the type of pulp processing (chemical vs. mechanical). Pitch is liberated from the pulp fibers at various points of time during processing particularly when there is a change of pH and/or temperature. Pitch can be deposited alone or with fibers, fillers, and defoamer components, coating binders from the broke, and insoluble inorganic salts. The formation of pitch deposits begins with chlorine during the bleaching process, when the double bonds of the pitch TGs are chlorinated. The chlorinated pitch is then released from the fibers and accumulates in the water used for paper manufacturing. Through largely unknown mechanisms, these chlorinated TGs form pitch deposits. Pitch problems are particularly severe for mechanical and sulfite pulping operations. During the Kraft process, the cooking liquors saponify TGs so that pitch problems due to these compounds are reduced. However, certain wood species such as aspen have high levels of nonsaponifiables that cause severe pitch problems for Kraft pulping operations.

5.2 Environmental Impact of Lipophilic Extractives

Besides causing a number of serious problems in the productions process, pitch is the major contributor to toxicity to aquatic life. Resin acids are environmentally significant because of their relative persistence and toxicity to fish and are responsible for a large part of the acute toxicity of pulp mill effluents (Sunito et al. 1988). The main resin acids contributing to toxicity in pulp mill effluents are dehydroabietic, abietic, isopimaric, and pimaric acids. Acute toxicity levels for these resin acids are between 0.4 and 1.8 mg/L (Chung et al. 1979). The works of Rabergh et al. (1992), Pesonen and Andersson (1992), Tana (1988), Oikari et al. (1983), and Oikari and Nakari (1982) give strong evidence that extractives such as resin acids are cytotoxic and enzyme inhibitors and that defensive responses would be expected in fish exposed to effluents containing high levels of extractives. Tana (1988) reported inhibition of the conjugation enzyme uranosylglucuronosyl-transferase (UDP-GT), which eliminates substances from the body, in rainbow trout exposed to 5 μg/L dehydroabietic acid (DHAA) for 60 days, and also reported decreased liver glycogen values in fish exposed to DHAA for 80 days. Pesonen and Andersson (1992) tested DHAA on primary cell cultures of rainbow trout hepatocytes. They found that this resin acid decreased the enzyme 7-ethoxyresorufin-*o*-deethylase (EROD) activity at concentrations between 0.1 and 40 μg/L. In a study by Rabergh et al. (1992), the cytotoxic action on isolated rainbow trout hepatocytes by resin acids was studied. Exposure to dehydroabietic and isopimaric acid inhibited bile acid uptake, confirming that resin acids cause impaired liver function in fish and can be major contributors to pulp mill effluent toxicity. The resin acids concentrations used by Rabergh et al. (1992) were much higher, 30–97 mg/L, than those used by Oikari and Nakari (1982) in a 3–11-day experiment with rainbow trout and simulated unbleached kraft mill effluent (70–150 μg/L).

Time-dependent stimulation and inhibition of the UDP-GT activity in rainbow trout simultaneously exposed to DHAA and trichlorophenol was observed by Tana (1988). Such response may be caused by compounds with different modes of action at the subcellular level so that some compounds such as trichlorophenol stimulate conjugation processes, whereas DHAA damages membranes and partly inhibits enzyme activities (Oikari and Nakari 1982). Resin and fatty acids levels are normally reduced to sublethal concentrations with the implementation of secondary treatment (Servizi et al. 1986). Besides being the major contributors to toxicity to aquatic life, these lipophilic extractives (pitch) cause a number of serious problems in the production process (Allen 1975). During the pulping process, these resinous materials are released from wood and later stick to tile and metal parts including the rolls and wires of the paper making machines. The pitch also stains the felts and canvas and eventually reaches the dryer section. This pitch accumulation can cause paper spotting and web breaks on the machine, which are severe problems in production. Wood extractive components such as TGs, resin acids, and steryl esters are major components of paper machine pitch deposits (Irie and Hata 1990; Fujita et al. 1992; Brush et al. 1994; Hata et al. 1996). In addition, pitch outbreaks are more common when resinous wood species are used and during seasons when wood resin content is particularly high. Pitch of pines, including loblolly, slash, and red pines, is known to cause serious problems. Hardwood pitch, particularly from tropical hardwood species and eucalyptus, can also be detrimental. In addition, these extractives are thought to increase the yellowing, i.e., brightness reversion of the pulp and paper products, to cause odor problems and to increase the necessity for chlorination of the pulp (Allen 1975).

Owing to the toxicity of these lipophillic wood extractives, it is necessary to have a pretreatment that reduces the liberation of these compounds, especially in mills lacking secondary effluent treatment.

5.3 Methods for Pitch Control

5.3.1 Conventional Treatment

Traditional methods to control pitch problems include deliberate storage of logs or wood chips in the mill before pulping (natural "seasoning") and adsorption or dispersion of the pitch particles with chemicals in the pulping and papermaking processes, which can be accomplished by adding alum, talc, ionic or nonionic dispersants, cationic polymers, and other types of additives (Allen 2000a, b; McLean et al. 2010). During wood storage, the content of extractives is decreased because some of them are subjected to hydrolytic or oxidative transformation by plant enzymes as well as by the action of wood colonizing microorganisms. The reactions of wood resin components during storage have been studied for several pulpwood species including spruce, pine, birch, aspen, and eucalypt (Ekman 2000; Gutiérrez et al. 1998; Silvério et al. 2008). However, prolonged storage cause decreases in pulp brightness and yield due to the uncontrolled action of microorganisms, therefore, the industrial practice today does not include overlong log or chip storage times. As an alternative, the use of selected fungi to accelerate and control the seasoning of wood, or enzymes to treat the pulp has been considered.

5.3.2 Biological Treatment

Biological treatments with microorganisms or enzymes have been suggested and tested in mill trials as an alternative to conventional treatments.

5.3.2.1 Use of Fungi

A variety of wood-inhabiting fungi including sapstain fungi, basidiomycetes, and molds are capable of degrading wood extractives (Farrell et al. 1997).

Sapstain fungi rapidly colonize the sapwood of logs and wood chips. These fungi grow mainly in ray and parenchyma cells and are capable of deeply penetrating logs and wood chips. These fungi can also grow within resin canals, tracheids and fiber cells and penetrate simple and bordered pits and sometime form boreholes through wood cell wall. Extractives and simple sugars found in the parenchymal cells are the major nutrient source for sapstain fungi. These fungi are not capable of degrading

Table 5.1 Extractive degradation by sap-stain fungi on nonsterile southern yellow pine

Fungal species	Control extractives (%)	Treated extractives (%)	Reduction (%)
C. adiposa	2.13	1.26	41
O. piliferum	3.34	2.27	32
C. adjuncti	1.98	1.44	27
C. minor	2.13	1.57	26
O. piceae	2.13	1.57	26
O. populina	2.13	1.62	24
O. abiocarpa	2.13	1.61	24
C. tremuloaurea	2.13	1.65	23
O. fraxinopennsylvanica	2.13	1.71	20
O. plurianulatum	2.13	1.73	19
E. aereum	1.98	1.61	19
C. ponderosa	2.19	1.86	18
C. penicullata	1.98	1.58	15
O. olivaceum	2.27	1.93	15
E. clavigerum	2.13	1.84	14
C. hunti	2.13	1.89	11
C. ambrosia	2.13	1.92	10
O. distortum	2.27	2.07	9
E. robustum	2.27	2.06	9
C. virescens	2.27	2.11	7
O. galeiformis	2.13	1.98	7
C. coerulescens	2.13	2.13	0
O. dryocetidis	2.24	2.27	0
O. stenorcerns	2.13	2.21	0
X. conudamae	2.44	3.16	0
X. hypoxylon	2.44	2.88	0

Farrell et al. (1997); reproduced with permission

cellulose and lignin. Hemicellulose is degraded to a very small extent. Common species of softwood sapstain include *Ophiostoma ips, O. piliferum, O. piceae, Aureobasidium pullulans, Leptographium lundbergii, Alternaria alternata, Cephaloascus fragrans, Cladosporium* spp., *Lasiodiplodia theobromae,* and *Phialophora* spp. and common species of sapstain fungi on hardwood include *Ophiostoma pluriannulatum, Ceratocystis moniliformis, L. theobromae, and Ceratocystis rigidum.* (Zabel and Morrell 1992). Many of these species are capable of degrading wood extractives. Extractive degradation by *Ophiostoma* spp. particularly *O. piliferum* and *O. piceae* has been most widely studied (Brush et al. 1994; Dorado et al. 2000a; Martínez-Íñigo et al. 1999; Gutiérrez et al. 1999; Rocheleau et al. 1999; Chen et al. 1994; Josefsson et al. 2006; Leone and Breuil 1998; Su et al. 2004).

A wide variety of sapstain fungi have been found to degrade wood extractives (Farrell et al. 1997; Breuil et al. 1998). *Ceratocystis adiposa, O piliferum, O. piceae* were found to be best species for extractive reduction. About 41, 32, and 26% reduction in extractives of southern yellow pine was obtained in 2 weeks at room temperature with *C. adiposa, O. piceae,* and *O. piliferum,* respectively (Table 5.1).

Table 5.2 Extractive content of sterile lodgepole pine and aspen treated with sap-stain fungi

Treatment	Aspen extractives (%)	Lodgepole pine extractives (%)
Control	3.09 ± 0.07	2.31 ± 0.03
Aged control	2.88 ± 0.04	2.26 ± 0.02
Cartapip 97	2.15 ± 0.01	1.94 ± 0.03
Strain A	2.13 ± 0.02	2.08 ± 0.01
Strain B	2.22 ± 0.01	NA
Strain C	2.07 ± 0.04	1.92 ± 0.01
Strain D	2.08 ± 0.05	1.92 ± 0.01

NA data not available
Farrell et al. (1997); reproduced with permission

Screening of nine different strains of *O. piliferum* on sterile southern yellow pine showed that 5 strains reduced dichloromethane extractives by 25–35%; 3 strains did not degrade the extractives. Screening of 45 different strains of *O. piceae* on sterile aspen chips showed that one strain reduced the extractives by 60%, 13 strains reduced extractives by 16–35%, 21 strains reduced extractives by 1–15% and 10 strains did not degrade the extractives.

Chen et al. (1994) also reported degradation of extractives in aspen and lodgepole pine with five different sapstain fungi. TGs were found to be the most abundant component of both aspen and softwood extractives. The wax and steryl ester content of aspen was about 3 times that of lodgepole pine. Fatty acids and resin acids were the second most common component of lodgepole pine extractives but were present in very small amounts in aspen. The extractives were reduced by 28–33% in aspen and 10–17% in lodgepole pine (Table 5.2). Analysis of extractive components showed that all the five fungi decreased TG content and that four of the five fungi increased free fatty acid content.

The fungus *O. piliferum* is marketed as Cartapip by Clariant, UK to the pulp and paper industry. It removes extractives by metabolizing them. The organism isolated from southern yellow pine wood chips is an albino strain of the staining fungus *Ophiostoma piliferum* that does not color wood (Blanchette et al. 1992) and often dominates in naturally seasoned piles. Cartapip is marketed as a dry powder. It is diluted with water to a 1–3% solids solution and sprayed on to wood chips. It rapidly proliferates and removes pitch/resin from wood chips within 4–10 days. Chip piles show good coverage of fungal growth, with pitch, as determined by dichloromethane extractables, being reduced 50% or more (Hoffman et al. 1982; Blanchette et al. 1992; Farrell et al. 1992, 1993). In general, fungal treatment results in an overall decrease in total fatty acids and total resin acids. TGs are almost completely hydrolyzed. The following resin acids are decreased by more than 50% after 2 weeks of treatment: levopimaric, abietic, palustric, pimaric and neoabietic. Linoleic and palmitic fatty acids are decreased by more than 50% after 2 weeks of treatment and oleic acid is decreased by 41%. The results of Cartapip treatment, such as pitch removal and maintenance of chip brightness and improved paper machine runnability, have been documented by use in mills. In a thermomechanical pulp mill, using southern yellow pine, a trial was performed comparing a 2-week period using the

Table 5.3 Use of a depitching organism in a TMP mill

DCM extractive content of secondary refiner pulp	−37.5%
Alum usage	−31.7%
Bleach usage	−36.9%
Brightness	+0.9%
Tensile index	+5.4%
Tear index	+3.4%
Burst index	+3.3%

Farrell et al. (1997); reproduced with permission

Cartapip product on their wood chips to a 2-week period without using Cartapip; the results are shown in Table 5.3. Reductions in the dichloromethane extractive content of secondary refiner pulp caused expected reductions in use of alum, a pitch control chemical. Use of Cartapip resulted in 36.9% reduction in bleach usage along with an increase in paper brightness of 0.9%. In addition, strength properties were increased due to lower extractive content of the paper.

A 2-month Cartapip 97 trial at a US Thermomechanical pulp mill using southern yellow pine showed significant reduction in the dichloromethane extractives of wood chips and an increase in burst index (Chen et al. 1994). A 1-week trial was performed at a mill in Northwestern USA using a blend of 60% lodgepole pine and spruce, and 40% fir and hemlock. Cartapip 97 treatment reduced the averaged dichloromethane extractive content of the chips reclaimed from the storage pile by 25% (Haller and Kile 1992). Both fungal treatment and natural microbial activity increased the free fatty acid content of the extractives. The increase in free fatty acid content results from initial hydrolysis of esterified fatty acids to free fatty acids. The free fatty acid content of the Cartapip treated chips is lower than that of the naturally aged chips, indicating further metabolism and removal of these components by the fungus. To date, the wood species that have been effectively treated for pitch removal include many different pine species, spruce, birch, eucalyptus, aspen, and mixed tropical hardwoods. The dosage of the fungus product and storage time in the chip pile may vary depending upon temperature and wood species. Cartapip is a unique solution to pitch problems before the wood enters the mills.

Albino strains of *O. piliferum* and related species also exert a biocontrol effect in addition to pitch reduction, preventing sapstaining and other wood-rotting fungi from growing on logs and wood chips (Held et al. 2003). This is due to the ability of *O. piliferum* as an early wood colonizer, which is specially noticeable when applied on freshly cut wood, and results in excluding not only late wood colonizers such as white-rot fungi but also other sapstaining species that are not able to start wood colonization due to the previous presence of the albino strain. When Cartapip is used as biocontrol agent it has the tradename Sylvanex™. Parrac Ltd., the current owners of the technology, improved production and storage qualities of the product in 2008 and announced the relaunching of Cartapip 97/Sylvanex 97 worldwide. Generally, this technology can refer to any albino *Ophiostoma* strain that can be used for resin decrease, also resulting in maintenance of brightness levels in transportation and storage of wood before pulping (Farrell 2007).

Table 5.4 Resin content (% of dry wood) of loblolly pine chips treated with *C. subvemispora* or *O. piliferum* after 1–4 weeks incubation

Incubation time	Control (no treatment)	*C. subvermispora*	*O. piliferum*
0	2.55	2.55	2.55
One week	2.62	2.04	2.39
Two weeks	2.64	1.93	2.16
Three weeks	2.55	1.75	1.81
Four weeks	2.63	1.75	1.70

Based on data from Fischer et al. (1994)

Table 5.5 Resin content of spruce chips treated with various fungi after 2 weeks incubation and kappa numbers after sulfite cooking

Fungus	Resin content (%)	Kappa number (% reduction)
Control	1.2	25.9
C. subvermispora	0.8	20.1 (22.39)
O. piliferum	0.9	25.5 (1.54)
P. chrysosporium	0.9	23.8 (8.1)

Based on data from Fischer et al. (1994)

Basidiomycetes also degrade extractives effectively (Farrell et al. 1997; Messner et al. 1992). These fungi extensively colonize wood. Brown rot fungi preferentially degrade wood polysaccharides including cellulose and cause rapid depolymerization. Brown rotted wood usually shows virtually no decrease in total lignin content. These fungi do not degrade lignin but modify it by oxidation and demethylation of methoxy groups. White rot fungi are the predominant degraders of lignin in nature. Some species of white rot fungi preferentially degrade lignin to wood polysaccharides and other species degrade all wood components simultaneously. The basidiomycetes have been shown to degrade pitch extractives. It has been reported that ethanol–benzene (1:2) extractive content of oak, treated with *Phaenerochaete chrysosporium*, decreased by 46% after 12 weeks (Lim and Cho 1992). Treatment of sterile southern yellow pine wood chips with *P. chrysosporium* for 2 weeks resulted in a 21% reduction in dichloromethane extractives (Farrell et al. 1997). Two biopulping fungi *Cereporiopsis subvermispora* and *Phanerochaete chrysosporium* have been examined to lower the resin content of wood chips (Fischer et al. 1994; Akhtar et al. 1998). The comparison was made with the commercial depitching fungus *O. piliferum*. *C. subvermispora* and *O. piliferum* lowered the resin content of loblolly pine (2.55–2.64%) by 18–27% in 2 weeks and 33–35% in 4 weeks. In spruce wood, all three fungi lowered the resin content from 1.2 to 0.8–0.9% in 2 weeks (Tables 5.4 and 5.5). Various basidiomycetes have shown significant reduction in extractive content (Farrell et al. 1997; Breuil et al. 1998) (Table 5.6). About 51% reduction in dichloromethane extractives in southern yellow pine was obtained with *P. chrysosporium* in 2 weeks at room temperature. *Hyphodontia setulosa*, *Perennipora subacida*, *P. gigantea*, and *Phlebia tremellosa* also performed well, reducing the extractives by 40%. Treatment with two brown rot fungi, *Coniophora puteana* and *Gloeophyllum saepiarium,* and one white rot fungus, *Phellinus igniarius,* resulted in large reductions in extractive content (Farrell et al. 1997).

Table 5.6 Extractive content of sterile southern yellow pine treated with various basidiomycetes

Fungal species	Control extractives (%)	Treatment extractives (%)	Reduction (%)
Phanerochaete chrysosporium	2.19	1.30	41
P. subacida	3.34	2.01	40
P. gigantea	3.34	2.03	39
P. tremellosa	1.98	1.21	39
H. setulosa	1.98	1.20	39
Coriolus versicolor	1.98	1.28	36
Inonotus rheades	3.34	2.18	34
Trichaptum abietinium	4.70	3.13	33
C. subvermispora	3.34	2.18	29
Trichaptum biforme	4.70	3.13	24
Schizophylum commune	2.50	2.03	17
Sistotrerma brinkmanii	2.44	2.17	11
Pleurotus ostreatus	2.44	2.30	6
Alurodiscus sp.	2.44	3.70	0
Ganoderma collosum	2.29	1.75	23
Phellinus igniarius	2.44	2.48	0

Farrell et al. (1997); reproduced with permission

Some studies report the fungal degradation of lipophilic extractives in sapwood and heartwood from Scots pine by white-rot fungi compared with sapstain fungi (Martínez-Íñigo et al. 1999). TGs, fatty acids, sterol esters, and waxes in pine sapwood were almost completely degraded by all the fungi assayed. Moreover, sterols and resin acids were extensively degraded by the white-rot strains, but very poorly removed by the sapstain fungi. This work shows that the fungal degradation of heartwood extractives was limited not only by the degradative ability of the various tested fungi but also by the inhibitory effect exerted by the extractive fraction. The white-rot fungus *Funalia trogii* was particularly inhibited on heartwood, whereas *Bjerkandera* sp. showed a higher tolerance to toxic extractives and was the most efficient fungus in degrading extractives in Scots pine heartwood and sapwood. Resin acids were reported to cause inhibition of other wood-inhabiting fungi (Eberhardt et al. 1994).

In another study, several white-rot fungi were tested for removal and detoxification of extractives from Scots pine sapwood and the time course of extractives degradation by two selected species namely *Bjerkandera* sp. and *Trametes versicolor* was monitored (Dorado et al. 2000b, 2001). These authors showed a fungal removal up to 90% of most lipophilic extractives that was accompanied by a 7- to 17-fold reduction in toxicity in the Microtox bioassay. Further studies were performed by these authors with *T. versicolor* to evaluate the effect of the fungal treatment of spruce chips on a laboratory scale, in terms of pulp and paper quality, thermomechanical pulping (TMP) parameters, and effluent toxicity, to identify advantages and potential problems before a pilot-scale trial (van Beek et al. 2007). On the contrary, several studies on lipophilic extractives removal from *E. globulus* wood showed that several white-rot fungi were able to remove up to 100% of all main lipophilic extractives present in this wood. These studies started with a screening of

Table 5.7 DCM extractive content of nonsterile southern yellow pine treated with various molds

Fungal species	Extractives (%) control[a]	Extractives (%) treatment	Reduction (%)
Phlebia roqueforti	3.34	2.16	35
Leptographium terrebrantis	2.27	1.92	15
Verticicladiella truncata	2.27	1.96	14
Diplodia pinea	2.27	2.01	11
Codinaea sp.	2.27	2.07	9
Aureobasidium pullulans	3.34	3.26	2

[a]The control was chips that had been frozen at –20°C since the start of the experiment
Farrell et al. (1997); reproduced with permission

a large number of fungal species (21 ascomycetes, 33 basidiomycetes, and 19 conidial fungi) including strains isolated from eucalypt wood (Martínez et al. 1999). Different patterns of lipophilic extractives degradation were analyzed (Gutiérrez et al. 1999) and several white-rot basidiomycetes, including *Phlebia* species, were selected for their ability to efficiently remove (up to 100%) both free and esterified sterols abundant in eucalypt wood unlike several ascomycete-type fungi including Cartapip™ that although they decreased the sterol ester content, they did not decrease the amount of free sterols. Further, the time course of the fungal removal of these compounds were followed to optimize the duration of the fungal treatment with selected fungal species, namely, *Phlebia radiata, Ceriporiopsis subvermispora, Bjerkandera adusta,* and *Pleurotus pulmonarius* (Martínez-Íñigo et al. 2000). Kraft pulping and TCF bleaching of the eucalypt chips treated with these four basidiomycetes, followed by papermaking evaluation of the pulps obtained, confirmed their potential for pitch biocontrol in hardwood pulping (Gutiérrez et al. 2000). Resin removal by wood chip treatment with white rot fungi often has additional benefits in biomechanical and biochemical pulping due to the preferential removal of lignin resulting in energy savings (Akhtar et al. 2000; Bajpai et al. 2001).

Molds are capable of degrading wood extractives to a lesser extent because molds grow less prolifically in wood than do other wood-inhabiting fungi. Molds grow best on wood that is very wet or that has been exposed to very high humidity for a long time. On softwoods, molds grow mainly on wood surfaces. On hardwoods, molds can enter the wood at exposed parenchyma, vessels, or ruptured cells and can move throughout the wood by penetrating pit membranes (Farrell et al. 1997; Breuil et al. 1998). *Penicillium roqueforti, Penicillium funiculosum, Rhizopus arrhizus,* and *Trichoderma lignorum* were found to degrade wood waxes in liquid culture (Nilsson and Asserson 1969). *R. arrhizus, Gliocladium viride, Penicillium rubrium, T. lignorum,* and *Aspergillus fumigatus* were found to reduce the ethanol/benzene (1:2) extractive content of wood chips (Farrell et al. 1997). The wood chips were stored at 35°C and sampled after 30 days. The type of wood chips tested was not given. Screening of various molds for their ability to degrade wood extractives revealed that the best fungus was *P. roqueforti* (Breuil et al. 1998). It reduced the dichloromethane extractive content by 33% in 2 weeks in southern yellow pine chips at room temperature (Table 5.7).

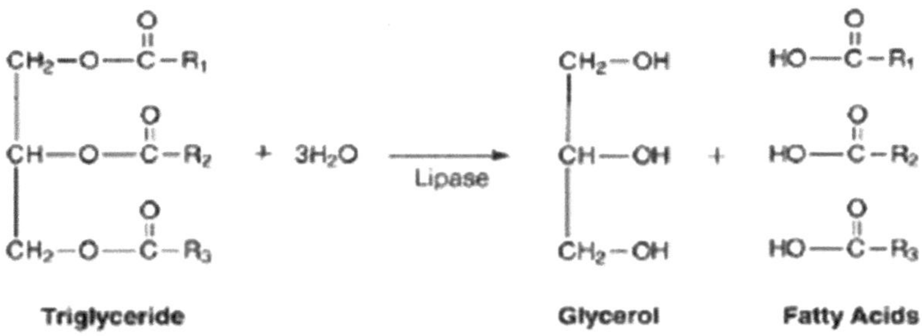

Fig. 5.1 Hydrolysis of pitch by lipase

5.3.2.2 Use of Enzymes

An increasing range of enzymes are available for treating the resins which occur naturally in wood fibers (Covarrubias 2009). Both hydrolytic and oxidative enzymes have been used.

Hydrolytic Enzymes

Use of hydrolytic enzyme (lipase) to treat the pulp was the first case in the world in which an enzyme was successfully applied in the papermaking process (Bajpai et al. 1999; Bajpai and Bajpai 2001; Fujita et al. 1991, 1992; Hata et al. 1996; Matsukura et al. 1990). This enzymatic pitch control technology was initially developed in the 1980s by Jujo Paper Co. (later Nippon Paper) in conjunction with the Japanese office of Novozymes. Lipase enzymes form a group of well-known hydrolytic enzymes (Bornscheuer et al. 2002; Hasan et al. 2006). These enzymes catalyze the hydrolysis of TGs which have been identified as one of the key components for pitch troubles (Fig. 5.1).

Lipases from *Aspergillus oryzae, Candida cylindraceae,* and *Pseudomonas* sp. have been used (Irie and Hata 1990; Fujita et al. 1992; Gibson 1991; Fischer and Messner 1992a, b; Fischer et al. 1993). Some thermostable lipases have been also evaluated (Fujita et al. 1992). Few commercial preparations of lipases for pitch removal are commercially available (Bajpai et al. 1999; Fujita et al. 1992).

Lipase from *Candida cylindraceae* was used for pitch removal from groundwood pulp from Japanese red pine (Irie and Hata 1990). On the lab-scale studies, when the resinous materials in the groundwood pulp slurry were treated with lipase, TG was hydrolyzed and the pitch deposits were remarkably reduced (Table 5.8). Since the effect of lipase on reducing pitch deposits was confirmed, the technology was applied to the actual paper making process (Irie and Hata 1990). The effects of enzyme concentration, reaction temperature, reaction time and agitation mode on the hydrolysis of TG were investigated to select optimum conditions of the lipase treatment for the mill trial (Irie and Hata 1990). It was found that it was necessary

Table 5.8 Effect of lipase treatment on pitch deposition

Resinous solid[1] Polar compounds/nonpolar compounds	9/1	7/3	5/5
Pitch-water suspension			
Control (mg)	76	137	204
Lipase treatment (mg)	49	53	79
Pitch-pulp suspension[2]			
Control (mg)	30	46	50
Lipase treatment (mg)	Trace	Trace	Trace

Based on data from Irie and Hata (1990)
[1]Contained 0.6 g of both compounds totally
[2]Pulp consistency: 1%

Table 5.9 Effect of lipase concentration on hydrolysis of trigycerides

Lipase concentration ppm/4% pulp slurry	Hydrolytic rate of TG (%)
1	12.9
3	50.0
5	74.2

Based on data from Irie and Hata (1990)

to have a strong mixing system to keep contact between enzyme and TG for the effective reaction. Under sufficient mixing conditions, lipase 5,000 U/kg groundwood pulp could hydrolyze more than 80% of the TG in the surface pitch within 2 h. No effect of the lipase treatment on the brightness and strength properties of the pulp was observed. The effect of lipase treatment on the pitch deposition was investigated at No.9 paper machine in Ishinomaki mill of Jujo Paper Company. The lipase was added to the groundwood pulping line and the hydrolytic rates were determined. In the three trials, the lipase was added into 4% groundwood pulp slurry at the rates of 5, 3, and 1 ppm respectively. The hydrolytic rates of TG were 74.2, 50 and 12.9% respectively (Table 5.9). During the trials, pitch deposits were rarely observed at the stock chest and the dryer felt rolls. Also, the frequency of pitch holes in the paper web was notably reduced and the amount of pitch control agents such as talc was reduced. A long run mill trial of 1 month was also conducted on the same paper machine. The lipase was added to the groundwood stock chest at the rate of 3 ppm. The content of surface pitch and TG in the pulp stock, amount of wet pitch (pitch deposit in the wire and press section), number of holes in the paper web caused by the pitch, and visible pitch deposition on the wall of the chest were measured during the production of yellow telephone-directory paper and newsprint. Lipase treatment was found to have a great effect on preventing pitch troubles during the long-run mill trial.

Lipases from *Aspergillus oryzae* for control of pitch were also used (Gibson 1991). It was found that lipases rapidly hydrolyzed TGs. Normally, dosages of 100–500 g enzyme preparation per ton of pulp added directly to the white water system were used for 90% destruction of TGs.

A commercial preparation of lipase (Resinase® A2X) for the control of pitch was also used (Fujita et al. 1992). The effect of lipase dose on rate of TG hydrolysis was investigated in laboratory experiments. More than 70% of TG was hydrolyzed at 500 ppm of lipase. At the Yatsushiro mill of Jujo paper Co., the plant-scale trial for enzymatic pitch control with Resinase® A2X revealed that the enzymatic treatment reduced TG content (Fujita et al. 1992). The contents of the surface pitch in groundwood pulp produced from unseasoned wood were 2.6–2.8% compared with 1.5–2.2% in groundwood produced from seasoned wood. TG content in the surface pitch of the seasoned wood was 7–9%. On the contrary, they were 18–25% in unseasoned wood, which was 2.5 times higher. The enzymatic treatment reduced TG content at the inlet of the mixing chest to less than 6% while 50% unseasoned was used. During the entire trial period, the removal frequency was kept at less than 0.3 times per day by the enzymatic treatment. So, the enzyme addition guaranteed a normal and stable operation at the paper machine with a relatively high unseasoned wood mixture. Finally, it was concluded that the enzymatic treatment would reduce woodyard space required for seasoning. Since the trial, enzymatic pitch control and unseasoned wood have been continuously used at the Yatsushiro mill. The groundwood pulp produced from fresh wood was found to give a brightness of 3.5 points higher than that from the wood which was seasoned to reduce the pitch. The mill has reduced the cost of seasoning and bleaching chemicals. At the Ishinomaki mill, the plant-scale trial with Resinase® A2X had a dramatic effect on all pitch related parameters. The amount of pitch was reduced to about 50% of the normal level. The significant decrease in the number of heavy and light defects resulted in less frequent recycling of substandard paper product. The use of micronized talc was reduced. During the trial, the paper dynamic friction coeffcient (DFC) increased and white carbon additions were reduced to 1%. DFC increased with enzyme addition but its effect leveled off with addition of more than 1,000 ppm. DFC of paper is normally affected by extractives, pigments and fillers. TG was found to be one of the key compounds for decreasing DFC since the enzyme hydrolyses only TG in pitch and the other compounds may remain unchanged.

In addition to Resinase®, other industrial lipases have been investigated for the enzymatic control of pitch (Gutiérrez et al. 2001b). In December 1999, Nanping Paper Mill in China started the world fastest newsprint paper machine at this time, whose main pulp supply came from groundwood and TMP pulp using local Masson pine as raw material. Owing to the very high content of lipophilic extractives of Masson pine, conventional pitch control programs were ineffective in preventing frequent pitch outbreaks on various parts of the paper machine during startup. In March 2000, the mill conducted a trial with a formulated pitch control treatment, which contained a variety of Novozymes enzymes including lipase products. The efforts focused not only on the operation of paper machine, but also on the pulp mills where the pitch had the highest concentration, and the pitch outbreaks were controlled successfully by the use of the enzyme–base treatment (Chen et al. 2001). More recently, Novozymes has developed by directed evolution a Resinase variant that is over 15°C more stable than the wild-type enzyme, in the frame of a European-Commission-funded RTD-project. The optimum temperature of this Resinase® HT ranges from 70 to 85°C. Currently, Resinase® A2X and Resinase® HT are used in mills from USA, Canada, China, Japan, and other countries in the Far East (that process wood species with very high resin content). Recently, it has been shown that combining a novel ecological surfactant with a lipase it is possible to reduce a broad range of extractives in a softwood TMP pulp (Dube et al. 2009).

Fujita et al. (1992) developed a thermostable fungal lipase which was found to be effective at 80°C. Mill trial with this enzyme at a concentration of 1,000 ppm resulted in 75% hydrolysis of TGs. The results were better than those obtained in the laboratory scale experiments.

The other common source of pitch deposits, sulfite pulp, also responds to lipase treatment. Fischer and Messner (1992a) assessed on the laboratory scale a method of TG removal from sulfite pulp that utilizes the *hydrolysis* of the TGs by lipases and the removal of the liberated fatty acids with sodium hydroxide. Pitch fractions with and without TGs and free fatty acids were chlorinated with chlorine water to simulate chlorination during the bleaching process. Chlorinated pitch without TGs and the fatty acids was found to be less adhesive than pitch containing free fatty acids or pitch that was treated with enzyme. It was found that (a) TGs content in pulp was a major contributor to pitch adhesiveness and (b) enzyme treatment followed by sodium hydroxide extraction reduced pitch adhesiveness. Fischer et al. (1993) also performed pilot-scale experiments concerning the application of lipases to unbleached pulp. Per day, twelve tons of pulp was treated with enzyme in a continuous process, the process conditions and the enzyme concentrations were varied, and the resins were analyzed for TGs and free fatty acids. The resin content of the pulp was reduced by about 60% during the process from 0.68 to 0.28% of the dry pulps. The hydrolysis of TG, C 57 fraction, which contains the main part of the most important unsaturated fatty acids (oleic acid and linoleic acid) was about 85%.

Lipases from *Candida cylindraceae* have been shown to be effective in hydrolyzing TGs in extractives of fresh birch and birch kraft pulp (Mustranta et al. 1995). The total amount of esterified compounds in fresh birch was decreased by 34%. The decrease was most significant in the fatty acids. About 50% of the esterified fatty acids and even 65% of the saturated fatty acids were hydrolyzed. The esters of fatty alcohols were also hydrolyzed to the same extent. The amounts of esterified sterols and terpenoids remained the same. For treatment of birch sulfate pulp, the lipases of *C. cylindraceae* hydrolyzed 30% of the esterified lipids as compared to 40% hydrolyzed by Resinase A (*Aspergillus*-lipase). Esters of saturated fatty acids, alcohols, and sterols were hydrolyzed by both lipases. The *C. cylindraceae* lipase was incapable of degrading esters of betulaprenols and triterpenoids, whereas the *Aspergillus* lipase hydrolyzed them to the some extent.

A combination of lipase and cationic polymers has been used in controlling pitch deposits during paper making (Sarkar et al. 1995; Sarkar and Finck 1993). Lipase was found to be effective in reducing the TG content of the stock slurry via hydrolysis to glycerol and fatty acids and cationic polymer reduced concentration of fatty acids released during hydrolysis.

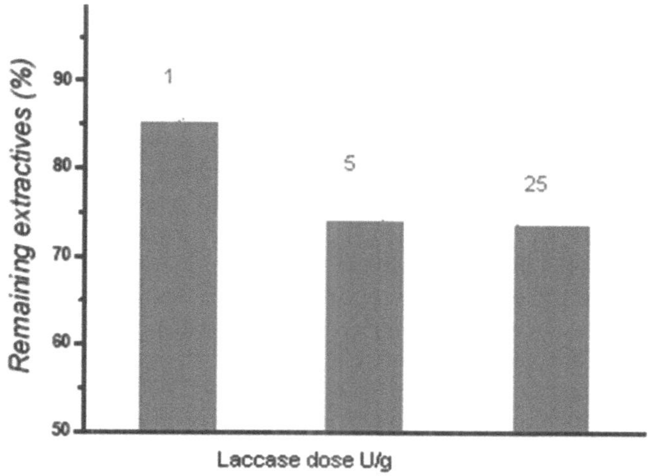

Fig. 5.2 Effect of Laccase treatment on removal of extractives from mechanical pulp, based on Paice (2005)

Oxidative Enzymes

Promising results have been reported by the use of oxidative enzymes, particularly laccases in the presence of redox mediators that are effective on several lipophilic extractives such as fatty acids, resin acids, free and conjugated sterols, and TGs. Laccases are characteristic of white-rot fungi and, as in the case of wood treatment with these lignin-degrading basidiomycetes, a double benefit can be obtained from their application on pulps in the presence of redox mediators. These compounds enable laccase removal of residual lignin, as well as extensive degradation of pulp extractives including the most recalcitrant compounds, such as sterols and resin acids. Karlsson et al. (2001) first reported reactivity of laccase (from a *Trametes* species) on polyunsaturated fatty acids (20% decrease after 4 h) and conjugated resin acids (29% decrease). Similar action (20–35% reduction after 3 h) of laccase on trilinolein was reported by Zhang et al. (2002). In the reaction of trilinolein, the dominant oxidation products detected were monohydroperoxides, bishydroperoxides, and epoxides. Likewise, a decrease over 30% of lipophilic extractives present in softwood pulp from TMP pulping and process waters was also reported (Buchert et al. 2002; Dube et al. 2009; Zhang et al. 2000, 2005). Paice (2005) reported about 85% removal of extractives from mechanical pulp by laccase treatment (Fig. 5.2).

Gutiérrez et al. (2006b) reported for the first time the high efficiency of the laccase-mediator system for the removal of lipophilic extractives present in pulps from different origins regardless of the pulping process, the raw material, or the chemical nature of the compound to be degraded, and a patent application was filed (Gutiérrez et al. 2006c). In these studies, the laccase from the basidiomycete *Pycnoporus cinnabarinus* in the presence of the mediator HBT was very efficient in removing free and conjugated sterols (95–100% decrease) from eucalypt kraft pulp; TGs, resin acids, and sterols (65–100% decrease) from spruce TMP pulp; and fatty alcohols, alkanes, and sterols (40–100% decrease) from flax soda pulp. The removal of lipids by laccase-HBT resulted in the formation of several oxidized derivatives that were absent or presented low abundances in the initial pulps. In spite of this, the total lipid content in pulps decreased significantly, and the most problematic compounds were completely removed. In another work, this enzymatic treatment was applied as an additional stage of an industrial-type TCF sequence for bleaching eucalypt kraft pulp (Gutiérrez et al. 2006a) showing the complete removal of free and conjugated sitosterol. Pulp brightness was also improved due to the simultaneous removal of lignin by the laccase-mediator treatment. Further investigations on the chemistry of the reactions of the laccase-mediator system with model lipids representative for the main lipophilic extractives present in hardwood, softwood, and nonwood paper pulps (including alkanes, fatty alcohols, fatty acids, resin acids, free sterols, sterol esters, and TGs) were carried out, and the reaction products were identified and quantified during the treatment, to better understand the degradation patterns observed in pulps (Molina et al. 2008). These studies evidenced that a 60–100% decrease of the initial amount of unsaturated compounds such as abietic acid, trilinolein, linoleic, and oleic acids, sitosterol, cholesteryl palmitate, oleate, and linoleate, was found at the end of 2-h laccase- HBT treatment. Likewise, a decrease of 20–40% of these unsaturated lipids was observed after treatment with laccase alone except in the cases of abietic acid that decreased 95%, and cholesteryl palmitate and sitosterol that were not affected. The above-mentioned study confirmed that laccase alone decreased the concentration of some unsaturated lipids (Karlsson et al. 2001; Zhang et al. 2002). However, the most rapid and extensive lipid modification was obtained with the laccase mediator system (Molina et al. 2008). Model unsaturated lipids were largely oxidized and the dominant products detected were epoxy and hydroxy-fatty acids from fatty acids, and free and esterified 7-ketosterols and steroid ketones from sterols and sterol esters. In the case of sterol linoleate, breakdown of the fatty acid chain is produced releasing the so-called core aldehydes. The enzymatic reaction on sterol esters largely depended on the nature of the fatty-acyl moiety, i.e., oxidation of saturated fatty acid esters started at the sterol moiety, whereas the initial attack of unsaturated fatty acid esters was produced on the fatty acid double bonds. By contrast, saturated lipids were not modified, although some of them decreased when the laccase-mediator reactions were carried out in the presence of unsaturated lipids suggesting participation of lipid peroxidation radicals.

In a study by Gutiérrez et al. (2007), unbleached eucalypt kraft pulp was treated with a fungal laccase in the presence of syringaldehyde, acetosyringone, and p-coumaric acid as mediators. The enzymatic treatment using syringaldehyde as mediator caused the highest removal (over 90%) of free and conjugated sitosterol, similar to that attained with HBT, followed by acetosyringone (over 60% removal), whereas p-coumaric acid was barely effective. Moreover, recalcitrant oxidized steroids surviving laccase-HBT treatment could be removed when using these natural mediators. Pulp brightness was also improved (from 57 to 66% ISO brightness) by the laccase treatment in the presence of the above phenols followed by the peroxide stage due to the simultaneous removal of lignin. The use of natural compounds as laccase mediators makes these enzymatic treatments more feasible to be applied in the pulp and paper industry. However, more work is required before this enzymatic treatment for simultaneous removal of pitch and lignin can be considered as a serious scheme to be implemented in the pulp and paper industry.

Finally, the use of lipoxygenases (EC 1.13.11.12) has been recently suggested by Zhang et al. (2007) for pitch control in softwood TMP pulp. The lipophilic extractives content of TMP pulp samples was reduced by more than 25% after a 2-h treatment with soybean lipoxygenase. In this work, the activity of lipoxygenase toward wood extractives was determined by using a mixture extracted from TMP. Lipoxygenase exhibited a significant activity toward these wood extractives. However, it was found that some of the extractives (such as resin acids) and lignin products may have inhibitory effects on lipoxygenase-catalyzed reactions with linoleic acid. Earlier work in a patent by Novozymes had suggested the possibility of using lipoxygenases to degrade a model wood "pitch" mixture containing linoleic acid, abietic acid, oleic acid, and olive oil (Borch et al. 2003). Additional studies on lipoxygenases for pitch biocontrol are in progress (Nguyen et al. 2007).

5.4 Advantages, Limitations, and Future Prospects

Biological treatment with enzymes or fungi helps to reduce pitch related problems to satisfactory level (Bajpai et al. 1999; Bajpai and Bajpai 2001). It reduces the downtime of the paper machine, by reducing defects on the paper web as well as the frequency of cleaning pitch deposits in the paper machine, and improves the pulp drainability and machine runnability and thus indirectly saves energy in subsequent operations. At the same time, it also offers several other advantages such as ecofriendly and nontoxic technology, improved pulp and paper quality, reduction in bleaching chemical consumption, reduction of effluent load, and space and cost savings in a mill wood yard by using unseasoned logs. By reducing the outside storage time of logs, this method reduces wood discoloration, wood yield loss, and the natural wood degradation, which occurs over longer storage time. The white carbon (talc) dosage is also reduced because the enzymatic treatment hydrolyses TGs and then increases the dynamic friction coefficient of the paper. With chemical pulps, the application of lipase improves the properties of the resins by lowering their adhesiveness.

Biotechnological processes based on the use of selective white-rot fungi could have some advantages compared to sapstain fungi. Some lipophilic extractives, such as TGs and fatty acids, are easily degraded by different fungi, and even an effective hydrolysis of sterol esters can be achieved by several ascomycetous fungi including sapstain species. However, free sterols and triterpenols as well as resin acids are more recalcitrant toward microbial degradation, and better removal has been obtained with white-rot basidiomycetes. In addition to the advantages directly resulting from the removal of wood extractives, white-rot fungi could accomplish other significant benefits. Wood chip pretreatment with white-rot fungi capable of degrading lignin (and recalcitrant extractives) selectively, a process known as biopulping, enables substantial savings in energy required for obtaining mechanical pulps. Unlike white-rot fungi, ascomycetes are unable to attack lignin. Therefore, such fungal treatments are not expected to reduce the energy requirements. Another limitation of sapstain ascomycetes is their poor ability to colonize the heartwood of most softwood species. Taking into account that the total amount of extractives in heartwood is generally much higher than that in sapwood, the low susceptibility of heartwood to biological colonization by sapstain fungi limits the effectiveness of wood depitching. The applicability of white-rot fungi in pulpwood depitching would greatly depend on their effects on pulp yield and properties. Fungal attack of cellulose in pulp fibers is highly undesirable as it causes reduction in yield and pulp strength properties. Therefore, these biodepitching/biopulping treatments need to be applied under controlled conditions to attain a maximal removal of lipids and lignin with a minimum deterioration of fibers and hydrolysis of cellulose.

The use of enzymes to remove extractives from pulp has advantages, compared to the use of fungal inocula to remove extractives from the wood before pulping, such as the shorter treatment times and the higher specificity in the removal of wood components. Commercial lipases are successful in the hydrolysis of TGs in softwood mechanical pulps, being currently used in different types of mills. Recently, some promising results have been reported by the use of oxidative enzymes, particularly laccases in the presence of redox mediators that are effective on several lipophilic extractives such as fatty acids, resin acids, free and conjugated sterols, and TGs. Laccases are characteristic of white-rot fungi and, as in the case of wood treatment with these lignin-degrading basidiomycetes, a double benefit can be obtained from their application on pulps in the presence of redox mediators. These compounds enable laccase removal of residual lignin, as well as extensive degradation of pulp extractives including the most recalcitrant compounds, such as sterols and resin acids.

Many mechanical and Kraft pulp mills are utilizing wood containing high content of extractives and are spending substantial amount of money on the chemicals for controlling the pitch. For these mills, the biological treatment processes seem to have a great potential.

References

Akhtar M, Blanchette RA, Myers G, Kirk TK (1998) An overview of biomechanical pulping research. In: Young RA, Akhtar M (eds) Environmentally friendly technologies for the pulp and paper industry. Wiley, New York, pp 309–340

Akhtar M, Scott GM, Swaney RE, Shipley DF (2000) Biomechanical pulping: a mill-scale evaluation. Resour Conserv Recycl 28:241–252

Allen LH (1975) Pitch in wood pulps. Pulp Paper Can 76(5):70–77

Allen LH (2000a) Pitch control in paper mills. In: Back EL, Allen LH (eds) Pitch control wood resin and deresination. TAPPI, Atlanta, pp 307–328

Allen LH (2000b) Pitch control in pulp mills. In: Back EL, Allen LH (eds) Pitch control wood resin and deresination. TAPPI, Atlanta, pp 265–288

Back EL (2000) The location and morphology of resin components in the wood. In: Back EL, Allen LH (eds) Pitch control wood resin and deresination. Tappi, Atlanta, pp 1–35
Back EL, Allen LH (2000) Pitch control wood resin and deresination. TAPPI, Atlanta
Bajpai P, Bajpai PK (2001) Solving pitch problems by biological treatment. Paper Int 5(2):9–12
Bajpai P, Bajpai PK, Kondo R (1999) Biotechnology for environmental protection in pulp and paper industry. Springer, Germany, pp 13–28 (Chapter 2)
Bajpai P, Bajpai PK, Akhtar M, Jauhari MB (2001) Biokraft pulping of eucalyptus with selected lignin-degrading fungi. J Pulp Paper Sci 27:235–239
Bergelin E, Moller R, Holmbom B (2005) Analysis of pitch and deposit samples in kraft pulp production. Pap Puu Pap Tim 87:399–403
Blanchette RA, Farrell RL, Burnes TA, Wendler PA, Zimmerman W, Brush TS, Snyder RA (1992) Biological control in pulp and paper production by *Ophiostoma piliferum*. Tappi J 75(12):102–106
Borch K, Franks N, Lund H, Xu H, Luo J (2003) Oxidizing enzymes in the manufacturing of paper materials. Patent (USA) US 2003/0124710 A1
Bornscheuer U, Bessler C, Srinivas R, Hari KS (2002) Optimizing lipases and related enzymes for efficient application. Trends Biotechnol 20:433–437
Breuil C, Iverson S, Yong G (1998) Fungal treatment of wood chips to remove extractives. In: Young RA, Akhtar M (eds) Environmentally friendly technologies for the pulp and paper industry. John Wiley & Sons, New York, pp 541–565
Brush T, Farrell RL, Ho C (1994) Biodegradation of wood extractives from Southern and Yellow pine by *Ophiostoma piliferum*. Tappi J 77:155–159
Buchert J, Mustranta A, Tamminen T, Spetz P, Holmbom B (2002) Modification of spruce lignans with Trametes hirsuta laccase. Holzforschung 56:579–584
Chen T, Wang Z, Gao Y, Breuil C, Hatton JV (1994) Wood extractives and pitch problems: analysis and partial removal by biological treatment. Appita 47:463–466
Chen T, Wang Z, Zhou Y, Breuil C, Aschim OK, Yee E, Nadeau L (1995) Using solid-phase extraction to assess why aspen causes more pitch problems than softwoods in kraft pulping. Tappi J 78:143–149
Chen S, Lin Y, Zhang Y, Wang XH, Yang JL (2001) Enzymatic pitch control at Nanping paper mill. Tappi J 84:44–47
Chung LTK, Meier HP, Leach JM (1979) Can pulp mill effluent toxicity be estimated from chemical analyses? Tappi J 62:71–74
Covarrubias R (2009) Enzymatic pitch control. Tissue World 38–39
del Río JC, Gutiérrez A (2006) Chemical composition of abaca (Musa textilis) leaf fibers used for manufacturing of high quality paper Pulps. J Agric Food Chem 54:4600–4610
del Río JC, Gutiérrez A, González-Vila FJ, Martín F, Romero J (1998) Characterization of organic deposits produced in the Kraft pulping of Eucalyptus globulus wood. J Chromatogr 823:457–465
del Río JC, Gutiérrez A, González-Vila FJ (1999) Analysis of impurities occurring in a totally-chlorine free-bleached Kraft pulp. J Chromatogr 830:227–232
del Río JC, Romero J, Gutiérrez A (2000) Analysis of pitch deposits produced in Kraft pulp mills using a totally chlorine free bleaching sequence. J Chromatogr A 874:235–245
Dorado J, Claassen FW, Lenon G, van Beek TA, Wijnberg JBPA, Sierra-Alvarez R (2000a) Degradation and detoxification of softwood extractives by sapstain fungi. Bioresource Technol 71:13–20
Dorado J, Claassen FW, van Beek TA, Lenon G, Wijnberg JBPA, Sierra-Alvarez R (2000b) Elimination and detoxification of softwood extractives by white-rot fungi. J Biotechnol 80:231–240
Dorado J, van Beek TA, Claassen FW, Sierra-Alvarez R (2001) Degradation of lipophilic wood extractive constituents in Pinus sylvestris by the white-rot fungi *Bjerkandera* sp. and Trametes versicolor. Wood Sci Technol 35:117–125
Dube E, Shareck F, Hurtubise Y, Beauregard M, Daneault C (2009) Enzyme-based approaches for pitch control in thermomechanical pulping of softwood and pitch removal in process water. EXFOR and Annual Meeting 2009, Montreal, QC, Canada, 3–4 Feb 2009, pp 69–74

Eberhardt TL, Han JS, Micales JA, Young RA (1994) Decay resistance in conifer seed cones – role of resin acids as inhibitors of decomposition by white-rot fungi. Holzforschung 48:278–284
Ekman R (2000) Resin during storage and its biological treatment. In: Back EL, Allen LH (eds) Pitch control wood resin and deresination. Tappi, Atlanta, pp 185–204
Ekman R, Holmbom B (2000) The chemistry of wood resin. In: Back EL, Allen LH (eds) Pitch control wood resin and deresination. TAPPI, Atlanta, pp 37–76
Farrell RL (2007) Cartapip/SylvanexTM: ophiostoma fungal product for commercial pulp and paper and solid wood applications. Proc 10 Intern Congr Biotechnology in the Pulp and Paper Industry, Madison, 10–14 June 2007, pp 63–64
Farrell RL, Blanchette RA, Brush TS, Gysin B, Hadar Y, Wendler PA, Zimmerman W, Cartapip® (1992) A biopulping product for control of pitch and resin acid problems in pulp and paper industry. Proceedings of 5th International Conference on Biotechnology in the Pulp and Paper Industry Kuwahara M, Shimada M, eds. Tokyo, Unipublishers, pp 23–32
Farrell RL, Blanchette RA, Brush TS, Hadar Y, Iverson S, Krisa K, Wendler PA, Zimmerman W (1993) Cartapip™ a biopulping product for control of pitch and resin acid problem in pulp mills. J Biotechnol 30:115–122
Farrell RL, Hata K, Wall MB (1997) Solving pitch problems in pulp and paper processes by the use of enzymes or fungi. Adv Biochem Eng Biotechnol 57:198–212
Fischer K, Messner K (1992a) Reducing troublesome pitch in pulp mills by lipolytic enzymes. Tappi J 75(2):130–134
Fischer K, Messner K (1992b) Biological pitch reduction of sulfite pulp on pilot scale. Proceedings of 5th International Conference on Biotechnology in Pulp and Paper Industry, Kyoto, Japan, pp 169–174
Fischer K, Puchinger L, Schloffer K, Kreiner W, Messner K (1993) Enzymatic pitch reduction of sulfite pulp on pilot scale. J Biotechnol 27:341–348
Fischer K, Akhtar M, Blanchette RA, Burnes TA, Messner K, Kirk TK (1994) Reduction of resin content in wood chips during experimental biological pulping process. Holzforschung; 48:285–290
Freire CSR, Silvestre AJD, Neto CP (2002) Identification of new hydroxy fatty acids and ferulic acid esters in the wood of Eucalyptus globulus. Holzforschung 56:143–149
Freire CSR, Silvestre AJD, Neto CP (2005) Lipophilic extractives in Eucalyptus globulus kraft pulps. Behavior during ECF bleaching. J Wood Chem Technol 25:67–80
Fujita Y, Awaji H, Matsukura M, Hata K (1991) Enzymic pitch control in papermaking process. Kami Pa Gikyoshi 45:905–921
Fujita Y, Awaji H, Taneda H, Matsukura M, Hata K, Shimoto H, Sharyo M, Sakaguchi H, Gibson K (1992) Recent advances in enzymatic pitch control. Tappi J 74(4):117–122
Gibson K (1991) Application of lipase enzmes in mechanical pulp production. Tappi Pulping Conference 1991, Book 2, Atlanta, pp 775–780
González-Vila FJ, Gutiérrez A, Martín F, Verdejo T (1997) Application of analytical pyrolysis to the characterization of Eucalyptus extractives and pitch deposits from a pulp mill. J Anal Appl Pyrolysis 40–41:501–510
Gutiérrez A, del Río JC (2001) Gas chromatography-mass spectrometry demonstration of steryl glycosides in eucalypt wood, kraft pulp and process liquids. Rapid Commun Mass Spectrom 15:2515–2520
Gutiérrez A, del Río JC (2003a) Lipids from flax fibers and their fate in alkaline pulping. J Agric Food Chem 51:4965–4971
Gutiérrez A, del Río JC (2003b) Lipids from flax fibers and their fate in alkaline pulping (Vol 51, pg 4965, 2003). J Agric Food Chem 51:6911–6914
Gutiérrez A, del Río JC (2005) Chemical characterization of pitch deposits produced in the manufacturing of high-quality paper pulps from hemp fibers. Bioresource Technol 96:1445–1450
Gutiérrez A, del Río JC, González-Vila FJ, Romero J (1998) Variation in the composition of wood extractives from Eucalyptus globules during seasoning. J Wood Chem Technol 18:439–446
Gutiérrez A, del Río JC, Martínez MJ, Martínez AT (1999) Fungal degradation of lipophilic extractives in Eucalyptus globules wood. Appl Environ Microbiol 65:1367–1371

Gutiérrez A, Martínez MJ, del Río JC, Romero J, Canaval J, Lenon G, Martínez AT (2000) Fungal pretreatment of Eucalyptus wood can strongly decrease the amount of lipophilic extractives during chlorine-free kraft pulping. Environ Sci Technol 34:3705–3709

Gutiérrez A, del Río JC, Martínez MJ, Martínez AT (2001a) The biotechnological control of pitch in paper pulp manufacturing. Trends Biotechnol 19:340–348

Gutiérrez A, Romero J, del Río JC (2001b) Lipophilic extractives from Eucalyptus globulus pulp during kraft cooking followed by TCF and ECF bleaching. Holzforschung 55:260–264

Gutiérrez A, Rodríguez IM, del Río JC (2004) Chemical characterization of lignin and lipid fractions in kenaf bast fibers used formanufacturing high-quality papers. J Agric Food Chem 52:4764–4773

Gutiérrez A, del Río JC, Ibarra D, Rencoret J, Romero J, Speranza M, Camarero S, Martínez MJ, Martínez AT (2006a) Enzymatic removal of free and conjugated sterols forming pitch deposits in environmentally sound bleaching of eucalypt paper pulp. Environ Sci Technol 40:3416–3422

Gutiérrez A, del Río JC, Rencoret J, Ibarra D, Martínez AT (2006b) Main lipophilic extractives in different paper pulp types can be removed using the laccase-mediator system. Appl Microbiol Biotechnol 72:845–851

Gutiérrez A, del Río JC, Rencoret J, Ibarra D, Speranza AM, Camarero S, Martínez MJ, Martínez AT (2006c) Sistema enzimamediador para el control de los depósitos de pitch en la fabricación de pasta y papel. Patent (International) PCT/ES06/070091

Gutiérrez A, Rodríguez IM, del Río JC (2006d) Chemical characterization of lignin and lipid fractions in industrial hemp bast fibers used for manufacturing high-quality paper pulps. J Agric Food Chem 54:2138–2144

Gutiérrez A, Rencoret J, Ibarra D, Molina S, Camarero S, Romero J, del Río JC, Martínez AT (2007) Removal of lipophilic extractives from paper pulp by laccase and lignin-derived phenols as natural mediators. Environ Sci Technol 41:4124–4129

Gutiérrez A, Rodríguez IM, del Río JC (2008) Chemical composition of lipophilic extractives from sisal (Agave sisalana) fibers. Ind Crops Prod 28:81–87

Gutiérrez A, del Río JC, Martínez AT (2009) Microbial and enzymatic control of pitch in the pulp and paper industry. Appl Microbiol Biotechnol 82:1005–1018

Haller T, Kile G (1992) Cartapip treatment of wood chips to reduce pitch and improve processing. Proceeding of Tappi Pulping Conference. Atlanta, Georgia, pp 1243–1252

Hasan F, Shah AA, Hameed A (2006) Industrial.applications of microbial lipases. Enzyme Microb Technol 39:235–251

Hata K, Matsukura M, Taneda H, Fujita Y (1996) Mill-scale application of enzymatic pitch control during paper production. In: Viikari L, Jeffries TW (eds) Enzymes for pulp and paper processing. ACS, Washington, pp 280–296

Held BW, Thwaites JM, Farrell RL, Blanchette RA (2003) Albino strains of Ophiostoma species for biological control of sapstaining fungi. Holzforschung 57:237–242

Hoffman GC, Brush TS, Farrell RL, Cartapip® (1982) A biopulping product for control of pitch and resin acid problems in pulp and paper mills. Naval Stores Rev 102(3):10–12

Irie Y, Hata K (1990) Enzymatic Pitch Control in papermaking system. Proceedings of 1990 Papermaking Conference, Atlanta, pp 1–10

Josefsson P, Nilsson F, Sundstrom L, Norberg C, Lie E, Jansson MB, Henriksson G (2006) Controlled seasoning of Scots pine chips using an albino strain of Ophiostoma. Ind Eng Chem Res 45:2374–2380

Karlsson S, Holmbom B, Spetz P, Mustranta A, Buchert J (2001) Reactivity of Trametes laccases with fatty and resin acids. Appl Microbiol Biotechnol 55:317–320

Leach JM, Thakore AN (1976) Toxic constituents in mechanical pulping effluents. Tappi 59:129–132

Leone R, Breuil C (1998) Filamentous fungi can degrade aspen steryl esters and waxes. Int Biodeterior Biodegrad 41:133–137

Lim CS, Cho NS (1992) Studies on the biological degradation of oak wood by white rot fungus *Phanerochaete chrysosporium*. J Tappi Korea 22:32–44

Liss SN, Bicho PA, Saddler JN (1997) Microbiology and biodegradation of resin acids in pulp mill effluents: a minireview. Can J Microbiol 43:599–611

Martínez MJ, Barrasa JM, Gutiérrez A, del Río JC, Martínez AT (1999) Fungal screening for biological removal of extractives from Eucalyptus globulus Labill. wood. Can J Bot 77: 1513–1522

Martínez-Íñigo MJ, Immerzeel P, Gutiérrez A, del Río JC, Sierra-Alvarez R (1999) Biodegradability of extractives in sapwood and heartwood from Scots pine by sapstain and white-rot fungi. Holzforschung 53:247–252

Martínez-Íñigo MJ et al (2000) Time course of fungal removal of lipophilic extractives from Eucalyptus globulus wood. J Biotechnol 84(2):119–126

Matsukura M, Fujita Y, Sakaguchi H (1990) On the use of Resinase™ A for pitch control. Novo A-6122:1–7

McLean DS, Stack KR, Richardson DE (2010) Evaluation of cationic polymers to control pitch deposition. Appita 63(3):199–205

Messner K, Marek S, Srebotnik E, Techt G (1992) Fungal pretreatment of wood chips for chemical pulping. Proceedings of 5th International Conference on Biotechnology in Pulp and Paper Industry, Kyoto, Japan, pp 9–13

Molina S, Rencoret J, del Río JC, Lomascolo A, Record E, Martínez AT, Gutiérrez A (2008) Oxidative degradation of model lipids representative for main paper pulp lipophilic extractives by the laccase-mediator system. Appl Microbiol Biotechnol 80:211–222

Morrison WHI, Akin DE (2001) Chemical composition of components comprising bast tissue in flax. J Agric Food Chem 49:2333–2338

Mustranta A, Fagernas L, Viikari L (1995) Effects of lipases on birch extractives. Tappi J 78(2): 140–146

Nguyen D, Zhang X, Paice MG, Tsang A, Renaud S (2007) Microplate enzyme assay for screening lipoxygenases to degrade wood extractives. Biocatal Biotransform 25:202–221

Nilsson T, Asserson A (1969) Treating wood chips with fungi to enhance enzymatic hydrolysis. US Patent 1969; 3, 486, 969; 5p

Oikari A, Nakari T (1982) Kraft pulp mill effluents cause liver dysfunction in trout. Bull Environ Contam Toxicol 28:266–270

Oikari A, Lonn BE, Castren M, Nakari T, Snickars-Nikinmaa B, Bister H, Virtanen E (1983) Toxicological effects of dehydroabietic acid (DHAA) on the trout, *Salmo gairdneri* Richardson, in fresh water. Water Res 17:81–89

Otero D, Sundberg K, Blanco A, Negro C, Holmbom B (2000) Effect of wood polysaccharides on the depositability of wood pitch. Nordic Pulp Paper Res J 15:607–613

Paice M (2005) Enzyme application in pulp and paper manufacturing. Lakehead University Symposium, 27 Sep 2005

Pesonen M, Andersson T (1992) Toxic effects of bleached and unbleached paper mill effluents in primary cultures of rainbow trout hepatocytes. Ecotoxicol Environ Safety 24:63–71

Rabergh CMI, Isomaa B, Eriksson KE (1992) The resin acids dehydroabietic acid and isopimaric acid inhibit bile acid uptake and perturb potassium transport in isolated hepatocytes from rainbow trout (*Oncorhynchus mykiss*). Aquat Toxicol 23:169–180

Rencoret J, Gutiérrez A, del Río JC (2007) Lipid and lignin composition of woods from different eucalypt species. Holzforschung 61:165–174

Rocheleau MJ, Sitholé BB, Allen LH, Noel Y (1999) Fungal treatment of aspen for wood resin reduction: effect on aged aspen wood chips at room temperature and at 5°C. Holzforschung 53:16–20

Sarkar JM, Finck MR (1993) Method for controlling pitch deposits using lipase and cationic polymer, US Patent 1993; 5, 256, 252; 25p

Sarkar JM, Tseng AM, Hartig ET (1995). Application of enzymes and polymers for controlling pitch in papermaking, Proceedings Papermakers Conference, Tappi Press, Atlanta, pp 175–182

Servizi JA, Martens DW, Gordon RW, Kutney JP, Singh M, Dimitriadis E, Hewitt GM, Salisbury PJ, Choi LSL (1986) Microbiological detoxification of resin acids. Wat Pollut Res J Can 21:119–129

Silvério FO, Barbosa LCA, Maltha CRA, Fidêncio PH, Cruz MP, Veloso DP, Milanez AF (2008) Effect of storage time on the composition and content of wood extractives in Eucalyptus cultivated in Brazil. Biores Technol 99:4878–4886

Silvestre AJD, Pereira CCL, Pascoal Neto C, Evtuguin DV, Duarte AC, Cavaleiro JAS, Furtado FP (1999) Chemical composition of pitch deposits from ECF Eucalyptus globulus bleached kraft pulp mill: its relationship with wood extractives and additives in process streams. Appita J 52:375–382

Su Y, Wang EI, Farrell R, Ho C-I, Chang H-M (2004) Screening of fungi for removal of wood extractives. Proc 58th Appita Ann Conf Exhibit, Canberra, 19–24 Apr 2004, pp 27–23

Sunito LR, Shiu WY, Mackay D (1988) A review of the nature and properties of chemicals present in pulp mill effluents. Chemosphere 17:1249–1290

Tana J (1988) Sublethal effects of chlorinated phenols and resin acids on rainbow trout (*Salmo gairdneri*). Water Science 2:77–85

van Beek TA, Kuster B, Claassen FW, Tienvieri T, Bertaud F, Lenon G, Petit-Conil M, Sierra-Alvarez R (2007) Fungal bio-treatment of spruce wood with Trametes versicolor for pitch control: Influence on extractive contents, pulping process parameters, paper quality and effluent toxicity. Bioresource Technol 98:302–311

Zabel RA, Morrell JJ (1992) Wood microbiology: decay and its prevention. Academic, New York, pp 476

Zhang X, Stebbing DW, Saddler JN, Beatson RP, Kruus K (2000) Enzyme treatments of the dissolved and colloidal substances present in mill white water and the effects on the resulting paper properties. J Wood Chem Technol 20:321–335

Zhang X, Eigendorf G, Stebbing DW, Mansfield SD, Saddler JN (2002) Degradation of trilinolein by laccase enzymes. Arch Biochem Biophys 405:44–54

Zhang X, Renaud S, Paice M (2005) The potential of laccase to remove extractives present in pulp and white water from TMP newsprint mills. J Pulp Paper Sci 31:175–180

Zhang X, Nguyen D, Paice MG, Tsang A, Renaud S (2007) Degradation of wood extractives in thermo-mechanical pulp by soybean lipoxygenase. Enzyme Microb Technol 40:866–873

Chapter 6
Bioretting

6.1 Introduction

Flax provides a variety of important industrial products such as textiles, oilseed, and paper/pulp. Fibers from flax, are the oldest textile known, dating back to early Egyptian times (Van Sumere 1992). Flax has been considered a prime source of natural fibers to replace glass in composites (Lepsch and Horal 1998; Van den Oever et al. 1999). Flax is also the source of linseed oil for industrial uses (Domier 1997) and nutritional flaxseeds (Westcott and Muir 2000). Fiber flax produced for textiles is grown under precise conditions to optimize fiber quality and harvested prior to full seed maturity (Van Sumere 1992). In contrast, seed flax is optimized for seed yield and quality, and the fiber, which is from straw residue, is generally shorter, coarser, and lower in quality. The large tonnage of straw residue from the linseed industry (Domier 1997; Euroflax Newsletter 2001), and perhaps future increases due to nutritional and health emphases of flaxseeds, constitutes an increasing environmental problem for disposal. Efforts are under way to use larger amounts of the linseed straw for high value fibers such as those in composites.

Fibers are obtained from flax stems by the process of retting. Retting is the separation or loosening of fiber bundles from the cuticularized epidermis and the woody core cells and subdivision to smaller bundles and ultimate fibers. Microbial activity during retting causes a partial degradation of the components that bind tissues together, thereby separating the cellulosic fibers from nonfiber tissues. Earlier work has clearly indicated the requirement of pectinases in flax-retting (Sharma 1987a; Van Sumere 1992).

P. Bajpai, *Biotechnology for Pulp and Paper Processing*,
DOI 10.1007/978-1-4614-1409-4_6,

6.2 Methods for Retting

Two primary methods for retting, namely water- and dew-retting, have been used traditionally to extract fibers for textile and other commercial applications. Despite the fact that the highest quality flax fiber is produced by water-retting, this practice has been largely discontinued in western Europe due to the high costs and the pollution and stench arising from fermentation of the plant material. Fermentation products absorbed by the fibers during water-retting impose an unpleasant odor. Dew-retting is now the most common practice for separating flax fibers, even though some water-retted fiber is still marketed (Van Sumere 1992).

Dew-retting is reportedly the oldest method of retting (Van Sumere 1992). Even though the flax fiber is of lower quality than that from water-retting, lower labor costs and high fiber yields make dew-retting attractive. Stalks are pulled, spread in uniform and thin, nonoverlapping swaths, and left in the field where the moisture and temperature encourage colonization and partial degradation of flax stems, primarily by saprophytic fungi. Dew-retting suffers from several disadvantages (Van Sumere 1992). In addition to poor and inconsistent fiber quality since dew-retting replaced water-retting (Sharma and Faughey 1999), dew-retting requires occupation of agricultural fields for several weeks and is restricted to geographical regions with appropriate moisture and temperature for effective microbial growth (Brown 1984; Van Sumere 1992). In Western Europe, which reportedly produces the highest quality linen due to a favorable climate for dew-retting (Hamilton 1986), substantial crop losses frequently occur. Too much rain and lack of sufficient time for drying further contribute to failed harvests of flax fiber. In other regions, dry weather after harvest prevents microbial growth and therefore retting. Considerable research has been undertaken to find a replacement for dew-retting. Chemical-retting has been evaluated using a variety of methods, including ethylenediaminetetraacetic acid (EDTA) or other chemical chelators at high pH, detergents, strong alkali, and steam explosion (Adamsen et al. 2002; Henriksson et al. 1998; Sharma 1987b, 1988; Tubach and Kessler 1994). Sharma (1987b) patented a chemical-retting method using chelating agents. Flash hydrolysis or steam explosion treatment, with or without impregnation before steam treatment, has been used to remove pectins and hemicelluloses from decorticated flax (Sotton and Ferrari 1989; Kessler et al. 1998); a fast decompression separates bundles to smaller bundles and ultimate fibers. Successful laboratory results have been reported with chemical-retting methods, but at times fiber properties are less satisfactory than those from other methods. Chemical treatment increases cost, but cost efficiencies are usually not reported. To date, no chemical-retting methods for flax are used commercially.

Enzymes have been considered for some time as a potential replacement for dew-retting of flax (Brown and Sharma 1984; Brown et al. 1986; Brown 1984; Sharma 1986; Sharma and Robinson 1983). Early work with water- and dew-retting microorganisms showed conclusively that pectinases were required for effective retting (Van Sumere 1992). Other matrix-degrading enzymes are often considered to be important for effective retting, possibly giving an appropriate balance of enzymes for degrading flax polysaccharides (Sharma 1987a; Sharma and Van Sumere 1992). Enzyme mixtures of polygalacturonase, pectinases, and hemicellulases have been found to give successful retting of flax in vitro (Gillespie et al. 1990; Sharma 1987a).

6.3 Enzymes Used in Flax-Retting

A wide range of enzyme mixtures containing polygalacturonases, pectin lyase, hemicellulases, and cellulases have been used for retting (Gillespie et al. 1990; Sharma 1987a; Bajpai 1999; Ebbelaar et al. 2001; Kenealy and Jeffries 2003; Hoondal et al. 2002; Antonov et al. 2007, 17, 25). These enzymes are commercially available. Enzymes, through selective biodegradation of the pectinaceous and matrix substances, facilitate the removal of fibers from the woody portion (i.e., shive) of a flax plant (Van Sumere 1992). Commercially available enzymes have been shown to promote a controlled retting of flax in a laboratory setting useful for developing improved retting methods (Van Sumere 1992; Akin et al. 1997, 2000; Akin 1998; Foulk et al. 2000).

6.4 Application of Enzymes in Flax-Retting

A chronological listing of research for enzyme-retting of flax beginning in 1932 has been reported by Van Sumere (1992). In the 1980s, extensive research was conducted to find an enzymatic replacement for dew-retting of flax for linen in Europe (Sharma and Van Sumere 1992). A wide range of commercially available enzyme mixtures containing polygalacturonases, pectin lyase, hemicellulases, and cellulases were screened for retting (Sharma 1987b) (Table 6.1). Novozym 249 was found to be the most suitable for retting, with conditions of 55°C for 20 h and a straw:liquid ratio of 1:11. Thermal stability of all four enzyme preparations was also tested, and it ranged between 45°C and 60°C. A semi-industrial retting of flax with Novozym 249 was carried out in tanks (2,000-L capacity) loaded with 80 kg of flax stems maintained at 49 ± 5°C for 26 h (Sharma 1987b). Novozym 249 was used at a concentration of 5 g/L at straw:liquor ratio of 1:11. The flax was well retted along the length of the stem. During enzyme-retting, the pH declined slowly from 6.2 to 4.6 after 20 h, and at the end of retting the pH increased slightly to 4.8. The population of bacteria steadily rose during the retting trial, peaking at the end to a density of 18×10^6 viable bacterial cells. The bacteria present during enzyme-retting were mainly *Actinobactor calcoaceticus*. The bacteria present may have contributed to retting of the flax by releasing further polysaccharide-degrading enzymes. At the end of retting, the enzyme levels of polygalacturonase, xylanase, and cellulase remained quite high. After the concentration of enzymes in the liquor was adjusted, the effluent from the retting trial could be reused for retting another batch of flax.

Table 6.1 Effect of enzymes on flax-retting

	Fiber fineness (decitex)			Fiber strength (g/denier)		
Enzyme	30°C	45°C	60°C	30°C	45°C	60°C
Novozym 249	33.4	29.5	28.2	5.3	4.6	4.2
Novozym 249+Pectinol AC	35.0	35.0	33.5	4.9	4.6	4.5
Novozym 249+Ultrazym	37.0	30.0	28.0	5.2	4.3	4.0
Novozym 249+Celluclast	36.8	30.2	28.0	5.2	2.7	2.0
Pectinol AC+Ultrazym	34.6	30.0	30.0	5.2	4.4	4.4
Pectinol AC+Ultrazym+Celluclast	35.6	32.0	30.2	5.1	3.0	2.9
Pectinol AC	42.0	36.0	35.6	4.8	4.2	4.1
Ultrazym	46.0	45.0	36.1	5.0	4.8	4.2
Celluclast	46.2	40.3	36.1	4.0	2.8	2.1
S.E.	3.14	3.32	2.05	0.33	0.24	0.12

Novozyme 249 contained 835 μg/mL/h polygalacturonase, 100 μg/mL/h pectin lyase, 350 μg/mL/h xylanase, and 100 μg/mL/h cellulase

Pectinol AC contained 850 μg/mL/h polygalacturonase, 100 μg/mL/h pectic lyase, and 205 μg/mL/h xylanase

Ultrazym contained 275 μg/mL/h polygalcaturonase and 2,100 μg/mL/h xylanase

Polygalacturonase and pectin lyase activities μg/mL/h galacturonic acid equivalent

Xylanase and cellulase activities = μg/mL/h glucose equivalent

Based on Sharma (1987b)

Table 6.2 Effects of enzyme-, chemical-, and water-retting on fiber yield and fiber properties

Retting	Fiber yield (%)	Fiber fineness (g denier^{-1})	Fiber strength (decitex)	Caustic weight loss (%)	Fluidity	Moisture regain (%)
Enzyme	15.4	46.1	7.4	28.4	0.6	11.1
Chemical	14.3	42.3	6.8	18.6	0.6	10.6
Water	14.9	40.9	6.7	24.7	0.7	10.6
S.E.	0.46	1.94	0.25	0.38	0.05	0.05

Based on Sharma (1987b)

During World War 11, unretted flax stems were processed to produce coarse green fibers, and the fibers were treated chemically (NaOH) to produce fine flax fibers. Preliminary research in the last 3–4 years on chemical-retting by chelating agents (EDTA, Trilon TB, BSF Ltd.) presented a fresh opportunity (Sharma 1987b) to examine the effects of chemical-retting on fiber yield and fineness (Table 6.2). Studies have shown that green flax fibers in the form of roving or in bundles can be treated with Novozym 249 to produce fine flax fibers.

Flaxzyme, a product from Novo Nordisk, was developed and was reported to produce fibers with good yield and quality (Van Sumere and Sharma 1991). The yield of enzyme-retted fibers was as good or even better than those from high-quality water-retting. A disadvantage, however, was potential lower fiber strength due to the continued activity of the cellulases in the mixtures. Treatment with an oxidizing agent, such as sodium hypochlorite, or reagents giving a high pH that denatured the enzymes prevented the continuing cellulolytic activity. Research on enzyme-retting led to a series of patents and to a semi-industrial scale trial (Van Sumere 1992), but no commercial system was developed.

Gillespie et al (1990) used the commercial enzyme Biopectinex, enzymes from *Penicillium capsulatum,* and an unidentified fungal isolate (possibly a mucor species) for retting of flax. Each of the enzyme preparation was found to affect the release of reducing sugars from flax straw. In the case of Biopectinex, the amounts of sugars released reflected the amount of enzyme used and up to about 20 h, also reflected the length of incubation. At 26 h, the amounts of reducing sugars in the supernatants were considerably less than that at 20 h because of the growth of contaminating microorganisms under the nonsterile conditions. The straw samples treated with the highest concentration of Biopectinex were fully retted only at 26 h, while those treated with lesser amounts of Biopectinex preparation were still poorly retted even at this time. By contrast, the *P. capsulatum* and preparation X-treated straws were well retted at 17 h and fully retted at 26 h, at which time the cellulose fibers were easily separated from the residue of the straw. The amounts of *P. capsulatum* and preparations used to obtain retting had considerably lower pectin-degrading and other enzyme activities than had the Biopectinex preparation. Yet, they released almost as many reducing sugars and effected retting at the same time. Some other enzyme activities besides pectinase-polygalacturonase, xylanase, arabinase, FPase, and CMCase might have contributed to the retting process. Each of the enzyme preparations affected the release of uronic acid-containing materials of various degrees of polymerization from the straw to the supernatant solution. The solubilized polymeric material eluting between fractions 12 and 17 had an apparent M value of 9.6 kDa. Much of the soluble uronic acid eluted at the same volumes as standard galacturonic acid (fractions 51–57), indicating that significant amounts of flax pectin (or other uronic acid-containing polymers in flax) had been converted to its monomeric constituents. For all the enzymes, the flax samples were adjudged visually to have been retted at 26 h. Treatment with preparation X and Biopectinex had removed the bulk of the pectin from the flax samples. Treatments with preparation X yielded 55 mg uronic acid per gm of straw, i.e., 145% on pectin basis. Scanning electron microscopic observation of the flax samples before and after treatment confirmed the effectiveness of the enzyme-retting.

Sharma (1987c) investigated the enzymatic degradation of residual noncellulosic polysaccharides present on dew-retted flax fibers with polysaccharide-degrading enzymes, and noted their effects on yarn quality number, strength, extension, caustic weight loss, and fluidity. Three commercial enzymes-Pectinol AC, Ultrazym, Novozym 249, and enzymes extracted from *Ceraceomyces sublaevis* and hybrid strains of *Pleurotus ostreatus* and *Pleurotus florida* were used Novozym 249 contains polygalacturonase, pectin lyase, xylanase, and traces of cellulase. In Pectinol AC, polygalacturonase, pectin lyase, xylanase, and traces of cellulase were present. Ultrazym contains xylanase as the main component along with traces of polygalacturonase. Laccase was not detected in any of the three commercial enzymes. Culture extracts of C. *sublaevis* contained mainly xylanase and laccase along with traces of polygalacturonase, pectin lyase, and cellulase. Xylanase, polygalacturonase, and laccase were detected in high concentrations along with traces of pectin lyase and cellulase in extracts of *Pleurotus*. All the enzymatic treatments gave yarns with high-quality numbers than that of the control. The coefficient of variation was also lower for yarns from the enzyme-treated rovings compared with control. A mixture of

Pectinol AC and Ultrazym was most effective in removing pectins and hemicellulases left on the dew-retted fibers. Of the two culture extracts, enzymes produced by C. *sublaevis* were the most effective in deploymerizing the residual noncellulosic materials left on the fiber. The quality number of yarn samples spun from rovings treated with an enzyme mixture of Pectinol AC and Ultrazym and NaOH at 95°C was marginally better than the quality number of the yarn spun from rovings treated with extracts of C. *sublaevis*. The reduction in weight loss was significantly lower for all the yarns spun from NaOH (high temperature) and enzyme-treated rovings compared to the control yarn. The enzyme extract of C. *sublaevis* was most effective in the removal of residual hemicelluloses, pectin, and lignin as shown by a low % caustic weight loss. The fluidity of the yarns spun from all NaOH- and enzyme-treated rovings revealed that the NaOH treatment and enzymatic depolymerization of the polysaccharides did not significantly reduce the chain lengths of cellulose and as a result, fluidities for the test yarns were low.

Interest has continued on use of enzymes for retting and bast fibers, including flax, with considerable interest at major international conferences (Gűbitz and Cavaco-Paulo 2001; Hardin et al. 2002; Kozlowski et al. 2005). The Agricultural Research Service of US Department of Agriculture began a project on enzyme-retting in the mid 1990s toward improving flax fibers that could be used in short staple spinning systems for the US textile industry. The objective of this work was, therefore, not long line fiber for traditional linen, but instead short staple fibers for blending with cotton and other fibers. The requirements to maintain long fiber length and other restrictions necessary for traditional linen could be avoided, and new methods could be explored to produce a total fiber product from diverse sources of flax. Results from several studies have been reviewed by Akin et al. (2004). From available enzyme sources evaluated at the time, Viscozyme L was found to be the most useful. This enzyme in combination with EDTA was extensively evaluated, and a spray enzyme-retting method was developed (Akin et al. 2000). Fibers produced from various formulations and flax sources were evaluated and ranked in test yarns (Akin et al. 2001). Use of Viscozyme and EDTA became the basis on which other products and protocols were compared. Akin et al. (2001, 2002) reported that 0.05–0.3% of the commercial product Viscozyme L with about 18–25 mM EDTA, pH 5.0, 40°C for 24 h resulted in an appropriate retting formulation and method. Progressively higher levels or longer incubation times with Viscozyme weakened the fibers. While washing and drying by the methods used prevented continued action of cellulases (Akin et al. 2004), their presence in the mixture initially weakened fibers. Eliminating cellulases and strength loss from commercial retting enzymes became a goal.

Advances in enzyme technology have resulted in new products and processes for textiles (Akin and Hardin 2003), including enzymes with potential to improve retting of flax (Antonov et al. 2005; Kozlowski et al. 2005). Further work involving these and other pectinolytic enzyme products (Brühlmann et al. 2000) could improve current retting methods. Foulk et al. (2008) screened new pectinolytic enzymes for retting efficiency, using a series of tests. The more likely candidates were further

Table 6.3 Properties of fibers from flax retted with different enzymes

Enzyme formulation	Strength (g/tex)	Elongation (%)	Fineness (SSI)
Unretted	42.0	1.9	5.20
2.0% Texazyme BFE + Mayoquest 200	36.7	1.8	4.69
5.0% Texazyme BFE only	34.6	1.6	4.27
0.1% Multifect FE + Mayoquest 200	21.6	1.3	4.33
0.2% Multifect FE + Mayoquest 200	17.8	0.5	4.12
0.1% Bioprep only	33.2	2.0	3.79
0.1% Bioprep + Mayoquest 200	34.9	2.3	2.95
0.1% Bioprep + Barapon + Clavodene	34.8	2.5	3.55
0/05% Viscozyme + Mayoquest 200	27.6	1.4	3.60

Based on Foulk et al. (2008)

evaluated by retting in larger amounts and processing in the USDA Flax Fiber pilot plant, using commercial type cleaning equipment. Fibers were assessed for fiber yield, fine fiber yield, and properties important for textile applications. Finally, chemical costs were compared based on available information. The Fried Test identified the most efficient enzymes and best retting conditions. All enzymes retted flax stems better in the presence of 18 mM EDTA. Pectinases that also contained cellulases reduced fiber strength, whereas those without cellulases effectively retted flax without substantial strength loss. Viscozyme, which has been used extensively in enzyme-retting research, and several pectinolytic enzymes were compared in pilot plant scale tests. Texazym BFE and Bioprep 3,000 L retted flax as well as Viscozyme in this system, and the fibers had higher tenacity. The monocomponent nature, commercial availability and price, and ability to ret flax in combination with EDTA at high pH indicated a potential advantage for Bioprep 3,000 L in these tests. Retting with different enzymes and formulations resulted in fibers with different properties, thereby leading to protocols for tailored fiber characteristics. Retting with Texazym BFE and Bioprep, which reportedly contain only pectinases, resulted in fibers with significantly higher strength than cellulase-containing mixtures, i.e., Multifect Pectinex FE and Viscozyme. Bioprep in all formulations resulted in higher elongation, although values were still small at 2.0–2.5% (Table 6.3). Retting with Bioprep and Viscozyme produced fibers that tended to be finer than those from other treatments.

Kawahara et al. (2007) studied enzymatic-retting of kudzu fibers with several commercial enzymes, and the effects on the retting were compared with respect to the smoothness of the surface of the fibers and the mechanical properties. The commercial enzymes were classified into two types, that is, cellulase and pectinase. For the cellulase type, enzymatic decomposition occurred almost topochemically because of the cooperation of cellulase with high activity, and then the retting was fully achieved, suppressing the damage to the intercellular matrix (middle lamella) by pectinase. For the pectinase type, decomposition predominately occurred in the middle lamella joining the adjacent fibrous cells. Therefore, the tensile properties of the retted fibers were lowered. In the retting of kudzu fibers, a topochemical process is promising for producing fibers with excellent luster that retain the tensile properties

6.5 Effect of Enzyme-Retting on Fiber Yield and Properties

In the retting trials carried out on a semi-industrial scale, flax stems retted with enzyme yielded more fibers than chemical- or water-retted stems. This may have been due to effective hydrolysis of the flax pectins and hemicelluloses by their respective enzymes without damage to celluloses (Sharma 1987b). The differences in fiber fineness and strength of chemical-, enzyme-, and water-retted stems were not significant. Fluidities of the samples from the three treatments were low, suggesting that cellulose chain lengths were not reduced by enzyme treatment.

6.6 Effect of Enzyme-Retting on Effluent Properties

The analysis of the effluents from chemical-, enzyme-, and water-retting revealed that biochemical oxygen demand (BOD) and chemical oxygen demand (COD) levels were highest in the effluents derived from the enzyme-retting trial, and lowest in the effluents resulting from the water-retting trial (Sharma 1987b) (Table 6.4). This may have been because of the presence of the by-products of enzymatic hydrolysis. Total nitrogen, sulfate, and potassium were present in higher levels in samples from enzyme-retting than those from water-retting. Continuous aeration of the liquor during enzyme-retting may help to reduce the levels of BOD and COD without denaturing the enzymes present in the liquor.

Since flax-retting by polysaccharide-degrading enzymes is quicker and relatively free from the bad odor usually associated with water-retting, this process would be preferable to either water- or chemical-retting. The cost-effectiveness of using complex enzyme mixtures would improve if the liquor could be used for more than one cycle. This would minimize the need for effluent treatment usually associated with other forms of retting (Sharma 1987b).

Table 6.4 Effect of enzyme-retting on effluent properties

Parameter	Enzyme	Chemical	Water
pH	4.8	8.9	6.0
Suspended matter	4.0	10.1	3.2
BOO	3,520	640	2,100
COD	5,950	2,160	2,700
Total organic N	187	160	32
Total phosphate	48	19	15
Chloride	815	156	208
Sulfate	170	103	125
Sodium	70	500	375
Potassium	450	120	134

Based on Sharma (1987b)

References

Adamsen AP, Akin DE, Rigsby LL (2002) Chemical retting of flax straw under alkaline conditions. Textile Res J 72:789–794

Akin D (1998) Enzyme retting of flax for linen fibers: recent developments. In: Book of Papers American Association Textile Chemists and Colorists. American Association Textile Chemists and Colorists, Research Triangle Park, NC, pp 273–280

Akin D, Hardin I (2003) The current state of the applications of biotechnology, in Proceedings of the Annual International Conference & Exhibition, American Association of Textile Chemists and Colorists, Research Triangle Park, NC, 10–12 Sep 2003, pp 189–195

Akin D, Morrison W, Gamble G, Rigsby L (1997) Effect of retting enzymes on the structure and composition of flax cell walls. Textile Res J 67(4):279–287

Akin D, Dodd R, Perkins W, Henriksson G, Eriksson K (2000) Spray enzymatic retting: a new method for processing flax fibers. Textile Res J 70(6):486–494

Akin D, Foulk J, Dodd R, McAlister D (2001) Enzyme-retting of flax and characterization of processed fibers. J Biotechnol 89(2–3):193–203

Akin D, Foulk J, Dodd R (2002) Influence on flax fibers of components in enzyme retting formulations. Textile Res J 72(6):510–514

Akin D, Henriksson G, Evans J, Adamsen A, Foulk J, Dodd R (2004) Progress in enzyme-retting of flax. J Nat Fibers 1(1):21–47

Antonov V, Maixner V, Vicenec R, Fishcer H (2005) How do enzymes contribute to bast fibres industry? In Proceedings of the 11th Conference for Renewable Resources and Plant Biotechnology, Institute of Natural Fibres, Poznan, Poland

Antonov V, Marek J, Bjelkova M, Smirous P, Fischer H (2007) Easily available enzymes as natural retting agents. Biotechnol J 2(3):342–346

Bajpai P (1999) Application of enzymes in pulp & paper industry. Biotechnol Prog 15(2): 147–157

Brown AE (1984) Epicoccirm nigrum, a primary saprophyte involved in the retting of flax. Trans Br Mycol Soc 83:29–35

Brown AE, Sharma HSS (1984) Production of polysaccharide-degrading enzymes by saprophytic fungi from glyphosate-treated flax and their involvement in retting. Ann Appl Biol 105:65–74

Brown AE, Sharma HSS, Black DLR (1986) Relationship between Pectin Content of Stems of Flax. Cultivars, Fungal Cell Wall Degrading Enzymes and Pre- harvest Retting. Ann Appl Biol 109:345–351

Brühlmann F, Leupin M, Erismann K, Fiechter A (2000) Enzymatic degumming of ramie bast fibers. J Biotechnol 76(1):43–50

Domier KW (1997) The current status of the field crop fibre industry in Canada. Euroflax Newslett 8:8–10

Ebbelaar M, van der Valk H, van Dam J, de Jong E (2001) Highly efficient enzymes for the production of natural fibres, 8th International conference on biotechnology in the pulp and paper industry, Helsinki, Finland, 4–8 June 2001 edited by Vahala P, Lantto R, p 240

Euroflax Newsletter (2001) Institute of natural fibres. Poznan, Poland

Foulk J, Dodd R, Akin D (2000) New low cost flax fibers for composites. Paper No. 2000-01-1133. Society of Automobile Engineers

Foulk JA, Akin DE, Dodd BR (2008) Pectinolytic enzymes and retting. Bioresources 3(1): 155–169

Gillespie AM, Keane D, Griffm TO, Thohy MG, Donaghy J, Haylock RW and Coughan MP (1990) In: Kirk TK, Chang HM (eds) Biotechnology in pulp and paper manzifacture. Butterworth-Heinmann, Boston, pp 211–219

G bitz G, Cavaco-Paulo A (2001) Biotechnology in the textile industry – perspectives for the new millenium. J Biotechnol 89(2,3):89–312

Hamilton IT (1986) Linen. Textiles 15:30–34

Hardin I, Akin D, Wilson S (eds) (2002) Advances in biotechnology for textile processing, department of textiles, mechandising and interiors. University of Georgia, Athens

Henriksson G, Eriksson KEL, Kimmel L, Akin DE (1998) Chemical physical retting of flax using detergent and oxalic acid at high pH. Textile Res J 68:942–947
Hoondal GS, Tiwari RP, Tewari R, Dahiya N, Beg QK (2002) Microbial alkaline pectinases and their industrial applications: a review. Appl Microbiol Biotechnol 59(4–5):409–418
Kawahara Y, Tsuda T, Minami H, Nishiuchi S, Endo R (2007) Enzymatic retting of Kudzu fibers. J Appl Polymer Sci 106(4):2759–2762
Kenealy WR, Jeffries TW (2003) Enzyme processes for pulp and paper: a review of recent developments, Wood deterioration and preservation: advances in our changing world, edited by Goodell B, Nicholas DD, Schultz TB, Chapter 12, pp 210–239 [ACS Symposium Series 845, Washington, DC, USA: American Chemical Society, 2003, 465pp]
Kessler RW, Becker U, Kohler R, Goth B (1998) Steam explosion of flax-a superior technique for upgrading fibre value. Biomass Bioenergy 14:237–249
Kozlowski R, Batog J, Konczewicz W, Mackiewicz-Talarczyk M, Muzyczek M, Sedelnik N, Tanska B (2005) Latest state-of-art in bast fibers bioprocessing, in Proceedings of the 11th Conference for Renewable Resources and Plant Biotechnology, Institute of Natural Fibres, Poznan, Poland
Lepsch D, Horal JW (1998) Development of an integrated modular plastic electrical carrier and flax/polypropylene shelf panel for a vehicle rear shelf system. In: Proceedings of the 1998 Society for Automotive Engineering International Congress and Exposition, Paper No. 980727, pp 87–94
Sharma HSS (1986) An alternative method of flax retting during dry weather. Ann Appl Biol 109:605–611
Sharma H (1987a) Screening of polysaccharide-degrading enzymes for retting flax stem. Int Biodeterior 23(3):181–186
Sharma H (1987b) Studies on chemical and enzyme retting of flax on a semi-industrial scale and analysis of the effluents for their physico-chemical components. Int Biodeterior 23(6):329–342
Sharma H (1987c) Enzymatic degradation of residual non-cellulosic polysaccharides present on dew-retted flax fibres. Appl Microbiol Biotechnol 26:358–362
Sharma H (1988) Chemical retting of flax using chelating agents. Appl Biol 113:159–165
Sharma HSS, Faughey GJ (1999) Comparison of subjective and objective methods to assess flax straw cultivars and fibre quality after dew-retting. Ann Appl Biol 135:495–501
Sharma HSS, Robinson K (1983) Tech Rep No. 2281 p 11
Sharma H, Van Sumere C (1992) Enzyme treatment of flax. Gen Eng Biotechnol 12:19–23
Sotton M, Ferrari M (1989) Le lin ultra-affine par le traitement hydrolyse flash. L'Industrie Textile 1197:58–60
Tubach M, Kessler RW (1994) Interdisciplinary approach for new flax products: examples of applied research at the IAF. In Proceedings World Fibre Flax Symposium, 7, 1–86. Connecticut Agricultural Experiment Station, New Haven, CT
Van den Oever MJA, Bos HL, Molenveld K (1999) Flax fibre physical structure and its effect on composite properties: impact strength and thermo-mechanical properties. Die Angew Makromol Chem 272:71–76
Van Sumere CF (1992) Retting of flax with special reference to enzyme-retting. In: Sharma HSS, Van Sumere CF (eds) The biology and processing of flax. M Publications, Belfast, pp 157–198
Van Sumere CF, Sharma HSS (1991) Analyses of fine flax fibre produced by enzymatic retting. Aspects Appl Biol 28:15–20
Westcott ND, Muir AD (2000) Medicinal lignans from flaxseed: isolation and purification. In: Shahidi F, Ho CT (eds) Phytochemicals and phytopharmaceuticals. AOCS Press, Champaign, pp 122–131

Chapter 7
Biopulping

7.1 Introduction

In the manufacture of paper from wood, the wood is first converted to pulp. Pulping involves treating wood to separate the cellulose fibers. Pulping processes are divided into two broad classes: chemical pulping and mechanical pulping. Chemical pulping involves the use of chemicals to solubilize the lignin in the wood cell wall and to release cellulose fibers. Lignin is a natural glue-like material that holds the wood cell wall together. Chemical pulping is a low yield process (about 50%) with significant waste treatment and chemical recycling costs; however, the pulp produced has extremely high strength properties. Mechanical pulping involves the use of mechanical force to separate cellulose fibers. Mechanical processes are high yield (up to 95%) but give paper with lower strength properties, high color reversion, and low brightness. Thus, currently available pulping processes offer a spectrum of pulp properties ranging from high yield, low strength mechanical pulps to low yield, high strength chemical pulp. A mixture of chemical pulp and mechanical pulp is used in many paper production processes to exploit these differences.

It has been suggested that biological systems can be also used to assist in the pulping of the wood. Attempts to improve primary pulp production processes by using isolated ligninolytic enzymes have so far been inhibited by the complex chemistry of the ligninolytic enzyme system, low yields in enzyme production, and the ultrastructure of wood itself. White-rot fungi, however, have great potential for this application. The concept of biopulping is based on the ability of some white-rot fungi to colonize and degrade selectively lignin in wood, thereby leaving cellulose relatively intact. There are certain process conditions and design requirements necessary to gain a biopulping effect (Akhtar et al. 1998). Biopulping can be carried out in bioreactors of different types, including open chip piles, depending on the requirements of the particular microorganism would have for optimal results. High moisture content (around 55–60%) should be kept in wood chips during the biotreatment step to ensure an optimal colonization and penetration of fungal hyphae. The degree of asepsis should be controlled to ensure a successful wood colonization by the

P. Bajpai, *Biotechnology for Pulp and Paper Processing*,
DOI 10.1007/978-1-4614-1409-4_7,

particular fungal strain used depending on its resistance against contamination and ability to compete with the microbial biota existing in the wood chips.

In mechanical pulping, biotreated wood allows energy savings during refining and provides stronger pulps. In chemical pulping, the wood biotreatment can increase pulp yield, reduce alkali requirement, and increase the cooking capacity during kraft pulping (Bajpai et al. 1999).

7.2 Pulping Processes

The manufacture of pulp for paper and paperboard employs mechanical (including thermomechanical), chemimechanical, and chemical pulping methods.

7.2.1 Mechanical Pulping

Mechanical pulping processes are mainly of four types – stone groundwood (SGW) pulping, refiner mechanical pulping (RMP), thermomechanical pulping (TMP), and chemithermomechanical pulping (CTMP) (Leask and Kocurek 1987). The groundwood pulping process was the first process used for the production of paper from wood. This process involves pressing a log against a grindstone to pull off fibers, which are continuously washed away by a water stream. High temperatures in the refining zone caused by friction soften the wood and ease fiberization. The yield is high (about 95%) but because most of the lignin remains, the fibers are stiff and bulky. Paper produced from groundwood pulp has low strength and high color reversion, but the opacity is excellent. Groundwood pulping is an energy-intensive process (Table 7.1). Groundwood pulp has been the quality leader in magazine papers, and it is predicted that this situation will remain (Arppe 2001). In the RMP process, chips are processed through a rotating disk refiner. The refiner plate is made up of three zones first to break the chips, then to produce intermediate size fragments, and finally to produce single fibers. This produces fibers with better bonding properties, and thus better paper strength than SGW pulp. However, opacity is reduced, color reversion is similar, and the energy expenditure is increased compared to SGW pulping (Table 7.1). RMP, as well as TMP and CTMP, is usually performed as a two-stage process. The first stage separates wood into individual fibers. The second stage loosens the structure of the fiber walls to increase fiber flexibility and fibrillation,

Table 7.1 Energy requirement in the production of mechanical pulps

Process	Net-energy requirement (kWh/ton)
Refiner mechanical pulping (RMP)	1,975–2,275
Groundwood pulping (SGW)	1,300–1,675
Thermomechanical pulping (TMP)	2,175–2,900
Chemithermomechanical pulping (CTMP)	1,375–1,775

improving fiber bonding and thus paper strength. The TMP process is a modification of the RMP process involving a steam pretreatment at 110–150°C to soften the wood followed by refining. In the first stage, the refiners are at elevated temperature and pressure to promote fiber liberation; in the second stage, the refiners are at ambient temperature to treat the fibers for papermaking. The higher temperature during refining in the first step, 110–130°C, softens the fibers and allows their recovery with minimal cutting and fines generation. The refining is performed just under the glass transition temperature of lignin, approximately 140°C, so that separation of fibers occurs at the S-1 cell wall layer. This improves fibrillation and access to hydroxyl groups for hydrogen bonding. The high strength of this pulp relative to the other mechanical pulps has made it the most important mechanical pulping process. Energy requirements are 2,175–2,900 kWh/ton (Table 7.1). Over two-thirds of this is used in the primary pressurized refining step, and less than one-third is used in the secondary atmospheric pressure refining step. The pulp yield is >93%. Solubilization of wood components results in relatively high BOD in mill effluents. The CTMP process is a further refinement of TMP which involves pretreatment of wood chips with sodium sulfite (about 2% on dry wood) at pH 9–10 or sodium hydroxide (with hydrogen peroxide in the alkaline peroxide method), then steaming at 130–170°C and finally refining. Liquor penetration is often achieved by a system that compresses the wood chips into a liquid-tight plug that is fed into the impregnator vessel where the chips expand and absorb the liquor. Chemical pretreatments of wood chips are used to enhance the strength properties of mechanical pulps. The addition of CTMP to pulp blends may reduce or eliminate the requirement for kraft pulp. Capital expenditures for a CTMP plant are one-fifth those of a kraft mill of comparable size (Karl 1990). The energy expenditures are decreased (Table 7.1) but the yield is also decreased to 85–91% by removing wood substance. The CTMP process generates more pollutants than other mechanical pulping processes and thus increases waste treatment costs.

Currently, mechanical pulps account for 20% of all virgin fiber material. It is foreseen that mechanical paper will consolidate its position as one major fiber supply for high-end graphic papers. The growing demand on pulp quality in the future can be achieved only by the parallel use of softwood and hardwood as a raw material. The largest threat to the future of mechanical pulp is its high specific energy consumption. In this respect, TMP processes are most affected due to their considerably higher energy demand than groundwood processes.

7.2.2 *Semichemical Pulping*

Semichemical pulping processes are characterized by a mild chemical treatment preceded by a mechanical refining step. Semichemical pulps, which apply to the category of chemical pulps, are obtained predominantly from hardwoods in yields of between 65 and 85% (average ca. 75%). The most important semichemical process is the neutral sulfite semichemical process (NSSC), in which chips undergo partial chemical pulping using a buffered sodium sulfite solution, and are then treated in disk refiners to complete the fiber separation. The sulfonation of mainly middle lamella lignin causes a partial dissolution so that the fibers are weakened for the subsequent mechanical defibration. NSSC pulp is used for unbleached products where good strength and stiffness are particularly important; examples include corrugating medium, grease-proof papers, and bond papers.

7.2.3 Chemical Pulping

Chemical pulps are made by cooking (digesting) the raw materials, using the kraft (sulfate) and sulfite processes.

7.2.3.1 Kraft Process

Kraft process produces a variety of pulps used mainly for packaging and high-strength papers and board. Wood chips are cooked with caustic soda to produce brownstock, which is then washed with water to remove cooking (black) liquor for the recovery of chemicals and energy. The Kraft process dominates the industry because of the advantages in chemical recovery and pulp strength. It represents 91% of chemical pulping and 75% of all pulp produced. A number of pulp grades are commonly produced, and the yield depends on the grade of product. Unbleached pulp grades, characterized by a dark brown color, are generally used for packaging products and are cooked to a higher yield and retain more of the original lignin. Bleached pulp grades are made into white papers. Nearly half of the Kraft production is in bleached grades, which have the lowest yields. The superiority of kraft pulping has further extended since the introduction of modified cooking technology in the early 1980s.

7.2.3.2 Sulfite Process

This process uses different chemicals to attack and remove lignin. Compared to Kraft pulps, sulfite pulps are brighter and bleach more easily, but are weaker. Sulfite pulps are produced in several grades but bleached grades dominate production. Yields are generally in the range of 40–50%, but tend toward the lower end of this range in bleached grades. Compared to the Kraft process, this operation has the disadvantage of being more sensitive to species characteristics. The sulfite process is usually intolerant of resinous softwoods, tannin-containing hardwoods, and any furnish-containing bark. Sulfite process produces bright pulp which is easy to bleach to full brightness and produces higher yield of bleached pulp which is easier to refine for papermaking applications. The sulfite process is characterized by its high flexibility compared to the kraft process, which is a very uniform method, which can be carried out only with highly alkaline cooking liquor. In principle, the entire pH range can be used for sulfite pulping by changing the dosage and composition of the chemicals. Thus, the use of sulfite pulping permits the production of many different types and qualities of pulps for a broad range of applications. The sulfite process can be distinguished according to the pH adjusted into different types of pulping. The main sulfite pulping processes are Acid (bi)sulfite, Bisulfite (Magnefite), Neutral sulfite (NSSC), and Alkaline sulfite.

7.3 Biomechanical Pulping

White-rot fungi have great potential for biotechnological applications. They not only produce the whole set of enzymes necessary for lignin degradation, but can also act as a transport system for these enzymes by bringing them into the depth of the wood chips and create the physiological conditions necessary for the enzymatic reactions. Fresh wood chips stored for pulp production are rapidly colonized by a variety of microorganisms, including many species of fungi. These organisms compete vigorously while easily assimilable foodstuffs last and then their population decreases. They are replaced by fungi that can degrade and gain nourishment from the cell wall structure polymers: cellulose, hemicelluloses, and lignin. Left unchecked, these last colonizers, mostly white-rot fungi, eventually decompose the wood to carbon dioxide and water. Some white-rot fungi selectively degrade the lignin component, which is what chemical pulping process accomplishes. It is these fungi which are useful for biopulping.

Fungal delignification of wood for biopulping was first seriously considered by industrial researchers of West Virginia Pulp and Paper Company (now Westvaco Corporation) in 1950s (Lawson and Still 1957). The researchers wondered whether wood chips could be inoculated with a lignin-degrading fungus during transport and storage, and thereby become partially pulped. They published a survey of 72 lignin-degrading fungi, summarizing knowledge about fungal degradation of lignin. In the 1970s, Eriksson's group at the Swedish Forest Products Laboratory (STFI) launched a more intensive investigation. A fungus isolated in Sweden, *Phanerochaete chrysosporium* was characterized by a high optimum temperature for growth, rapid growth, and selective lignin degradation in incipient stages of birch wood decay. This fungus was proposed to be a useful "wood defibrator" in the pulping process. A U.S. patent was obtained by STFI for the process (Eriksson et al. 1976). Considerable efforts at STFI were directed toward developing cellulase-less mutants of selected white-rot fungi to improve the selectivity of lignin degradation and thus the specificity of biopulping (Johnsrud and Eriksson 1985). In one study, using spruce and pine wood, up to 23% energy savings and an increase in tensile index were noticed. On a large scale, success was achieved on bagasse (Johnsrud et al. 1987) while the results using wood chips were less encouraging. An energy requirement of 4,800 kWh/ton for producing chemimechanical pulp (CMP) of 70°SR according to the Cuba-9 process (6% NaOH treatment at 90°C for 10–20 min) was

Table 7.2 Energy requirement for chemimechanical pulp (CMP) and biochemimechanical pulp (BCMP) from bagasse

	Energy input (kWh/ton)	
Refining equipment	CMP	BCMP
Defibrator and PFI mill	4,800	1,700
Disk refiner	3,100	2,100

Based on Johnsrud et al. (1987)

decreased to 1,700 kWh/ton by pretreating the bagasse with fungi as shown in Table 7.2. The strength properties of biochemimechanical pulp (BCMP) were better than those of CMP but there was a small drop in the yield of BCMP due to fungal degradation of bagasse. STFI's work has been summarized in a number of publications on biomechanical pulping and related aspects (Eriksson et al. 1976, 1980; Johnsrud and Eriksson 1985; Johnsrud et al. 1987; Ander and Eriksson 1975; Eriksson and Vallander 1980, 1982; Eriksson 1985; Setliff et al. 1990).

Preliminary research on biopulping was conducted at Forest Products Laboratory (FPL-USDA) at Madison, Wisconsin in 1970s (Kirk 1993). Kirk et al. (1994) at FPL showed that aspen wood chips treated with *Rigidoporus ulmaris* consumed less energy during pulping and produced stronger paper (FPL internal report 1972). Bar-lev et al. (1982) showed that treatment of a coarse mechanical pulp with *P. chrysosporium* decreased the energy requirement (25–30%) for further fiberization and improved the paper strength properties. Akamatsu et al. (1984) found that treatment of wood chips with white-rot fungi decreased the mechanical pulping energy and increased paper strength.

A comprehensive evaluation of biomechanical pulping was launched in 1987 at the FPL at Madison, Wisconsin after the establishment of Biopulping Consortium I, which involved the FPL, the Universities of Wisconsin and Minnesota and 20 pulp and paper and related companies. The overall goal was to establish the technical feasibility of using fungal treatment with mechanical pulping to save energy and/or improve paper strength. It was assumed that fungal pretreatment would have less environmental impact than would chemical pretreatments, which turned out to be the case. The consortium research was conducted by seven closely coordinated research teams: fungal, pulp and paper, enzyme, molecular genetics, economics, engineering and scale-up, and information. The scientists of the consortium investigated all fields of research relating, to biopulping (Kirk et al. 1994; Lawson and Still 1957; Otjen et al. 1987; Blanchette et al. 1988, 1992a, b; Myers et al. 1988; Sachs et al. 1989, 1990, 1991; Leatham and Myers 1990; Akhtar et al. 1992a, b, 1993, 1998; Leatham et al. 1990a, b; Wegner et al. 1991). The first report of Biopulping consortium I, a 5-year research and information program, was published in 1993 (Kirk 1993). Biopulping Consortium II was established in 1992 and extended until June 1996, mainly for the scale-up of the process and other important aspects. Several white-rot fungi were screened for their biopulping performance using aspen wood chips (Myers et al. 1988; Leatham and Myers 1990; Leatham et al. 1990a, b; Akhtar et al. 1996). Based on energy savings and improvements in paper strength properties, six fungi – *P. chrysosporium*, *Hypodontia setulosa*, *Phlebia brevispora*, *Phlebia subserialis*, *Phlebia tremellosa*, and *Ceriporiopsis subvermispora* – were selected. The energy-saving potentials of these fungi on biomechanical pulping of loblolly pine are given in Table 7.3.

Table 7.3 Energy savings from biomechanical pulping of loblolly pine chips with different white-rot fungi (4-week incubation)

Fungus	Energy savings (%)
Phanerochaete chrysosporium	14
Hyphodontia setulosa	26
Phlebia brevispora	28
Phlebia subserialis	32
Phlebia tremellosa	36
Ceriporiopsis subvermispora	42

Based on Leatham et al. (1990a, b)

Out of about 200 strains, two fungi exhibiting a great deal of intraspecific variation (Akhtar et al. 1992a and Blanchette et al. 1992b) seem to be especially useful for biopulping: *P. chrysosporium* for hardwoods and *C. subvermispora* for hardwoods and softwoods. Various reactor types including rotary drums (Myers et al. 1988), stationary trays (Akhtar et al. 1992a, 1996) and a static bed bioreactor (Akhtar et al. 1992b) were tested on a 2–5 kg scale. The best results were obtained with strains of *C. subvermispora* on aspen and loblolly pine (Akhtar et al. 1992b). On aspen, energy savings of 48% were accompanied by increases in burst and tear indices of 40 and 162% respectively. The effects on loblolly pine amounted to 37% energy savings and 41 and 54% increase in burst and tear indices, respectively. The optical properties deteriorated with both types of wood. After 4 weeks of treatment, a weight loss of 6% for aspen and 5% for loblolly pine was measured. *C. subvermispora* proved to be superior to other selective white rotters (Leatham et al. 1990a, b). However, *P. chrysosporium* has the advantage of competitiveness at temperature between 35 and 40°C. When different strains of *C. subvermispora* were tried on pine, the energy savings ranged from 21 to 37% (Kirk 1993). Adding nutrient nitrogen to the chips as a defined source (L-glutamate or ammonium tartrate) increased energy savings and improved strength properties but led to a high weight loss. Addition of a chemically undefined N source to aspen chips gave large biopulping benefits with low weight loss, using both *P. chrysosporium* and *C. subvermispora.* Wood batch was found to have little influence on the outcome of biopulping and chip storage method (fresh, air dried or frozen) and inoculum age and form (spore, mycelial suspension or colonized chips) were without significant influence.

Another American group has also reported that aspen chips treated with *C. subvermispora* for 17 days required 20% less energy for pulping, while the refining energy of Norway spruce was reduced by 13% (Setliff et al. 1990). Strength properties were increased with aspen and spruce, but no increase was found with eucalyptus.

In Japan, biopulping research has been conducted mainly in industrial laboratories. Kobe steel and Oji paper seem to be the major industrial players. Kobe steel has obtained a broad US patent on the use of white-rot fungi particularly NK-1148, for the treatment of primary mechanical pulp to save energy (Kobe Steel 1988). Applications for four Japanese patents were filed: (1) an inoculum method (Kojima 1988), (2) two improved biopulping strains (Kobe Steel 1988), (3) a silo type bioreactor (Akamatsu et al. 1988), and (4) treatment of chips during transport in ships with a white-rot fungus to enhance pulping (Heden et al. 1988). Nishida (1989) and Nishida et al. (1988) have developed a screening method for selective lignin degradation which was used to identify the strains used in two of the above patents.

Biomechanical pulping of nonwood fibers-straw, kenaf, and jute was also successful (Martinez et al. 1994; Sabharwal et al. 1994, 1995). The energy consumption in refining was substantially lower and the strength properties were higher for the fungal-treated bast strands (Sabharwal et al. 1994, 1995). The opacity and drainage properties were also superior for biomechanical bast pulps, but the brightness level was lower. Scanning electron microscopy of fungus-treated bast strands after refining showed that fibers appeared to separate more readily from adjacent fibers than in noninoculated treatments. Italian researchers studied treatment of nonwoody raw materials with a mixture of various type of enzymes for saving energy and reducing chemical consumption while maintaining good properties of CTM pulp (Giovannozzi-Sermanni et al. 1997). The level of energy savings was found to depend on the type of raw material, ranging from 21% for rice straw up to 40% for kenaf bast. Enzyme treatment significantly improved tear index regardless of the cellulose source whereas the tensile index decreased in wheat straw and kenaf bast samples. Burst index was slightly improved in all the biotreated samples, except kenaf. Pulp yields of the biotreated samples were, without exception, significantly higher than those of the corresponding control samples. This was apparently due to the lower chemical charge needed for biotreated samples.

One of the major costs foreseen during the scale-up of biopulping was for inoculum production. Akhtar et al. (1996, 1997a, b) discovered that the amount of inoculum could be lowered to 5 kg/ton wood chips (dry weight basis) or less by adding an inexpensive and commercially available nutrient source, corn steep liquor (CSL), to the mycelial suspension. Subsequent studies have also identified a better strain of *C. subvermispora* that gave up to 38% energy savings and improved tear index by 51% compared to the control in the presence of CSL (Akhtar et al. 1996). After the practical and economical feasibility of biopulping was proved on the laboratory scale, accurate kinetic data were needed to determine the potential for the biopulping process on a large scale. Techniques for monitoring dry weight loss and growth rate as functions of time using carbon dioxide production data have been developed (Wall et al. 1993). Other aspects of biomechanical pulping like prediction of energy saving and brightness stability were also studied (Akhtar 1994; Akhtar et al. 1995a, b; Sykes 1993). Based on the practical and economical feasibility study of biopulping, a chip pile-based system has been proposed (Fig. 7.1) (Akhtar et al. 1997a).

Biopulping Consortium conducted successful 5 and 100 ton trials in outdoor chip piles at the FPL-USDA, Madison, WI. The results obtained were similar to those obtained in the laboratory scale bioreactors. The fungal treatment saved 32% electrical energy (Akhtar et al. 1997a). Equipment and techniques for the pilot scale treatments and scale-up to mill scale were discussed by Scott et al. (1997). Contaminating microorganisms on the chip surfaces are controlled by brief exposure of the chips to steam, before addition of the fungal inoculum fortified with CSL. Metabolic heat from the growth of the fungus on the wood chips is removed by forced ventilation. The fungus *P. subserialis* showed operational advantages over *C. subvermispora*, because less mycelial growth outside the chips and lower compressibility of the treated chips resulted in lower resistance to air flow (Scott et al. 1997). Economic analysis of a mill-scale design suggested net savings of about

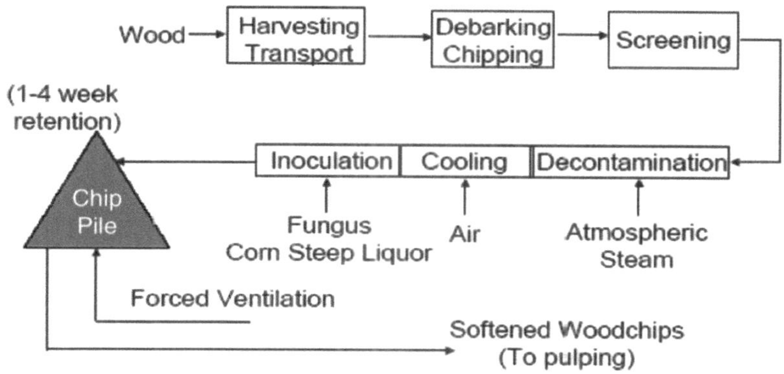

Fig. 7.1 Biopulping process can be fitted into an existing mill's wood handling system

$10/ton and a 25% annual return on investment (Scott et al. 1997). The Biopulping consortium has obtained several patents on the process (Blanchette et al. 1991; Akhtar et al. 1995a, b; Akhtar 1997).

Brazilian researchers biotreated *Eucalyptus grandis* wood chips with *C. subvermispora* in a 50-ton chip pile and evaluated for TMP and CTMP processing on a mill scale (Guerra et al. 2004, 2005, 2006). Biotreatment on the 50-ton chip pile was performed after a series of scale-up procedures starting with precolonized wood chips prepared in 20 L bioreactors. The first step included 760 kg of decontaminated wood chips and 40 kg of the start-up precolonized wood chips. Second scale-up used the 800 kg colonized wood chips to prepare an 8-ton pile. A final scale-up was conducted using the wood chips precultured in the 8-ton pile as inoculum seed to build a 50-ton pile. After 60 days of biodegradation, the wood chips from the last pile were refined on a mill-scale by using a two-stage thermomechanical process (Guerra et al. 2006). In this trial, the wood weight loss was 9% based on basic wood density values of untreated and biotreated samples: 413 and 376 kg/m^3, respectively. The average energy consumption for producing TMP pulps with 450–470 CSF was 913 and 745 kWh/ton for control and biotreated wood chips, respectively (18% of energy saving in the pulping process). In the case of CTMP pulps with similar CSF, energy consumption was 1,038 and 756 kWh/ton for control and biotreated wood chips, respectively (27% of energy saving in the pulping process). Tensile indexes of biomechanical pulps were higher in comparison to reference pulps (Table 7.4). However, chip pile contamination with opportunist fungi has been observed when the process was initiated by wood chip inoculation with blended mycelium and CSL as a co-substrate (Ferraz et al. 2008).

Brazilian researchers further investigated biopulping of *E. grandis* wood chips with *P. chrysosporium* RP-78 under nonaseptic conditions in laboratory and mill wood-yard (Masarin and Ferraz 2008; Masarin et al. 2009). The ability of *P. chrysosporium* to compete with indigenous fungi present in fresh wood chips was notorious under controlled laboratory experiments. A subsequent step involved an industrial test performed with 10 ton of fresh wood chips inoculated and maintained

Table 7.4 Tensile indexes of biomechanical pulps

Properties	Laboratory-prepared 2-stage TMP/RMP (blank)	Laboratory-prepared 2-stage BioTMP/RMP	Mill-prepared 2-stage TMP/RMP (blank)	Mill-prepared 2-stage BioTMP/RMP	Mill-prepared 2-stage CTMP	Mill-prepared 2-stage BioCTMP
Tensile index (Nm/g)	5.2	11.3	12±2	11±2	16±3	16±2
CSF (mL)	402	390	472±31	455±26	485±16	409±23

Based on Ferraz et al. (2008)

Table 7.5 Properties of mill-refined pulps prepared from *Eucalyptus* wood chips treated with *Phanerochaete chrysosporium*

Samples	Consistency, first disk refiner (%, w/v)	Shive content of the pulps (%, w/w)	Refining degree (degree SR)	Specific volume (cm^3/g)	Average fiber length (mm)	Tensile index (Nm/g)	Delamination strength (kPa)
Control	29.4	0.12	22	2.9	557	25±1	217±19
Fungal-treated	22.5	0.04	22	2.5	562	33.6±0.5	295±30

Based on Masarin et al. (2009)

at 37±3°C for 39 days in a biopulping pilot plant. Biotreated wood chips were pulped in a CTMP mill. Net energy consumption during refining was 745 and 610 kWh/ton of processed pulp for control and biotreated wood chips, respectively. Accordingly, 18.5% net energy saving could be achieved. Biopulps contained lower shive content and had improved strength properties compared to control pulps. Tensile index improved from 25±1 to 33.6±0.5 Nm/g and delamination strength from 217±19 to 295±30 kPa (Table 7.5).

Mechanical pulping processes release organic materials (sugars, low molecular weight lignins, extractives) from wood, and these materials appear in the effluent stream (Eriksson 1985). Mill effluents could contain as many as 100 potentially toxic constituents. Moreover, even low levels of several benign compounds could interact synergistically to produce a toxic effluent (Johnson and Butler 1991). CTMP processes produce effluents of high color and BOD which may be difficult to treat; environmental considerations have kept this process from being used in many locations. Canadian CTMP mills have been reported to generate effluents having pollution loads of 35–65 kg BOD/ton pulp and 70–200 kg COD/ton pulp, depending on the pulping conditions and process as well as wood species used (Environment Canada Report 1988). Lo et al. (1991) based on the characterization of effluents from 17 different sources in a Canadian CTMP pulp and paper mill, showed that the concentrations and loadings of BOD, COD, resin and fatty acids (RFA), and other polluting constituents in the effluent from CTMP washing were very much higher than those in other effluents. The approximate pollutant amounts sent to the receiving water from the primary clarifier are BOD 63.3 kg, COD 90 kg, and RFA 0.43 kg/ton pulp. RFA constituents of wood are extracted during mechanical pulping/refining, and are major contributors to effluent toxicity (Leach and Thakore 1976).

Table 7.6 Characteristics of bleached CTMP wastewater

Parameter	Value
pH	6.5
Acetic acid (mg/L)	1,360
COD-total (mg/L)	9,300
COD-soluble (mg/L)	5,030
TSS (g/kg)	2.45
VSS (g/kg)	1.98
Resin acids (mg/L)	36–40

Based on Kennedy et al. (1992)

Table 7.7 Composition of resin acids in bleached CTMP wastewater

Resin acid	Percentage
Pimaric	7
Sandaracopimaric	6
Isopimaric	12
Levopimaric	21
Dehydroabietic	22
Abietic	26
Neoabietic	6
Total	*100*

Based on Kennedy et al. (1992)

CTMP wastewater characteristics are given in Table 7.6 and the composition of resin acids is represented in Table 7.7 (Kennedy et al. 1992). It has also been observed that the CTMP wastewater can be toxic to anaerobic bacteria (anaerobic treatment is sometimes given to CTMP effluents to reduce their pollution load) because of the presence of resin acids (Welander 1988; Hall and Cornacchio 1988). However, Kennedy et al. (1992) reported that resin acids were toxic to anaerobic bacteria but were not responsible for all the toxicity in bleached CTMP effluent. Furthermore, resin acid shocks were found to be inhibitory to the batch anaerobic system. However, short-term shocks of resin acids up to concentrations of 400 mg/L (38 mg resin acid/g VSS.d) had little effect on UASB (Upflow anaerobic sludge blanket) reactor efficiency. Long-term continuous exposure to abietic acid up to a concentration of 600 mg/L (60 mg resin acid/g VSS.d) did not significantly affect UASB reactor performance. In other words, anaerobic bacterial biomass can be acclimatized, to some extent, to the resin acid containing wastewater from CTMP plants.

The environmental impact of a new pulping process is a critical factor in assessing its viability. Ideally, a new process should be more environmentally compatible than existing processes. One question about biopulping is whether its effluents are more detrimental to the environment than are effluents from traditional mechanical pulping and refining. Specifically, do fungal metabolites or fungal degradation products of wood introduce additional toxicity to the effluent or significantly increase the BOD and COD load?

The microtox method of toxicity test has been used satisfactorily as a rapid screening method in the evaluation of acute toxicity of pulp mill effluents (Firth and

Table 7.8 BOD, COD, and toxicity of nonsterile aspen chips after treatment with *C. subvermispora*

Pulp[a]	BOD (g/kg pulp)	COD (g/kg pulp)	EPA toxicity[b] (100/EC_{50})
Raw chips	18	40	33
Control			
No nutrients	10	30	5
Nutrients added	12	33	9
Fungus treated			
No nutrients	10	33	6
Nutrients enriched	11	35	4

Based on Sykes (1994)
[a]All pulps incubated for 4 weeks at 27°C, except raw chips
[b]EC_{50} is a measure of toxicity

Backman 1990). Microtox assay uses luminescent bacteria to measure acute toxicity. Samples of wastewater from the first refiner passes of aspen chips pretreated with either *P. chrysosporium* or *C. subvermispora* were analyzed for BOD, COD, and microtox toxicity (Kirk 1993; Sykes 1994). Effluents from pulping fungus-treated chips were substantially less toxic than effluents from pulping raw chips (Table 7.8) (Sykes 1994). BOD values for effluents from fungus-treated pulps were slightly higher than for RMP of raw chips. The COD values for effluents from fungus-treated pulps were considerably higher than for RMP of raw chips probably because of the release of lignin degradation products (Sykes 1994). Addition of nutrients to the aspen chips also affected the BOD and COD loads of effluents. BOD decreased after 4-week incubation with nutrient-enriched *P. chrysosporium*, while BOD remained unchanged following incubation with nutrient-enriched *C. subvermispora*. COD remained unchanged after 4-week incubation with nutrient-enriched *P. chrysosporium*, while COD increased in effluents from chips pretreated for 4 weeks with nutrient-enriched *C. subvermispora*. As different white-rot fungi used for pulp treatment have differing nutrient requirements, an important factor in optimizing the biopulping process is to establish the minimum amount of nutrient required to assure fungal growth. The BOD and COD data indicate that *C. subvermispora* does not require much added nutrients. It has been concluded that the aspen biopulp effluents were less toxic, and probably contained considerably lower BOD and COD levels, than aspen TMP or CTMP at comparable yields (Sykes 1994) although no data on CTMP were available for comparison. It is likely that the environmental impact of CTMP effluent will be more than that of RMP or TMP effluents because of more extractives in the CTMP effluent. As the resin content of the wood chips decreases on fungal treatment (Fischer et al. 1994) it is expected that the amount of these extractives will be less in the effluent after mechanical pulping/refining of these chips, resulting in reduced toxicity (Table 7.9).

Published values for BOD load, based on the EPA survey for establishing effluent guidelines, are 20 kg/ton of o.d. pulp for TMP and 95 kg/ton of o.d. pulp for CTMP (Springer 1986). A commercial CTMP mill reported a BOD load of 45 kg/air dry ton, or approximately 52 kg/ton of o.d. pulp, for aspen CTMP (Jakko Poyry Inc 1985). *C. subvermispora* biopulping effluents contained approximately 24 kg/ton pulp and *P. chrysosporium* biopulping effluents contained 20 kg/ton pulp, at

Table 7.9 Effect of fungal treatment on resin content (% of dry wood) of loblolly pine and spruce chips

Time	Loblolly pine		Spruce		
(weeks)	Control	*C. subvermispora*	Control	*C. subvermispora*	*P. chrysosporium*
0	2.55	2.55	1.2	1.2	1.2
1	2.62	2.04	–	–	–
2	2.64	1.93	1.2	0.8	0.9
3	2.55	1.75	–	–	–
4	2.63	1.75			

Based on Fischer et al. (1994)

comparable yield (Sykes 1994) less than half the commercial CTMP values. *P. chrysosporium* has been used to remove color in the MyCoR (mycelial color removal) system (Eaton et al. 1982; Joyce and Pellinen 1995) and it was later discovered that this also decreased effluent toxicity. The decrease of toxicity observed for the fungus-treated biopulps is consistent with biotreatment of mill effluents by MyCoR system.

7.4 Biochemical Pulping

Fungal pretreatment of wood before chemical pulping has received relatively little attention. However, the fungi that are effective in biomechanical pulping have been tested as pretreatments for both kraft and sulfite pulping in a few studies (Messner et al. 1997).

Kraft pulps prepared from chips of aspen or red oak pretreated with *P. chrysosporium* for 10–30 days cooked faster and gave higher yields at a given kappa number (residual lignin content) than pulps from untreated chips (Oriaran et al. 1990, 1991). The improved cooking properties of the fungus-treated chips were attributed to enhanced penetration of cooking liquor and a lower lignin content. The fungus-pretreated pulps were more responsive to beating and gave higher tensile strength than the control pulps (Oriaran et al. 1990). The environmental consequences of the fungal treatments were not addressed in these studies, but the substantial wood weight loss (up to 17%) during the fungal pretreatment would negate the reported yield improvement, and the darkening of the wood by the fungus would probably require application of more bleaching chemicals.

Pretreatment of mixed hardwood chips with the Cartapip® 97 fungus, *Ophiostoma piliferum*, for 21 days improved kraft pulping efficiency; the kappa number decreased by 29% or the active alkali concentration in the pulping liquor decreased by 20% (Wall et al. 1996). Pulp yield was unaffected and viscosity, an indicator of pulp strength, was increased. The improvements were attributed to enhanced liquor penetration resulting from removal of ray parenchyma cells, resin deposits, and pit membranes. The Cartapip pretreatment could be used to reduce the environmental impact of kraft pulp production by decreasing bleach chemical usage or by decreasing cooking chemical usage (Wall et al. 1996).

Table 7.10 Biokraft pulping of eucalyptus with *C. subvermispora* at reduced active alkali charge

	AA charge (%)			
	17	14	14	14
Parameter	Control	Treated	Control	Treated
(a) Pulp properties				
P. No.	13.50	15.86	16.28	15.86
Lignin (%)	1.55	1.57	–	–
Unbleached brightness (% ISO)	27.3	28.3	25.9	28.3
Unbleached pulp yield (%)	45.67	45.53	47.15	45.53
Final brightness (% ISO)	87.0	88.3	87.6	89.1
Bleach chemical consumption (kg/ton)				
Elemental Cl_2	37.5	37.5	46.1	46.1
NaOH	19.1	19.1	18.9	18.9
Hypo	13.5	13.5	12.8	12.8
Chlorine dioxide	6.0	6.0	6.0	6.0

	Unbleached			Bleached	
	Control		Treated	Control	Treated
Parameter	17% AA	14% AA	14% AA	17% AA	14% AA
(b) Strength properties					
Wetness (SR)	16.5	17.0	17.5	35.0	35.0
Beating time (min)	–	–	–	29.0	22.5
Tensile index (Nm/g)	33.68	34.10	40.75	66.25	72.26
Breaking length (m)	3,435	3,478	4,157	6,757	7,364
Burst index (kN/g)	1.38	1.62	1.89	4.30	4.85
Tear index (mNm^2/g)	5.45	5.77	6.81	7.68	7.88
Double fold (No.)	5	8	10	58	80

Fungal treatment for 2 weeks; Inoculum level, 5 g/T wood
From Bajpai et al. (2001, 2003)

Screening of 283 basidiomycetes for ability to improve the efficiency of kraft pulping of pine chips revealed *Coriolus versicolor, Pycnoporus sanguineus,* and *Stereum hirsutum* as the most promising species (Wolfaardt et al. 1996). Pretreatment of the pine chips with *S. hirsutum* for 9 weeks reduced the cooking time required to reach a kappa number of 28, but increased alkali consumption and lowered pulp yield and viscosity. The fungal pretreatment did not seem to give economical or environmental benefits.

Bajpai et al. (2001, 2003) observed that extractive content reduced by 17–39% and the AA requirement reduced by 18%, when eucalyptus chips treated with the fungus *C. subvermispora* for 2 weeks were subjected to Kraft pulping (Table 7.10). Brightness and strength properties of biopulps were better than the control and the pulps were easier to bleach and easier to refine – requiring less energy (by 18–30%).

Çöpür and Tozluoğlu (2007) examined the effects on pulp and paper properties of pretreating Brutia pine wood chips with *C. subvermispora* and adding AQ and sodium borohydrate to the white liquor. Results showed that, compared with the control kraft method, pulp rejects were lower for biokraft pulp and a significant reduction was observed when the biokraft process was modified by sodium borohydride. An increase in pulp yield and reduction in kappa number was also seen with

Table 7.11 Soda pulping of wheat straw with *C. subvermispora* strains 1 and 2 at reduced alkali charges

Parameters	Control	Strain 1	Strain 2	Strain 3
EA (%)	12	10	10	9.0
Kappa no.	28.6	28.6	26.7	28.3
Yield (%)	45.9	48.1	45.9	46.7
Brightness (% ISO)	34.6	34.6	35.7	34.7
Residual alkali (g/L)	1.9	2.1	2.2	1.9

Based on Bajpai et al. (2004a)

biokraft pulp. The addition of AQ and sodium borohydride into biopulping gave positive results in terms of pulp yield and kappa number compared with biokraft pulp. There was a significant increase in pulp brightness when biokraft pulping was modified with 2% sodium borohydrate. Biotreatments resulted in pulps that were easier to refine and the refined bio-kraft-AQ pulps had the highest tensile index. However, biopulps had a lower tear but higher burst index compared to the control kraft pulp.

Recently research has been carried out by Garmaroody et al. (2011) in which poplar chips were pretreated by *Trametes versicolor* for 1, 2, and 3 weeks and then after washing, the chips were air-dried for kraft pulping to achieve pulp kappa number of around 20. Analysis of the pulp samples indicated that fungi pretreatment of chips can degrade lignin and carbohydrates and affect kraft pulping and fiber characteristics. In pretreated pulp samples, higher chemical charge in pulping was observed, together with lower fine and higher long fiber fraction. Increasing pretreatment time increased the fiber length, cross-sectional area, width, cell wall thickness, and volume index, while fine length, fiber coarseness, and curl were reduced. It has been recommended that 2-week pretreatment of chips would produce acceptable overall fiber properties in kraft pulping.

Bajpai et al. (2004a) also studied pretreatment of wheat straw with lignin-degrading fungi to study its effect on chemical pulping. Treatment with *C. subvermispora* reduced the lignin and extractive content of wheat straw by 16.5 and 44.3%, respectively. For chemical pulping, pretreatment reduced the kappa number by 22–27% at the same alkali charge, reduced the alkali charge by 30 kg/ton of raw material for the same kappa number (Table 7.11), or reduced the cooking time by up to 30% for the same kappa number. The biopulp had higher brightness and whiteness than the pulp obtained from feedstock without any fungal treatment (Table 7.12). The COD load in the effluent was lower for biopulping than for conventional pulping. Biopulping benefits obtained with other white-rot fungal cultures, *P. subserialis* and *P. brevispora*, were less pronounced than the benefits obtained with *C. subvermispora*.

Bajpai et al. (2004b) also studied the pretreatment of bagasse with *C. subvermispora* strains and its effect on chemical pulping. Treatment of depithed bagasse with different strains of *C. subvermispora* reduced the kappa number by 10–15% and increased unbleached pulp brightness by 1.1–2.0 ISO points on chemical pulping at the same alkali charge. Bleaching of biopulps at the same chemical charge increased final brightness by 4.7–5.6 ISO points and whiteness by 10.2–11.4 ISO points. Fungal treatment did not result in any adverse effect on the strength properties of pulp.

Table 7.12 Effect of cooking time on soda pulping of *C. subvermispora*-treated wheat straw

	Cooking time (min)	Kappa no.	Yield (%)	Brightness (% ISO)
Control	60	28.1	45.9	34.1
	45	30.1	46.5	33.9
	30	31.5	47.1	33.1
	15	Ellipsis	–	–
C. subvermispora, Strain 2	60	21.9	46.1	38.2
	45	22.5	47.2	37.6
	30	24.1	47.8	37.1
	15	26.1	48.1	36.2

Based on Bajpai et al. (2004a)

Treatment of birch, maple, oak, sycamore, or pine chips with an enzyme mixture containing cellulase and hemicellulase for 24 h disrupted the pit membranes blocking pores between cells and increased the longitudinal and transverse diffusion rates of sodium hydroxide in the chips (Jacobs-Young et al. 1998). Kraft pulping experiments with enzyme-treated sycamore chips confirmed the expected improvement in delignification; addition of pectinase to the enzyme mixture further lowered kappa numbers without reducing yield (Jacobs et al. 1998). The pulps produced from enzyme-treated chips were easier to bleach with chlorine dioxide than control pulps, and had comparable strength properties. Reduced bleaching chemical use should result in lower effluent BOD, COD, and chloroorganic loadings.

Franco et al. (2006) studied the potential of *Drimys winteri* for the conventional kraft and biokraft pulp production. For biokraft pulping, wood chips were biotreated with the white-rot fungus *Ganoderma australe*. During the biotreatment, a selective pattern of biodelignification was observed and the wood chips biotreated for 15, 30, and 45 days were submitted to kraft cooking. At low cooking severity (*H*-factor below 1,500/h, 15% active alkali and 25% sulfidity), all biopulps presented lower kappa numbers than control pulps and approximately the same screened pulp yield. Biopulps were easily refined in a PFI mill, requiring less PFI revolutions to achieve the same fibrillation degree. The strength properties of the biopulps were similar to those of the control pulps.

Eucalyptus nitens requires more severe cooking conditions to produce bleachable kraft pulps. Mardones et al. (2006) attempted to find out whether a pretreatment of *E. nitens* with *C. subvermispora* would improve its performance during kraft pulping and improve the pulp properties. The biotreatment of the chips carried out for a period of 15 days resulted in 13.3% lignin loss and a limited glucan degradation (2%). The pulping of biotreated samples required lower active alkali charge to reach the target kappa number compared to the control untreated sample and exhibited better pulping selectivity. The pulp yield increased by 3 and 1.5% for the pulps of 22 and 16 kappa numbers, respectively. The biotreated pulp's strength properties were improved and were similar to those of *Eucalyptus globulus* reference pulp (Table 7.13).

Table 7.13 Properties of kraft pulps[a] prepared from *Eucalyptus nitens* and *Eucalyptus globulus*

Sample	Burst index (kPa m²/g)	Tensile index (Nm/g)	Tear index (m Nm²/g)
(a) Physical strength properties			
Control			
E. nitens	5.9	110.2	6.6
Biotreated			
E nitens	6.2	113.8	7.1
E. globulus	6.5	114.6	7.4

Sample	Opacity (%)	Brightness (%)
(b) Brightness and opacity		
Control		
E. nitens	96.2	39.4
Biotreated		
E nitens	94.2	41.3
E. globulus	93.4	39.5

Sample	Coarseness (mg/100 m)	Average fiber length (mm)	Fines content (%)
(c) Fiber properties			
Control			
E. nitens	5.79	0.63	6.08
Biotreated			
E. nitens	5.39	0.64	7.06
E. globulus	5.81	0.68	5.98

Based on Mardones et al. (2006)
[a]Kappa number 16 ± 1

Pretreatment of chips with while-rot fungi increases the rate of their delignification in sulfite pulping also (Messner et al. 1997). In magnesium-based sulfite pulping, treatment of birch and spruce chips with *P. tremellosa, P. brevispora, Dichomitus squalens,* and especially *C. subvermispora* for 2–4 weeks significantly reduced pulp kappa number. However the fungal treatments also reduced pulp strength and brightness. Pretreatment of loblolly pine chips with *C. subvermispora* for 2 weeks increased the rates of lignin and yield loss to the same extent in sodium bisulfite pulping, but preferentially enhanced delignification in calcium-acid sulfite pulping, and decreased shives production. The fungal treatment darkened the chips, so that equal amounts of bleaching chemicals were needed to brighten the treated and control pulps. BOD and COD levels were the same in effluents from the fungus-treated and control pulps, but the Microtox toxicity in the effluent from the fungus-treated chips was less than half that of the control. The reduced toxicity was attributed to biodegradation of RFAs by the fungus (Messner et al. 1997).

7.5 Biopulping with Laccase Mediator System

The bulk of biopulping research on lignocellulosics has focused on the utilization of the white-rot fungi. Some researches have also used laccase mediator system to study biopulping (Widsten and Kandelba 2008; Dyer and Ragauskas 2004; Petit-Conil et al. 2002; Vaheri et al. 1991). Softwood (pine) chips were treated with laccase together with an ABTS, HBT, or VA mediator by Dyer and Ragauskas (2004) before kraft pulping. Laccase/HBT was found to be the most beneficial LMS in terms of enhancing delignification and pulp yield. Petit-Conil et al. (2002) treated softwood (spruce) chips with laccases obtained from three fungi with a mediator (HBT) prior to TMP. Laccase/HBT saved refiner energy with two of the laccases by up to 20%, but the third laccase increased it. The effect on pulp properties in terms of mechanical strength and brightness of handsheets was mostly positive. The improved pulp properties were attributed to a modification of the fiber surface chemistry, and increased external fibrillation and bonding potential. A decrease of 15% in peroxide consumption during subsequent bleaching to equal brightness was achieved with one of the laccases without mediator compared to bleaching without laccase pretreatment. The patent by Vaheri et al. (1991) describes the use of laccase pretreatment for reducing energy consumption during mechanical pulping. The treatment also boosts the pulp strength properties and blue reflectance factor.

7.6 Mechanism of Biopulping

Mechanism of biopulping is not completely understood though extensive investigation has been made over the last decade. Akhtar et al. (1998) examined at the microscopic level the fungal growth patterns of *P. chrysosporium* and *C. subvermispora* in aspen wood chips to gain insight into the mechanism of biopulping. *P. chrysosporium* grew well both across the chip surfaces and throughout the cell walls. The hyphae penetrated the chips through the lumens of wood vessels and fiber cells, as well as through natural wood cell pits and fungal bore holes. Partial degradation of the cell lumen walls was evident. Erosion troughs and localized wall fragmentation or thinning were clearly visible, as was a generalized swelling and relaxing of the normally rigid wood cell wall structure. *C. subvermispom*-treated aspen chips showed packed hyphae within the ray cells. Numerous crystals of calcium oxalate were found on the hyphae, during both the incipient and advanced stages of growth. Physical basis for the biopulping efficacy of the fungal treatment is likely to involve an overall softening and swelling of the cell walls, as well as thinning and fragmentation in localized areas (Sachs et al. 1989). Use of scanning electron microscopy revealed increased fibrillation on biomechanical pulp as compared with RMP (control). The biomechanical pulp fibers appeared more woolly, looser, and more uniform in length than the conventional mechanical and chemimechanical pulp fibers. Fiber bonding in handsheets produced from biomechanical pulp fibers appeared to be similar to that observed in handsheets produced from chemical pulps. Handsheets made from mechanically processed pulps showed uncollapsed fibers, leading to poor conformability and reduced bonding. The kraft pulps yielded handsheets that exhibited fibers of enhanced compressibility and conformability. Handsheets prepared from biomechanical pulps visually resembled the kraft handsheets, exhibiting good compressibility and conformability of the fibers.

Beneficial effects of wood pretreatment with white-rot fungi are obtained in the initial stages of biodegradation when weight losses are lower than 5% (Akhtar et al. 1998). Within biodegradation periods as short as 1 week, biotreated wood chips become softer and easier to disrupt along the fiber axes. This softening effect has been the basis for initial proposal of biological pretreatment of wood chips for mechanical pulping. In mechanical pulping, wood chips are disrupted in disk refiners to produce fibers suitable for papermaking. The process is energy intensive and pulp quality depends on several variables including the refiner design, wood species, and the desired refining level. In this way, wood chips that disrupt along fibers by requiring less mechanical strain would save energy and provide stronger pulps simply because fibers would suffer less damage during refining steps.

Guerra et al. (2002, 2003, 2004) have pointed out for two different types of wood transformations considering the chemical changes induced by the fungus in wood. One of them involves intense lignin depolymerization in short biotreatment periods, while the other indicates that esterification of oxalate secreted by the fungus on the polysaccharides chains increases the water saturation point of the fibers (Hunt et al. 2004). Both transformations are expected to affect the fiber–fiber bonding and, consequently, the physical resistance of wood. Obviously, these changes in wood structure are not isolated events and the overall change in the wood structure and ultrastructure would affect the behavior of biodegraded wood chips during mechanical or chemical pulping processes. Lack of correlation is found between the biopulping benefits and the extent of wood weight or component losses (Leatham et al. 1990a, b; Hunt et al. 2004; Ferraz et al. 2000; Mendonca et al. 2002). For example, the extent of lignin removal during fungal pretreatment is not related to the energy savings in biomechanical pulping or to the increase in delignification rates observed in kraft pulping.

Extensive removal of extractives during wood biodegradation by some white-rot fungi should provide alkali savings in kraft cooking. Extractive removal can result in unobstructed resin canals, facilitating the liquor penetration and reducing the active alkali consumption by nonlignin components. Actually, this benefit has been reported by Fischer et al. (1994) and Kohler et al. (1997) for seasoning of wood chips as well as for wood chips biotreatment by the nonlignin degrading fungus *O. piliferum*. Pulping experiments with extractive-free wood chips have been useful to evaluate the benefits of extractives removal during biopulping. The residual lignin contents in pulps prepared from extractive-free samples are midway between the undecayed controls and the fungal-treated samples (Mendonca et al. 2002), showing that extractives removal facilitates the subsequent kraft pulping but it cannot explain all the benefits observed in biokraft pulping, since even a sample without extractives is not delignified as easily as the fungal-treated samples. The extent of extractives removal during biopulping confirms this conclusion, since, similar to lignin losses, extractive losses are progressive with biodegradation time, whereas the benefits of the fungal treatment are not (Mendonca et al. 2002). The changes induced in the wood chips that provide benefits for mechanical pulping processes are not necessarily the same ones required for chemical pulping. For example, although correlating with the benefits in biomechanical pulping (Hunt et al. 2004), esterification of oxalate to the fibers is not expected to present a clear benefit for kraft pulping because the oxalate esters would consume part of the active alkali used in the cooking process.

Schwanninger et al. (2004) have reported that some near infra-red (NIR) bands from wood change significantly in biodegradation periods as short as 4 days. These changes in NIR band intensities prove that structural changes in wood components initiate very early during biodegradation. More relevant is that NIR bands can reflect not only changes in the covalent bonds of wood components, but also changes in fiber–fiber interactions such as hydrogen bonds. It is possible that minor changes in hydrogen bonding between fiber surfaces would be responsible by the softening effect observed in wood chips biotreated by white-rot fungi that facilitates disruption of the lignocellulosic matrix by disk refiners. To arrive at a comprehensive description of the biopulping chemistry, wood transformations occurring during biodegradation need to be explored in detail.

7.7 Advantages of Biopulping

Biomechanical pulping saves substantial amount of electrical energy or increases mill throughput significantly. It also improves paper strength compared to conventional RMP. Studies suggest that fungal treatment is also effective for depitching wood chips. It decreases dichloromethane extractable resin by about 30% (Fischer et al. 1994) including a 60% reduction in triglycerides which are responsible for sticky deposits on the paper machine (Fischer and Messner 1992). The cost of incorporating the fungal treatment process into existing mills is minimal. It is a relatively simple process that can be carried out in any woodyard.

Biochemical pulping reduces the amount of cooking chemicals, increases the cooking capacity, or enables extended cooking, resulting in lower consumption of bleaching chemicals. Increased delignification efficiency results in an indirect energy saving for pulping, and reduces pollution (Kirk et al. 1994). The waste load produced by biopulping should be considerably lower and more benign than effluents currently produced by commercial CTMP mills. In fact, the effluents from fungus-treated mechanical pulps have been found to be less toxic (Sykes 1994) although sometimes they may contain slightly higher BOD and COD than effluents from untreated pulp. These findings suggest that biopulping is environmentally compatible. Biopulping technology has advanced rapidly within recent years and pilot mill trials have been started worldwide (Reid et al. 2010).

7.8 Limitations and Future Prospects

A biopulping process would require inoculum on a regular basis for commercial scale applications, which would involve additional work and expense. Large-scale production of basidiomycetes is usually difficult. The fungal treatment is lengthy; a minimum of 2-week incubation is required to get the desired benefits. At first glance, the long reaction time needed for the fungal process seems to be a great disadvantage. However, considering that wood chips are often stored at the mill for at least 2 weeks, time and space should be available in the pulp mill to introduce this process. In fact, this type of bioprocess is already familiar to the pulp and paper industry as a similar process based on Cartapip® has been used commercially in many US mechanical pulp mills since 1990 (Farrell et al. 1992; Brush et al. 1994; Wall et al. 1994). The success of the Cartapip process shows that mills are able and willing to insert a biological step into their existing operations. Nevertheless it is desirable to apply classical or molecular genetic methods to improve the effectiveness of the biopulping fungi, leading to shorter reaction treatment times.

Although a chip pile-based biopulping system has been designed and evaluated on a pilot scale, the process requires demonstrated long-term operation at mill scale.

Fungal treatment reduces the brightness of resulting mechanical pulps by as much as 15–20 Elrepho brightness points in 4 weeks and 8–10 points in 2 weeks. However, aspen biorefiner mechanical pulp (BRMP) could be readily bleached to 60% Elrepho brightness with 1% sodium hydrosulfite, and 80% brightness with a two-step bleach sequence using sodium hydrosulfite and alkaline hydrogen peroxide. Based on the accelerated thermal and photoaging tests, the brightness stability of BRMP was found to be slightly lower than that of RMP but slightly higher than that of CTMP (Sykes 1993).

Although biomechanical pulping seems to have a high potential to reduce pollution problems, very few data on the environmental performance of biomechanical pulping vis-à-vis CTMP and other high yield pulping processes are available. These data need to be generated from systematic studies comparing biomechanical pulping to other pulping processes, for equivalent quality and yield of pulp.

References

Akamatsu I, Yoshihara K, Kamishima H (1984) Influence of white-rot fungi on poplar chips and thermo-mechanical pulping of fungi-treated chips. Mokuzai Gakkaishi 30:697–702

Akamatsu IH, Ueshima KY, Umeda TA (1988) Biological pulping apparatus for wood chips. Japanese Patent Application 63/83537

Akhtar M (1994) Biomechanical pulping of aspen wood chips with three strains of *Ceriporiopsis subvermispora*. Holzforschung 48:199–202

Akhtar M (1997) Method of enhancing biopulping efficiency. US Patent 5,620,564

Akhtar M, Attridge NC, Myers GC (1992a) Biomechanical pulping of loblolly pine with different strains of the white-rot fungus *Ceriporiopsis subvermispora*. Tappi J 75(2):105–109

Akhtar M, Attridge MC, Blanchette RA (1992b) The white-rot fungus *Ceriporiopsis subvermispora* saves electrical energy and improves strength properties during biomechanical pulping of both hardwood and softwood chips. In: Kuwahara M, Shimada M (eds) Biotechnology in the pulp and paper industry. UNI Publishers, Kyoto, pp 3–8

Akhtar M, Attridge MC, Myers GC (1993) Biomechanical pulping of loblolly pine chips with selected white-rot fungi. Holzforschung 47:36–40

Akhtar M, Blanchette RA, Burnes T (1995a) Using Simons stain to predict energy savings during biomechanical pulping. Wood Fiber Sci 27(3):258–264

Akhtar M, Attridge MC, Koning JW et al (1995b) Method of pulping wood chips with a fungi using sulfite salt-treated wood chips. US Patent 5,460,697

Akhtar M, Blanchette RA, Kirk TK (1996) Biopulping: an overview of consortia research. In: Srebotnik E, Messner K (eds) Biotechnology in the pulp and paper industry. Recent advances in applied and fundamental research. Facultas-Universitatsverlag, Vienna, pp 187–192

Akhtar M, Blanchette RA, Kirk TK (1997a) Fungal delignification and biomechanical pulping of wood. Adv Biochem Eng Biotechnol 57:159–195

Akhtar M, Lentz MJ, Blanchette RA (1997b) Corn steep liquor lowers the amount of inoculum for biopulping. Tappi J 80(6):161–164

Akhtar M, Blanchette RA, Myers G (1998) An overview of biomechanical pulping research. In: Young RA, Akhtar M (eds) Environmentally friendly technologies for the pulp and paper industry. Wiley, New York, pp 309–340

Ander P, Eriksson KE (1975) Influence of carbohydrates on lignin degradation by the white-rot fungus *Sporotrichum pulverulentum*. Sven Papperstid 78:643–652

Arppe M (2001) Mechanical pulp: has it got a future or will it be discontinued? Int Papwirtsch 10:45–50

Bajpai P, Bajpai PK, Kondo R (1999) Biotechnology for environmental protection in pulp and paper industry. Springer, Berlin, pp 141–170

Bajpai P, Bajpai PK, Akhtar M (2001) Biokraft pulping of *Eucalyptus* with selected lignin-degrading fungi. J Pulp Pap Sci 27(7):235–239

Bajpai P, Bajpai PK, Akhtar M (2003) Eucalyptus biokraft pulping process. US Patent 6,613,192

Bajpai P, Mishra SP, Mishra OP, Kumar S, Bajpai PK, Singh S (2004a) Biochemical pulping of wheat straw. Tappi J 3(8):3–6

Bajpai P, Mishra SP, Mishra OP, Kumar S, Bajpai PK (2004b) Biochemical pulping of bagasse. Biotechnol Prog 20(4):1270–1272

Bar-Lev SS, Kirk TK, Chang H-M (1982) Fungal treatment can reduce energy requirements for secondary refining of TMP. Tappi J 65(10):111–113

Blanchette RA, Burnes TA, Leatham GF (1988) Selection of white-rot fungi for biopulping. Biomass 15:93–101

Blanchette RA, Leatham GF, Attridge M (1991) Biomechanical pulping with *C. subvermispora*. US Patent 5,055,159

Blanchette RA, Akhtar M, Attridge MC (1992a) Using Simons stain to evaluate fiber characteristics of biomechanical pulps. Tappi J 75(11):121–124

Blanchette RA, Burnes TA, Eerdmans MM (1992b) Evaluating isolates of *Phanerochaete chysosporium* and *Ceriporopsis subvermispora* for use in biological pulping processes. Holzforschung 46:109–115

Brush TS, Farrell RL, Ho C (1994) Biodegradation of wood extractives form southern yellow pine by *Ophiostoma piliferum*. Tappi J 77(1):155–159

Çöpür Y, Tozluoğlu A (2007) The effect of AQ and NaBH4 on bio-kraft delignification (*Ceriporiopsis subvermispora*) of brutia pine chips, International Biodeterioration & Biodegradation 60(2):126–131

Dyer TJ, Ragauskas AJ (2004) Laccase: a harbinger to kraft pulping. ACS Symp Ser 889:339–362

Eaton DC, Chang H-M, Joyce TW (1982) Method obtains fungal reduction of the color of extraction stage kraft bleach effluent. Tappi J 65(6):89–92

Environment Canada Report (1988) WTC Bio-07-1988

Eriksson KE (1985) Swedish developments in biotechnology related to the pulp and paper industry. Tappi J 68(7):46–55
Eriksson KE, Vallander L (1980) Biomechanical pulping. In: Kirk TK, Higuchi T, Chang H-M (eds) Lignin biodegradation: microbiology, chemistry, and potential applications, vol II. CRC Press, Boca Raton, pp 213–233
Eriksson KE, Vallander L (1982) Properties of pulps from thermomechanical pulping of chips pretreated with fungi. Sven Papperstid 85:R33–R38
Eriksson KE, Ander P, Henningsson B (1976) Method for producing cellulose pulp. US Patent 3,962,033
Eriksson KE, Grunewald A, Vallander L (1980) Studies of growth conditions in wood for three white-rot fungi and their cellulaseless mutants. Biotechnol Bioeng 22:363–376
Farrell RA, Blanchette RA, Brush TH (1992) In: Kuwahara M, Shimada M (eds) Biotechnology in the pulp and paper industry. UNI Publishers, Kyoto, pp 27–32
Ferraz A, Mendonca R, Silva FT (2000) Organosolv delignification of white- and brown-rotted *Eucalyptus grandis* hardwood. J Chem Technol Biotechnol 75:18–24
Ferraz A, Guerra A, Mendonca R, Masarin F, Vicentim MP, Aguiar A (2008) Technological advances and mechanistic basis for fungal biopulping. Enzyme Microb Technol 43:178–185
Firth B, Backman C (1990) Comparison of Microtox testing with rainbow trout (acute) and *Ceriodaphnia* (chronic) bioassays in mill wastewaters. Tappi J 73(12):169–174
Fischer K, Messner K (1992) Reducing troublesome pitch in pulp mills by lipolytic enzymes. Tappi J 75(2):130–134
Fischer K, Akhtar M, Blanchette RA, Burnes TA, Messner K, Kirk TK (1994) Reduction of resin content in wood chips during experimental biological pulping processes. Holzforschung 48:285–290
Franco H, Freer J, Rodrıguez J, Baeza J, Elissetche JP, Mendon R (2006) Kraft pulping of *Drimys winteri* wood chips biotreated with *Ganoderma australe*. J Chem Technol Biotechnol 81:196–200
Garmaroody ER, Resalati H, Fardim P, Hosseini SZ, Rahnama K, Saraeeyan AR, Mirshokraee SA (2011) The effects of fungi pre-treatment of poplar chips on the kraft fiber properties. Bioresour Technol 102(5):4165–4170
Giovannozzi-Sermanni G, Cappelletto PL, D'Annibale A (1997) Enzymatic pretreatment of non-woody plants for pulp and paper production. Tappi J 80(6):139–144
Guerra A, Mendonca R, Ferraz A (2002) Characterization of the residual lignins in *Pinus taeda* biodegraded by *Ceriporiopsis subvermispora* by using in situ CuO oxidation and DFRC methods. Holzforschung 56:157–160
Guerra A, Mendonca R, Ferraz A (2003) Molecular weight distribution of wood components extracted from *Pinus taeda* biotreated by *Ceriporiopsis subvermispora*. Enzyme Microb Technol 33:12–18
Guerra A, Ferraz A, Lu F, Ralph J (2004) Structural characterization of lignin during *Pinus taeda* wood treatment with *Ceriporiopsis subvermispora*. Appl Environ Microbiol 70:4073–4078
Guerra A, Mendonca R, Ferraz A (2005) Bio-chemimechanical pulps from *Eucalyptus grandis*: strength properties, bleaching, and brightness stability. J Wood Chem Technol 25:203–216
Guerra A, Pavan PC, Ferraz A (2006) Bleaching, brightness stability and chemical characteristics of *Eucalyptus grandis*-bio-TMP pulps prepared in a biopulping pilot plant. Appita J 59:412–415
Hall E, Cornacchio LA (1988) Anaerobic treatability of Canadian pulp and paper mill wastewaters. Pulp Pap Can 89:T188–T192
Heden CG, Eriksson KE, Johnsrud K (1988) Japanese Patent Application 152/380
Hunt C, Kenealy W, Horn E, Houtman C (2004) A biopulping mechanism: creation of acid groups on fiber. Holzforschung 58:434–439
Jacobs CJ, Vendetti RA, Joyce TW (1998) Effect of enzyme pretreatments on conventional kraft pulping. Tappi J 81(2):143–147
Jacobs-Young CJ, Venditti RA, Joyce TW (1998) Effect of enzymatic pretreatment on the diffusion of sodium hydroxide in wood. Tappi J 81(1):260–266
Jakko Poyry Inc. (1985) Multiclient report. Lindingo, Sweden, p 2

Johnson I, Butler R (1991) Paper mill effluents: a move to toxicity-based consents. Pap Technol 32(6):21–25

Johnsrud SC, Eriksson KE (1985) Cross-breeding of selected and mutated homokaryotic strains of *Phanerochaete chrysosporium* K-3: new cellulase deficient strains with increased ability to degrade lignin. Appl Microbiol Biotechnol 21:320–327

Johnsrud SC, Fernandez N, Lopez P (1987) Properties of fungal pretreated high yield bagasse. Nord Pulp Pap Res J (Special Issue) 2:47–52

Joyce TW, Pellinen J (1995) White rot fungi for the treatment of pulp and paper industry wastewater. In: Proceedings of the Tappi environmental conference, Seattle

Karl W (1990) The 1990's could be the decade for CTMP. Tappi J 73(Suppl 2000 and Beyond):90–92

Kennedy KJ, McCarthy PJ, Droste RL (1992) Batch and continuous anaerobic toxicity of resin acids from chemithermomechanical pulp wastewater. J Ferm Bioeng 73(3):206–212

Kirk TK (1993) Biopulping: a glimpse of the future? Res Rep FPL-RP-523. Forest Products Laboratory, Madison

Kirk TK, Akhtar M, Blanchette RA (1994) Biopulping: seven years of consortia research. In: Proceedings of the Tappi biological science symposium. Tappi Press, Atlanta, pp 57–66

Kobe Steel (1988) Lignin degrading microorganisms having high activity and selectivity (for lignin). Japanese Patent EP295063

Kohler LJF, Dinus RJ, Malcolrn EV, Rudie AW, Farrell RL, Brush TS (1997) Improving softwood mechanical pulp properties with *Ophiostoma piliferum*. Tappi J 80:135–140

Kojima Y (1988) Inoculation of lignocellulosic materials with microbes in delignification. Japanese Patent Application 63/91077

Lawson LR Jr, Still CN (1957) The biological decomposition of lignin – literature survey. Tappi J 40(9):56A–80A

Leach JM, Thakore AN (1976) Toxic constituents in mechanical pulping effluents. Tappi J 59(2):129–132

Leask RA, Kocurek MJ (1987) Mechanical pulping. Joint Textbook Committee of the Paper Industry, Montreal

Leatham GF, Myers GC (1990) A PFI mill can be used to predict biomechanical pulp strength properties. Tappi J 73(4):192–197

Leatham GF, Myers GC, Wegner TH (1990a) Biomechanical pulping of aspen chips: paper strength and optical properties resulting from different fungal treatments. Tappi J 73(3):249–255

Leatham GF, Myers GC, Wegner TH (1990b) Biomechanical pulping of aspen chips: energy savings resulting from different fungal treatments. Tappi J 73(5):197–200

Lo SN, Liu HW, Rousseau S (1991) Characterization of pollutants at source and biological treatment of a CTMP effluent. Appita 44(2):133–138

Mardones L, Gomide JL, Freer J, Ferraz A, Rodnguez J (2006) Kraft pulping of *Eucalyptus nitens* wood chips biotreated by *Ceriporiopsis subvermispora*. J Chem Technol Biotechnol 81:608–613

Martinez AT, Camarero S, Guillén F (1994) Progress in biopulping of non-woody materials: chemical, enzymatic and ultrastructural aspects of wheat straw delignification with ligninolytic fungi from the genus *Pleurotus*. FEMS Microbiol Rev 13:265–274

Masarin F, Ferraz A (2008) Evaluation of *Eucalyptus grandis* biopulping with *Ceriporiopsis subvermispora* under nonaseptic conditions. Holzforschung 62:1–7

Masarin F, Pavan PC, Vicentim MP, Souza-Cruz PB, Loguercio-Leite C, Ferraz A (2009) Laboratory and mill scale evaluation of biopulping of *Eucalyptus grandis* Hill ex Maiden with *Phanerochaete chrysosporium* RP-78 under non-aseptic conditions. Holzforschung 63:259–263

Mendonca R, Guerra A, Ferraz A (2002) Delignification of *Pinus taeda* wood chips treated with *Ceriporiopsis subvermispora* for preparing high-yield kraft pulps. J Chem Technol Biotechnol 77:411–418

Messner K, Koller K, Wall MB (1997) Fungal treatment of wood chips for chemical pulping. In: Young RA, Akhtar M (eds) Environmentally friendly technologies for the pulp and paper industry. Wiley, New York, pp 385–419
Myers GC, Leatham GF, Wegner TH (1988) Fungal pretreatment of aspen chips improves strength of refiner mechanical pulp. Tappi J 71(5):105–108
Nishida T (1989) Lignin biodegradation by wood-rotting fungi. V. A new method for evaluation of the ligninolytic activity of lignin-degrading fungi. Mokuzai Gakkaishi 35(7):675–677
Nishida T, Kashino Y, Mimura A (1988) Lignin biodegradation by wood-rotting fungi. I. Screening of lignin-degrading fungi. Mokuzai Gakkaishi 34:530–536
Oriaran TP, Labosky P Jr, Blankenhorn PR (1990) Kraft pulp and papermaking properties of *Phanerochaete chrysosporium* degraded aspen. Tappi J 73(7):147–152
Oriaran TP, Labosky P Jr, Blankenhorn PR (1991) Kraft pulp and papermaking properties of *Phanerochaete chrysosporium* degraded red oak. Wood Fiber Sci 23:316–327
Otjen L, Blanchette R, Effland M (1987) Assessment of 30 white rot basidiomycetes for selective lignin degradation. Holzforschung 41:343–349
Petit-Conil M, Semar S, Niku-Paavola M-L, Sigoillot JC, Asther M, Anke H (2002) Potential of laccases in softwood-hardwood high-yield pulping and bleaching. Prog Biotechnol 21:61–71
Reid ID, Bourbolnnais R, Paice MG (2010) Biopulping and biobleaching. In: Heitner C, Dimmel DR, Schmidt JA (eds) Lignin and lignans: advances in chemistry. CRC Press, Boca Raton, pp 521–554
Sabharwal HS, Akhtar M, Blanchette RA (1994) Biomechanical pulping of kenaf. Tappi J 77(12):105–112
Sabharwal HS, Akhtar M, Blanchette RA (1995) Refiner mechanical and biomechanical pulping of jute. Holzforschung 49:537–544
Sachs IB, Leatham GF, Myers GC (1989) Biomechanical pulping of aspen chips by *Phanerochaete chrysosporium*: fungal growth pattern and effects on wood cell walls. Wood Fiber Sci 21:331–342
Sachs IB, Leatham GF, Myers GC (1990) Distinguishing characteristics of biomechanical pulp. Tappi J 73(9):249–254
Sachs IB, Blanchette RA, Cease KR (1991) Effect of wood particle size on fungal growth in a model biomechanical pulping process. Wood Fiber Sci 23:363–375
Schwanninger M, Hinterstoisser B, Gradinger C, Messner K, Fackler K (2004) Examination of spruce wood degradation by *Ceriporiopsis subvermispora* using near and mid infrared spectroscopy. J Near Infrared Spectros 12:397–409
Scott GM, Akhtar M, Lentz MJ (1997) Engineering, scale-up, and economic aspects of fungal pretreatment of wood chips. In: Young RA, Akhtar M (eds) Environmentally friendly technologies for the pulp and paper industry. Wiley, New York, pp 341–383
Setliff EC, Marton R, Granzow SG (1990) Biomechanical pulping with white-rot fungi. Tappi J 73(8):141–147
Springer AM (1986) Industrial environmental control. Pulp and paper industry. Wiley-Interscience, New York
Sykes M (1993) Bleaching and brightness stability of aspen biomechanical pulps. Tappi J 76(11):121–126
Sykes M (1994) Environmental compatibility of effluents of aspen biomechanical pulps. Tappi J 77(1):160–166
Vaheri M, Salama N, Ruohoniemi K (1991) Procedure for the production of pulp. European Patent Application EP429422, 29 May 1991
Wall MB, Cameron DC, Lightfoot EN (1993) Biopulping process design and kinetics. Biotechnol Adv 11:645–662
Wall MB, Brecker J, Noel Y et al (1994) Cartapip 97® treatment of wood chips to improve chemical pulping efficiency. In: Proceedings of the Tappi biological sciences symposium, Madison, pp 67

Wall MB, Stafford G, Noel Y (1996) Treatment with *Ophiostoma piliferum* improves chemical pulping efficiency. In: Srebotnik E, Messner K (eds) Biotechnology in the pulp and paper industry. Recent advances in applied and fundamental research. Facultas-Universitatsverlag, Vienna, pp 205–210

Wegner TH, Myers GC, Leatham GF (1991) Biological treatments as an alternative to chemical pretreatment in high-yield wood pulping. Tappi J 74(3):189–193

Welander T (1988) An anaerobic process for treatment of CTMP effluent. Wat Sci Technol 20:143–147

Widsten P, Kandelba A (2008) Laccase applications in the forest products industry: a review. Enzyme Microb Technol 42:293–307

Wolfaardt JF, Bosman JL, Jacobs A (1996) Bio-kraft pulping of softwood. In: Srebotnik E, Messner K (eds) Biotechnology in the pulp and paper industry. Recent advances in applied and fundamental research. Facultas-Universitatsverlag, Vienna, pp 211–216

Chapter 8
Biobleaching

8.1 Introduction

Pulp and paper industry is a capital and resource-intensive industry that contributes to many environmental problems, including global warming, human toxicity, ecotoxicity, photochemical oxidation, acidification, nutrification, and solid wastes. The most significant environmental impacts of the pulp and paper manufacture result from the pulping and bleaching processes: some pollutants are emitted to the air, others are discharged to the wastewaters, and solid wastes are generated as well. Much research has focused upon the bleaching technology employed because this component of the production process has historically been associated with the formation of chlorinated dioxins and other chlorinated organic chemicals. These pollutants are toxic, nonbiodegradable and have the tendency to contaminate food chains through bioaccumulation. The dioxins are known for their extreme toxicity and are believed to be carcinogenic. Pulp mills are being forced to reduce their use of chlorine and chlorine dioxide in the bleaching process because of market and environmental pressures. There are several options for mills to do this, but most require the expenditure of large sums of capital. Biobleaching is one of the promising alternatives for reducing/eliminating chlorine-based chemicals in pulp bleaching process. This process requires little capital investment and is thus attractive to the industry. So far, two enzyme-based approaches have been investigated; one uses xylanase enzymes and the other uses ligninolytic enzymes. Significant effort has also been made to study the potential of white-rot fungi for the bleaching of chemical pulps.

8.2 Xylanase Enzymes

The use of xylanase enzymes to enhance the bleaching of the pulp was first reported by Viikari et al. (1986). The concept of enzyme bleaching is based on the fact that limited removal of hemicellulose in kraft pulps by hemicellulases increases the

P. Bajpai, *Biotechnology for Pulp and Paper Processing*,
DOI 10.1007/978-1-4614-1409-4_8, © Springer Science+Business Media, LLC 2012

extractability of lignin in subsequent chemical bleaching stage (Viikari et al. 1994). Xylanase bleaching of chemical pulp is the most widely used and best established biotechnical application in pulp bleaching. This technology has become one of the solutions considered by the pulp and paper industry to give an innovative, environmentally and economically acceptable answer to the pressures exerted on chlorine bleaching by regulatory authorities in Western countries and by more demanding, environmentally minded consumers (Bajpai 2004; Viikari et al. 2002, 2009; Reid et al. 2010). There has been a significant application in kraft mills, due in part to the rapid advances in biotechnology and molecular biology, resulting in more effective, cheaper enzymes.

Today more than 10% of all kraft pulp is manufactured with xylanase prebleaching. In North America, Iogen Corp. based in Ottawa has established a market leadership position. Globally, other suppliers such as Novozymes, Genencor, AB Enzymes, and more recently Diversa are also selling to this market. In Japan, Oji Paper is unique in manufacturing xylanase on-site at its Yonago mill (Paice and Zhang 2005). The enzyme is produced from a bacterial fermentation of pulp side stream which results in a xylanase/pulp mixture. This mixture is then fed to the main pulp brownstock storage tank.

8.2.1 Production and Properties of Xylanases

Several articles have been published on the production and properties of xylanases (Biely 1985; Viikari et al. 1993; Wong et al. 1988; Suominen et al. 1992; Viikari et al. 1990; Polizeli et al. 2005; Bajpai 1997a; Yang et al. 2005). These enzymes are produced by fungi, bacteria, yeast, marine algae, protozoans, snails, crustaceans, insects, seeds, etc. Filamentous fungi are particularly interesting producers of this enzyme from an industrial point of view, due to the fact that they excrete xylanases into the medium. Furthermore, xylanase levels from fungal cultures are generally much higher than those from yeasts or bacteria. In addition to xylanase, fungi typically produce several accessory xylanolytic enzymes, which are necessary for debranching substituted xylans. Fungi known to produce xylanase include *Aspergillus, Disporotrichum, Penicillium, Neurospora, Fusarium, Neocallimastix, Trichoderma, Coniothyrium, Thermomyces lanuginosus,* etc. Xylanases are also produced from several species of bacteria, some from extreme environments (hot, alkaline, etc.) which makes them more suitable for industrial environments.

The cost of enzymes is one of the factors determining the economics of a biocatalytic process and can be reduced by finding optimum conditions for their production by the isolation of hyperproducing mutants and (possibly) by the construction of efficient producers using genetic engineering. A rational approach to these goals requires knowledge of the regulatory mechanisms governing enzyme production. Studies of the regulation of xylanolytic enzymes have largely focused on the induction of enzyme activities under various conditions rather than on gene regulation. In addition, the studies have mainly focused on xylanase and xylosidase. Xylanases appear to be inducible. They are produced in high amounts during growth on xylan, and the synthesis of the enzyme is catabolite repressed by easily metabolized carbon sources such as glucose or xylose. Xylan can not enter the cells, so the signal for accelerated synthesis of xylanolytic enzymes must involve low-molecular-weight fragments, mainly xylobiose and xylotriose. The oligosaccharides are formed by the hydrolysis of xylan in the medium by small amounts of enzymes produced constitutively. Induction can also be achieved by various synthetic alkyl and aryl β-D-xylosides in a *Streptomyces* sp. and by methyl β-D-xyloside in yeast. These compounds enable the production of xylanolytic enzymes in the absence of xylan and xylooligosaccharides. In the yeast *Cryptococcus albidus*, only methyl β-D-xyloside induced the xylanolytic system. Other alkyl and aryl β-D-xylosides were unable to do so; they could induce a nonspecific β-glucosidase that hydrolyzed aryl β-D-xylosides but not xylooligosaccharides. Xylanolytic systems of yeast can also be induced by positional isomers of xylobiose (Bajpai 2009). Induction with 1,2-β-xylobiose is analogous to the induction of cellulases in filamentous fungi by sophorose. However, the slow response of cells to 1,2-β-xylobiose compared to 1,4-β-xylobiose, as well as the evidence for its transformation into 1,4-β-xylobiose, the natural inducer, suggested that, in yeast, the isomeric diasaccharide is a precursor of the natural inducer. The nature of the regulation in filamentous fungi has not been established. Generalization will perhaps never be possible due to the diversity of cellular control mechanisms. Fungal xylanases appear to be inducible or under derepression control, which includes enzyme production on carbon sources that are used slowly. Regulatory studies in fungi are often complicated by the concurrent production of xylanase and cellulase and by substrate cross-specificity of cellulases and xylanases. There are different types of cellulases and xylanases, the substrate specificities of which range from absolute for one polymer to about equal affinity for both polymers. The xylanolytic and cellulolytic systems in some filamentous fungus are likely to be under separate regulatory control. During growth on xylan, several species produce specific xylanases with little or no cellulase. However, when grown on cellulose, cellulases are produced together with xylanases. The reason for the production of specific xylanases on cellulose is unclear. Perhaps, it results from the presence of xylan remnants in cellulose or depression on cellulose, a carbon source that is used slowly. Experiments with defined low-molecular-weight inducers in *Trichoderma reesei* afforded similar results. Sophorose induced both specific and nonspecific endo-1,4-β-glucanases, cellobiohydrolase I, and very little xylanase. Induction with xylobiose produced only specific xylanases. Therefore, the strategy for xylanolytic systems free of cellulases might be simply to grow cells on xylan uncontaminated by cellulose. However, this strategy could not be applied to all fungi. In *Schizophyllum commune*, high xylanase production is linked strictly to cellulase production. The fungus grows poorly on xylan in the absence of cellulose. The possibility of producing xylanolytic systems free of cellulase should be clarified in strains of *Aspergillus* because they belong to the best xylanase and xylosidase producers. An alternative and promising approach to the production of xylanolytic systems free of cellulases is the isolation of cellulase-deficient mutants. Another possibility is the construction of appropriate recombinants by genetic engineering.

Recombinant DNA techniques offer opportunities for the construction of microbial strains with selected enzyme machinery. In xylan bioconversion, the main objectives for recombinant DNA technology are the construction of producers of xylanolytic systems free of cellulolytic enzymes and the improvement of the fermentation characteristics of industrially important xylose-fermenting organisms by introducing genes for xylanase and xylosidase so that the direct fermentation of xylan is possible.

Most of the published cloning work has been restricted to bacterial genes. The isolation of xylanase genes from *Bacillus* sp. and their expression in *Escherichia coli* have been reported. In only one case does the expressed enzyme appear to be secreted from the host cells. Further biochemical studies of xylanase-secreting and -nonsecreting transformants could lead to a better understanding of the secretary process and to the development of cloning strategies that would guarantee the secretion of the desired products. Further difficulties with cloning genes from eukaryotic microorganisms can be expected. In addition to permitting the introduction of novel genes, cloning techniques could enable the amplification of the expression genes already present. For instance, the production of xylanase in *Bacillus subtilis* was enhanced successfully using a plasmid vector carrying the *Bacillus pumilus* genes. The transformant produced approximately 3 times more extracellular xylanase than the donor strain. Moreover, the enzyme was produced constitutively, suggesting that regulatory elements of the donor organism were absent in the vector used for the transformation.

Several fungal xylanases as well as bacterial xylanases have been extensively studied (Bajpai 1997a). A large number of the xylanases that have been purified are rather small (molecular weight 20 kDa) monomeric proteins with basic isoelectric points (pI 8–9.5). They also show great homology at the molecular level and belong to the family G (or 11) of glycosyl hydrolases. The other xylanases with high molecular mass (molecular mass >40 kDa) and lower pI values belong to the other identified endoxylanase family, F (or 10). The optimum pH for xylan hydrolysis is around 5 for most fungal xylanases, and they are normally stable between pH values of 2 and 9. The pH optima of bacterial xylanases are generally slightly higher than those of fungal xylanases. Alkalophilic *Bacillus* species and alkalophilic actinomycetes produce xylanases with high activity at alkaline pH values. Most of the fungi and bacteria produce xylanases that tolerate temperatures below 40–50°C. Xylanases have also been characterized from thermophilic organisms. These strains include *Thermoascus aurantiacus, Talaromyces emersonii, T. lanuginosus, Melanocarpus albomyces,* and *Sporotrichum thermophile.* The most thermostable xylanase described is that from an extremely thermophilic species of *Thermotoga* with a half-life of 20 min at 105°C. Xylanases with half-lives from a few minutes up to 90 min at 80°C are produced by *T. aurantiacus, Bacillus stearothermophilus, Caldoceum saccharolyticum, Clostridium stercorarium,* and *Thermomonospora* sp. (Gruninger and Fiechter 1986; Luthi et al. 1990; Mathrani and Ahring 1992; McCarthy et al. 1985; Perttula et al. 1993; Tan et al. 1987).

The most important characteristics of xylanases are their pH and temperature stability and activity (Viikari et al. 2009). A number of enzymes produced by extremophilic organisms have been characterized. Particularly, enzymes originating from thermophilic bacteria have shown superior characteristics but, surprisingly, are not available for commercial use, obviously due to problems related to their efficient production in heterologous host strains.

Xylanases show the highest activity against polymeric xylan; the rate of the hydrolysis reaction normally decreases with the decreasing chain length of oligomeric substrates (Biely 1985). They do not hydrolyze xylobiose, and the hydrolysis of xylotriose is in most cases negligible or at least limited. The main products formed from the hydrolysis of xylan are xylobiose, xylotriose, and substituted oligomers of two to four xylosyl residues. The length and type of the substituted products depend on the mode of action of the individual xylanases. Most of the enzymes studied cleave the xylan backbone, leaving the substituent on the nonreducing end of the xylosyl chain of the oligosaccharide. Some xylanases leave the substituent on the nonreducing end and in the middle of the oligosaccharide chain of the end products; xylotriose has been reported to inhibit the action of xylanases. In addition to hydrolytic activity, transferase activity has been detected in several xylanases.

Of the xylanases produced by *Trichoderma* species, two main groups can be identified (Polizeli et al. 2005). Both types have low molecular weights, but the isoelectric points are different. Whereas xylanases with high pI have been quite extensively studied, the other type of xylanase (with pI near to pH 5) has been purified and characterized only from *T. lignorum* and *T. reesei.* Basic xylanases have been isolated from *T. harzianum, T. koningii,* and *T. longibrachiatum.* However, only one of these enzymes made a major contribution to the total xylanase activity in the culture filtrate. The pI 9.0 and 5.5 xylanases of *T. reesei* have been shown to be different gene products and were both classified as belonging to the family G. Multiple xylanases, have also been purified from culture filtrates of *Aspergillus niger, A. oryzae, A. kawachii, A. awamori,* and other fungi, as well as from bacteria. Xylanases with high pH and temperature optima have been isolated and tested for improving the bleachability of kraft pulps. Several alkali-tolerant strains of *Bacillus* have been used for the production of xylanases with pH optima of around 9.0. *T. lanuginosus* has been found to be an excellent producer of thermostable β-xylanase. A wild-type strain has been found to produce about 60,000 nkat/mL xylanase activity in 6 days. The most thermophilic xylanases described are produced by an extremely thermophilic bacterium, *Thermotoga* sp. *T. maritima* produces at least two hyperthermophilic xylanases; one has its temperature optimum at 90°C and the other at 105°C. However, microorganisms living in extreme conditions are often difficult to grow in the laboratory and the productivity of xylanases is usually low. The amino acid homology of xylanases has opened up the possibility of using novel genetic techniques to screen for better xylanases for industrial applications. Thus, several xylanase genes encoding proteins active at temperatures from 75 up to 95°C (pH 6 to 8) have been isolated from the extremely thermophilic bacteria *Thermotoga* and *Dictyoglomus* without the laborious production and purification of the enzymes.

The research progress and trend in the structure correlating with the important properties of xylanases has been reviewed (Yang et al. 2005). Analyses of three-dimensional structures and properties of mutants have revealed that glutamine and aspartic acid residues are involved in the catalytic mechanism. The thermostability of xylanases correlated with many factors, such as disulfide bridges, salt bridges, aromatic interactions, content of arginine and proline, and some multidomain xylanases have thermostability domains in N or C terminal. But no single mechanism is responsible for the remarkable stability of xylanases. The isoelectric points and reaction pH of xylanase are influenced by hydrophobicity and content of electric charges. Many researchers have demonstrated that aromatic amino acid, histidine, and tryptophan play an important role in improving enzyme-substrate affinity. The research on structures and functions of xylanases are of great significance in understanding the catalytic mechanism and directing the improvement of properties of xylanases to meet the application requirement.

8.2.2 Performance of Xylanases in Bleaching

The use of xylanases in different bleaching sequences of kraft pulp consistently leads to a reduction in chemical consumption (Viikari et al. 1994; Bajpai 1999, 2004, 2009). However, the benefits obtained by enzymes are dependent not only on the type of pulp but also on the chemical bleaching sequence used as well as the final target brightness and environmental goals of the mill. Originally, xylanases were applied in order to reduce the consumption of chlorine chemicals, especially elemental chlorine. Later enzymes have been combined with various elemental chlorine free (ECF) and totally chlorine-free (TCF) bleaching sequences to improve the otherwise lower final brightness value of pulp or to decrease the bleaching cost. Results from laboratory studies and mill trials show about 35–41% reduction in active chlorine at the chlorination stage for hardwoods and 10–20% for softwoods, whereas savings in total active chlorine were found to be 20–25% for hardwoods and 10–15% for softwoods (Tolan and Canovas 1992; Werthemann 1993; Skerker et al. 1992; Yang and Eriksson 1992; Allison et al. 1993a, b; Bajpai et al. 1993, 1994; Bajpai 1999; Bim and Franco 2000; Valchev et al. 1998, 2000; Zhan et al. 2000; Awakaumova et al. 1999; Senior and Hamilton 1991, 1992a, b, c, 1993; Senior et al. 1992, 1999, 2000; Thibault et al. 1999; Atkinson et al. 1993; Saleem et al. 2009; Ko et al. 2011).

In the elementary chlorine-free bleaching sequences, the use of xylanase increases the productivity of the bleaching plant when the production capacity of ClO_2 is a limiting factor. This is often the case when the use of chlorine gas has been abandoned. In TCF bleaching sequences, the addition of xylanase increases the final brightness value, which is a key parameter in the marketing of chlorine-free pulps. In addition, the savings in TCF bleaching are important with respect to both costs and the strength properties of the pulp.

For the batch kraft pulp, xylanase treatment decreased the demand for active chlorine by 15% but decreased active chlorine of pulp from a continuous process by only 6–7%. It has been shown that the prebleaching effect on black spruce pulp is associated with a drop in the degree of polymerization, even though the xylan content decreases slightly. Prebleaching thus appears to be associated with xylan

Table 8.1 Plant-scale trial results with xylanase

Parameter	No enzyme	Enzyme
(a) Effect on bleach chemical requirement		
Production rate (a.d. metric tons/day)	796	789
Unbleached pulp		
Soda loss (as Na_2SO_4, kg/ton)	9.7	10.1
Kappa number	31.1	31.4
C_D stage		
Active chlorine multiple	0.23	0.21
Chlorine (%)	4.58	3.75
Chlorine dioxide (%)	0.7	0.82
Chlorine dioxide substitution (%)	28.5	36.4
Total equivalent Cl_2 (kg/a.d. metric tons)	64.1	59.0
EOP stage		
Hydrogen peroxide (%)	0.4	0.4
Oxygen (%)	0.8	0.8
D_1 stage		
Chlorine dioxide (%)	1.67	1.58
Total equivalent Cl_2 (kg/a.d. metric tons)	43.9	41.7
D_2 stage		
Chlorine dioxide (%)	0.23	0.20
Total equivalent Cl_2 (kg/a.d. metric tons)	6.0	5.2
Brightness (% ISO)	89.2	90.3
Overall total equivalent Cl_2 (kg/a.d. metric tons)	114.0	105.8
Total ClO_2 demand (%)	2.6	2.6
(b) Effect on physical strength properties[a]		
PFI revolutions	2,280	2,333
Tear index (mN m²/g)	137±7	130±6
Tensile strength (km)	9.2±0.4	9.6±0.3
Burst index (kPa m²/g)	77±4	78±3
Strength factor (km mN m²/g)	1,260±57	1,248±54
Viscosity (cps)	23.5±2.3	24.5±2.1

Based on data from Manji (2006)
[a]At a freeness of 500 mL CSF

depolymerization, even though not necessarily with solubilization of the xylan-derived hemicellulose components. Canadian researchers showed that xylanase treatment and extraction change the reactivity of the pulp by enabling a higher chlorine dioxide substitution to achieve a target brightness and that they raise the brightness ceiling of fully bleached pulps. Xylanase enzyme was used in a Canadian pulp mill to reduce the consumption of chlorine and chlorine dioxide in the bleaching processes (Manji 2006). The total equivalent chlorine decreased by 8 kg/air dried (a.d.) metric ton of pulp during enzyme treatment (Table 8.1a). Consequently, the substitution in the front end of the bleach plant increased from 28.5 to 36.4%. In addition, the active chlorine multiple decreased from 0.23 to 0.21 without a loss in pulp production rate and an insignificant change in the physical properties of pulp. During the enzyme treatment period, the AOX being discharged into the receiving waters decreased from 2.4 to 2.2 kg/a.d. metric ton. The pulp quality results showed no significant difference in the strength factor during the enzyme treatment period (Table 8.1b).

Latorre et al. (2008) reported that in Jacarei mill (Votorantim Celulose e Papel), xylanase treatment enabled a decrease of 1.5 points of kappa number and an increase of 2.5% ISO in brightness. In addition, no differences in viscosity were observed between pulps treated enzymatically and those without enzymatic treatment.

Valls et al. (2010) used two new bacterial xylanases from families 11 to 5 to obtain modified fibers with high-cellulose content. When these xylanases were applied separately or simultaneously in a complete ECF bleaching sequence, both xylanases were found to improve delignification and bleaching during the sequence, while a synergistic effect of the enzymes was observed on several pulp and paper properties. The xylanases enhanced the release of xylooligosaccharides branched with hexeneuronic acids (HexA), producing fibers with a reduced HexA and xylose content. On the other hand, these effects were found to be dependent on the xylanase used, with the family 11 enzyme being more efficient than the family 5 xylanase. Effluent properties were affected by the enzymatic sequences, due to the dissolution of lignin and xylooligosaccharides, while some changes in the fiber morphology were also produced without affecting the final paper strength properties.

The most conventional method is to add xylanase to the brownstock pulp prior to the high density (HD) tower (Tolan 1992, 2001; Tolan and Canovas 1992; Tolan et al. 1996). The enzyme reaction takes place in the tower and the treated pulp then passes into the bleach plant. Various ways to add enzymes have been used including: (1) spraying on the decker pulp mat, (2) adding to either the decker repulper or discharge chute, (3) adding into the stock of medium consistency pulp leading to the HD tower, and (4) adding directly into the HD tower. Xylanase has also been added later in the bleaching sequence rather than to the brownstock pulp. The latest generation of alkali-tolerant enzymes requires little, if any, addition of acid to adjust the pH. Earlier generation of enzymes had pH optima ranging from 5 to 6.5 and required acid addition to brownstock pulp. Instances of corrosion problems were seen when acid was incorrectly applied. New xylanases have higher pH optima and function optimally without pH adjustment.

The acid of preference by far has been sulfuric acid. However, with the development of alkaline xylanases, noncorrosive carbon dioxide is an excellent choice and also improves washer performance. The addition of acid prior to the D_0 stage in ECF bleaching has also been shown to improve the performance of D_0 stage. This is because the higher acidity in the stage prevents decomposition of chlorine dioxide to chlorate and because the chemistry of delignification with chlorine dioxide favors an acidic environment. Typical site of acidification are also indicated in Fig. 8.1. Acid added to the low consistency pulp prior to the washer vat provides the benefits of reducing pitch deposits; however, acid charges here tend to be much higher due to the large volume that must be treated. Acid can also be added on the washer shower, bars shower or in the repulper discharge section. Experience has shown that prevention of corrosion must be a priority. Xylanase pretreatment has been shown to be easily applicable with existing industrial equipment, which is a considerable advantage of this technology.

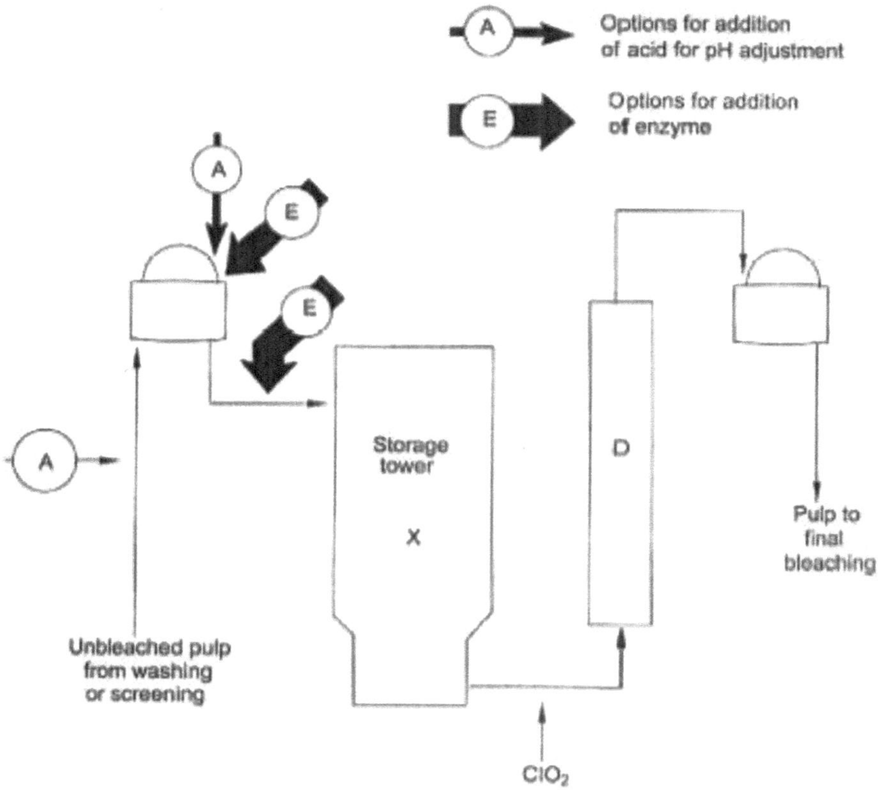

Fig. 8.1 Typical xylanase and acidification sites (based on Bajpai 2004)

Detailed laboratory work is generally needed to optimize and adapt the enzymatic treatment to individual existing mill conditions. Interestingly, however, xylanase bleaching has been scaled up directly from laboratory scale to the large industrial scale (1,000 TP/day) without intermediate pilot stages. It has also been observed that even higher brightness values can be reached on the full scale than those attainable in the laboratory, which is due to the more efficient mixing systems and higher pulp consistencies. No expensive capital investments have generally been necessary for full-scale runs. The most significant requirement is the addition of pH adjustment facilities. Xylanase pretreatment has been shown to be easily applicable.

Bleaching with xylanase requires proper control of pH, temperature, as well as retention time (Tolan and Guenette 1997). The optimum pH and temperature for enzyme treatment varies among enzymes. Generally, xylanases derived from strains of bacterial origin are most effective between pH 6 and 9, while those derived from strains of fungal origin should be used within the pH range of 4–6. The optimum temperature ranges from 35 to 60°C with different enzymes. To obtain the best results from enzyme use, enzyme dosage must be optimized in each single case. In addition, the pulp consistency must be optimized to obtain effective dispersion of enzyme and improve the efficiency of enzyme treatment. Screw conveyers and static mixers are examples of efficient mixing systems. Most of the bleaching effect is obtained after only 1 h of treatment. Usually, the reaction time is set to 2–3 h. Long reaction time must be avoided if cellulases are present. Commercial xylanase enzyme preparations consist mainly of endoxylanases. Most of the enzymes are active at acidic or neutral pH, although some of them function under alkaline conditions. Xylanases are sold as concentrated liquids and the amount required per metric ton of pulp is very low, less than a liter. The cost of enzyme per ton of pulp varies and depends on the dosage required and the supplier. The approximate cost of enzyme treatment is around $1.2–2.0/TP. Due to the low enzyme price and low capital costs of enzyme stage, the potential economic benefits of enzyme bleaching are significant. Eiras et al. (2009) have recently reported that the application of the enzymes yielded savings of USD3/t of bleached pulp, regardless of sequence and enzyme type.

Mill operations also affect the performance of the xylanase enzyme. The effect of raw material, pulping process, brown stock washing, and bleaching sequence should be assessed by laboratory testing prior to mill usage of enzymes (Tolan and Guenette 1997).

Among raw materials, the important distinction is between hardwoods and softwoods. The percentage of the bleaching chemicals saved by xylanase treatment is thus greater on hardwoods than on softwoods. At good treatment conditions, the decrease in chlorine chemicals is about 20% on hardwoods and 15% on softwoods. The digestor operation affects the xylan content of the pulp significantly. For example, sulfite pulping destroys most of the xylan and thus sulfite pulp is not suitable for enhanced bleaching by enzyme treatment. In conventional kraft pulping, the xylan content depends strongly on the effective alkalinity. The lower the alkalinity, the higher the xylan content and the benefits of using xylanase enzymes. At high alkalinity (19–22%), much of the xylan is solubilized which decreases the benefit of xylanase treatment. At low alkalinity (less than about 18%), the xylan structure is more stable, and the bleaching enhancement by xylanase is greater by two- to threefold over the high alkalinity pulp. This is often the case for pulp cooked to higher kappa number. Kraft pulping at severe conditions, such as conventional cooking of softwood to kappa number less than 23, also destroys much of the hemicellulose that is accessible to the enzyme. On the other hand, MCC or oxygen-delignified pulps with low unbleached kappa number respond well to enzyme treatment. Much smaller enzyme benefit has been reported for batch-cooked pulp at kappa number 21 than for MCC and oxygen-delignified pulp at the same kappa number. The MCC and oxygen-delignified pulps have hemicellulose structures that are similar to that for conventional, high kappa number pulps. Enzyme benefits have been achieved in mills with conventional, MCC, and O_2 delignification systems. The brownstock black liquor properties vary greatly from mill to mill. In some mills black liquor can inhibit enzyme performance due to the presence of highly oxidizing compounds. This effect differs significantly among enzymes and should always be checked before proceeding with full-scale enzyme use. It is important to note that it is not necessary to wash the pulp after enzyme treatment (before chlorination) to achieve the enhanced bleaching. Identical enzyme benefits with and without a postenzyme washing have been obtained. The bleaching sequence and brightness target influence the enzyme's benefits to the mill. The enzyme benefit is greater at higher brightness targets, especially near the brightness ceiling, and lower at lower brightness targets.

Xylanase pretreatment of pulps prior to bleach plant reduces bleach chemical requirements and permits higher brightness to be reached (Viikari et al. 2009). The reduction in chemical charges can translate into significant cost savings when high levels of chlorine dioxide and hydrogen peroxide are being used. A reduction in the use of chlorine chemicals clearly reduces the formation and release of chlorinated organic compounds in the effluents and the pulps themselves. The ability of xylanases to activate pulps and increase the effectiveness of the bleaching chemicals may allow new bleaching technologies to become more effective. This means that for expensive chlorine-free alternatives such as ozone and hydrogen peroxide, xylanase pretreatment may eventually permit them to become cost effective. Traditional bleaching technologies also stand to benefit from xylanase treatments. Xylanases are easily applied and require essentially no capital expenditure. Because chlorine dioxide charges can be reduced, xylanase may help eliminate the need for increased chlorine dioxide generation capacity. Similarly, the installation of expensive oxygen delignification facilities may be avoided. The benefit of a xylanase bleach boosting stage can also be taken to shift the degree of substitution toward higher chlorine dioxide levels while maintaining the total dosage of active chlorine. Use of high chlorine dioxide substitution dramatically reduces the formation of AOX.

In TCF-bleaching sequences, the addition of enzymes increases the final brightness value, which is a key parameter in marketing chlorine free pulp. In addition, savings in TCF bleaching are important with respect to both costs and the strength properties of the pulp. The production of TCF pulp has increased dramatically during recent years. Several alternative new bleaching techniques based on various chemicals such as oxygen, ozone, peroxide, and peroxyacids have been developed. In addition, an oxygen delignification stage has already been installed at many kraft mills. In the bleaching sequences in which only oxygen-based chemicals are used, xylanase pretreatment is generally applied after oxygen delignification to improve the otherwise lower brightness of the pulp or to decrease bleaching costs. The TCF sequences usually also contain a chelating step in which the amount of interfering metal ions in pulp is decreased. It has been observed that the order of metal removal (Q) and enzymatic (X) stages is important for an optimal result. When aiming at the maximal benefit of enzymatic treatment in pulp bleaching, the enzyme stage must be carried out prior to or simultaneously with the chelating stage. In fact, the neutral pH of enzyme treatment is optimum in many cases for chelation of magnesium, iron, and manganese ions that must be removed before bleaching with hydrogen peroxide. The TCF technologies applied today are usually based on bleaching of oxygen-delignified pulps with enzymes and hydrogen peroxide.

A survey of mill usage of xylanase revealed that the mills have spent most of their efforts in decreasing AOX (by decreasing chlorine usage), followed closely by meeting customer demands (which in many cases was decreasing chlorine usage), and eliminating chlorine gas (Tolan et al. 1996). These objectives were followed in effort by decreasing off-grade pulp, decreasing BOD, and cutting costs. The least effort was devoted to increasing throughput eliminating dioxin and converting to TCF. The most widely reported benefit of enzyme treatment is a savings in bleaching chemicals. The chemical savings was 8–15% with an average of 11% of the total chemical across the bleach plant. The other widespread benefits were in improved effluent including decreases in AOX of 12–25%, decreases in effluent color and other improvements to the effluent. Other benefits of enzyme treatment reported increased bleached brightness (1 point gain), tear strength (5% gain), and pulp throughput (10% increase). Xylanase enzymes can cut bleaching related energy usage by 40%. This would result in carbon dioxide emissions savings of between 155,000 and 270,000 tons annually in European paper industry.

The most common problems with xylanase treatment cited in a mill survey have been corrosion of equipment and maintaining the brownstock residence time. Sulfuric acid corrosion of mild steel has been encountered in several mills. The brownstock residence time must be maintained for as long as possible, but usually at least 1–2 h to obtain the maximum benefits of enzyme treatment. This sometimes means that the mills must maintain the storage tower nearly full, which curtails its ability to act as a buffer between the pulping mill and the bleach plant. Other problems reported with enzyme treatment included difficulties in application and in bleach plant control. These relate to subtle action of enzymes, which is not easily observed on-line or in rapid testing. A decreased tear strength and pitch formation were also reported in some mills.

8.2.3 Effect of Xylanases on Pulp and Effluent Quality

The xylanase-treated pulps show unchanged or improved strength properties. Also, these pulps are easier to refine than the control pulps. The viscosity of the pulp is improved as a result of xylanase treatment. However, the viscosity of the pulp is adversely affected when cellulase activity is present. Therefore, the presence of cellulase activity in the enzyme preparation is not desirable. Xylanase pretreatment leads to reductions in effluent AOX and dioxin concentrations due to reduced amount of chlorine required to achieve a given brightness. The level of AOX in effluent is significantly lower for xylanase-pretreated pulps as compared to conventionally bleached control pulps.

8.2.4 Mechanism of Bleaching

Xylanases such as endoxylanases are xylan-specific enzymes. They catalyze the hydrolysis of xylose–xylose bonds within the xylan chain and solubilize only a fraction of the total xylan present. However, the actual enzymatic mechanism in bleaching is not yet well understood. One hypothesis suggests that precipitated xylan blocks or occludes extraction and that xylanase increases the accessibility (Kantelinen et al. 1993b). This model is based on reports that xylan reprecipitates on the fiber surfaces (Yllner et al. 1957). It has been reported that no extensive relocation of xylan to the outer surface occurs during pulping, so the occlusion model might not be a sound premise (Suurnakki et al. 1997). Another possible explanation for xylanase action in bleaching is that the disruption of xylan chain by xylanase interrupts lignin-carbohydrate bonds, improves the accessibility of the bleaching chemicals to the pulps, and facilitates easier removal of solubilized lignin in bleaching (Paice et al. 1992). Skjold-Jorgensen et al. (1992) found that xylanase treatment decreased the demand for active chlorine for a batch kraft pulp by 15% but decreased active chlorine of pulp from a continuous process by only 6–7%. They also showed that DMSO extraction of residual xylan does not lead to an increase in bleachability but that xylanase treatment does. This shows that DMSO-extractable xylan is not involved in bleach boosting. Paice et al. (1992) have shown that the prebleaching effect on black spruce pulp is associated with a drop in the degree of polymerization, even though the xylan content decreases slightly. Prebleaching thus appears to be associated with xylan depolymerization, even though not necessarily with solubilization of the xylan-derived hemicellulose components. Senior and Hamilton (1993) have shown that xylanase treatment and extraction change the reactivity of the pulp by enabling a higher chlorine dioxide substitution to achieve a target brightness and that they raise the brightness ceiling of fully bleached pulps.

8.2.5 Conclusion and Future Prospects

Xylanase enzymes have proven to be a cost effective way for mills to realize a variety of bleaching benefits including: Reducing AOX discharges, primarily by decreasing chlorine gas usage, Debottlenecking mills limited by chlorine dioxide generator capacity, Eliminating chlorine gas usage for mills at high chlorine dioxide substitution levels, Increasing the brightness ceiling, particularly for mills contemplating ECF and TCF bleaching sequences, and decreasing cost of bleaching chemicals, particularly for mills using large amounts of peroxide or chlorine dioxide. These benefits are achieved over the long term when the enzymes are selected and applied properly in the mill.

Xylanase-aided bleaching has been identified as a future technology. The development is focusing on improved enzyme properties and improved enzyme performance. Improved properties include higher pH and temperature tolerance of the enzymes, to make the enzyme treatment operations more compatible with existing mill operations. Improved enzyme performance is being approached by tailoring the enzyme action more closely to the hemicellulose structure of the pulp, to result in a greater bleaching benefit or higher pulp yield.

8.3 Lignin-Oxidizing Enzymes

Lignin-oxidizing enzymes are mainly produced by white-rot fungi. The most important lignin-oxidizing enzymes are lignin peroxidases, manganese peroxidases, and laccases. Lignin peroxidase and manganese peroxidase appear to constitute a major component of the ligninolytic system. The lignin peroxidases are able to catalyze the oxidation of non-phenolic aromatic rings in lignin to cation radicals in the presence of hydrogen peroxide. Manganese-dependent peroxidases oxidize phenolic units in lignin. These require Mn^{2+}, which is oxidized to Mn^{3+} in the presence of chelators and hydrogen peroxide. Mn^{3+} is the real oxidizing agent attacking the lignin molecule. Laccase uses molecular oxygen as a cosubstrate. The enzyme oxidizes phenolic subunits in lignin and simultaneously reduces oxygen to water. The substrate range of laccases can be extended to non-phenolic subunits by adding readily oxidized substrates.

The crystallographic structure of lignin peroxidase from *Phanerochaete chrysosporium* has been studied at different resolutions (Edwards et al. 1993; Poulos et al. 1993). The model comprises all 343 amino acids, 1 heme molecule, and 3 sugar residues. The crystal structure reveals that the enzyme consists mostly of helical folds with separate domains on either side of the catalytic heme. The enzyme also contains four disulfide bridges formed by the eight cysteine residues and two structural calcium ions, which appear to be important for maintaining the integrity of the active site (Poulos et al. 1993). Manganese peroxidase presents a good sequence homology with lignin peroxidase.

8.3.1 Performance of Lignin-Oxidizing Enzymes in Bleaching

Several reports in the literature suggest that the enzymes – lignin peroxidases (LiP), manganese peroxidases (MnP), and laccases (L) could prove useful in bleaching of pulps. Several patents have been filed on the use of ligninases for bleaching (Call 1994a, b; Farrell 1987; Farrell et al. 1987a, b; Olsen et al. 1989, 1991; Vaheri and Miiki 1991; Vaheri and Piirainen 1992; Gysin and Griessmann 1991). With ligninase 118 from *P. chrysosporium* followed by alkaline extraction, Egan (1985) reported kappa number reductions of 24 and 26%. The pulp viscosity did not change but no brightness data were reported. Arbeloa et al. (1992) used lignin peroxidase enzyme from *P. chrysosporium* to improve the bleachability of hardwood and softwood Kraft pulps. Enzyme treatment prior to chemical bleaching increased brightness and decreased lignin content in the pulp. The final brightness of pulp was found to be higher by about 0.8–0.9 points as compared to control. Other researchers have not been able to achieve much bleaching effects from either pure or crude enzyme preparations from *P. chrysosporium*.

At Paprican, manganese peroxidase from *Trametes versicolor* was examined for bleaching of Kraft pulp (Paice et al. 1993, 1995a, b). Studies on the effect of enzyme treatment on delignification of hardwood Kraft pulp showed about 14% reduction in kappa number. The delignification was found to be accompanied by the release of methanol from phenolic methoxyl groups in Kraft pulp lignin. With softwood Kraft pulp, delignification was observed over a wide range of initial lignin contents (kappa number). Only the pulp of lowest initial kappa number gave a higher brightness after enzyme treatment. Subsequent treatment with alkaline hydrogen peroxide resulted in pulps up to 7–8 points brighter than those obtained without enzyme. In the absence of added manganese and malonate buffer, comparable brightness gain of 6 points could be achieved.

Moreira et al. (2001) isolated the MnP of Bjerkadera and tested in vitro with eucalyptus oxygen-delignified kraft pulp (ODKP) based on measuring the reduction in kappa number as an indicator of lignin oxidation. The MnP preparation applied at 60 U/g pulp for 6 h caused a significant decrease of 11–13% in the kappa number in the ODKP under optimal conditions compared to parallel-incubated controls lacking enzyme. Hatakka et al. (1997) reported degradation of wheat straw with enzymes from *P. chrysosporium* and *Ceriporiopsis subvermispora*. Enzymes from both the fungi decreased the amount of lignin and hemicellulose and increased the relative amount of cellulose.

Combination of manganese peroxidase and xylanase has pulp-bleaching effects that are far superior to those of the individual enzymes used sequentially (Bermek et al. 2000). The bleaching effect was a synergistic action between the enzymes rather than an addition of the brightening abilities of each individual enzyme. Bermek et al. (2002) have reported the effect of unsaturated fatty acids; thiol-containing compounds and various other organic compounds in pulp bleaching experiments with MnP. Thiol-containing compounds did not improve the pulp bleaching effect by MnP. Some unsaturated fatty acids, linoleic acid, and linolenic acid provided a better pulp bleaching effect than Tween 80. A combination of Tween 80 and a carboxylic acid resulted in higher pulp brightness than that obtained with Tween 80 alone. A laccase mediator, 3-hydroxy-1,2,3-benzotriazin-4 (3H)-one, could also enhance the MnP pulp bleaching effect.

Ducka and Pekarovicova (1995) used crude ligninases from *P. chrysosporium* for bleaching of softwood Kraft pulp. The pulp after oxygen bleaching with kappa number 19.7 and 36.5% MgO brightness was pretreated with ligninases (L) or xylanases (X) and bleached in QEOPDP sequence. The brightness of pulp bleached in this sequence was 87.8%, which is about 3.2 points higher than the control and is approximately the same as with the use of commercial xylanases.

Martinez et al. (2000) have reported delignfication of wheat straw with enzyme from *Pleurotus eryngii*. Enzyme treatment did not have any adverse effect on physical strength properties of the pulp. Kantelinen et al. (1993a) studied the capability of lignin-modifying enzymes of *Phlebia radiata* to improve the bleachability of Kraft and peroxyformic acid pulps. The effect of lignin modifying enzymes alone on Kraft pulps was insignificant. Bleachability of the Kraft pine pulp was found to be improved only when laccase was used after hemicellulase treatment (Kantelinen et al. 1993b).

Niku-Paavola et al. (1994) treated oxygen delignified pine Kraft pulps with lignin modifying enzymes – laccase, manganese dependent peroxidases, lignin peroxidase – and xylanase in order to improve bleachability. Pulps were treated either with a purified single enzyme or alternatively enzymes were successively added in different combinations. The residual pulps were delignified with alkaline hydrogen peroxide and analyzed for kappa number and brightness. Lignin modifying enzymes did not improve the bleachability when acting alone. Xylanase treatment increased the brightness by 1–2.5 ISO units and xylanase combined with lignin modifying enzymes increased the brightness by a further 1 unit. By extending the alkaline peroxide step, the final brightness increased in all samples, whereas the relative difference between reference and enzyme treated samples remained constant (Niku-Paavola et al. 1994).

Machii et al. (2004) have examined the biobleaching of manganese-less oxygen-delignified hardwood kraft pulp (E-OKP) by the white-rot fungi *Phanerochaete sordida* YK-624 and *P. chrysosporium* in the solid-state fermentation system. *P. sordida* YK-624 possessed a higher brightening activity than *P. chrysosporium*, increasing pulp brightness by 13.4 points after 7 days of treatment. In these fermentation systems, lignin peroxidase (LiP) activity was detected as the principle ligninolytic enzyme, and manganese peroxidase and laccase activities were scarcely detected over the course of treatment of E-OKP by either fungus. Moreover, a linear relationship between brightness increase and cumulative LiP activity was observed under all tested culture conditions with *P. sordida* YK-624 and *P. chrysosporium*. These results indicated that LiP is involved in the brightening of E-OKP by both white-rot fungi.

Kadimaliev et al. (2003) have studied the Lignin consumption and synthesis of lignolytic enzymes by the fungus Panus (Lentinus) tigrinus cultivated on solid phase (modified and unmodified birch and pine sawdusts). The fungus grew better and consumed more readily the birch lignin than the pine wood. Peroxidase activity was higher in the case of pine sawdust; laccase; and lignolytic activities, in the case of birch sawdust. Treatment with ammonia or sulfuric acid decreased lignin consumption by the fungus cultivated on either medium. Modification of sawdust by ultrasound increased lignin consumption and may be recommended for accelerating biodegradation of lignocellulose substrates.

Milagres et al. (1995) studied the effect of number of stages on the treatment of hardwood Kraft pulp with xylanase (X) alone and sequentially, with xylanases (X) and laccases (L). They also studied the effect of various orders of sequential treatment with these enzyme preparations. It was noted that a multistage process achieved greater delignification than a one-stage process. About 23 and 11.2% delignification was obtained using the sequence X-X-X-L and L-L-L-X, respectively, whereas about 7.9 and 5% delignification was obtained with the sequence X-L and L-X, respectively. Sequence order L-L-X, L-X-L did not have much effect on the lignin content (12.9% delignification). The absolute number of X stages was found to have a greater impact.

Three laccases, a natural form and two recombinant forms obtained from two different expression hosts, were characterized and compared for paper pulp bleaching

(Sigoillot et al. 2004). Laccase from *Pycnoporus cinnabarinus*, a well-known lignolytic fungus, was selected as a reference for this study. The corresponding recombinant laccases were produced in *A. oryzae* and *A. niger* hosts using the lacI gene from *P. cinnabarinus* to develop a production process without using the expensive laccase inducers required by the native source. It was observed that treatment of wheat straw Kraft pulp using laccases expressed in *P. cinnabarinus* or *A. niger* with 1-hydroxybenzotriazole (HBT) as redox mediator achieved a delignification close to 75%, whereas the recombinant laccase from *A. oryzae* was not able to delignify pulp.

Vaheri and Piirainen (1992) showed that the oxidizing enzyme laccase can be used in conjunction with manganese ions to reduce the consumption of chlorine chemicals applied in the later stages of bleaching. Viikari et al. (1993) were unable to demonstrate any bleaching effects on pine Kraft pulp with ligninases and oxidases from the white-rot fungus *P. radiata* or purified ligninase from *P. chrysosporium*. When the ligninases were applied after treatment with hemicellulases, there were slight reductions in kappa number. However, these were within the accuracy limits of the analytical method. An extensive delignification and brightening observed with the fungus were not achieved with the isolated manganese peroxidase (Paice et al. 1993) or lignin peroxidase and laccase (Kantelinen et al. 1993b; Arbeloa et al. 1992).

These results show that single enzymes are not able to mimic the complete biological system. Small improvements can be achieved by addition of low molecular weight aromatic compounds like veratryl alcohol or other substances such as 2,2′-azinobis (3-ethylbenzthiazoline-6-sulfonate) (ABTS) and Remazol blue (Bourbonnais and Paice 1990; Olsen et al. 1989). Lignozym GmbH Germany, continued work with enzymes plus chemical mediators, which create a redox system throughout the pulp treatment period (Call 1994a, b; Call and Mücke 1994a, b; Call and Mücke 1995a, b). Their idea was to find a system, which is a good mimic of the natural situation. Starting in 1987 with the enzyme mediator concept, Lignozym improved the performance of the mediator system for the laccase of *T. versicolor* by changing and further fine tuning the chemical nature of the component. The mediator is HBT (Call and Mücke 1995a, b). The treatment of pulp with laccase alone does not result in any degradation of lignin but just in a structural change or repolymerization, the laccase-mediator system (LMS) causes a significant kappa number reduction. According to the present understanding, the laccase while oxidizing the chemical mediator is generating a strongly oxidized co-mediator, which is the real bleaching agent (Fig. 8.2).

The LMS provides a broad flexibility with respect to the pulp substrate, the technical requirements for application and the final quality of the pulp. LMS is compatible with all other bleaching sequences. The performance of the LMS system has been proven in pilot plant trial. The summary of the results is presented in Table 8.2. A degree of delignification of >50% could be obtained in a single step, even if the mediator dosage is reduced by factor 0.6.

Table 8.3 illustrates the conceptual difference between the xylanase and direct enzymatic lignin attack.

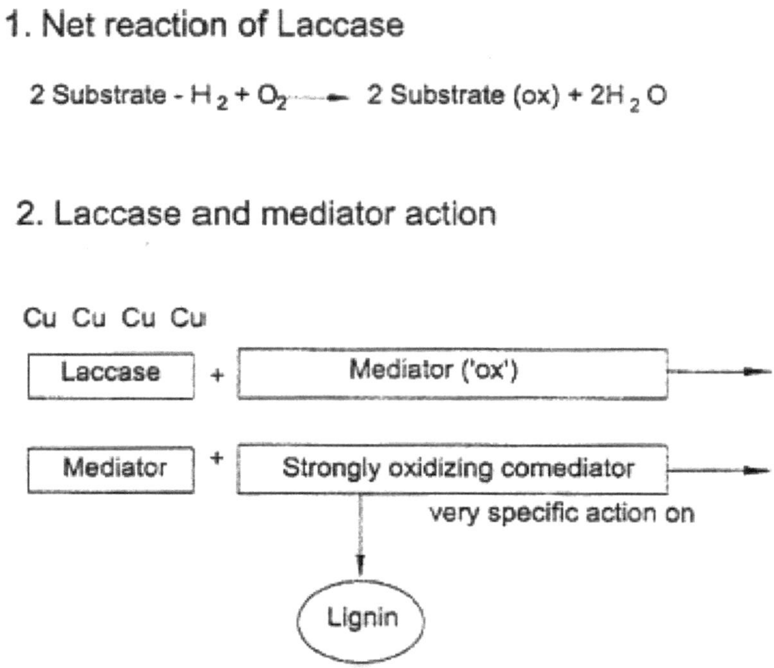

Fig. 8.2 Possible mechanism of laccase and mediator action on lignin (based on Call and Mücke 1995a, b, 1997)

Table 8.2 Summary of results from the pilot plant trial with laccase-mediator system (LMS)

Sequence	Pulp	Dosage of enzyme/ mediator (kg/TP)	Degree of delignification (%)	Maximum brightness (% ISO)
L-E-Q-P	A	2/13	56.6	76.5
L-E-L-E-Q-P	A	2X2/2X8	50.6/67.7	82.7
L-E-Q-(P)	B	2/8	44.2	

Conditions

Parameter	L stage	E stage	Q stage	P stage
Consistency (%)	10	10	5	10
Temperature (°C)	45	60	60	75
pH	4.5	11.5	5	11.2
Residence time (min)	120	60	30	210
Pressure (bar)	2	–	–	–
Dosage	Enzyme: 2 kg/T, Mediator: variable	NaOH	0.2% DTPA	3% Peroxide

Based on data from Call and Mücke (1997)

Table 8.3 Conceptual difference between the xylanase and laccase/mediator treatment

Xylanase
No or very poor kappa reduction
Moderate bleaching effect
Saving of bleaching chemicals
Laccase/mediator
Very good kappa reduction
Good bleaching effect
Saving of bleaching chemicals
Totally chlorine-free (TCF) pulp production possible

Herpoel et al. (2002) reported enzymatic delignification of wheat straw pulp by a sequential xylanase and laccase mediator treatment. It was found that sequential treatment of wheat straw chemical pulp with xylanase and laccase followed by alkaline extraction lowered the kappa number by about 60%. The enzymatic treatment reduced the chlorine consumption and resulted in higher final brightness.

Poppius-Levlin et al. (1997) have reported 60% reduction in kappa number after laccase-mediator treatment and alkaline extraction. Pulp yield and viscosity were found to be high with negligible changes in hemicellulose and hexenuronic acid. Variations in treatment temperature, laccase charge and mediator charges are found to have a pronounced effect on lignin reactions; their effect on pulp carbohydrates, however, was insignificant.

Sealey et al. (1997) reported that with oxygen reinforced alkaline extraction; laccase-HBT biobleaching could obtain over 70% delignification in one stage. Further treatment of L (EOP) pulp with another laccase HBT stage increased the delignification to about 80%. It seems that laccase HBT biobleaching is capable of reacting with the last vestiges of residual lignin, which are typically very unreactive.

Sigoillot et al. (2005) have studied the ability of feruloyl esterase (from *A. niger*) and Mn^{2+}–oxidizing peroxidases (from *P. chrysosporium* and *P. eryngii*) to decrease the final lignin content of flax pulp. Laccase from *P. cinnabarinus* (without mediator) also caused a slight improvement of pulp brightness that was increased in the presence of aryl-alcohol oxidase. However, the best results were obtained when the laccase treatment was performed in the presence of a mediator, 1-HBT, enabling strong delignification of pulps. The enzymatic removal of lignin resulted in high-final brightness values that are difficult to attain by chemical bleaching of this type of pulp.

Lignozym introduced a new mediator, *N*-hydroxyacetinalide (NHAA) that is biodegradable and has been claimed to be cost-effective (Amann 1997; Call and Mücke 1997). Biobleaching with NHAA also allows the enzyme to maintain about 80% of its original activity after an hour treatment, whereas biobleaching with HBT causes a severe loss of enzyme activity. Wacker Chemie GmbH took the worldwide exclusive license of the Lignozym technology for its commercialization.

Fu et al. (2000) bleached *Eucalyptus urophylla* Kraft pulp with laccase in the presence of NHAA and obtained 43% reduction in kappa no. after alkali extraction. The addition of surfactant improved the dissolution of lignin and hence improved the pulp brightness and also improved the activity of the laccase. The effectiveness of HBT and NHAA in LMS has been confirmed (Chakar and Ragauskas 2000). Higher levels of delignification were achieved with HBT compared to NHAA. The benefits of oxidative reinforced alkaline stages were demonstrated. NMR analysis of lignin samples demonstrated the advantages of LMS/HBT vs. LMS/NHAA systems. Residual lignin isolated following LMS/HBT treatment was more enriched in lignin carboxylic acid than that from LMS/NHAA.

The effect of ABTS on laccase catalyzed oxidation of Kraft pulp was reported by Bourbonnais and Paice (1992). It was found that demethylation (diagnostic for delignification) was enhanced and kappa number was decreased with ABTS. Bourbonnais et al. (1995) compared the activities of induced laccases I and II (from *T. versicolor*) with and without ABTS, on hardwood and softwood Kraft pulps. It was noted that in the absence of ABTS, the two purified laccases were not able to reduce the kappa number of either pulp, but both produced small amounts of methanol with the hardwood pulp only. When ABTS was present, the two enzymes were equally effective in delignifying and demethylating either pulp. The enhancement of delignification by the dye ABTS is not fully understood, it may be that the ABTS radical cation acts as an electron carrier between residual lignin in the Kraft pulp fiber wall and the large laccase molecule. Significant differences in reactivity between various fungal laccases for pulp delignification in the presence of mediators ABTS and HBT were found (Bourbonnais et al. 1997).

Geng and Li (2002) synthesized and investigated a number of hydroxamic acids as laccase-mediators for pulp bleaching. As compared with *N*-hydroxyacetanilide (NHA), one of the most effective laccase-mediators reported so far, *N*-(4-cyanophenyl)acetohydroxamic acid (NCPA), resulted in the highest brightness and lowest kappa number of hardwood kraft pulp of all the laccase-mediators studied. The bleaching efficacy of a laccase/7-cyano-4-hydroxy-2H-1,4-benzoxazin-3-one system was also comparable with that of a laccase/NHA system.

Arias et al. (2003) have demonstrated that application of the laccase from *Streptomyces cyaneus* in the presence of ABTS to biobleaching of eucalyptus kraft pulps resulted in a significant decrease in the kappa number (2.3 U) and an important increase in the brightness (2.2%) of pulps, showing the suitability of laccases produced by *Streptomycetes* for industrial purposes. A comparison of *T. versicolor* laccase with various mediators, including ABTS, HBT, remazol blue, nitrosonapthols and phenothiazines has shown that HBT gave the most extensive delignification, but deactivated the enzyme and, therefore, required a higher enzyme dosage. The properties of the fully bleached pulps were not affected by the LMS treatments. Three different chemical pulps, i.e., a pine Kraft pulp, a two-stage oxygen-delignified pine Kraft pulp, and birch formic acid/peroxyformic acid (MILOX) pulps were subjected to HBT- and ABTS-mediated laccase treatments. HBT was more effective than ABTS in the presence of laccase in delignification and gave higher pulp brightness. All the laccase/HBT treated pulps showed an increased response to alkaline hydrogen peroxide bleaching, and as a consequence oxygen-delignified pulp and MILOX pulp reached full brightness in one and two stages, respectively. Even the pine Kraft pulp reached a final brightness of 83%. Moreover, a xylanase stage before or together with laccase/HBT slightly improved the effect of laccase/HBT and gave a higher final brightness after peroxide bleaching than without the xylanase treatment.

Kandioller and Christov (2001) used hydroxybenzotriazole (HBT), *N*-hydroxyacetanilide (NHA), violuric acid (VA), and potassium-octacyanomolybdate (IV) (PCM) as mediators in combination with the laccase (L) of *T. versicolor* at various dosages to delignifiy and bleach the pulps. Kappa number reductions between 5.6 and 64.3%, depending on pulp type, enzyme and mediator charge, were obtained following alkaline extraction (E). On all pulps, VA was the most efficient mediator in terms of kappa number reduction. In terms of brightness gain after LE treatment, VA was most efficient on bagasse soda pulp (up to 1.0 points) and hardwood sulfite dissolving pulp (up to 7.0 points). HBT was the most efficient mediator in terms of brightness boost on bagasse soda pulp (4.0 points) and hardwood soda-AQ pulp (1.5 points), whereas VA was most efficient on hardwood sulfite dissolving pulp (1.1 points). Chlorine dioxide savings were achieved on all the three pulps: hardwood sulfite dissolving pulp (55% savings), hardwood soda-AQ pulp (50%), and bagasse soda pulp (50%).

Camarero et al. (2004) have compared three fungal laccases (from *P. cinnabarinus*, *T. versicolor*, and *P. eryngii*) and two mediators, ABTS and 1-HBT. *P. cinnabarinus* and *T. versicolor* laccases in the presence of HBT gave the best results in terms of high brightness and low lignin content (kappa number). The former laccase also resulted in the best preservation of cellulose and the largest removal of residual lignin as revealed by analytical pyrolysis, and was selected for subsequent TCF bleaching. Up to 90% delignification and strong brightness increase were attained after a laccase-mediator treatment followed by H_2O_2 bleaching. This TCF sequence was further improved by applying H_2O_2 under pressurized O_2.

Chandra et al. (2001) have reported that the bleaching of high kappa Kraft pulps with a LMS provided 42.6–61.1% delignification following an E+P stage when violuric acid was used as the mediator. The pulps yield after the LMS(E+P) treatment were +99.9%. A comparison of mediator efficiency indicated that violuric acid was a superior reagent for LMS delignification of high lignin content pulps in comparison to *N*-hydroxybenzotriazole or *N*-acetyl-*N*-phenyl hydroxylamine. Molecular modeling of these three mediators indicates an elevated impaired electron density for violuric acid over the other mediators, perhaps accounting for its improved performance. Structural analysis of the residual lignin after the LMS treatments indicated that the LMS-stage oxidizes, primarily the phenolic structures of lignin. Full sequence ECF bleaching of high- and low-kappa SW Kraft pulps after a LMS(EOP) or (LMS)(LMS)(EOP) indicated the pulps could be readily bleached to +85 TAPPI brightness.

Call et al. (2004) reported that by combining two enzyme based systems, the overall performance can be strongly enhanced. Therefore, combined enzymatic pulp treatment methods such as an extended HOS (hydrolase mediated oxidation system) and an improved LASox system (laccase oxidation system) were used (on mixed hardwood pulp consisting mainly of oak and beech). The main target of these treatments was to evaluate the chlorine dioxide saving potential of the mentioned enzymatic approaches. It was demonstrated, that a saving of ca. 50% of the chlorine dioxide charge at good strength properties and at comparable costs is possible. A drawback was some loss in hemicellulose content probably due to impurities in the hydrolase (lipase) formulation used.

In Paprican, the use of transition metal complexes as catalytic mediators in the enzymatic delignification and bleaching of Kraft pulps has been investigated (Bourbonnais et al. 2000; Paice et al. 2001). An oxygen-delignified softwood Kraft pulp was treated with laccase in the presence of a transition metal complex – potassium octacynomolybdate. At all charges of mediator, pulp delignification exceeded that of the control pulps. Pulp viscosity did however suffer a slight loss at the highest dosage of mediator. Treatment of hardwood Kraft pulp at the same reaction conditions produced similar results. The molybdenum mediator can be recycled after pulp delignification and reused with the same efficiency as a fresh solution of mediator. LMS is also found to be effective in removing hexenuronic acid from Kraft pulp (Fagerström et al. 2001).

Gamelas et al. (2005) have applied a series of manganese substituted polyoxotungstates, $[SiW_{11}Mn^{III}(H_2O)O_{39}]^{5-}$ and $[PW_{11}Mn^{III}(H_2O)O_{39}]^{4-}$, and $[SiW_{11}V^{V}O_{40}]^{5-}$ as catalysts for the oxygen delignification of unbleached eucalypt kraft pulp with laccase of *T. versicolor*. Unlike to modest results obtained in the LMS at 45–60°C (lignin oxidation and catalyst re-oxidation occurred at the same stage), a sustainable delignification with removal of about 50% of residual lignin was achieved with $[SiW_{11}Mn^{III}(H_2O)O_{39}]^{5-}$ and $[SiW_{11}V^{V}O_{40}]^{5-}$ when the kraft pulp treatment was carried out with polyoxometalate at 110°C (lignin oxidation stage) and with at 45°C (catalyst re-oxidation stage) in separate stages. The use of $[PW_{11}Mn^{III}(H_2O)O_{39}]^{4-}$ in this multistage process was limited by the low reoxidation rate with laccase. The best selectivity on the pulp delignification was found with polyoxoanion $[SiW_{11}Mn^{III}(H_2O)O_{39}]^{5-}$, whereas $[SiW_{11}V^{V}O_{40}]^{5-}$ was the most effective in the oxidative delignification. The influence of process factors on the POMs re-oxidation, such as the amount of laccase, oxygen pressure, and temperature has been studied. UV–vis and ^{51}V NMR studies indicated that POMs maintained stable after redox turnovers during the pulp delignification.

Ehara et al. (2000) investigated the role of Tween 80 in the biobleaching of unbleached hardwood Kraft pulp with manganese peroxidase. The effect of the surfactant on the brightness of the pulp and the manganese peroxidase activity was compared with the effect of two other surfactants (Tween 20). Tween 80 and Tween 20 both dispersed degraded lignin and activated manganese peroxidase, although these effects did not explain the significantly higher brightness increase achieved with Tween 80.

Vares et al. (1997) studied delignification of semichemical wheat straw pulp with LMS and manganese peroxidase in combination with Tween 80. Differences between the enzymatically treated pulp and untreated control were rather small. However, some favorable or prominent effects were caused by Tween 80 alone or in combination with manganese peroxidase and by laccase-HBT treatment followed by a chelating step with ethylene diamine tetraacetic acid.

Tavares et al. (2004) have studied the catalytic oxygen bleaching of kraft pulp using the heteropolyanion $[SiW_{11}Mn^{III}(H_2O)O_{39}]^{5-}$ ($SiW_{11}Mn^{III}$) and laccase of *T. versicolor*. The oxidation of the residual lignin in pulp with the heteropolyanion was followed by the catalyst re-oxidation by laccase in a separate stage. The alternative treatment in a multistage process with $SiW_{11}Mn^{III}$ at 110°C and with laccase at 45°C allowed a sustainable eucalypt kraft pulp delignification with removal of about 50% of the residual lignin with minimal polysaccharides damage.

Fillat and Blanca Roncero (2009) undertook research to examine the performance of a laccase mediator treatment (L) in the biobleaching of flax pulp, together with the influence of the treatment time and of supplying the medium with oxygen on various properties of the resulting pulp and effluent. The enzyme used was laccase from *Trametes villosa* and the L samples were subjected to an alkaline extraction (E) stage. It was found that the biotreatment involves two distinct stages in both L and LE sequences. Initially, the pulp exhibited a fast delignification and a slow viscosity decrease, which was followed by slow delignification in the second stage. Pulp brightness initially decreased with respect to the initial pulp after the L stage and then increased rapidly, eventually leveling off. After the LE sequence, brightness initially increased rapidly and more gradually afterward. It was found that supplying the medium with oxygen and increasing the oxygen concentration in it affected the kinetics of the process. The CIE L*a*b* color coordinates study revealed that the enzyme treatment not only removed lignin, but also altered the structure of the pulp by causing the formation of chromophoric groups giving color. In an E stage, such groups were removed.

Moldes et al. (2010) studied a new biobleaching sequence involving two enzymatic stages based on the application of LMS, in order to increase the effectiveness of enzyme delignification on eucalypt kraft pulp. Synthetic and natural mediators were utilized in the laccase stages and the biobleached pulp was compared in terms of chemical, optical, and physico-mechanical properties. It was found that the pulp bleached with 1-hydroxybenzotriazole (HBT) or violuric acid (VA) exhibited similar delignification (64.1 and 65.9%, respectively) and optical properties (86.4%, 86.1% ISO brightness, respectively) as an industrial TCF pulp. Syringaldehyde was found to improve these properties to a lower extent, 56.71% delignification and 80.52% ISO brightness. The bioleaching sequence had no negative effect on physico-mechanical properties of the pulp and in very specific cases, some slight improvements were observed.

Valls and Roncero (2009) reported that xylanase pretreatment would effectively reduce the laccase and mediator doses or the reaction time in the XLE sequence. As the xylanase pretreatment assisted access to cellulose fibers, this enhanced the effect of the LMS in reducing the content of a residual lignin and releasing more HexA. A similar brightness and smaller kappa number could be obtained by using 30% less laccase, 80% less HBT, and a 45% shorter reaction time.

Xu et al. (2010a, b) studied the use of Recombinant manganese peroxidise (rMnP) from *Pichia pastoris* for kraft pulp delignification and TCF and ECF bleaching.

Three sequential rMnP treatments of hardwood kraft pulp combined with alkaline extraction yielded pulp with the kappa number reduced by 61% and brightness increased by 26 points. In TCF bleaching, the inclusion of the rMnP treatment in the bleaching sequence produced a significant improvement in pulp brightness, while in ECF bleaching, inclusion of the three-stage rMnP treatment combined with alkaline extraction reduced the consumption of chlorine dioxide by 41% for HWKP and 30% for SWKP. Moreover, the pulp strength properties of the biobleached pulp were better or comparable with those of conventionally bleached pulp.

8.3.2 Effect of Lignin-Oxidizing Enzymes on Pulp and Effluent Quality

Lignin-oxidizing enzymes do not affect the viscosity, strength properties, and yield of the pulp due to the specific action on lignin. Due to reduced chlorine consumption, effluent properties are also expected to improve. Bourbonnais and Paice (1992) have reported that the treatment of Kraft pulp with laccase and ABTS did not affect the pulp viscosity and zero span breaking length. In a pilot plant trial with LMS, the strength properties and viscosity of the pulp with L-E-Q-P sequence were not found to be affected (Call and Mücke 1995a, b). Three enzyme preparations (crude cellulase, laccase, and proteinase) were evaluated for their potential to improve the papermaking properties of mechanical pulp (Wong et al. 2000). After treating a long fiber-rich fraction of the pulp with enzyme, the fibers were recombined with untreated fines for handsheet making and testing. None of the enzymes altered the retention of fines or the consolidation of the furnish mix during handsheet formation. All three enzymes increased tensile stiffness index, which is a measure of the initial resistance of the handsheets to strain. Only the laccase preparation, an enzyme that modifies pulp lignin, consistently increased fiber bonding to enhance other strength properties of the handsheets. Bourbonnais and Paice (1996) reported the effect of laccase-ABTS treatment on viscosity and strength properties of an oxygen delignified softwood Kraft pulp. However, the tear index was found to be decreased by enzyme-mediator action. Egan (1985) reported that pulp viscosity did not change after treatment of pulp with Ligninase 118. Camarero et al. (2004) have reported that when high-quality flax pulp was bleached in a TCF sequence using a LMS, the pulp properties obtained by using this system, could not be attained by conventional TCF bleaching of flax pulp. Arbeloa et al. (1992) reported that when softwood TMP was treated with lignin peroxidase enzyme, some of the strength properties viz. burst index, tear index, and breaking length slightly increased. Kondo et al. (2001) have reported that manganese peroxide treatment is quite effective for improving the strength properties of softwood and hardwood bleached Kraft pulps and deinked pulp. In L-E-L-E sequence, the total viscosity loss was 19% compared with the initial pulp. This viscosity loss was found to be comparable with that found following alkaline extraction alone. The viscosity of laccase-ABTS treated sulfite pulp following the hot alkaline extraction was higher as compared to that of untreated pulp because of hemicellulose removal. Herpoel et al. (2002) reported higher tear strength when wheat straw chemical pulp was treated with xylanase and laccase followed by alkaline extraction. Vaheri and Miiki (1991) reported that treatment of pulp with laccase produced an effluent with lower content of chloroorganics.

8.3.3 Mechanism of Bleaching

Lignin peroxidase and manganese peroxidase appear to constitute a major component of the ligninolytic system. These enzymes have been crystallized from *P. chrysosporium* and their tertiary structure examined. The crystallographic structure of lignin peroxidase from *P. chrysosporium* has been studied by Edwards et al. (1993) and Poulos et al. (1993) at different resolutions. The model comprises all 343 amino acids, one heme molecule, and three sugar residues. The crystal structure reveals that the enzyme consists mostly of helical folds with separate domains on either side of the catalytic heme. The enzyme also contains four disulfide bridges formed by the eight cysteine residues and two structural calcium ions which appear to be important for maintaining the integrity of the active site (Poulos et al. 1993). Manganese peroxidase presents a good sequence homology with lignin peroxidase (Sundaramoorthy et al. 1994).

Lignin peroxidase catalyzes a large variety of reactions e.g., cleavage of β-0-4 ether bonds and Cα–Cβ bonds in dimeric lignin model compounds – the basis for the deploymerization reactions catalyzed by lignin peroxidase. The enzyme also catalyzes decarboxylation of phenyl-acetic acids, oxidation of aromatic Cα-alcohols to Cα-oxocompounds, hydroxylation, quinone formation, and aromatic ring opening (Higuchi 1989, 1990, 1993). Lignin peroxidase oxidizes its substrate by two consecutive one-electron oxidation steps, with intermediate cation-radical formation. Due to its high redox potential, lignin peroxidase can also oxidize non-phenolic methoxy substituted lignin subunits. The enzyme can deploymerize dilute solutions of lignin, oxidize and degrade a variety of dimers and oligomers structurally related to lignin in vitro, and catalyze the production of activated oxygen species (Barr et al. 1992; Hammel and Moen 1991; Higuchi 1990). Studies of lignin peroxidase oxidation of DHPs confirm the involvement of lignin peroxidase in the initial degradation of lignin (Higuchi 1993; Umezawa and Higuchi 1989). Lignin peroxidase has a similar catalytic cycle to that of the horseradish peroxidase. The five redox states of lignin peroxidase have been characterized, and its catalytic cycle has been investigated (Cai and Tein 1989, 1992).

Although much of the research has focused on lignin peroxidases, these enzymes are not necessarily involved in lignin degradation and may not be secreted by all lignin-oxidizing fungi (Paice et al. 1995b). *Coriolus versicolor* produces laccases as well as lignin and manganese – dependent peroxidases; however, Archibald (1992a) found that lignin peroxidases secreted by *C. versicolor* did not appear to play an important role in lignin degradation. This enzyme could not be detected in pulp bleaching culture and, moreover, addition of a large excess of lignin peroxidase inhibitor did not interfere bleaching (Archibald 1992a). Also, fungal bleaching was not found to be enhanced by addition of exogenous enzyme. *Dichomitus squalens* and *Rigidoporus lignosus* produce both laccase and a manganese-dependent peroxidase but do not produce lignin peroxidase (Paice et al. 1995b).

Manganese peroxidase acts exclusively as a phenol oxidase on phenolic substrates using Mn^{2+}/Mn^{3+} as an intermediate redox couple. Manganese peroxidase has also been shown to produce H_2O_2 in the oxidation of glutathione, NaDPH, and dihydroxymalic acid (Paszczynski et al. 1985). However, H_2O_2 production is not the major function of manganese peroxidase. Instead, this enzyme is involved in the oxidation of phenols and phenolic lignin structures. It oxidizes Mn^{2+} to Mn^{3+} in the presence of a proper chelating agent and Mn^{3+} must form a complex with the chelators before it oxidizes phenolic substrates (Wariishi et al. 1992). Organic acids are good chelators and white-rot fungi are producers of oxalic acid, malonic acid, pyruvic acid, and malic acid. Mn^{3+}/oxalate and Mn^{3+}/malonate form very stable chelators, which probably also function in vivo. Malonate facilities Mn^{3+} dissociation from the enzyme and has a relatively low Mn^{2+} binding constant (Wariishi et al. 1992). Manganese peroxidase has a catalytic cycle very similar to that of lignin peroxidase.

A Mn^{3+} complex can oxidize phenolic lignin substructures by acting as a mediator between the enzyme and the polymer, and leads to the formation of phenoxy radicals as intermediates (Gold et al. 1989). Subsequently, Cα–Cβ cleavage or alkyl-phenyl cleavage would yield deploymerized fragments including quinones and hydroxyquinones. The purified manganese peroxidase has been reported to deploymerize DHP and also to degrade high molecular mass chlorolignins (Lackner et al. 1991; Wariishi et al. 1991). Several lines of evidence suggest that manganese peroxidase is a key enzyme in fungal bleaching: (1) catalase, an enzyme that destroys H_2O_2 inhibits the bleaching; (2) mutant of *C. versicolor* deficient in manganese peroxidase do not bleach, and bleaching activity is partially restored by addition of manganese peroxidase; and (3) isolated manganese peroxidase produces partial delignifiction of pulps when supplied with H_2O_2, Mn^{2+} and chelator (Paice et al. 1993).

Laccase appears to play an important role in degradation of lignin. Both constitutive and induced forms of laccases are known (Eggert et al. 1996; Garzillo et al. 1992; Vasdev and Kuhad 1994; White and Body 1992). All laccasses are glycoprotein and they generally contain four copper ions (Reinhammer 1984; Sariaslani 1989). These are distributed among three different binding sites and each copper ion appears to play an important role in the catalytic mechanism. Laccase is a true phenol oxidase with broad specificity toward aromatic compound containing hydroxyl and amine group. The enzyme oxidizes phenols and phenolic substructures by one electron abstraction with formation of radicals that can either repolymerize or lead to depolymerization (Higuchi 1993).

Laccase catalyzes demethoxylation reactions of terminal phenolic units. It can also degrade β-dimers and β-0-4 dimers via Cα oxidation alkyl-aryl cleavage and Cα–Cβ cleavage (Kawai et al. 1988). Laccase has also been shown to catalyze the cleavage of aromatic rings in a similar way to lignin peroxidase (Kawai et al. 1988). It was observed that laccase from *C. subvermispora* transformed 4,6-di-tert-butyl-

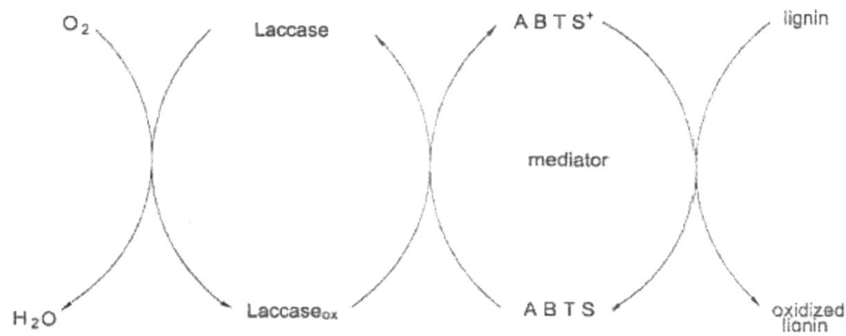

Fig. 8.3 Oxidative pathway for catalytic action of laccase on lignin (based on Bajpai 1997b)

guaiacol into a ring opening product, the muconolactone derivative 2,4-di-(tert-butyl), 4-(methoxy-carbonylmethyl)-2-buten-4-olide. This work clearly demonstrated that ^{18}O incorporated into muconolactone was from $^{18}O_2$ but not from $H_2{}^{18}O$. It is obvious from the results that laccase, like L_1P, can cleave aromatic rings. Studies of side chain cleavage and ring opening of lignin model compounds have shown that both laccase and lignin peroxidase which catalyze one-electron oxidation of either phenolic or non-phenolic compounds, are involved in the initial degradation of lignin substructure model compounds (Higuchi 1993). Until 1990, laccase had been considered to be able to degrade only phenolic lignin model compounds (Higuchi 1990). However, Bourbonnais and Paice (1990) reported that oxidation of non-phenolic lignin substructures by laccase from *T. versicolor* took place in the presence of a suitable redox mediator i.e., the dye ABTS. Laccase along with ABTS has been shown to also delignify kraft pulp (Bourbonnais and Paice 1992). The presence of the mediator prevented reploymerization of kraft lignin by *T. versicolor* laccase (Bourbonnais et al. 1995). Other phenolic reagents have also been proposed and found to work as mediators (Call 1993). Brightening and delignification of kraft pulp has been shown with another mediator 1-HBT (Call 1994a, b). Moreover, the production of a fungal metabolite, which acts as a physiological lacasse/redox mediator has been shown to be produced by the white-rot fungus *P. cinnabarinus* in University of Georgia.

Paice et al. (1995a) have suggested that Cellobiose quinone oxidoreductase (CBQase), a quinone-reducing enzyme also plays a number of roles in delignification of kraft pulp. Studies with CBQase have shown that the enzyme can reduce reaction products of laccase or peroxidases such as quinones, radicals, and Mn^{2+} (Bao et al. 1994). Concurrently, cellobionate, a potential chelater of Mn^{3+} is formed. However, cellobionate is less efficient than malonate or oxalate as a chelater in the manganese peroxidase catalytic cycle, probably because it has a higher binding constant for Mn^{2+}.

Figures 8.3 and 8.4 show the oxidative pathways for catalytic action of laccase and manganese peroxidase on lignin. Jurasek et al. (1994) attempted to develop a model of the three-dimensional structure of lignin to provide a framework for interpretation and prediction of interactions between the enzymes and lignin on the

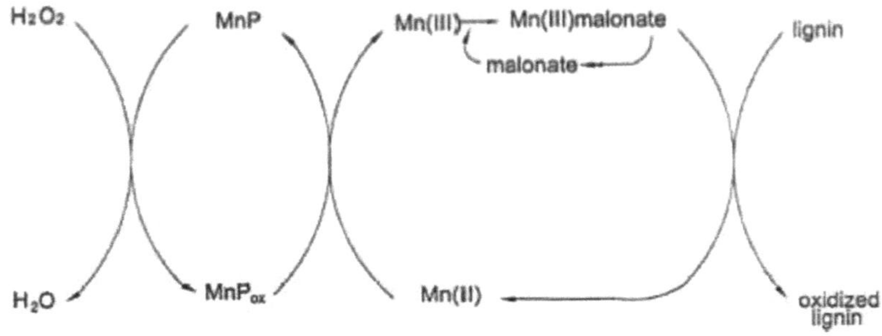

Fig. 8.4 Oxidative pathway for catalytic action of manganese peroxidase on lignin (based on Bajpai 1997b)

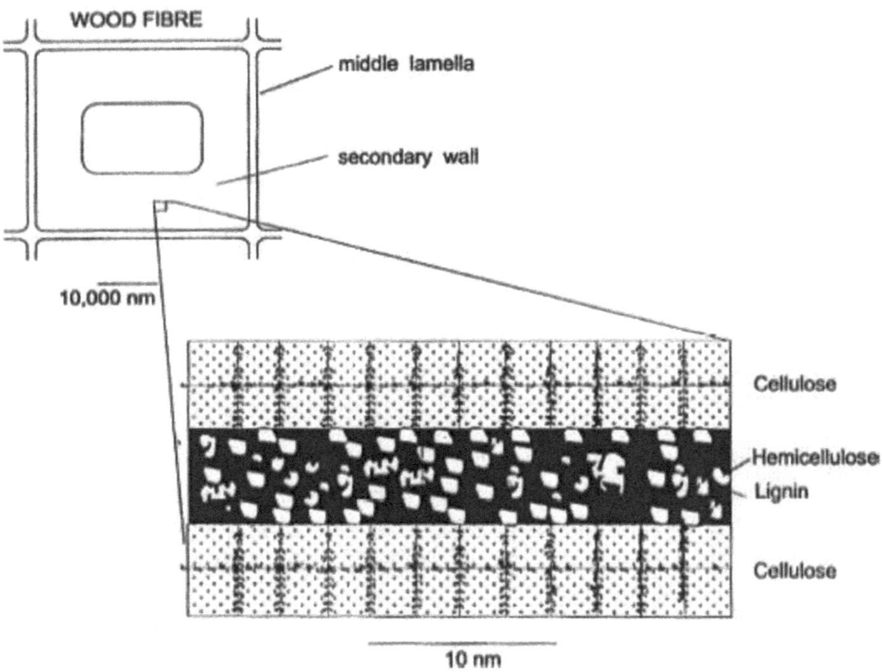

Fig. 8.5 Model of a cross-section of a small portion of secondary wall of wood fiber (based on Jurasek et al. 1994)

molecular level. Inspection of the model of lignified secondary wall showed the accessibility restrictions upon lignin-oxidizing enzymes and points toward a concept of an enzyme factory outside the cell wall, producing chemicals which have access to and break down the lignin and allow its fragments to leave the network (Figs. 8.5 and 8.6). Once out of the wall, these fragments may interact with enzymes directly, resulting in further reduction of the fragment size and finally mineralization.

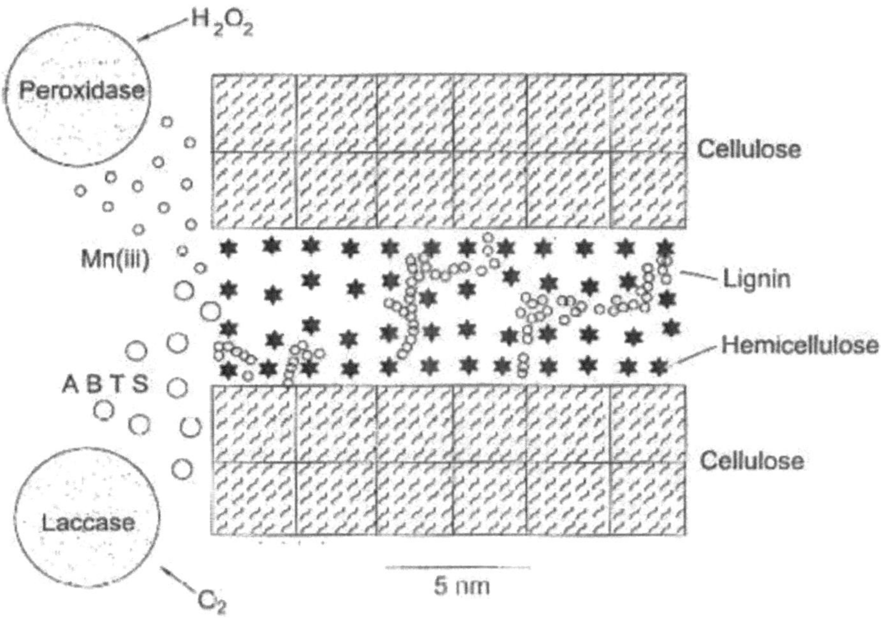

Fig. 8.6 Model of a cross-section area of kraft fiber shown in comparison with some enzyme molecules (based on Jurasek et al. 1994; Paice 2005)

8.3.4 *Advantages, Limitations, and Future Prospects*

Lignin-oxidizing enzymes are highly specific toward lignin; there is no damage or loss of cellulose (Bajpai et al. 2006). This results in better strength and yield of bleached pulp. As compared to oxygen delignification, treatment with lignin-oxidizing enzymes results in more removal of lignin. This translates into substantial savings of energy and bleaching chemicals which in turn would lead to lower pollution load. Lignin-oxidizing enzymes are not currently available in sufficient quantity for mill trials and scale-up of enzyme production from fungal cultures is costly. Cloning of genes for lignin-oxidizing enzymes has been reported and may provide an alternative production route. The laccase-mediators are expensive and alternatives are less effective. There are two ways of improving LMS for pulp bleaching. One is to discover new laccases that have extraordinarily high redox potential, the other is to find a very effective laccase-mediator. Since the laccase-mediators reported so far have very wide structural variations, it should be possible to discover inexpensive and more effective laccase-mediators than what has been done to date. Chelating agents for Mn^{3+} in the manganese peroxidase reaction may also be a signficant cost item. Manganese peroxidase require hydrogen peroxide but are inactivated by concentrations above about 0.1 mmol/L. Even when hydrogen peroxide is kept below 0.1 mmol/L, manganese peroxidase becomes inactive relatively rapidly.

Experience with xylanase and with other enzymes has shown that enzymes can be successfully introduced in the plant. Thus, oxidative enzymes which can be regarded as catalysts for oxygen and hydrogen peroxide driven delignification, may also find a place in the bleach plant in coming years.

8.4 White-Rot Fungi

Several fungi, which attack wood in nature, are capable of degrading lignin. So, there appears to be an opportunity to exploit them for direct attack on lignin in pulps, an attack that can be specific and effective even if the amount of residual lignin is low. The fungi that most effectively biodegrade lignin, are basidiomycete fungi. Lignin-oxidizing species are called white-rot fungi, because they typically turn wood white as they decay. This effect is another indication that these fungi may be useful in pulp bleaching.

8.4.1 Performance of White-Rot Fungi in Bleaching

Significant effort has been also made to study the potential of white-rot fungi for bleaching of chemical pulps. Kirk and Yang (1979) found that *P. chrysosporium* and some other white-rot fungi could reduce the kappa number of unbleached softwood kraft pulp by upto 75%, leading to reduced requirement for chlorine during subsequent chemical bleaching. The pulp was incubated with the fungi in shallow stationary layers for several days, and then extracted with alkali. Kappa number reduction was inhibited by added nutrient nitrogen and enhanced by the oxygen enrichment of the atmosphere, as is lignin degradation by *P. chrysosporium*. Attack on the cellulose of the pulp was severe unless alternative carbohydrate sources were added to the cultures; even in the presence of glucose, the pulp showed a 60% drop in viscosity. Other fungi tested, including *T. versicolor*, had lesser effects. In contrast to the results reported by Kirk and Yang (1979), Pellinen et al. (1989) found that *P. chrysosporium* failed to delignify unbleached softwood kraft pulp in stationary culture, but did so in agitated cultures. Tran and Chambers (1987) reported that the effects of culture conditions on delignification of unbleached hardwood kraft pulp by *P. chrysosporium* were similar to those observed by others with synthetic lignin or lignin in wood. They did not determine the effects of the fungal treatment on the bleachability or the paper-making properties of the pulp.

Researchers at Nippon Paper Industries (Iimori et al. 1994), carried out extensive screening for pulp bleaching fungi. By plating samples of decayed wood or fruit-bodies directly on agar containing unbleached kraft pulp, they isolated 1,758 cultures that produced decolorization zones; 266 of these bleached oxygen-delignified hardwood pulp to 70–81% ISO brightness within 7 days of incubation under solid-state fermentation conditions. The most active isolate, SKB-1152 also brightened agitated suspensions (2% consistency) of O_2-delignified hardwood pulp to 80% ISO. Optimization studies, with the aim of shortening treatment time, showed that dilute pulp suspensions in the range of 0.5–1% consistency were brightened faster than those with higher consistency, inoculum-to-pulp ratio of 6% was found to be adequate. Incubation of the mycelial suspension used as inoculum in a nutrient medium for 24 h before adding it to the pulp eliminated the lag period before the onset of brightening (Iimori et al. 1996).

Paprican researchers reported that *T. versicolor* could markedly increase the brightness of hardwood kraft pulp (Paice et al. 1989). The fungal treatment was carried out in agitated, aerated cultures for 5 days. Under these conditions, *T. versicolor* performed better than *P. chrysosporium*. The kappa number was decreased from 12 to 8, and the brightness increased from 34 to 48%, it could be further increased to 82% with DED bleaching. In initial experiments, Paice et al. (1989) did not observe similar brightening of softwood pulp. Subsequently, *T. versicolor* was found to delignify and brighten softwood kraft pulps (Reid et al. 1990). After 14 days of treatment with the fungus followed by alkaline extraction, the pulp had kappa number 8.5 and could be bleached to 61% ISO brightness with a DED sequence. Whereas without fungal treatment, the pulp had kappa number 24 and was bleached to only 33% brightness. Mycelium of *T. versicolor* immobilized in polyurethane foam was also able to delignify the pulp. *T. versicolor* caused a much smaller direct increase in the brightness of softwood kraft pulp than hardwood kraft pulp. To determine the contributions of higher residual lignin contents (kappa numbers) and structural differences in lignins to recalcitrance of softwood kraft pulps to biobleaching, Reid and Paice (1994a, b) tested softwood and hardwood kraft pulps cooked to the same kappa numbers 26 and 12. A low lignin content (over cooked) softwood pulp resisted delignification by *T. versicolor*, but a high-lignin-content (lightly cooked) hardwood pulp was delignified at the same rate as a normal softwood pulp. The longer time taken by *T. versicolor* to brighten softwood kraft pulp than hardwood kraft pulp was found to result from the higher residual lignin content of the softwood pulp. Softwood pulps, whose lignin contents were decreased by extended modified continuous cooking or oxygen delignification to kappa numbers as low as 15 were delignified by *T. versicolor* at the same rate as normal softwood pulp (Reid and Paice 1994a, b). More intensive O_2 delignification, like overcooking, decreased the susceptibility of the residual lignin in the pulps to degradation by *T. versicolor* (Reid and Paice 1994a, b). This fungus was found to delignify and brighten kraft pulp in dilute (1–2% consistency) agitated suspensions. Delignification of both hardwood and softwood pulps with *T. versicolor* was found to be accompanied by a moderate decrease in viscosity indicating some cellulose depolymerization (Reid et al. 1990; Ho et al. 1990). Addition of excess glucose to repress cellulose biosynthesis did not prevent the loss in viscosity (Kirkpatrick et al. 1990a, b). However, the damage to cellulose was not severe enough to cause important strength losses. Paper sheets made from pulp delignified with *T. versicolor* were slightly stronger than those made from unbleached pulp (Paice et al. 1989; Reid et al. 1990). Delignification by *T. versicolor* was not restricted to conditions of secondary metabolism. The fungus degraded lignin while growing vigorously (Roy and Archibald 1993). Ho et al. (1990)

Table 8.4 Bleaching conditions and optical properties of conventionally bleached and fungal bleached hardwood kraft pulp

	Dosage (% on pulp)						Brightness (% ISO)		
Bleaching sequence	C	E_1	D_1	E_2	D_2	As effective chlorine	Before aging	After aging[a]	PC number
CEDED (Conventional process)	5.0	3.6	0.8	0.2	0.3	7.89	88.8	84.2	0.78
FCED (Fungal bleaching)	1.4	0.8	0.3	0.2	0.3	2.19	88.1	85.3	0.46

Based on data from Fujita et al. (1991)
[a]At 105°C for 1 h

Table 8.5 Bleaching conditions and optical properties of conventionally bleached and fungal bleached softwood kraft pulp[b]

	Dosage (% on pulp)						Brightness (% ISO)			
Bleaching sequence	C	E_1	D_1	E_2	D_2	As effective chlorine	Before aging	After aging[a]	PC number	Yield (%)
CEDED (Conventional bleaching process)	9.0	6.4	1.0	0.5	0.5	12.95	84.2	82.4	0.38	91.4
FCED (Fungal bleaching process)	2.4	1.7	0.4	0.5	0.5	3.45	84.6	83.4	0.25	91.4

Based on data from Fujita et al. (1993)
[a]At 105°C for 1 h
[b]Commercial SWKP was treated with fungus IZU-154 for 6 days, with kappa number falling from 40.0 to 14.9

reported that when kraft pulp was added to inoculum of *T. versicolor*, the lag before the onset of brightening in subsequent pulp treatment was reduced. The presence of pulp induced production of manganese peroxidase and accumulation of organic acids that can chelate Mn^{3+}.

Addleman and Archibald (1993) investigated the ability of 10 dikaryotic and 20 monokaryotic strains of *T. versicolor* to bleach and delignify hardwood and softwood kraft pulps. The isolates were found to vary in their bleaching ability. Dikaryotic strains produced brightness changes ranging from −2 to +22 points in hardwood kraft pulp; monokaryons tended to give a higher brightness increase, with a maximum of 26 points. A monokaryon, 52 J, isolated by protoplasting the dikaryon originally used at Paprican, gave slightly higher brightening than its parents and was adopted for further work. In addition to better bleaching, the monokaryon had the advantage of less biomass production, no dark pigment formation, and a simpler genome (Addleman and Archibald 1993).

Fujita et al. (1991) found that a 5 day fungal (F) treatment of hardwood kraft pulp with IZU-154 replaced a CE_1DE_2D sequence with an FCED sequence yielding a target brightness of 88% ISO with 72% less chlorine, 79% less NaOH, and 63% less ClO_2 (Table 8.4). With softwood kraft pulp, similar chemical savings were achieved (Table 8.5) (Fujita et al. 1993). The yield and burst strength of the pulp bleached

Table 8.6 Optical properties of conventionally bleached and fungal bleached pulps

Bleaching sequence	Brightness (% ISO) Before aging	After aging	Post color number	Yield (%)
OCED	87.9	85.7	0.36	97.4
OPP	73.8	71.8	0.89	97.4
OFEP, 3 days				
4% H_2O_2	85.0	83.7	0.26	97.7
5% H_2O_2	86.3	84.9	0.26	97.8
OFEP, 5 days				
2% H_2O_2	84.3	83.5	0.17	97.1
4% H_2O_2	87.3	85.6	0.29	97.0

Based on data from Fujita et al. (1993)

with a CED sequence after delignification with IZU-154 were equivalent to those of a control bleached with CEDED sequence. However, the tensile index and tear index of the fungus-treated pulp were decreased by 9 and 2.6%, respectively.

Murata et al. (1992) applied the fungus IZU-154 to the delignification and brightening of oxygen-bleached hardwood kraft pulp to establish an absolutely chlorine-free bleaching process. The fungus brightened the pulp and simultaneously decreased its kappa number. Brightness was increased by 17 and 22 points by 3- and 5-day treatments respectively, and kappa number was decreased from 10.1 to 6.4 by a 5-day treatment (Murata et al. 1992). The combination of the 3-day fungal treatment, alkaline extraction (2% NaOH charge), and hydrogen peroxide bleaching with 5% charge of H_2O_2 gave a pulp of 86.3% ISO brightness (Table 8.6). The 5-day-treated pulp was brightened to 87.3% ISO brightness by 4% H_2O_2 bleaching after alkaline extraction. Optical and strength properties of OFEP-bleached pulp were comparable to those of conventional OCED-bleached pulp.

Tsuchikawa et al. (1995) have reported biobleaching of hardwood kraft pulp with lignin-oxidizing fungi *P. sordida* YK-624. When the hardwood kraft pulp was treated with the fungus for 10 days, the kappa number was decreased from 14.4 to 5.75 and the brightness was increased to 61% ISO. If the fungal incubation was interrupted after 5 days and the pulp was extracted with alkali and treated with fungus for another 5 days, the kappa number was lowered to 4.8, and the brightness reached 80% ISO. The intermediate alkaline extraction, which was more effective than water washing, seemed to reactivate the lignin toward degradation as it does with chemical bleaching reagents. The tensile and burst strengths of the fungus treated pulp were almost as high as those of control pulp, but the tear strength was 34% lower.

Nishida et al. (1995) investigated the biobleaching of hardwood unbleached kraft pulp by *P. chrysosporium* and *T. versicolor* in the solid state and liquid state fermentation systems with four different culture media (low nitrogen-high carbon, low nitrogen-low carbon, high nitrogen-high carbon, and high nitrogen-low carbon). In the solid state fermentation system with low nitrogen and high carbon culture medium, pulp brightness increased by 15 and 30 points after 5 days of treatment with *T. versicolor* and *P. chrysosporium*, respectively. The pulp kappa number decreased with the increasing brightness, and a positive correlation between the kappa number decrease and brightness increase of the fungus-treated pulp was observed.

Laccase and manganese peroxidase enzymes were detected during biobleaching of softwood kraft pulps with *P. chrysosporium* and *T. versicolor* (Katagiri et al. 1997). However, the enzyme lignin peroxidase was not detected. *T. versicolor* did not delignify softwood kraft pulp but produced laccase, whereas *P. chrysosporium* did the reverse.

Ultraviolet irradiation was used to prepare 120 strains of mutants of *P. sordida* YK-624 (Ishimura et al. 1998). The role of the reductive enzyme system was explored. The strains were screened using as indicators the tetrazolium salts for reducing enzymes and Poly R for lignin-oxidizing enzymes. The two mutants obtained had weak tetrazolium salts reducing activity and normal Poly R decolorization activity, and strong reducing activity, and normal oxidizing activity, respectively. The two mutants and a wild strain were compared focusing on treatment of unbleached hardwood, softwood, and oxygen bleached softwood kraft pulps. Mutant 1 was observed to secrete manganese peroxidase under low manganese (11) concentration. Uptake of manganese (11) by mutant 1 was faster than that of the wild-type.

Wroblewska and Zielinski (1995) examined biodelignification of beech and birch pulp wood by selected white-rot fungi. One of the strains designated as DL-Sth-4, was found to be best for selective delignification of beech wood. About 25% of lignin was lost with very little loss in cellulose content. Pazukhina et al. (1995) used the culture filtrate of several white-rot fungi – *Pitcairnia sanguinea, C. versicolor, Ganoderma applanatum,* and *Trichoptum biforma* – for bleaching hardwood kraft pulp. *P. sanguinea* showed the highest selectivity in lignin degradation.

Moreira et al. (1997) tested the ability of 25 white-rot fungal strains to bleach *Eucalyptus globulus* oxygen delignified kraft pulp. Under nitrogen-limited culture conditions, eight outstanding biobleaching strains were identified that increased the brightness of OKP by more than 10 ISO units compared to pulp incubated in sterile control medium. The highest brightness gain of approximately 13 ISO units was obtained with *Bjerkandera* sp. strain BOS55, providing a high final brightness of 82% ISO. This strain also caused the greatest level of delignification, decreasing the kappa number of OKP by 29%. When the white-rot fungal strains were tested in nitrogen-sufficient medium, the extracellular activities of laccase and peroxidases increased in many strains; nonetheless, the pulp handsheets were either destroyed or brightness gains were lower than those obtained under nitrogen-limitation. The titre of ligninolytic enzymes was not found to be indicative of biobleaching potential. However, the best biobleaching strains were generally characterized by a predominance of manganese dependent peroxidase activity compared to other ligninolytic enzymes and by a high decolorizing activity toward the polyanthraquinone ligninolytic indicator dye-Poly R-478.

Moreira et al. (1998b) investigated the manganese requirement for biobleaching by *T. versicolor, P. chrysosporium, P. radiata, Stereum hirsutum,* and *Bjerkandera* sp. strain BOS55. When manganese was present in the medium, the kraft pulp was bleached by all five strains. In the absence of manganese, only *Bjerkandera* sp. strain BOS55 bleached the kraft pulp only in the presence of organic acids.

Bjerkandera sp. was also the only fungus which produced manganese peroxidase in the absence of manganese. Fungus *Bjerkandera* sp. strain BOS55 can be used to bleach EDTA-extracted eucalyptus oxygen delignified kraft pulp independent of manganese (Moreira et al. 1998a). Adding simple physiological organic acids at 1–5 mM produced 2–3 times the brightness and pulp delignification compared to control cultures. Inorganic acids improve the manganese-free biobleaching by enhancing the production of manganese peroxidase and lignin peroxidase. Increased physiological concentration of veratryl alcohol and oxalate were also a factor. The resulting improvement in production of superoxide anion radicals might explain the more extensive biobleaching. Results indicate that manganese peroxidase from *Bjerkandera* is deliberately produced in the absence of manganese and might function independently of manganese in OKP delignification. Lignin peroxidase might also be a contributing factor.

P. chrysosporium and *C. versicolor* have been successfully immobilized on polyurethane foam (Kirkpatrick et al. 1990a, b; Ziomek et al. 1991; Kirkpatrick and Palmer 1987). Immobilized and free cultures of *C. versicolor* have been found to bleach hardwood and softwood kraft pulp at a comparable rate and to a similar extent (Reid et al. 1990; Kirkpatrick et al. 1990a, b). The results showed that intimate contact between the fungal hyphae and pulp fibers was not required as long as the media was renewed through contact with the fungus (Kirkpatrick et al. 1990a, b; Archibald 1992b). Immobilization enabled the pulp to be separated from the mycelia. Another advantage of immobilization was that the same fungal biomass could be reused to treat other batches of pulp either immediately or after storage at 4°C (Kirkpatrick et al. 1990a, b).

Cell-free filtrates from bleaching cultures of *T. versicolor* did not cause detectable pulp brightening (Archibald 1992b). These filtrates contain laccase and manganese peroxidase, which can, under appropriate conditions, delignify and brighten pulp (Paice et al. 1993). The failure to detect pulp brightening by culture filtrates can be attributed to lack of H_2O_2 and tween 80 to support brightening by manganese peroxidase and the absence of mediator for laccase. The ability of intact *T. versicolor* cultures to extensively delignify and brighten kraft pulps in the absence of a mediator for laccase or an analogue of tween 80 suggests that the fungus produces one or more unknown enzymes that contribute to its pulp-bleaching ability. *P. sordida* YK-624 could delignify and brighten hardwood kraft pulp separated from the mycelium by a membrane filter with 0.1 μm pores; a thin polycarbonate membrane supported more bleaching than a thicker cellulose nitrate membrane (Kondo et al. 1994). *P. chrysosporium* and *T. versicolor* could also be used through the membrane although to a smaller extent than can YK-624.

Iimori et al. (1998) have investigated the potential to biobleach unbleached (UKP) and oxygen bleached (OKP) hardwood kraft pulp using culture filtrate containing manganese peroxidase and lignin peroxidase from *P. chrysosporium* in a short-term treatment of several hours. The brightness increase following biotreatment with manganese peroxidase, lignin peroxidase, and cofactors was greater than that by manganese peroxidase alone. There was a pronounced 7–8 points increase in brightness for UKP and OKP during the first 3 h. When the 3 h treatment was repeated, the brightness increase of OKP was halted at 78%.

Kondo et al. (2000) reported biobleaching of hardwood kraft pulp by a marine fungus and its enzymes. Marine fungi isolated from Japanese mangrove stands, were screened based on their delignification ability. Strain MG-60 was used to bleach hardwood kraft pulp in vivo. The pulp brightness bleached by MG-60 was found to be considerably higher than that by *P. chrysosporium*. The higher manganese peroxidase activity of MG-60 was observed compared to that of *P. chrysosporium* at the same conditions. Also, the manganese peroxidase secreted from MG-60 demonstrated good stability against high sea salt concentrations. It was also found that unbleached hardwood kraft pulp was biobleached by crude enzymes secreted by MG-60 at 0 and 3% sea salts conc. in vitro.

8.4.2 Effect of White-Rot Fungi on Pulp and Effluent Quality

In order to be adopted by the pulp and paper industry, biological bleaching must not compromise with the quality of the pulp. Results from lab scale fungal bleachings indicate an improvement in strength characteristics in both hardwood and softwood pulps. It has been suggested that the lignin may become more flexible and hydrophilic as a result of fungal enzyme action, resulting in a softer pulp with improved bonding and stronger paper characteristics (Jurasek and Paice 1988). Reduced color reversion was another benefit noted with the fungus IZU-154 (Fujita et al. 1991). Some viscosity loss, indicating limited cellulose depolymerization, has been reported as a result of fungal bleaching (Reid et al. 1990; Fujita et al. 1991). However, based upon experiments done with free and immobilized cultures, Kirkpatrick et al. (1990a, b) reported that upto 25% of the reduction in the pulp viscosity may be due to the presence of fungal mycelia, rather than cellulose cleavage. Although fungal bleaching is primarily an oxidative process, it appears to be more selective than oxygen bleaching at high pH and at kappa number less than 17 because there is a better retention of pulp viscosity.

The impact of fungal bleaching on effluent quality has not been much studied. Fujita et al. (1991) reported that the COD and color loading in the bleach plant waste water were reduced by 50 and 80%, respectively in FCED bleaching sequence. These authors suggested that higher reductions could be obtained with an FED or FE_1DE_2D sequence, although there may be slight loss in pulp yield.

8.4.3 Advantages, Limitations, and Future Prospects

Pretreatment with fungi has been shown to replace upto 70% of the chemicals needed to bleach kraft pulp. The usual specificity of biological reactions and their mild reaction conditions make biological delignification an interesting alternative to bleaching with chemicals such as pressurized oxygen or ozone.

A serious shortcoming of the fungal bleaching process is the long incubation time required for contact with the biomass. Typical contact periods range from 5 to 14 days for different types of pulps. So the size of the fungal bioreactor would have to be very large considering that daily production could range from 200 to over 1,000 a.d. tones of pulp. Most fungal bleaching studies have been performed at low pulp consistencies. Only few studies (Fujita et al. 1991) have been conducted at high consistency, which would allow for a small reactor. Fungal treatment can be conducted in the unbleached storage itself with minor modifications provided the treatment time is reduced to a practically feasible duration. The reaction time, required at its current stage of development, makes it economically unattractive. There is a need to identify/develop fast growing white-rot fungal cultures that could do the job in less time.

References

Addleman K, Archibald FS (1993) Kraft pulp bleaching and delignification by dikaryons and monokaryons of *Trametes versicolor*. Appl Environ Microbiol 59:266–273

Allison RW, Clark TA, Wrathall SH (1993a) Pretreatment of radiata pine kraft pulp with a thermophillic enzyme Part I. Effect on conventional bleaching. Appita 46(4):269–273

Allison RW, Clark TA, Wrathall SH (1993b) Pretreatment of radiata pine kraft pulp with a thermophillic enzyme Part II. Effect on oxygen, ozone and chlorine dioxide bleaching. Appita 46(5):349–353

Amann A (1997) The Lignozym process coming closer to the mill. In: Proceedings fo the 9th ISWPC, Montreal, QC, pp F4-1–F4-5

Arbeloa M, de Leseleuc J, Goma G, Pommier JC (1992) An evaluation of the potential of lignin peroxidases to improve pulps. TAPPI J 75(3):215–221

Archibald FS (1992a) Lignin peroxidase is not important in biological bleaching and delignification of kraft brownstock by *Trametes verisicolor*. Appl Environ Microbiol 58:3101–3109

Archibald FS (1992b) The role of fungus fiber contact in the biobleaching of kraft brownstock by *Trametes versicolor*. Holzforschung 46:305–310

Arias ME, Arenas M, Rodriguez J, Soliveri J, Ball AS, Hernandez M (2003) Kraft pulp biobleaching and mediated oxidation of a nonphenolic substrate by laccase from *Streptomyces cyaneus* CECT 3335. Appl Environ Microbiol 69(4):1953–1958

Atkinson D, Moody D, Gronberg V (1993) Enzymes make pulp bleaching faster. Invest Tech Pap 35(136):199–209

Awakaumova AV, Nikolaeva TV, Vendilo AG, Kovaleva NE, Sinitzyn AP (1999) ECF bleaching of hardwood kraft pulp: new aspects. In: 13th International papermaking conference – progress-99, Cracow, Poland, 22–24 Sept 1999, pp IV-5-1–IV-5-13

Bajpai P (1997a) Microbial xylanolytic enzyme system: properties and applications. In: Neidleman S, Laskin A (eds) Advances in applied microbiology, vol 43. Academic Press, New York, NY, pp 141–194

Bajpai P (1997b) Enzymes in pulp and paper processing. Miller Freeman, San Francisco, CA

Bajpai P (1999) Application of enzymes in pulp & paper industry. Biotechnol Prog 15(2): 147–157

Bajpai P (2004) Biological bleaching of chemical pulps. Crit Rev Biotechnol 24(11):1–58

Bajpai P (2009) Xylanases. In: Schaechter M, Lederberg J (eds) Encyclopedia of microbiology, vol 4, 3rd edn. Academic Press, San Diego, CA, pp 600–612

Bajpai P, Bhardwaj NK, Maheshwari S, Bajpai PK (1993) Use of xylanase in bleaching of eucalypt kraft pulp. Appita 46(4):274–276

Bajpai P, Bhardwaj NK, Bajpai PK, Jauhari MB (1994) The impact of xylanases in bleaching of eucalyptus kraft pulp. J Biotechnol 36(1):1–6
Bajpai P, Ananad A, Bajpai PK (2006) Bleaching with lignin oxidizing enzymes. Biotechnol Annu Rev 12:349–378
Bao WL, Fukushima Y, Jensen KA, Moen MA (1994) Oxidative degradation of non-phenolic lignin during lipid peroxidation by fungal manganese peroxidase. FEBS Lett 354:297–300
Barr DP, Shah MM, Grover TA, Aust SD (1992) Production of hydroxyl radical by lignin peroxidase from *Phanerochaete chrysosporium*. Arch Biochem Biophys 298:480–485
Bermek H, Li K, Eriksson KE (2000) Pulp bleaching with manganese peroxidase and xylanase: a synergistic effect. TAPPI J 83(10):69
Bermek H, Li K, Eriksson KE (2002) Studies on mediators of manganese peroxidase for bleaching of wood pulps. Bioresour Technol 85(3):249–252
Biely P (1985) Microbial xylanolytic systems. Trends Biotechnol 3:286–290
Bim MA, Franco TT (2000) Extraction in aqueous two-phase systems of alkaline xylanase produced by *Bacillus pumilus* and its application in kraft pulp bleaching. J Chromaogr 743(1): 346–349
Bourbonnais R, Paice MG (1990) Oxidation of non-phenolic substrates. An expanded role for laccase in lignin biodegradation. FEBS Lett 267:99–102
Bourbonnais R, Paice MG (1992) Demethylation and delignification of kraft pulp by *Trametes versicolor* laccase in the presence of 2,2′-azinobis-3-ethylbenzthiazoline-6-sulphonate. Appl Microbiol Biotechnol 36:823–827
Bourbonnais R, Paice MG (1996) Enzymatic delignification of kraft pulp using laccase and a mediator. TAPPI J 76(6):199–204
Bourbonnais R, Paice MG, Reid ID, Lanthier P, Yaguchi M (1995) Lignin oxidation by laccase isozymes from *Trametes versicolor* and role of the mediator 2, 2′-Azinobis (3-ethylbenzothiazoline-6-sulfonate) in kraft lignin depolymerization. Appl Environ Microbiol 61(5):1876–1880
Bourbonnais R, Leech D, Paice MG, Freiermuth B (1997) Reactivity and mechanism of laccase-mediators for pulp delignification. In: Proceedings of the TAPPI biological science symposium. TAPPI Press, Atlanta, GA, pp 335–338
Bourbonnais R, Rochefort D, Paice MG, Renaud S, Leech D (2000) Transition metal complexes: a new class of laccase-mediators for pulp bleaching. TAPPI J 83(10):68
Cai D, Tein M (1989) On the reactions of lignin peroxidase compounds III (isozyme H8). Biochem Biophys Res Commun 162:464–470
Cai D, Tein M (1992) Kinetic studies on the formation and decomposition of compound II and III. Reactions of lignin peroxidase with hydrogen peroxide. J Biol Chem 267:11149–11155
Call HP (1993) Process for producing cellulose from lignin containing raw materials using an enzyme or microorganism while monitoring and maintaining the redox potential. US Patent 5,203,964
Call HP (1994a) Multicomponent bleaching system. WO 94/29425
Call HP (1994b) Process for modifying, breaking down or bleaching lignin, materials containing lignin or like substances. PCT World patent application WO 94/29510
Call HP, Mücke I (1994a) Enzymatic bleaching of pulps with the laccase-mediator-system. In: Pulping conference AlChE session, San Diego, CA, pp 38–52
Call HP, Mücke I (1994b) State of the art of enzyme bleaching and disclosure of a breakthrough process In: International non-chlorine bleaching conference, Amelia Island, FL
Call HP, Mücke I (1995a) Further improvements of the laccase-mediator-system (LMS) for enzymatic delignification and results from large scale trials. In: International non-chlorine bleaching conference, Amelia Island, FL, 5–9 March 1995, p 16
Call HP, Mücke I (1995b) The laccase-mediator-system (LMS). In: Srebotnk E, Messner K (eds) Biotechnology in the pulp and paper industry: recent advances in applied and fundamental research. Proceedings of the 6th international conference on biotechnology in the pulp and paper industry, Vienna, Austria, pp 27–32
Call HP, Mücke I (1997) History, overview and applications of mediated lignolytic systems, especially laccase-mediator-systems (Lignozym-process). J Biotechnol 53:163–202

Call H-P, Call S, Garcia-Lindgren C, Marklund A (2004) Extended lab trials: combined enzymatic delignification and bleaching systems. In: 9th International conference on biotechnology in the pulp and paper industry, Durban, South Africa, 10–14 Oct 2004
Camarero S, García O, Vidal T, Colom J, Del río JC, Gutiérrez A, Gras JM, Monje R, Martínez MJ, Martínez AT (2004) Efficient bleaching of non-wood high-quality paper pulp using laccase-mediator system. Enzyme Microb Technol 35(2–3):113–120
Chakar FS, Ragauskas AJ (2000) The effects of oxidative alkaline extraction stages after laccase HBT and laccase NHAA treatments – an NMR study of residual lignins. J Wood Chem Technol 20(2):169–184
Chandra RP, Chakar FS, Allison L, Kim DH, Ragauskas AJ, Elder T (2001) Delving into the fundamental LMS delignification of high kappa kraft pulps. In: 8th International conference on biotechnolgy in the pulp and paper industry, Helsinki, Finland, 4–8 June 2001, pp 54
Ducka I, Pekarovicova A (1995) Ligninases in bleaching of softwood kraft pulp. In: 6th International conference on biotechnology in the pulp and paper industry, Vienna, Austria, 11–15 June 2005
Edwards SL, Raag R, Wariishi H, Gold MH, Poulos TL (1993) Crystal structure of lignin peroxidase. Proc Natl Acad Sci USA 90:750–754
Egan M (1985) Proceedings of the second annual pulp and paper chemical outlook conference, Corpus Information Services Ltd., Montreal, QC, 12–13 Nov 1985
Eggert C, Temp V, Eriksson K-EL (1996) The ligninolytic system of the white-rot fungus *Pycnoporus cinnabarinus:* purification and characterization of the laccase. Appl Environ Microbiol 62:1151–1158
Ehara K, Tsutsumi Y, Nishida T (2000) Role of tween 80 in biobleaching of unbleached hardwood kraft pulp with manganese peroxidase. J Wood Sci 46(2):137–142
Eiras KM, Milanez AF, Colodette L (2009) Biobleaching of eucalyptus pulp. In: 42nd Pulp and paper international congress and exhibition, Sao Paulo, Brazil, 26–29 Oct 2009, pp 8
Fagerström R, Tenkanen M, Kruus K, Buchert J (2001) Removal of hexenuronic acid side groups from kraft pulp by laccase/mediator treatment. In: 8th International conference on biotechnology in the pulp and paper industry, Helsinki, Finland, 4–8 June 2001, pp 225–230
Farrell R (1987) Use of rldmtm 1–6 and other ligninolytic enzymes. PCT Int. Appl. WO 87/00, 564
Farrell RL, Gelep P, Anillouis A, Javaherian K, Malone TE, Rusche JR, Sadownick BA, Jackson JA (1987a) Sequencing and expression of ligninase cDNA of *Phanerochaete chrysosporium.* EP 0216080
Farrell RL, Kirk TK, Tien M (1987b) Novel enzymes for degradation of lignin. WO 87/00550
Fillat U, Blanca Roncero M (2009) Biobleaching of high quality pulps with laccase mediator system: influence of treatment time and oxygen supply. Biochem Eng J 44(2–3):193–198
Fu S, Zhan H, Yu H (2000) Preliminary study on biobleaching of *Eucalyptus urophylla* kraft pulp with laccase-mediator system. China Pulp Pap 19(2):8–15
Fujita K, Kondo R, Sakai K, Kashino Y, Nishida T, Takahara Y (1991) Biobleaching of kraft pulp using white-rot fungus IZU-154. TAPPI J 74(11):123–127
Fujita K, Kondo R, Sakai K (1993) Biobleaching of softwood kraft pulp with white-rot fungus IZU-154. TAPPI J 76(1):81–84
Gamelas JAF, Tavares APM, Evtuguin DYV, Xavier AMB (2005) Oxygen bleaching of kraft pulp with polyoxometalates and laccase applying a novel multi-stage process. J Mol Catal B Enzym 33:57–64
Garzillo AMV, Dipaolo S, Burla G, Buonocore V (1992) Differently-induced extracellular phenol oxidases from *Pleurotus ostreatus*. Phytochemistry 31:3685–3690
Geng X, Li K (2002) Degradation of non-phenolic lignin by the white-rot fungus *Pycnoporus cinnabarinus*. Appl Microbiol Biotechnol 60(3):342–346
Gold MH, Wariishi H, Walli K (1989) Extracellular peroxidases involved in lignin degradation by the white-rot basidiomycete *Phanerochaete chrysosporium*. ACS Symp Ser 389:127
Gruninger H, Fiechter A (1986) A novel, highly thermostable D-xylanase. Enzyme Microb Technol 8:309–314
Gysin B, Griessmann T (1991) Bleaching wood pulp with enzymes. EP 0418201 A2
Hammel KE, Moen MA (1991) Depolymerization of a synthetic lignin in vitro by lignin peroxidase. Enzyme Microb Technol 13:15–18

Hatakka AI, Bocchini P, Vares T, Galletti GC (1997) Production of lignin-degrading enzymes on solid straw medium by *Phanerochaete chrysosporium* and *Ceriporiopsis subvermispora* and degradation of the lignocellulosic substrate. In: 1997 Biological sciences symposium, San Francisco, CA, 19–23 Oct 1997, pp 19–23
Herpoel I, Jeller H, Fang G, Petit-Conil M, Bourbonnair R, Robert J-L, Asther M, Sigoillot J-C (2002) Efficient enzymatic delignification of wheat straw pulp by a sequential xylanase-laccase mediator treatment. J Pulp Pap Sci 28(3):67–71
Higuchi T (1989) Mechanism of lignin degradation by lignin peroxidase and laccase of white-rot fungi. In: Lenis NG, Paice MG (eds) Biogenesis and biodegradation of plant cell polymers. ACS Symposium No. 399, American Chemical Society, Washington, DC, pp 482–502
Higuchi T (1990) Lignin biochemistry: biosynthesis and biodegradation. Wood Sci Technol 24:23–63
Higuchi T (1993) Biodegradation mechanism of lignin by white-rot basidiomycetes. J Biotechnol 30(1):1–11
Ho C, Jurasek L, Paice MG (1990) The effect of inoculum on bleaching of hardwood kraft pulp with *Coriolus versicolor*. J Pulp Pap Sci 16:J78–J83
Iimori T, Kaneko R, Yoshikawa H, Machida M, Yoshioka H, Murakami K (1994) Screening of pulp bleaching fungi and bleaching activity of newly isolated fungus SKB-1152. Mokuzai Gakkaishi 40(7):733–737
Iimori T, Yoshikawa H, Kaneko R, Miyawaki S, Machida M, Murakami K (1996) Effects of treatment conditions on treatment times for biobleaching by SKB-1152. Mokuzai Gakkaishi 42: 313–317
Iimori T, Miyawaki S, Machida M, Murakami K (1998) Biobleaching of unbleached and oxygen-bleached hardwood kraft pulp by culture filtrate containing manganese peroxidase and lignin peroxidase from *Phanerochaete chrysosporium*. J Wood Sci 44(6):451–456
Ishimura D, Kondo R, Sakai K, Hirai H (1998) Biobleaching of kraft pulp with mutants from white-rot fungus *Phanerochaete sordida* YK-624. In: Proceedings of the 7th international conference on biotechnology in the pulp and paper industry, vol B, Vancouver, BC, 16–19 June 1998, p B237
Jurasek L, Paice MG (1988) Biological treatments of pulps. Biomass 15:103–108
Jurasek L, Archibald FS, Bourbonnais R, Paice MG, Reed ID (1994) Prospects for redox enzymes to enhance Kraft pulp bleaching. In: Proceedings of the biological sciences symposium, Minneapolis, MN, 3–6 Oct 1994, p 239
Kadimaliev DA, Revin VV, Atykian NA, Samuilov VD (2003) Effect of wood modification on lignin consumption and synthesis of lignolytic enzymes by the fungus *Panus (Lentinus) tigrinus*. Prikl Biokhim Mikrobiol 39(5):555–560
Kandioller G, Christov L (2001) Efficiency of *Trametes versicolor* laccase-mediator systems in pulp delignification and bleaching. In: 8th International conference on biotechnology in the pulp and paper industry, Helsinki, Finland, 4–8 June 2001, pp 223
Kantelinen A, Hortling BO, Ranua M, Viikari L (1993a) Effects of fungal and enzymatic treatments on isolated lignins and pulp bleachability. Holzforschung 47:29–35
Kantelinen A, Hortling B, Sundquist J, Linko M, Viikari L (1993b) Proposed mechanism of the enzymatic bleaching of kraft pulp with xylanases. Holzforschung 47:318–324
Katagiri N, Tsutsumi Y, Nishida T (1997) Biobleaching of softwood kraft pulp by white-rot fungi and its related enzymes. J Jpn Wood Res Soc 46(8):678–685
Kawai S, Umezawa T, Shimada M, Higuchi T (1988) Aromatic ring cleavage of 4,6-di(tert-butyl) guaiacol, phenolic lignin model compound by laccase of *Coriolus versicolor*. FEBS Lett 236:309–311
Kirk TK, Yang HH (1979) Partial delignification of unbleached kraft pulp with ligninolytic fungi. Biotechnol Lett 1:347–352
Kirkpatrick N, Palmer JH (1987) Semi-continuous ligninase production using foam-immobilized *Phanerochaete chrysosporium*. Appl Microbiol Biotechnol 27:129–133
Kirkpatrick N, Reid ID, Ziomek E, Paice MG (1990a) Biological bleaching of hardwood kraft pulp using *Trametes versicolor* immobilized in polyurethane foam. Appl Environ Microbiol 33:105–108

Kirkpatrick N, Reid ID, Ziomek E, Paice MG (1990b) Physiology of hardwood kraft pulp bleaching by *Coriolus versicolor* and use of foam immobilization for the production of mycelium-free bleached pulps. In: Kirk TK, Chang HM (eds) Biotechnology in pulp and paper manufacture. Butterworth-Heinemann, Boston, MA, pp 131–136

Ko C-H, Tsai C-H, Tu J, Yang B-Y, Hsieh D-L, Jane W-N, Shih T-L (2011) Identification of *Paenibacillus* sp. 2S-6 and application of its xylanase on biobleaching. Int Biodeterior Biodegr 65(2):334–339

Kondo R, Kurashiki K, Sakai K (1994) In vitro bleaching of hardwood kraft pulp by extracellular enzymes secreted from white-rot fungi in a cultivation system using a membrane filter. Appl Environ Microbiol 60:921–926

Kondo R, Li X, Sakai K (2000) Biobleaching of hardwood kraft pulp by a marine fungus and its enzyme. In: Pulp and paper research conference, Tokyo, Japan, 28–29 June 2000, pp 12–17

Kondo R, Tsuchikawa K, Sakai K (2001) Application of manganese peroxidase to modification of fibers. In: 8th International conference on biotechnology in the pulp and paper industry, Helsinki, Finland, 4–8 June 2001, p 70

Lackner R, Srebotnik E, Messner K (1991) Oxidative degradation of high molecular weight chlorolignin by manganese peroxidase of *Phanerochaete chrysosporium*. Biochem Biophys Res Commun 178:1092

Latorre UF, Sacon VM, Bassa A (2008) Selection of commercial xylanases to improve pulp bleaching in Jacarei mill (Votorantim Celulose e Papel). Influence of pH and COD in process efficiency. In: International pulp bleaching conference, Quebec City, QC, 2–5 June 2008, pp 265–266

Luthi E, Jasmat NB, Berquist P (1990) Xylanase from the extremely thermophilic bacterium *"Caldocellum saccharolyticum"*: overexpression of the gene in *Escherichia coli* and characterization of the gene product. Appl Environ Microbiol 56:2677–2683

Machii Y, Hirai H, Nishida T (2004) Lignin peroxidase is involved in the biobleaching of manganese-less oxygen-delignified hardwood kraft pulp by white-rot fungi in the solid-fermentation system. FEMS Microbiol Lett 233(2):283–287

Manji AH (2006) Extended usage of xylanase enzyme to enhance the bleaching of softwood kraft pulp. TAPPI J 5(1):23–26

Martinez AT, Camarero S, Ruiz-Duenas FJ, Heinfling A, Martinez MJ (2000) Studies on microbial and enzymatic applications in paper pulp manufacturing from non-woody plants based on white-rot fungi from the genus Pleurotus. In: 2000 Pulping/process and product quality conference, Boston, MA, 5–8 Nov 2000, 10 pp

Mathrani IM, Ahring BK (1992) Thermophilic and alkalophilic xylanases from several *Dictyoglomus* isolates. Appl Microbiol Biotechnol 38:23–27

McCarthy AJ, Peace E, Broda P (1985) Studies on the extracellular xylanase activity of some thermophilic actinomycetes. Appl Microbiol Biotechnol 21:238–244

Milagres AMF, Medeiros MB, Borges LA (1995) Sequential treatment of eucalyptus kraft pulp with *Penicillium janthinellium* xylanase and *Pleurotus ostreatus* laccase. In: 6th International conference on biotechnology in the pulp and paper industry, Vienna, Austria, 11–15 June 1995

Moldes D, Cadena EM, Vidal T (2010) Biobleaching of eucalypt kraft pulp with a two laccase-mediator stages sequence. Bioresour Technol 101(18):6924–6929

Moreira MT, Feijoo G, Sierra-Alvarez R, Lema J, Field JA (1997) Biobleaching of oxygen delignified kraft pulp by several white-rot fungal strains. J Biotechnol 53:237–251

Moreira MT, Feijoo G, Merter T, Mayorga P, Sierra-Alvarez R, Field JA (1998a) Role of organic acids in the manganese-independent biobleaching system of *Bjerkandera* sp. strain BOS 55. Appl Environ Microbiol 64(7):2409–2417

Moreira MT, Sierra-Alvarez R, Feijoo G, Field JA (1998b) Evaluation of the manganese requirement for biobleaching by white-rot fungi. In: Proceedings of the 7th international conference on biotechnology in the pulp and paper industry, vol B, Vancouver, BC, 16–19 June 1998, pp B229

Moreira MT, Sierra-Alvarez R, Lema JM, Feijoo G, Field JA (2001) Oxidation of lignin in eucalyptus kraft pulp by manganese peroxidase from *Bjerkandera* sp. strain BOS55. Bioresour Technol 78(1):71–79

Murata S, Kondo R, Sakai K, Kashino Y, Nishida T, Takahara Y (1992) Chlorine-free bleaching process of kraft pulp using treatment with the fungus IZU-154. TAPPI J 75(12):91–94
Niku-Paavola ML, Ranua M, Suurnakki A, Kantelinen A (1994) Effects of lignin-modifying enzymes on pine kraft pulp. Bioresour Technol 50:73–77
Nishida T, Katagiri N, Tsutsumi Y (1995) New analysis of lignin-degrading enzymes related to biobleaching of kraft pulp by white-rot fungi. In: 6th International conference on biotechnology in the pulp and paper industry, Vienna, Austria, 11–15 June 1995
Olsen WL, Slocomb JP, Gallagher HP, Kathleen BA (1989) Enzymatic delignification of lignocellulosic material. EP 0,345,715 A1
Olsen WL, Gallagher HP, Burris AK, Bhattacharjee SS, Slocomb JP, Dewitt DM (1991) Enzymatic delignification of lignocellulosic material. EP 406, 617
Paice M (2005) Enzyme application in pulp and paper manufacturing. In: Lakehead University symposium, 27 September 2005
Paice M, Zhang X (2005) Enzymes find their niche. Pulp Pap Can 106(6):17–20
Paice MG, Jurasek L, Ho C, Bourbonnais R, Archibald FS (1989) Direct biological bleaching of hardwood kraft pulp with the fungus *Coriolus versicolor*. TAPPI J 72(5):217–221
Paice MG, Gurnagul N, Page DH, Jurasek L (1992) Mechanism of hemicellulose directed prebleaching of kraft pulp. Enzyme Microb Technol 14:272–276
Paice MG, Reid ID, Bourbonnais R, Archibald FS, Jurasek L (1993) Manganese peroxidase produced by *Trametes versicolor* during pulp bleaching, demethylates and delignifies kraft pulp. Appl Environ Microbiol 59:260–265
Paice MG, Bourbonnais R, Reid ID (1995a) Bleaching kraft pulps with oxidative enzymes and alkaline hydrogen peroxide. TAPPI J 78(9):161–170
Paice MG, Bourbonnais R, Reid ID, Archibald FS, Jurasek L (1995b) Oxidative bleaching enzymes. J Pulp Pap Sci 21:J280–J284
Paice MG, Bourbonnais R, Renaud S, Amann M, Candussio A, Rochefort D, Leech D, Labonte S, Sacciadis G (2001) Laccase/mediator catalysed delignification: trials with new mediators. In: 8th International conference on biotechnology in the pulp and paper industry, Helsinki, Finland, 4–8 June 2001, p 48
Paszczynski A, Huynh V-B, Crawford R (1985) Enzymatic activities of an extracellular manganese-dependent peroxidase from *Phanerochaete chrysosporium*. FEMS Microbiol Lett 29:37–40
Pazukhina GA, Soloviev VA, Malysheva ON (1995) Bleaching of kraft pulp with filtrates of white-rot fungi. In: 6th International conference on biotechnology in the pulp and paper industry, Vienna, Austria, 11–15 June 1995
Pellinen J, Abuhasan J, Joyce TW, Chang HM (1989) Biological delignification of pulp by *Phanerochaete chrysosporium*. J Biotechnol 10:161–170
Perttula M, Ratto M, Konradsdottir M, Kristijansson JK, Viikari L (1993) Xylanases of thermophilic bacteria from Icelandic hot springs. Appl Microbiol Biotechnol 38:592–595
Polizeli ML, Rizzatti AC, Monti R, Terenzi HF, Jorge JA, Amorim DS (2005) Xylanases from fungi: properties and industrial applications. Appl Microbiol Biotechnol 67(5):577–591
Poppius-Levlin K, Wang W, Ranua M, Niku-Paavola ML, Viikari L (1997) Biobleaching of chemical pulps by laccase/mediator systems. In: Proceedings of the TAPPI biological science symposium. TAPPI Press, Atlanta, GA, pp 329–333
Poulos TL, Edwards SL, Wariishi H, Gold MH (1993) Crystallographic refinement of lignin peroxidase at 2 Å. J Biol Chem 268(6):4429–4434
Reid ID, Paice MG (1994a) Biological bleaching of kraft pulps by white-rot fungi and their enzymes. FEMS Microbiol Rev 13:369–376
Reid ID, Paice MG (1994b) Effect of residual lignin type and amount on bleaching of kraft pulp by *Trametes versicolor*. Appl Environ Microbiol 60(5):1395–1400
Reid ID, Paice MG, Ho C, Jurasek L (1990) Biological bleaching of softwood kraft pulp with the fungus *Trametes versicolor*. TAPPI J 73(8):149–153
Reid ID, Bourbolnnais R, Paice MG (2010) Biopulping and biobleaching. In: Heitner C, Dimmel DR, Schmidt JA (eds) Lignin and lignans: advances in chemistry. CRC Press, Boca Raton, FL, pp 521–554

Reinhammer B (1984) Laccase. In: Lontie R (ed) Copper proteins and copper enzymes. CRC, Boca Raton, FL, p 1
Roy BP, Archibald F (1993) Effects of Kraft Pulp and Lignin on *Trametes versicolor* Carbon Metabolism. Appl Environ Microbiol 59(6):1855–1863
Saleem M, Tabassum MR, Yasmin R, Imran M (2009) Potential of xylanase from thermophilic *Bacillus* sp. XTR-10 in biobleaching of wood kraft pulp. Int Biodeterior Biodegr 33(8):1119–1124
Sariaslani FS (1989) Microbial enzymes for oxidation of organic molecules. Crit Rev Biotechnol 9:171–257
Sealey JE, Ragaukas AJ, Runge TM (1997) Biobleaching of kraft pulps with laccase and hydroxybenzotriazole. In: Proceedings of the TAPPI biological science symposium. TAPPI Press, Atlanta, GA, pp 339–342
Senior DJ, Hamilton J (1991) Use of xylanase to decrease the formation of AOX in kraft pulp bleaching. In: Proceedings of the environment conference of the technical section, Canadian Pulp and Paper Association, Quebec, Canada, 8–10 Oct 1991, pp 63–67
Senior DJ, Hamilton J (1992a) Bleaching with xylanases brings biotechnology to reality. Pulp Pap 66(9):111–114
Senior DJ, Hamilton J (1992b) Reduction in chlorine use during bleaching of kraft pulp following xylanase treatment. TAPPI J 75(11):125–130
Senior DJ, Hamilton J (1992c) Use of xylanase to decrease the formation of AOX in kraft pulp bleaching. J Pulp Pap Sci 18(5):J165–J168
Senior DJ, Hamilton J (1993) Xylanase treatment for the bleaching of softwood kraft pulps: the effect of chlorine dioxide substitution. TAPPI J 76(8):200–206
Senior DJ, Hamilton J, Bernier RL Jr (1992) Use of *Streptomyces lividans* xylanase for biobleaching of kraft pulps. In: Visser J, Beldmann G, Kusters-van Someren MA, Voragen AGJ (eds) Xylans and xylanases. Progress in biotechnology, vol 7. Elsevier, Amsterdam, The Netherlands, pp 555
Senior DJ, Hamilton J, Taiplus P, Torvinen J (1999) Enzyme use can lower bleaching costs, aid ECF conversions. Pulp Pap 73(7):59–62
Senior DJ, Bernhardt SA, Hamilton J, Lundell R (2000) Mill implementation of enzymes in pulp manufacture. In: Biological science symposium, San Francisco, CA, 19–23 Oct 2000, pp 163
Sigoillot C, Record E, Belle V, Robert JL, Levasseur A, Punt PJ, van den Hondel CA, Fournel A, Sigoillot JC, Asther M (2004) Natural and recombinant fungal laccases for paper pulp bleaching. Appl Microbiol Biotechnol 64(3):346–352
Sigoillot C, Camarero S, Vidal T, Record E, Asther M, Pérez-Boada M, Martínez MJ, Sigoillot JC, Asther M, Colom JF, Martínez AT (2005) Comparison of different fungal enzymes for bleaching high-quality paper pulps. J Biotechnol 115(4):333–343
Skerker PS, Labbauf MM, Farrell RL, Beerwan N, McCarthy P (1992) Practical bleaching using xylanases: laboratory and mill experience with Cartazyme HS-10 in reduced and chlorine free bleach sequences. In: TAPPI pulping conference, Boston, 1–5 Nov 1992. TAPPI press, Atlanta, GA, p 27
Skjold-Jorgensen S, Munk N, Pederson LS (1992) Recent progress within the application of xylanases for boosting the bleachability of kraft pulp. In: Kuwahara M, Shimada M (eds) Biotechnology in the pulp and paper industry. Uni Publishers, Tokyo, Japan, pp 93–99
Sundaramoorthy S, Kishi K, Gold MH, Poulos TL (1994) Preliminary crystallographic analysis of manganese peroxidase from *Phanerochaete chrysosporium*. J Mol Biol 238(5):845–856
Suominen P, Mantyla A, Saarelainen R, Paloheimo M, Fagerstrom P, Parkkinen E, Nevalainen H (1992) Genetic engineering of *Trichoderma reesei* to produce suitable enzyme combinations for applications in the pulp and paper industry. In: Proceedings of the 5th international conference on biotechnology in the pulp and paper industry, Kyoto, Japan, 27–30 May 1992, p 439
Suurnakki A, Tenkanen M, Buchert J, Viikari L (1997) Hemicellulases in the bleaching of chemical pulps. In: Scheper T (ed) Advances in biochemical engineering/biotechnology, vol 57. Springer, Berlin, Germany, pp 261–287

Tan LUL, Mayers P, Saddler JN (1987) Purification and characterization of a thermostable xylanase from a thermophilic fungus *Thermoascus aurantiacus*. Can J Microbiol 33:689–694

Tavares APM, Gamelas JAF, Gaspar A, Evtuguin DV, Xavier AMB (2004) A novel approach for the oxidative catalysis employing polyoxometalate-laccase system: application to the oxygen bleaching of kraft pulp. Catal Commun 5:485

Thibault L, Tolan J, White T, Yee E, April R, Sung W (1999) Use of an engineered xylanase enzyme to improve ECF bleaching at Weyenhaeuser Prince Albert. In: 85th Annual meeting, Montreal, QC, 26–29 Jan 1999, pp B263

Tolan JS (1992) Mill implementation of enzyme treatment to enhance bleaching. In: Proceedings of the 78th CPPA annual meeting, Montreal, QC, 28–29 Jan 1992, pp A163–A168

Tolan JS (2001) How a mill can get more benefit out of its xylanase treatment. In: 8th International conference on biotechnology in the pulp and paper industry, Helsinki, Finland, 4–8 June 2001, pp 81

Tolan JS, Canovas RV (1992) The use of enzymes to decrease the chlorine requirements in pulp bleaching. Pulp Pap Can 93(5):39–42

Tolan JS, Guenette M (1997) Using enzymes in pulp bleaching: mill applications. In: Scheper T (ed) Advances in biochemical engineering/biotechnology, vol 57. Springer, Berlin, Germany, pp 288–310

Tolan JS, Olson D, Dines RE (1996) Survey of mill usage of xylanase. In: Jeffries TW, Viikari L (eds) Enzymes for pulp and paper processing. ACS Symposium Series 655, American Chemical Society, Washington, DC, pp 25–35

Tran AV, Chambers RP (1987) Delignification of an unbleached hardwood pulp by *Phanerochaete chrysosporium*. Appl Microbiol Biotechnol 25:484–490

Tsuchikawa K, Kondo R, Sakai K (1995) Application of ligninolytic enzymes to bleaching of kraft pulp II: totally chlorine-free bleaching process with the introduction of enzyme treatment with crude enzymes secreted from *Phanerochaete sordida* YK-624. Jpn TAPPI J 49:1332–1337

Umezawa T, Higuchi T (1989) Cleavage of aromatic ring and β-4-O-bond of synthetic lignin (DHP) by lignin peroxidase. FEBS Lett 242:325–330

Vaheri M, Miiki K (1991) Redox enzyme treatment in multistage bleaching of pulp. EP 0,408,803 A1

Vaheri M, Piirainen O (1992) Bleaching of pulp in presence of oxidizing enzyme and transition metal compound. WO 92/09741

Valchev V, Valchev V, Christova E (1998) Introduction of an enzyme stage in bleaching of hardwood kraft pulp. Cellul Chem Technol 32(5–6):457–462

Valchev I, Valchev I, Ganev I (2000) Improved elemental chlorine free bleaching of hardwood kraft pulp. Cellul Chem Technol 33(1–2):61–66

Valls C, Roncero MB (2009) Using both xylanase and laccase enzymes for pulp bleaching. Bioresour Technol 100(6):2032–2039

Valls C, Gallardo O, Vidal T, Pastor FIJ, Diaz P, Roncero MB (2010) New xylanases to obtain modified eucalypt fibres with high-cellulose content. Bioresour Technol 101(19):7439–7445

Vares T, Almondros G, Galletti GC, Hatakka A, Dorado J, Bocchini P, Martinez AT (1997) Effect of ligninolytic enzymes and mediators on paper. In: 1997 Biological sciences symposium, San Francisco, CA, 19–23 Oct 1997, pp 405–412

Vasdev K, Kuhad RC (1994) Decolourization of poly R-478(polyvinylamine sulfonate anthrapyridone) by *Cyathns bulleri*. Folia Microbiol 39(1):61–70

Viikari L, Ranua M, Kantelinen A, Sundquist J, Linko M (1986) Bleaching with enzymes. In: Proceedings of the 3rd international conference on biotechnology in the pulp and paper industry, Stockholm, Sweden, pp 67–69

Viikari L, Kantelinen A, Poutanen K, Ranua M (1990) Characterization of pulps treated with hemicellulolytic enzymes prior to bleaching. In: Kirk TK, Chang HM (eds) Biotechnology in pulp and paper manufacture. Butterworth-Heinemann, Boston, MA, pp 145

Viikari L, Tenkanen M, Buchert J, Ratto M, Bailey M, Siika-aho M, Linko M (1993) Hemicellulases for industrial applications. In: Saddler JN (ed) Bioconversion of forest and agricultural plant residues. CAB International, Wallingford, UK, pp 131–182

Viikari L, Kantelinen A, Sundquist J, Linko M (1994) Xylanases in bleaching: from an idea to industry. FEMS Microbial Rev 13:335–350

Viikari L, Poutanen K, Tenkanen M, Tolan JS (2002) Hemicellulases. In: Flickinger MC, Drew SW (eds) Encyclopedia of bioprocess technology: fermentation, biocatalysis, and bioseparation. Wiley, Chichester, West Sussex (Update. Electronic release)

Viikari L, Suurna kki A, Gronqvist S, Raaska L, Ragauskas A (2009) Forest products: biotechnology in pulp and paper processing. In: Schaechter M (ed) Encyclopedia of microbiology, 3rd edn. Academic Press, New York, NY, pp 80–94

Wariishi H, Valli K, Gold MH (1991) In vitro depolymerization of lignin by manganese peroxidase of *Phanerochaete chrysosporium*. Biochem Biophys Res Commun 176:269–275

Wariishi H, Valli K, Gold MH (1992) Manganese(II) oxidation by lignin peroxidase from the basidiomycete *Phanerochaete chrysosporium*. Kinetic mechanism and role of chelators. J Biol Chem 267:23688–23699

Werthemann D (1993) Prebleaching of *Pinus radiata* pulp using enzymes – technology to reduce AOX. Jpn J Pap Technol 10:15–17

White NA, Body L (1992) Differential extracellular enzyme production in colonies of *Coriolus versicolor, Phlebia radiata* and *Phlebia rufa*: effect of gaseous regime. J Gen Microbiol 138(12):2589–2595

Wong KKY, Tan LUL, Saddler JN (1988) Multiplicity of β-1,4-Xylanase in microorganisms: functions and applications. Microbiol Rev 52:305–315

Wong KKY, Richardson JD, Mansfield SD (2000) Enzymatic Treatment of Mechanical Pulp Fibers for Improving Papermaking Properties. Biotechnol Prog 16(6):1025–1029

Wroblewska H, Zielinski MH (1995) Biodelignification of beech and birch pulpwood by selected white-rot fungi. In: 6th International conference on biotechnology in the pulp and paper industry, Vienna, Austria, 11–15 June 1995

Xu H, Scott GM, Jiang F, Kelly C (2010a) Recombinant manganese peroxidise (rMNP) from *Pichia pastoris*. Part 1: kraft pulp delignification. Holzforschung 64(2):137–143

Xu H, Scott GM, Jiang F, Kelly C (2010b) Recombinant manganese peroxidise (rMnP) from *Pichia pastoris*. Part 2: application in TCF and ECF bleaching. Holzforschung 64(2):145–151

Yang JL, Eriksson K-EL (1992) Use of hemicellulolytic enzymes as one stage in bleaching of kraft pulps. Holzforschung 46(6):481–488

Yang HM, Yao B, Fan YL (2005) Recent advances in structures and relative enzyme properties of xylanase. FEMS Microbial Rev 21(1):6–11

Yllner S, Ostberg K, Stockmann L (1957) A study of the removal of the constituents of pine wood in the sulphate process using a continuous liquor flow method. Sven Papperstidn 60:795–802

Zhan H, Yue B, Hu W, Huang W (2000) Kraft reed pulp TCF bleaching with enzyme treatment. Cellul Chem Technol 33(1–2):53–60

Ziomek E, Kirkpatrick N, Reid ID (1991) Effect of polydimethylsiloxane oxygen carriers on the biological bleaching of hardwood kraft pulp by *Trametes versicolor*. Appl Microbiol Biotechnol 35:669–673

Chapter 9
Biodeinking

9.1 Introduction

Growing environmental awareness, robust overseas markets, and domestic demand are driving increased paper recycling activities. Recycling technologies have been improved by developments in pulping flotation deinking, cleaning, screening, and bleaching, as well as by efforts to boost overall yield, that will encourage developments of products based on recycled materials. Wastepaper recycling enables substitution of virgin pulp with recycled fibers, reducing the exploitation of old forests and reduces disposal problems. Producing recycled paper uses less energy than virgin paper, less water, and releases fewer pollution emissions to air and water. Wastepaper pulp also requires less refining than virgin pulp and can be corefined with other pulps. In addition, deinked pulp provides special properties to the finished papers compared with those made from wood pulp, including opacity, less curling tendency, less fuzziness, and improved formation.

A significant difficulty in dealing with secondary fiber is the removal of contaminants, particularly ink. The difficulty of ink removal depends primarily on the ink type, printing process, and fiber type. Some paper grades, such as newspapers printed with oil-based inks, can be deinked with relative ease by conventional deinking processes. Nonimpact-printed papers are more difficult to deink and the quantity of these papers continues to grow as a proportion of total recovered paper (Vidotti et al. 1992). Similarly, color printing via offset lithography is expanding in the US; other countries are also expected to follow. The cross-linking inks used in this process are also difficult to remove by conventional methods. Mixed office waste (MOW) is a large, virtually untapped source of high-quality fiber that can be used for fine papers and many other products, if the deinking process can be improved. Ink removal continues to be a major technical obstacle to greater use of recycled paper. Many of the conventional deinking processes require large quantities of chemicals, resulting in high wastewater treatment costs to meet environmental regulations. Deinking processes create substantial amounts of solid and liquid waste. Disposal is a problem, and deinking plants would benefit from more effective

P. Bajpai, *Biotechnology for Pulp and Paper Processing*,

DOI 10.1007/978-1-4614-1409-4_9,

and less polluting processes. Enzyme-assisted deinking has been shown to represent a potential environmentally friendly alternative to conventional alkaline deinking processes (Anon 2010; Bajpai 1999; Bajpai and Bajpai 1998; Bajpai 2006; Ma and Jian 2002; Mohammed 2010; Puneet et al. 2010).

9.2 Enzymes Used in Deinking

Different enzymes have been used for deinking (Bajpai and Bajpai 1998). These enzymes include lipases, esterases, pectinases, hemicellulases, cellulases, amylases, and ligninolytic enzymes. Lipases and esterases degrade vegetable-oil-based inks. Pectinases, hemicellulases, cellulases, and ligninolytic enzymes alter the fiber surface or bonds in the vicinity of the ink particles, thereby freeing ink for removal by washing or flotation. Many patent applications have been filed or granted concerning the use of enzymes in deinking. Several patents specify the use of cellulases, particularly alkaline cellulases for deinking. Few patents claim that esterases can be used while others specify the use of lipases or pectinases. One patent application mentions the use of laccase enzyme from white rot fungi. Most of the published literature on deinking deals with cellulases and hemicellulases.

9.3 Mechanisms of Enzyme Deinking

Different mechanisms for ink removal by enzymes have been proposed (Welt and Dinus 1995). Korean researchers pointed out that enzymes partially hydrolyze and depolymerize cellulose between fibers, freeing them from one another (Kim et al. 1991). Ink particles are dislodged as the fibers separate during pulping. Researchers also believe that enzyme treatment weakens the bonds, probably by increasing fibrillation or removing surface layers of individual fibers (Eom and Ow 1990). Woodward et al. (1994) suggested that catalytic hydrolysis may not be essential, since enzymes can remove ink under nonoptimal conditions. Mere cellulase binding alone may be enough to disrupt the fiber surface to an extent sufficient to release ink during pulping. It is also reported that cellulases peel fibrils from fiber surfaces, thereby freeing ink particles for dispersal in suspension (Eom and Ow 1990). Enzymatic effects may be indirect, removing microfibrils and fines and thereby improving freeness and facilitating washing or flotation (Jeffries et al. 1994). Fines content, however, is not always reduced during enzymatic deinking (Putz et al. 1994). Enzymatic treatment of nonimpact-printed paper has been reported to remove material from ink particles, thereby increasing particle hydrophobicity and facilitating separation during flotation (Jeffries et al. 1994). Mechanical action is supposed to be critical and a prerequisite to enzymatic activity (Zeyer et al. 1994), and some authors propose that it distorts cellulose chains at or near fiber surfaces, thereby increasing vulnerability to enzymatic attack. However, research conducted by Putz

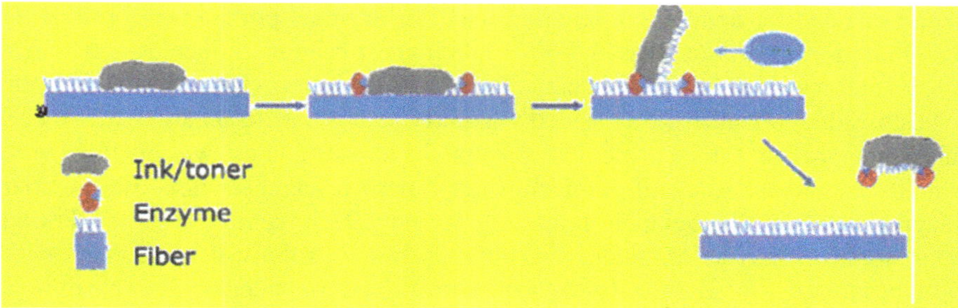

Fig. 9.1 Schematic diagram showing mechanism of Cellulase action on fiber. Mohammed (2010); Reproduced with permission

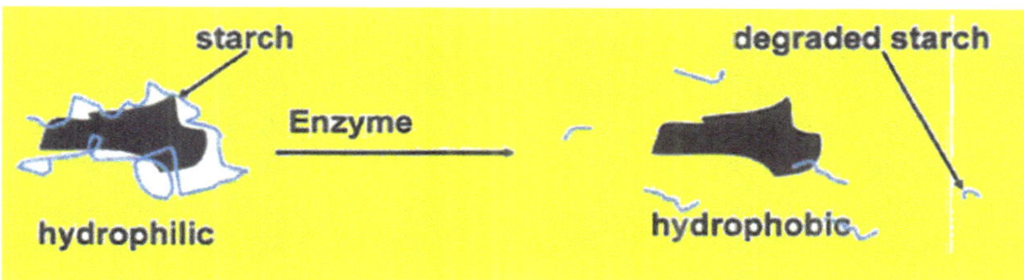

Fig. 9.2 Schematic diagram showing mechanism of Amylase action on fiber. Mohammed (2010); Reproduced with permission

et al. (1994) disputes the importance of mechanical action. It is likely that a particular deinking system would involve more than one of these mechanisms. However, the relative importance of each mechanism would be dependent on fiber substrate, ink composition, and enzyme mixture. Probable mechanism of cellulose and amylase action on fiber is presented in Figs. 9.1 and 9.2 (Mohammed 2010).

9.4 Application of Enzymes in Deinking

Several research groups have examined the application of enzymes in deinking of different types of wastepapers (Zhang and Hu 2004; Spiridon and de Andrade 2005; Bajpai and Bajpai 1998; Wang and Kim 2005; Zuo and Saville 2005; Spiridon and Belgacem 2004; Leenen and Tausche 2004; Xu et al. 2004; Gu et al. 2004; Tausche 2002, 2005a, b, 2007; Magnin et al. 2001; Prasad et al. 1992a, b; Prasad 1993; Heise et al. 1996; Paik and Park 1993; Kim et al. 1991; Rushing et al. 1993; Putz et al. 1994; Baret et al. 1991; Jeffries et al. 1994, 1995; Sykes et al. 1995; Rutledge-Cropsey et al. 1994; Franks and Munk 1995; Yang et al. 1995; Ow et al. 1996; Zeyer et al. 1994, 1995; Woodward et al. 1994; Floccia 1988; Eriksson and Adolphson 1997; Heitmann et al. 1992). Low-pH cellulase and hemicellulase mixtures have been evaluated for deinking of letterpress and color offset-printed newsprint at pH 5.5 (Prasad et al. 1992a, Prasad 1993). The highest brightness increase for letterpress paper was obtained with a hemicellulase preparation (xylanase). However, the lowest residual ink areas as measured via image analysis were achieved with a cellulase preparation. For colored offset papers, the best brightness was obtained with a mixture of cellulases and hemicellulases. These researchers also used similar enzymes to deink flexographic-printed newspaper (Heitmann et al. 1992; Prasad et al. 1992b). Enzyme treatment and flotation removed the water-based ink with ease, resulting in brightness levels well above those obtained with conventional deinking. When the hemicellulases from *Aspergillus niger* and cellulases from *Trichoderma virdie* were evaluated for deinking, brightness increased with increasing enzyme dosage and reaction time (Paik and Park 1993).

Cellulases and hemicellulases have a significant effect on the enzymatic deinking of old newsprint (ONP), improving deinking efficiency and fiber modification (Wang and Kim 2005). Compared to DIPs from conventional chemical materials, the enzymatically deinked pulps exhibit better bleachability. The enzymatically bleached pulp showed a brightness of 59.1% ISO, which was 9% higher than unbleached pulp. Spiridon and de Andrade (2005) studied the effects of three enzymatic preparations – mixed cellulase and xylanase, cellulose alone, and lipase – on the properties of ONP that was 7 months old before deinking using a conventional flotation technique. There was a considerable improvement in drainage rates for the treated fiber suspensions. The enzymatic treatments affected the handsheet mechanical properties. Pulps treated with cellulase/xylanase and pulps treated with lipase had constant tensile and burst indices, but their tear indices decreased. All showed improved optical properties: opacity, brightness, and effective residual ink content (ERIC). The treatments using cellulase/xylanase and lipase produced the best properties.

Zhang and Hu (2004) studied the enzymatic deinking of postconsumer printing paper using cellulase and compared the deinking efficiency of the enzymatic process with the conventional process using chemicals. The enzymatically deinked pulp showed superior drainage, improved physical properties, and better bleachability than the chemically deinked pulp. Gu et al. (2004) used mixtures of lipase, cellulose, and xylanase for deinking ONP. An equal mixture of cellulase and xylanase, itself mixed at a ratio of 60:40 with lipase, gave the best deinking performance. The breaking length, the burst index, and the tear index of handsheets from the deinked pulp were increased by 3.2%, 7.4%, and 7.1%, respectively, compared to pulp deinked with the cellulase/xylanase mixture alone. Higher pulp yield and improved pulp drainage were also obtained.

Zuo and Saville (2005) studied the efficacy of immobilized cellulase for deinking MOW. When they used immobilized-enzyme treatment, they found the residual ink levels were lower than with soluble-enzyme treatment. Their results suggest that immobilized cellulase could be useful for deinking MOW. Spiridon and Belgacem (2004) investigated the effectiveness of enzymes in deinking office papers. They took recycled fibers from office papers and gave them enzymatic pretreatment using cellulase alone or a mixture of cellulase and xylanase. Then they observed the effects on freeness, sheet strength properties, sheet optical properties, and paper surface properties. The fiber suspensions showed a significant improvement in drainage rates, and the mechanical properties of the handsheet showed a substantial increase in burst strength. The tensile strength remained almost constant for pulps treated with the mixture of cellulase and xylanase. For all treatments, the tear index decreased significantly, but the brightness and ERIC improved. The mixture of cellulase and xylanase was the most suitable treatment for laser-printed paper.

The deinked pulp obtained after deinking of sorted office waste with hydrolytic enzymes showed higher brightness (1.0–1.6 points) and whiteness (2.7–3.0 points) and lower residual ink as compared to chemically deinked pulp (Bajpai et al. 2004). It was possible to obtain pulp of <10 ppm dirt count with combination of cellulase and alpha-amylase enzymes resulting in reduced chemical consumption. COD and color loads were lower in case of effluents generated during enzymatic deinking.

Treatment of 100% multiprints furnishes with cellulase and amylase enzymes at pH 7–7.5 improved the pulp brightness by 2 ISO points in the laboratory investigation as a part of EUREKA Enzyrecypaper Project (Gill et al. 2007). The ink particles released on treating with amylase enzymes appeared to be more hydrophobic than ink particles released on treating with cellulose. During the mill trial using highly specific amylases, the brightness was significantly improved up to 8 points. The ash content also reduced to a great extent after flotation and washing, resulting in a change of the final pulp characteristics.

MOW often contains a large variety of dyed papers. The color must be removed to make the pulp suitable for reuse. For this reason, it is frequently underutilized source of wastepapers. Usually, several chemical bleaching agents like ozone, oxygen, hydrogen peroxide, or sodium hydrosulfite have been used to bleach secondary fibers. Now, there is an alternative color stripping process for secondary fibers – the laccase-mediator system. In a study by Arjona et al. (2007), a bleaching sequence included an enzyme stage called laccase-mediator system stage (L), a hydrogen peroxide stage (P), and a sodium hydrosulfite stage (Y) on a mixture of different colored writing & printing papers. After the application of L-P-Y sequence, a pulp with optical properties near to eucalyptus totally bleached pulp was obtained. The L-P-Y sequence reaches a color removal of 90% and saves chemicals in the final stages.

A mill-scale enzymatic deinking project was begun in 2001 by Van Houtum Papier (VHP) of the Netherlands and Enzymatic Deinking Technologies (EDT) of the US, with project subsidies from the Dutch government (Leenen and Tausche 2004). EDT analyzed the mill and the product development process. In a laboratory trial, a Blue Print analysis indicated poor dirt removal and brightness development, and incorrect dilution. A mixture of enzymes was tested on VHP recycled paper furnishes. A short mill trial of 2 weeks was then conducted in which furnish and stock preparation were optimized, giving a brightness gain of 2.7 points and furnish savings. In a long mill trial of 2 months, further optimization was carried out, leading to reduced enzyme dosage and changes in stock preparation. Enzymatic deinking can have a significant effect on the mill's performance but it is necessary to customize the treatment to suit the mill's situation.

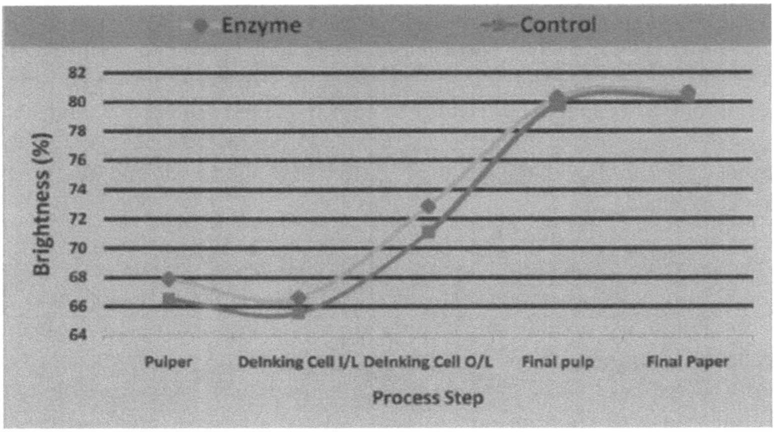

Fig. 9.3 Effect of enzyme on brightness. Mohammed (2010); Reproduced with permission

Xu et al. (2004) studied the deinking of old newspaper (ONP) using cellulase or hemicellulase in conjunction with a laccase-mediator system. The synergistic use of the two enzymes led to the production of pulps with superior brightness and strength compared to those prepared using only one of the enzymes. ONP deinked using cellulase and the laccase-mediator system had a brightness after bleaching with hydrogen peroxide of 55.9% ISO, a breaking length of 2.13 km and a tear index of 6.43 mN m^2/g. The respective increases in brightness were 2.4 and 3.8% points compared to the use of cellulase alone and laccase system alone. The breaking length was 30% higher, a pulp brightness (after hydrogen peroxide bleaching) of 60.4% ISO, a breaking length of 1.94 km, and a tear index of 6.54 mN m^2/g. The respective increases in brightness were 2.7 and 8.3% points compared to the use of hemicellulase alone and laccase system alone. The breaking length was 20% higher than obtained using hemicellulase alone.

Mill-scale results show that enzymatic deinking gives a 50% reduction in visible and subvisible dirt (Tausche 2002). Effective residual ink concentrations have been reduced by 35% in old newsprint/old magazine (ONP/OMG) mills using enzymatic deinking. Stickies reductions have reached 30–50% in mills that use tracking systems. There are also optical benefits of cleaner pulp for tissue and towel production. Yield improvements have averaged 2%. Some mills using wastepaper mix have achieved a 15% decrease in furnish costs by using enzymatic deinking. Mohammed (2010) reported that in one of the Indian Paper mill, use of enzymatic deinking on multigrade furnish like CBS, MOW, and ONP for producing writing and printing paper reduced the residual ink count, increased brightness and gave cost benefit by almost 50% saving in sodium hydroxide, 37% saving in sodium silicate and complete elimination of hydrogen peroxide (Fig. 9.3–9.5).

Magnin et al. (2001) conducted pilot-scale trials to compare enzymatic deinking with conventional alkaline deinking on a typical wood-containing paper composition and on a typical wood-free paper composition. Work showed promising results, particularly a reduction in the number and area of specks in the final deinked pulp.

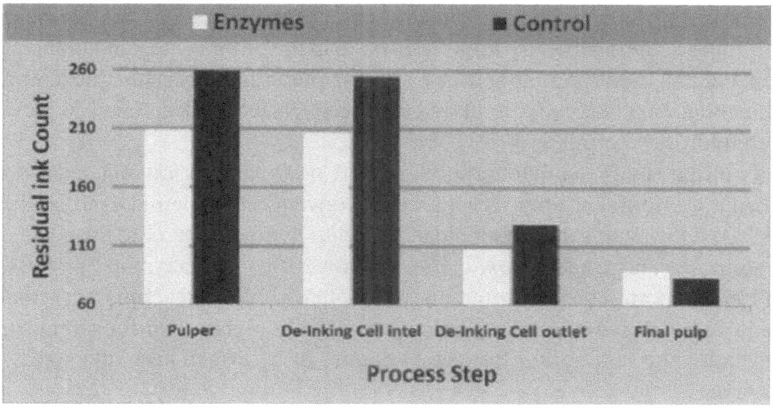

Fig. 9.4 Effect of enzyme on residual ink count. Mohammed (2010); Reproduced with permission

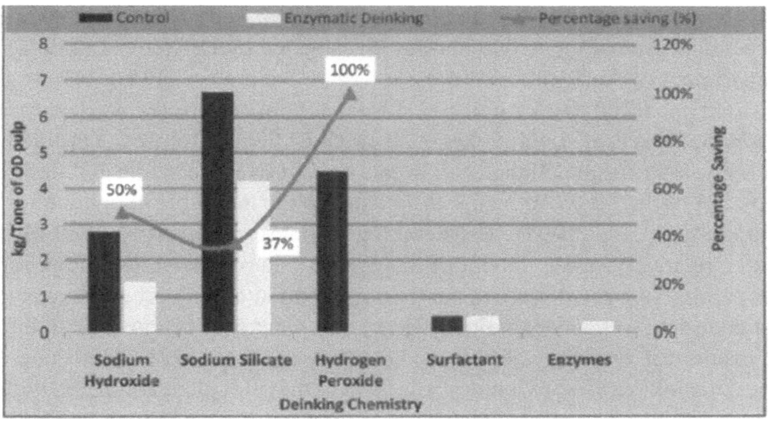

Fig. 9.5 Effect of enzyme on chemical consumption. Mohammed (2010); Reproduced with permission

Full-scale enzymatic deinking was then performed at a mill producing wood-free deinked pulp from 100% printed coated wood-free papers. The results showed that good ink removal and lower specks contamination were obtained by enzymatic treatment in neutral conditions.

Soaking with enzyme before pulping was beneficial. but prolonged soaking reduced ink particle size, lowered flotation effectiveness, and reduced brightness. An optimal blend of cellulase and hemicellulase gave higher brightness gains than conventional deinking. Regardless of ink type or printing process, enzyme treatment tends to reduce ink particle size. It has been reported that reduction in particle size varied with pulping time in the presence of cellulases; overall reduction was greater than in conventional deinking (Kim et al. 1991). Prasad et al. (1992a, b) and Rushing et al. (1993) reported reductions in particle size from 16 to 37%, depending on ink type. Kim et al. (1991) have reported that newspaper pulps bleached after being deinked by enzymatic and conventional means had similar brightness values. Conventional deinking uses hydrogen peroxide in the pulping step and in the bleaching step, but enzymatic deinking uses hydrogen peroxide only in the bleaching step. Consequently, enzymatically deinked pulps were easier to bleach and required half as much hydrogen peroxide.

In a similar study with letterpress-printed newspaper, enzymatically deinked pulps had lower initial brightness values than conventionally deinked pulps (Rushing et al. 1993). However, subsequent bleaching with hydrogen peroxide produced similar brightness values, and peroxide use was lowest for the enzymatic process. Putz et al. (1994) reported that brightness levels obtained after bleaching enzymatically deinked offset-printed newspaper pulp were slightly higher than for pulp produced by conventional deinking, for the same quantity of hydrogen peroxide applied during pulping.

The benefits of neutral cellulase for deinking MOW were exploited by a French group (Baret et al. 1991). Using a neutral cellulase as a post-treatment to a standard alkaline chemical treatment, they reported additional brightness and greater ink removal. Baret et al. (1991) did not consider the use of neutral cellulase without any other chemical pretreatment, but this was investigated by Jeffries et al. (1994) at the US Department of Agriculture's Forest Products Laboratory (FPL). FPL researchers reported enhanced deinking of wood-free, nonimpact-printed wastepapers with cellulases (Jeffries et al. 1994; Sykes et al. 1995; Rutledge-Cropsey et al. 1994). Not surprisingly, the neutral cellulases showed a benefit over the acidic cellulases, even when the pH of the MOW furnish was adjusted to the initial pH region with sulphuric acid (Jeffries et al. 1994). The deinking response observed by the FPL group required a relatively small amount of enzyme to achieve the optimum ink removal responses, although the dose–response curve reported by this group is unusual and has not been satisfactorily explained. Their pilot plant results agreed with the laboratory results for two enzymes (Rutledge-Cropsey et al. 1994). With one of the enzymes, the ink removal efficiency was 94% in the pilot plant compared with 96% in laboratory trials. They reported that in continuous processing of 2,300 kg batches of 100% toner-printed office papers, cellulases greatly reduce the residual particle count while increasing brightness and freeness (Jeffries et al. 1995). Strength properties and fiber length were essentially unchanged.

Novo-Nordisc, Denmark, also performed several lab flotation runs using a neutral cellulase (Novozym 342) it had developed for use in deinking copier paper (Franks and Munk 1995). The brightness improvement was similar to the improvement seen by the FPL investigators. Treatment with a pure alkaline cellulase significantly improved brightness levels of photocopied and laser-printed papers relative to pulping in water without enzymes (Prasad 1993). A brightness improvement of 4 ISO units was observed. Residual ink area (dirt count) was reduced by 94%. Enzyme treatment also affected fiber length distributions. These results might be expected, as the papers typically contain bleached softwood chemical pulp, and cellulases are more likely to affect fiber distributions of chemical pulps. Enzyme-treated pulps showed a similar increase in freeness and in strength properties (breaking length and burst index) relative to control pulps.

Heise et al. (1996) reported the results of three industrial-scale trial runs to evaluate enzymatic deinking of nonimpact-printed toners. Increased ink removal was achieved using a low level of a commercially available enzyme preparation in combination with a surfactant. The brightness of enzymatically deinked pulp was 2 percentage points higher than the brightness of the control pulp. The enzyme trials also had improved drainage and comparable strength when compared with the control. There were no significant differences in the quality and treatability of the process water, although the effluents from these trials had lower oxygen demand and toxicity than the effluents from the control. Enzymatic deinking of mixed wastepapers (laser-printed and UV-coated papers) and ONP/OMG eliminated or substantially reduced the use of chemicals in the deinking process (Yang et al. 1995). The brightness of enzymatically deinked MOW papers containing 90% laser copies, 3% colored paper, and 7% other papers was significantly greater than for an MOW furnish deinked using the chemical method. Enzymatic deinking achieved a 94% lower dirt count (visible) as well as 82% lower total dirt count.

The treatment of US ONP pulp with a blended cellulase in a Korean newsprint mill also gave about a brightness improvement of about 2 percentage points (Ow et al. 1996). They also conducted a mill trial to evaluate enzymatic deinking of white ledger-grade paper pulp at one of the largest Korean tissue mills. The trial ran for several days, and the results showed a reduction of residual ink count. The total ink removal efficiency increased from 93.9 to 98.3% by using the blended cellulase deinking. Zeyer et al. (1995) studied the performance of enzymes for deinking ONP and found that the arrangement of unit operations was important. No deactivation of enzymes by shear stress was observed. Statistical investigation of particles on handsheets demonstrated that many ink particles were probably still at their original location.

Novo researchers used monocomponent cellulases SP-476 and SP-613 to deink MOW (Franks and Munk 1995). The response using SP-476 was similar to the response for the multicomponent cellulase preparation. But SP-613 gave a dose response closer to what would be expected for a typical enzyme system. The concomitant increase in brightness and decrease in ink count helped to confirm that the use of brightness as an assessment method could provide a rough measure of response in these systems. This study shows that the monocomponent portion of the multicomponent cellulase system plays a major role by enhancing deinking of the wood-free nonimpact-printed wastepapers. Paper sizing and other additives can prevent or limit direct physical contact between enzyme and substrate. This inhibits the effectiveness of the enzymes, since contact is a prerequisite to activity.

The implications of sizing effects have been studied by many researchers (Zeyer et al. 1994, 1995; Rutledge-Cropsey et al. 1994). The literature shows that paper sizing reduces enzymatic deinking efficiency and that the effect may vary with sizing agents. For nonimpact-printed papers, deinking efficiency was lowest for papers sized with rosin and alum (Rutledge-Cropsey et al. 1994). These papers have the greatest resistance to wetting and the highest fiber hydrophobicity. Papers sized with alkyl succinic anhydride are less resistant to wetting but are almost as difficult to deink. Alkaline lipases are claimed to facilitate the removal of oil-based inks. Nakano (1993) has reported that an alkaline lipase efficiently removed offset printing inks.

Enzymes that catalyze the removal of surface lignin may hold promise for deinking of newsprint that contains a proportion of lignin-rich mechanical pulp. This approach has been evaluated using the white-rot fungus *Phanerochaete chrysosporium* and with lignin-degrading enzymes (Call and Strittmatter 1992).

Ink removal by a laccase preparation proved comparable to conventional chemical deinking. However, the enzyme-treated pulps showed higher brightness and were easier to bleach. A novel deinking process that couples separation technology with cellulose treatment has been described by Woodward et al. (1994). They reported that ink particles dislodged from newsprint, presumably by cellulase activity, readhered to smaller fibers originally present or created by enzymatic action. The smaller fibers and adhered ink were then separated from longer deinked fibers. The longer deinked fibers are usable without further treatment. Since ink adhered to the shorter fibers, conventional washing or flotation would be unnecessary, and ink would not be released into the effluents. The reason for strong association between ink and short fibers could not be identified. The separation of such fibers is technically feasible (Floccia 1988).

Centre Technique du Papier (CTP) performed a pilot scale test, employing an alkaline cellulase in a neutral deinking of mixed office paper (Gallo et al. 2004). The recovered paper used in the test was made up of 35% MOW, 35% toner, and 30% wood-free magazines. Two cellulases were used: a well-known crude cellulase (Novozyme 342) containing cellulase and hemicellulase activities, and a monocomponent alkaline cellulase (Novozyme 613). Preliminary trials showed that the deinking of this type of recovered paper was more effective using cellulases rather than hemicellulases or amylases. Use of the Novozymes typically increased efficiency of deinking from 95% to around 99%. Overall results confirmed that alkaline cellulases can be used in a neutral deinking process, enhancing the performance of conventional alkaline deinking. Mild alkaline conditions used could impact positively on stickies problems at mill scale. By optimizing the process in a complete deinking line, enzyme use could lead to a reduction in pulping time, thus saving energy and potentially increasing production.

The potential of combining cellulase enzymes with sulfite deinking to achieve a superior natural deinking strategy for deinking of old news print (ONP)/old magazine paper (OMG) was also examined by Zhang et al. (2008). They reported substantial improvement in the deinking performance of ONP/OMG in 70:30 ratio as compared to either cellulase enzyme or sulfite deinking.

The exposure of recovered newsprint to high temperature in closed containers accelerates the oxidation and polymerization of ink particles (summer effect) (Haynes 2000). Sulphite deinking has been shown to give relatively better deinking performance as compared to alkaline deinking (Chezick et al. 2004) but no improvement in the brightness. However, combining enzymes with sulphite chemistry significantly enhances sulphite deinking to achieve deinking of aged newsprint at neutral pH.

Commercial use of enzymes for deinking has started in many countries. The Enzynk process developed at the University of Georgia by Eriksson's group has been commercialized by EDT. The process uses a mixture of enzymes in combination with surfactants and a few other chemicals (Eriksson and Adolphson 1997). The

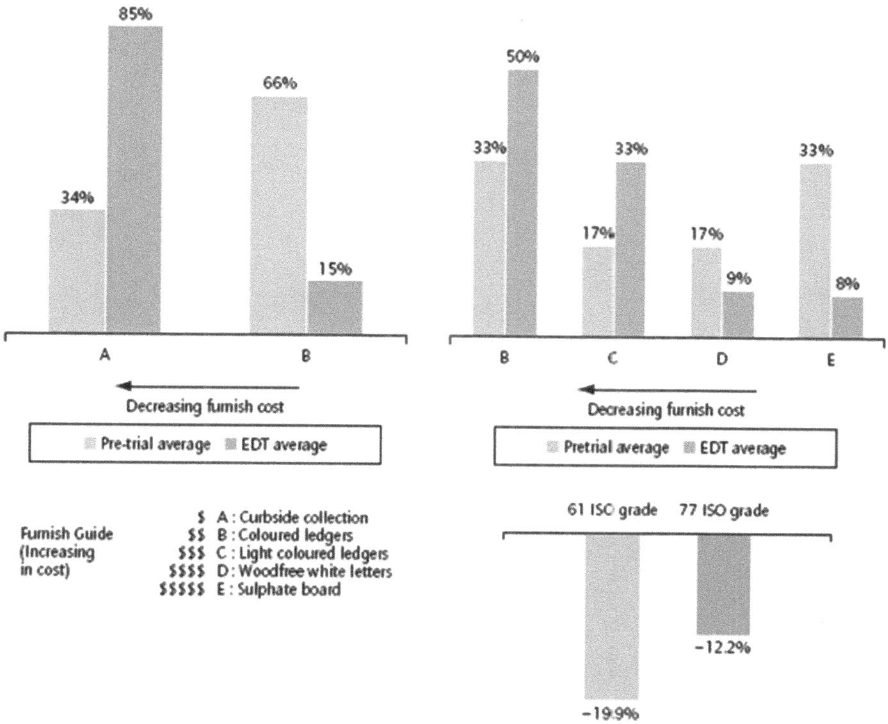

Fig. 9.6 Enzymatic deinking (**a**) furnish composition of tissue with ISO brightness 61 (**b**) furnish composition of tissue with ISO brightness 77 (**c**) net cost change in total raw materials (furnish plus all chemistry) by using enzymatic deinking. Tausche (2005a, b); Reproduced with permission

enzyme mixture very much depends on the choice of furnish. The Stora Dalum deinking plant used an enzymatic deinking process developed by EDT (Knudsen et al. 1998). Trials showed that dirt specks can be reduced by up to 35% and stickies by up to 50%, brightness levels can be increased by 1.2% ISO% before bleaching and by 2.2 ISO% after bleaching. The usage of certain chemicals was also reduced and the mill experienced a 1.8% higher yield with the use of enzymes. The average daily production of the mill increased by over 8 tons per day.

EDT is exhibiting its Enzynk technology for use in deinking mills to improve quality and reduce total production costs (Tausche 2002, 2005a, b, 2007). EDT's mill-specific enzyme blend tailors a treatment based on the mill's furnish mix, deinking plant configuration, key operating conditions, and desired results from the treatment. The company says that this patented technology provides superior ink and contaminant detachment from the fibers such that deinking plant equipment can be more efficient across flotation, cleaning, and washing stages. In one case, a mill wishing to reduce furnish costs was able to achieve a 12–20% reduction in net furnish costs (Fig. 9.6). A second mill was able to achieve a 5% improvement in yield with a 30% reduction in sludge generation. Figures 9.7–9.9 show improvement in dirt reduction and brightness gain and increased fiber yield from Enzynk use.

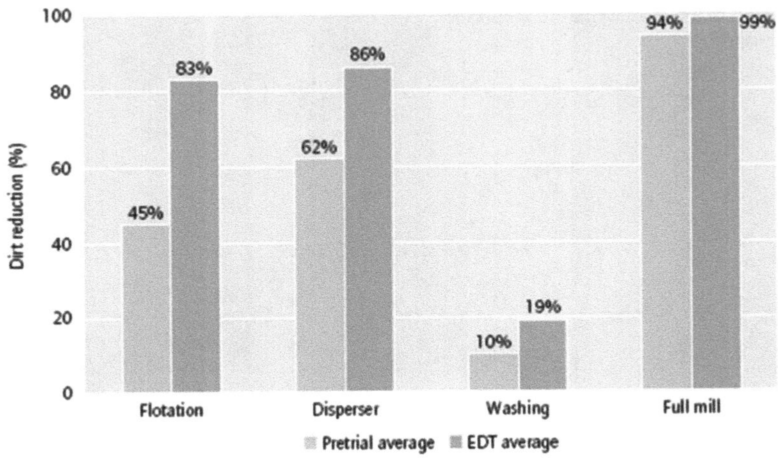

Fig. 9.7 Enzymatic deinking: Tappi dirt reductions. Tausche (2005a, b); Reproduced with permission

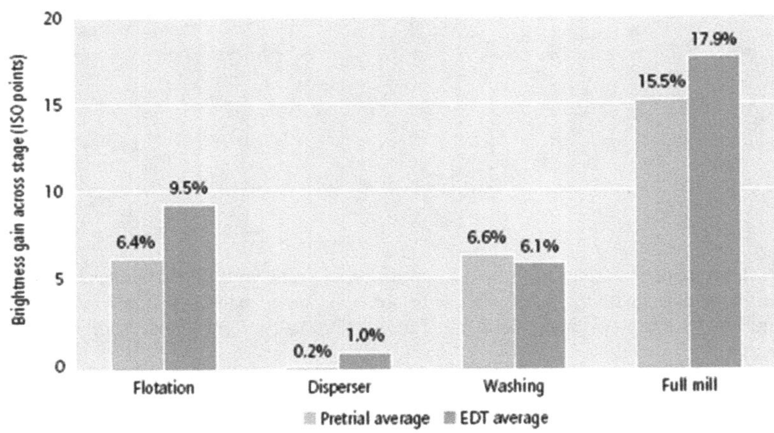

Fig. 9.8 Enzymatic deinking: Brightness gains. Tausche (2005a, b); Reproduced with permission

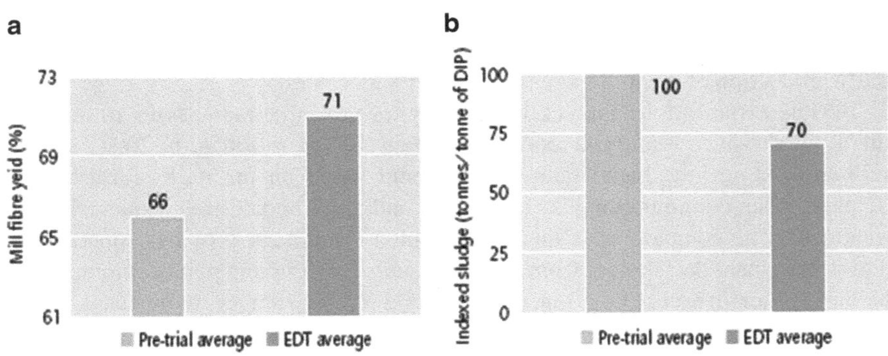

Fig. 9.9 Enzymatic deinking: (**a**) mill fiber yield (**b**) indexed sludge generation. Tausche (2005a, b); Reproduced with permission

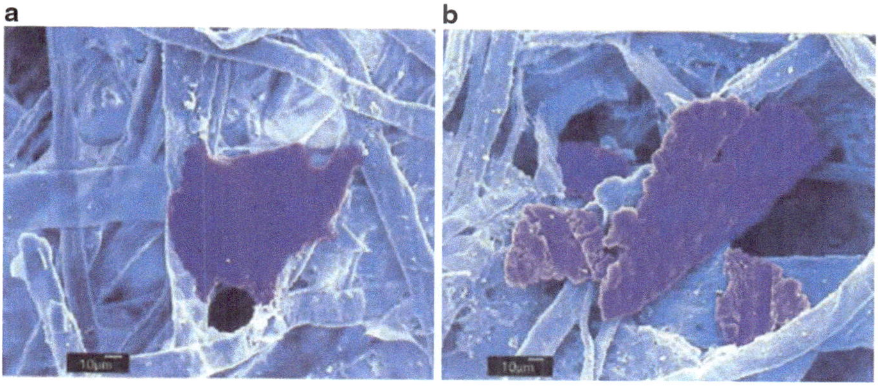

Fig. 9.10 SEM of toner particle detachment (**a**) conventional deinking (**b**) enzymatic deinking. Tausche (2005a, b); Reproduced with permission

Figure 9.10 compares scanning electron micrographs of toner ink particles treated with a conventional chemical deinking approach and EDT's enzymatic deinking process; it clearly shows the effect of improved ink detachment. Toner inks are essentially melted plastic that impregnate the sheet, forming large ink–fiber complexes. Normal pulping fails to destroy these "hairy particles," hampering the removal and yield control across flotation, cleaning, and washing. The enzymatic process detaches inks more effectively from fibers to facilitate their removal and close the gap between virgin and recycled quality.

The use of enzymes for deinking, has been employed with a variety of strategies to reduce energy costs, both in the deink plant and on the machine. Tausche (2007) reported five such strategies to reduce energy costs per ton: increased tonnage output, enhanced pulper defiberization, reduced dispersion, reduced refining, and reduced drying energy. Enzymatic deinking has helped mills increase production rates while maintaining good deinked pulp quality, overcoming the challenges resulting from running faster or at higher consistencies. One major benefit of increased tonnage is the reduction in specific energy cost per ton produced. Enzymatic strategies have also focused on fiber modification to enhance drainage and increase machine output. In one application where drainage in a dewatering screw was a limitation, an enzyme treatment was developed to successfully debottleneck this stage. Enzymatic treatments have helped mills reduce pulping times, and hence energy. In one application, high consistency pulping time was reduced by 23%, from 22 to 17 min. This method can also increase the running rate for mills that are bottlenecked at the pulper. The disperger energy may be reduced or in some cases eliminated by using enzymes. Certain enzymatic treatments have been found to "clean" fibers such that more hydrogen bonding sites present themselves to provide strength in the sheet. In mill applications where strength enhancement was a goal, sheet strength increases from 5 to 13% were attained. While mills leverage these increases in different ways, a common strategy is to decrease refining energy such that the same sheet strength is achieved. For example, one mill reduced refining from 125 kWh/ton to the 70–90 kWh/ton range and still achieved the desired final sheet properties (Tausche 2007). Enzymatic treatments can enhance the drainability of stock on the paper machine. Certain treatments have enabled the steam to escape more readily and have reduced drying energy per ton. Mills using such enzymatic approaches have been able to reduce total energy usage by over €10 per tissue ton.

9.5 Effect of Enzyme on Fiber and Paper Quality

Enzymatically deinked pulp typically has higher brightness, greater freeness, and superior paper quality strength properties compared with chemically deinked pulp. Yang et al. (1995) reported that the freeness of enzymatically deinked MOW pulp was 32% higher than that of control pulp. Heise et al. (1996) found that enzyme treatment significantly increased pulp freeness from 510 to 570 ml Canadian Standard Freeness (CSF) and Prasad (1993) found an increase from 440 to 490 ml. Prasad (1993) observed that freeness increased in all the enzyme-treated samples compared with the control. The freeness increase varied from 50% for cellulase-treated colored flexo-printed newsprint to 14% for black and white printed newsprint treated with a hemicellulase preparation. Enzyme-deinked pulp also had better runnability on paper machines. It is mainly the enhanced drainage and oxygen wet web strength that contributes to enhanced runnability. Baret et al. (1991) claimed to see a physical property enhancement even when applying as much as 4 dm^3 cellulase per ton of pulp. A hemicellulase preparation showed the largest strength increase with the smallest improvement in freeness.

9.6 Effect of Enzyme on Pulp Yield

Yield results are inconclusive. Some yield reduction appears to arise from losses of fines and other small particles through the action of the applied enzyme. More precise control over enzyme dosages and reaction times are expected to minimize these losses. Kim et al. (1991) have reported that reducing sugars were released during enzymatic deinking of ONP, but yield losses were immaterial. Relatively short reaction times were thought to have restricted enzyme attack of fibrils on fiber surfaces. In another study with ONP, sugar release increased with enzyme dosage and reaction time (Paik and Park 1993). The yield was reduced by 5%, but freed sugars did not explain all the loss. Microfibrils freed from fibers by enzymatic activity were said to have been lost during flotation. Even then, yields from enzymatic deinking were higher than from conventional deinking.

9.7 Effect of Enzyme on Effluent Characteristics

In comparison to alkaline deinking process, the chemical oxygen demand is lower in enzymatic deinking; this reduces the load on wastewater treatment systems (Yang et al. 1995). Wastewater effluent from enzymatic deinking was reported to have a 20–30% lower COD than wastewater from chemical deinking processes (Kim et al. 1991). There may be extra environmental advantages by avoiding the use of high alkalinity in the pulping stage (Jeffries et al. 1994). Another report indicated that the COD load after enzyme treatment was 50% lower than for conventional deinking (Putz et al. 1994).

The following observations on effluent characteristics are based on the results of Heise et al. (1996), who reported on the industrial-scale enzyme deinking of nonimpact-printed toners. Process water entering the clarifier from the enzyme runs contained lower total and dissolved biochemical oxygen demand (BOD) and higher COD than the comparable control (Table 9.1). However, the dissolved air flotation (DAF) cell readily clarified the process water, and the water exiting the DAF contained lower BOD and COD than the clarified control water (Table 9.2). The best quality reprocessed water was achieved with enzyme run 2, which was also the best trial for ink removal. Enzyme run 2 also had the lowest BOD and COD in the reject effluent stream (Table 9.3). There was no detectable difference in the BOD of

Table 9.1 Quality of water entering dissolved air flotation clarifier

		Enzyme runs	
Parameter*	Control run	1	2
Total BOD_5	441	327	253
Dissolved BOD_5	234	168	138
Total COD	1,080	1,190	1,455

*All values in mg/l
Based on data from Heise et al. 1996

Table 9.2 Quality of water exiting dissolved air flotation clarifier

		Enzyme runs	
Parameter*	Control run	1	2
Total BOD_5	298	192	115
Dissolved BOD_5	275	165	123
Total COD	510	565	415

*All values are mg/l
Based on data from Heise et al. 1996

Table 9.3 Quality of reject stream

Trial	Dissolved BOD_5 (mg/l)	Total COD (mg/l)
Control run	219	440
Enzyme run 1	210	427
Enzyme run 2	180	346

Based on data from Heise et al. 1996

different samples from each trial. The reject streams had comparable toxicities. However, if the conventional chemical control were used for comparison, the enzyme runs would undoubtedly be less toxic than the conventional run, as previously observed on effluents collected from bench-scale experiments (Jeffries et al. 1994; Sykes et al. 1995).

In deinking offset newsprint, the COD load in the white water from the pulp suspension after reaction was lowest, at 5 kg per ton of deinked pulp, when no chemicals were added (Putz et al. 1994). With the total chemical reference formulation, a COD load of 22 kg per ton was generated, because of the alkaline pulping environment. However, for the enzyme-treated pulp, the COD load depends on the amount of enzymes as well as on the enzyme type. On average, the following CODs were obtained (Putz et al. 1994): 11 kg per ton for 0.2% enzyme addition and 20 kg per ton for 1.0% enzyme addition. Compared with the common alkaline deinking procedure, the COD load of a treatment with 0.2% enzymes was 50% lower.

Knudsen et al. (1998) reported that COD coming in to the biological treatment plant was not statistically different between chemical and enzymatic deinking. Bioplant outlet COD and therefore treatment plant efficiency were also unchanged. SVI was not impaired by the transition to enzymatic deinking. Ash content almost doubled probably due to increased transfer of calcium ions from the deinking plant to the biological treatment plant. This changed the dewatering properties of the biosludge necessitating a minor modification of the dewatering equipment. Then significantly higher solids content could be achieved compared to reference.

9.8 Benefits and Limitations

Conventional deinking is a chemical-intensive process that requires extensive wastewater treatment, which is expensive and becoming highly regulated. Enzyme-based deinking offers a potential means for reducing the amount of chemicals in the deinking process, hence reducing the load on wastewater treatment systems. Conventional methods are relatively ineffective in deinking MOW, which presents technical and economic challenges to the paper recycler. MOW contains a wide variety of fibers and contaminants as well as toners and other nonimpact polymeric inks from laser printing, which are the most difficult to deal with. Toners and laser printing inks do not disperse readily during a conventional repulping process and are not readily removed during flotation or washing. Conventional deinking uses surfactants to float toners away from fibers, high temperatures to make toner surfaces form aggregates, and vigorous high-intensity dispersion for size reduction.

Most of the deinking chemicals and high-energy dispersion steps are expensive. Microbial enzymes enhance the release of toners from office wastepapers. The size distribution and the shape of the ink removed can be effectively controlled using the enzymatic process to maximize the efficiency of the flotation process, which relies heavily on particle size. This can be accomplished by selectively varying enzyme composition, dose and residence time, and by varying other additives and the pH of the system to effectively dislodge the normally large, flat and rigid ink particles into much finer and nonplatelet forms. Enzymes may also retard redeposition of ink particles onto the fibers. The most promising implication of high deinking efficiency from enzyme-enhanced deinking is that the dewatering and dispersion steps – as well as subsequent reflotation and washing – may not be essential. This should save capital expenses in construction of deinking plants while also reducing consumption of electrical energy for dewatering and dispersion.

Less bleaching chemicals are usually needed for enzymatic deinking than for conventional chemical deinking. Lower chemical use would reduce waste treatment costs and reduce the impact on the environment. Lower bleaching costs and less pollution can also be anticipated, since enzymatically deinked pulps have proved easier to bleach and require less chemicals than pulps deinked by conventional methods. Enzymatically deinked pulp also displays improved drainage, superior physical properties, higher brightness, and lower residual ink compared with chemically deinked recycled pulps. Improved drainage results in faster machine speed, which yields significant energy savings, hence overall cost savings. In addition, the use of recycled fiber reduces the need for virgin pulp. This brings great savings in the energy required for pulping, bleaching, refining, etc., which will also reduce pollution problems. Introducing enzymatic deinking technology in a mill-scale operation will require extensive customization of the enzyme formulation and process variables to achieve optimal effectiveness. After extensive experience with US mill-based trials (Yang et al. 1995), it is clear that the enzyme formulations will vary widely and will depend on the furnish, process water, equipment configuration, and desired specifications of the deinked pulp. Moreover, an enzyme-based deinking process will naturally lead to a new chemical balance throughout the mill's entire water system. If enzymatic deinking is to be effectively introduced into the pulp and paper industry, the costs and risks of conversion must be minimized.

9.9 Conclusions

Enzymes for deinking are now commercially available and at lower cost than in the past. Several pilot plant and mill-scale trials have been conducted and promising results have been obtained. Several mills in the world are regularly using enzymes for deinking. EDT has been one of the most active companies. Increased usage and advances in fermentation technology are expected to lower the production costs of enzymes. Alternatively, genetic engineering techniques can be used to identify the gene for a specific enzyme and transfer it to another organism, e.g., *Escherichia coli*, that normally does not produce the enzyme. Transfer and expression of cellulase genes have also been accomplished and several firms are now producing individual cellulases.

References

Anon (2010) Enzymes: the new approach to minimizing fiber costs? Perini J 99(35):94–96

Arjona I, Vidal T, Roncero MB, Torres AL (2007) A new color stripping sequence for dyed secondary fibres. In: 10th international congress on biotechnology in the pulp and paper industry (June 10–15, 2007), Madison, Wisconsin, USA, PS LPA 3.2, p 127

Bajpai P (1999) Application of enzymes in pulp & paper industry. Biotechnol Prog 15(2): 147–157

Bajpai P (2006) Advances in recycling and deinking, PIRA international, U.K. Chapter 6, p.75–88

Bajpai P, Bajpai PK (1998) Deinking with enzymes: a review. Tappi J 81(12):111

Bajpai P, Mishra OP, Mishra SP, Kumar S, Bajpai PK (2004) Enzyme assisted deinking of sorted non-impact white office paper. In: 9th international conference on biotechnology in the pulp and paper industry (October 10–14, 2004), Durban, South Africa, P1.5, p 121

Baret JL, Leclerc M, Lamort JP (1991) Enzymatic Deinking Process. Int Appl, PCT DK91/00090

Call HP, Strittmatter G (1992) Application of ligninolytic enzymes in the paper and pulp industry – recent results. Papier 46(10A):V32

Chezick C, Allen J, Hill G, Lapierre L, Dorris G, Merza J, Haynes RD (2004) A 10-day mill trial of near-neutral sulfite deinking, part i: deinked pulp optical and physical properties. Pulp and Paper Canada 104(4):33–38

Eom TJ, Ow SSK (1990) Process for Removing Printing Ink from Wastepaper. German Patent GB 3,934,772

Eriksson K-EL, Adolphson RB (1997) Pulp bleaching and deinking pilot plants use chlorine-free process. Tappi J 80(6):80

Floccia L (1988) Fractionation and separate bleaching of wastepaper, tappi international pulp bleaching conference, Orlando, FL, US

Franks NE, Munk N (1995) Alkaline Cellulases and the enzymatic deinking of mixed office waste, tappi pulping conference, Chicago IL

Gallo I; Mosele G; Elegir G (2004). Pilot scale evaluation of alkaline cellulases for enzymatic deinking of mixed office paper, 9th international conference on biotechnology in the pulp and paper industry, book of abstracts, Durban, South Africa, 10–14 Oct. 2004, pp 69–70

Gill R, Hillerbrand M, Keijsers E, Kessel LV, Loosvelt I, Lund H, Luo J, Muller E, Pedersen HH, Snelders A, Valk Hvd, Westenbroek A, Willemsen J (2007) Enzymatic deinking of recycled paper: from laboratory to mill scale. In: 10th international congress on biotechnology in the pulp and paper industry (June 10–15, 2007), Madison, Wisconsin, USA, IndusAPP 2.1, p 52

Gu QP, You JX, Yong Q, Yu SY (2004) Enzymatic deinking of ONP with lipase/cellulase/xylanase. China Pulp Paper 23(2):7

Haynes RD (2000) The impact of the summer effect on ink detachment and removal. Tappi J 83(3):56–65

Heise OU, Unwin JP, Klungness JH, Fineran WG Jr, Sykes M, Abubakr S (1996) Industrial Scale-up of enzyme-enhanced deinking of non-impact printed toners. Tappi J 79(3):207

Heitmann JA, Joyce TW, Prasad DY (1992) Enzyme deinking of newsprint waste, international conference on biotechnology in the pulp and paper industry, OZEPA, Kyoto, Japan

Jeffries TW, Sykes MS, Cropsey KR, Klungness H, Abubakr S (1995) Enhanced removal of toners from office waste papers by microbial cellulases, sixth international conference on biotechnology in the pulp and paper industry, OZEPA, Vienna

Jeffries TW, Klungness JH, Sykes MS, Rutledge-Cropsey KR (1994) Comparison of enzyme-enhanced with conventional deinking of xerographic and laser-printed paper. Tappi J 77(4):173

Kim TJ, Ow SSK, Eom TJ (1991). Enzymatic deinking method of waste paper, Tappi pulping conference, Orlando FL

Knudsen O, Young JD, Yang JL (1998). Mill experience of a new technology at Stora Dalum deinking plant. PTS-CTP Deinking Symposium 1998, Munich, Germany, 5–7 May 1998, pp 17

Leenen M, Tausche J (2004) Principles of enzymatic deinking (EDT) and practical implementation in a paper mill, eighth pira international conference on paper recycling technology, Prague 2004

Ma JH, Jian C (2002) Enzyme applications in the pulp and paper industry. Progress in Paper Recycling 11(3):36–46

Magnin L, Lantto R, Delpech P (2001) Potential of enzymatic deinking, 8th international conference on biotechnology in the pulp and paper industry, VTT Biotechnology, Helsinki 2001

Mohammed SH (2010) Enzymatic deinking: a bright solution with a bright future. IPPTA 22(3):137–138

Nakano J (1993) Recent research trends of pulping chemistry. J Korea Tappi 25(1):85

Ow SK, Park JM, Han SH, Srebotnik E, Messner K (1996) Effects of enzyme on ink size and distribution during the enzymatic deinking process of old newsprint, 6th international conference on biotechnology in the pulp and paper industry', OZEPA, Vienna

Paik KH, Park JY (1993) Enzyme deinking of newsprint waste i: effect of Cellulase and Xylanase on brightness, yield and physical properties of deinked pulps. J Korea Tappi 25(3):42

Prasad DY (1993) Enzymatic deinking of laser and xerographic office wastes. Appita J 46(4):289

Prasad DY, Heitman JA, Joyce TW (1992a) Enzymatic deinking of flexographic printed newsprint: black and colored inks. Papiripar, Papir-es Nyomdaipari Mueszaki Egyesuelet 36(4):122

Prasad DY, Heitman JA, Joyce TW (1992b) Enzyme deinking of black and white letterpress printed newspaper waste. Progress in Paper Recycling 1(3):21

Puneet P, Bhardwaj NK, Singh AK (2010) Enzymatic deinking of office waste paper: an overview. IPPTA 22(2):83–88

Putz HJ, Renne K, Gottsching L, Jokinen O (1994). Enzymatic deinking in comparison with conventional deinking of offset news, Tappi pulping conference, San Diego CA

Rushing W, Joyce TW, Heitmann JA (1993) Hydrogen peroxide bleaching of enzyme deinked old newsprint, seventh international symposium on wood and pulping chemistry, Beijing

Rutledge-Cropsey K, Jeffries T, Klungness JH, Sykes M (1994) preliminary results of effect of sizings on enzyme-enhanced deinking, Tappi recycling symposium, Boston MA

Spiridon I, Belgacem MN (2004) Enzymatic deinking of laser printed papers. Progress in Paper Recycling 13(4):12

Spiridon I, de Andrade AM (2005) Enzymatic deinking of old newspaper (ONP). Progress in Paper Recycling 14(3):14

Sykes M, Klungness J, Abubakr S, Rutledge-Cropsey K (1995) Enzymatic deinking of sorted mixed office waste: recommendations for scale-up, Tappi recycling symposium, New Orleans, LA

Tausche J (2002). Mill-scale benefits in enzymatic deinking, 7th pira international recycling technology conference, Brussels

Tausche J (2005a) Furnishing better deinking: tailoring enzyme to suit your recycling needs. Pulp & Paper International 47(7):20

Tausche J (2005b) Deinking mills dodge financial crunch with customized enzymes. Pulp Paper 79(10):49–51

Tausche (2007). More tons, less energy, and reduced total costs via enzymatic deinking, TissueWorld nice

Vidotti RM, Johnson DA, Thompson EV (1992). Comparison of bench-scale and pilot plant floatation of photocopied office waste paper', Tappi pulping conference, Boston MA

Wang S and Kim M (2005). Study on old newsprint deinking with Cellulases and Xylanase, 59th appita annual conference and exhibition, ISWFPC, Auckland

Welt T, Dinus RJ (1995) Enzymatic deinking – a review. Progress in Paper Recycling 4(2):36

Woodward J, Stephan LM, Koran LJ Jr, Wong KKY, Saddler JN (1994) Enzymatic separation of high-quality uninked pulp fibers from recycled newspaper. Biotechnology 12(9):905

Xu QH, Qin MH, Shi SL, Zhang AP, Xu Q (2004) Synergistic deinking of ONP by Cellulase/ Hemicellulase combined with laccase-mediator system. China Pulp Paper 23(8):6

Yang JL, Ma J, Eriksson K EL, Srebotnik E, Messner K (1995) Enzymatic deinking of recycled fibers – development of the enzynk process. 6th international conference on biotechnology in the pulp and paper industry, Vienna

Zeyer C, Joyce TW, Heitmann JA, Rucker JW (1994) Factors influencing enzyme deinking of recycled fiber. Tappi J 77(10):169

Zeyer C, Heitmann JA, Joyce TW and Rucke JW (1995). Performance study of enzymatic deinking using Cellulase/Hemicellulase blends, 6th International conference on biotechnology in the pulp and paper industry, Vienna

Zhang SF, Hu XG (2004) Enzymatic deinking of postconsumer printing paper. China Pulp Paper 23(2):10

Zhang X, Renaud S, Paice M (2008) Cellulase deinking of fresh and aged ONP/OMG. Enzyme Microb Technol 42(2):103–108

Zuo Y, Saville BA (2005) Efficacy of immobilized cellulase for deinking of mixed office waste. J Pulp and Paper Science 31(1):3

Chapter 10
Fiber Modification

10.1 Introduction

Wood fibers are composed mainly of cellulose and hemicellulose microfibrils encrusted in lignin–carbohydrate matrices. They are multilayered structures that can have internal delamination and external fibrillation after chemical and/or mechanical processing. Wood pulp can be treated with enzymes, and some of the cellulose in the fiber is hydrolyzed. This biochemical treatment reduces the amount of mechanical treatment needed to reach the desired fiber properties. Less mechanical action and less energy are required. Since refining requires significant energy input as well as capital investment for equipment, helping the refining process could provide numerous benefits. In the last few years, interest in the use of enzymes as a way of modifying fiber properties to improve the beatability/refinability and drainage of pulps has increased (Bhardwaj et al. 1995, 1996; Pastor et al. 2002; Garcia et al. 2002; Torres et al. 1999; Scartazzini et al. 1995; Wong et al. 1999a; Bajpai and Bajpai 2001; Bajpai et al. 2004, 2005, 2006; Michalopoulos et al. 2005; Covarrubias 2006, 2007; Gill 2008; Loosvelt 2008, 2009; Lecourt et al. 2010). The use of commercially produced enzymes in papermaking is relatively new. Typical enzymes include amylases, proteases, lipases, xylanases, and cellulases. Produced in nature by fungi, bacteria, and protozoans, cellulases break down the cellulose walls of plant fiber. There are several kinds of cellulases which differ structurally and mechanistically. These enzymes can be classified into two broad groups according to the specific function they perform. Endo-cellulases break internal bonds to disrupt the crystalline structure of cellulose to expose individual polysaccharide chains. Exo-cellulases on the other hand cleave two to four units from the ends of the cellulose chains resulting in much smaller tetra- or disaccharide molecules. In papermaking, the enzymes used are principally of the endo-cellulase type. When used to treat papermaking pulps, fiber modification enzymes deliver a number of beneficial effects on the manufacturing process and paper properties. The main process-related impacts are seen in reduction in refining energy, substitution of expensive pulps by more cost-effective ones, increase in dewatering, lowering of drying energy, and

P. Bajpai, *Biotechnology for Pulp and Paper Processing*,

DOI 10.1007/978-1-4614-1409-4_10,

reduction in starch use. In some cases increases in machine speeds and therefore productivity have been realized. Effects on paper quality include increase in tensile strength, higher bulk, porosity, and tissue softness. Successful applications result in delivering substantial return on investment (ROI) for the paper makers. There are a number of running applications using these fiber modification enzyme products in the paper industry.

10.2 Enzymes Promoting Beatability/Refinability

The use of enzymes to modify wood pulp is not new. In 1942 a patent claimed that microbial hemicellulases from *Bacillus* and *Aspergillus* species could aid refining and the hydration of pulp fibers (Diehm 1942). In 1959, Bolaski et al. patented the use of cellulases from *Aspergillus niger* to separate and fibrillate pulps, mainly in cotton linters and other nonwood pulps. In 1968, cellulases from a white rot fungus, which were applied at a concentration of 0.1–1% by weight, reduced the beating or refining time (Yerkes 1968). While enhancing beating, the enzyme also facilitated drainage by removing fines. In other applications cellulases have been used to remove fines from pits and felts in the papermaking machinery. French researchers employed xylanase enzymes from mutants of *Sporotrichium pulverulentum* and *S. diorphosphorum* to fibrillate pulps while suppressing the cellulase activity (Comtat et al. 1984; Mora et al. 1986; Noe et al. 1986; Barnoud et al. 1986). The enzyme treatment increased the °SR of the pulp. When the enzyme-treated pulps were compared with untreated controls, the time required to obtain the same degree of freeness decreased by about 60%. Along with the slower drainage, the water retention increased by about 40% following enzyme treatment, and more than doubled following refining. The tensile strength and the zero-span breaking length of the enzyme-treated refined pulp also increased. Comtat et al. (1984) claimed similar results using xylanases produced by cloning the DNA for the enzyme into a bacterium. In addition to the increase in water retention, Mora et al. (1986) showed that the mean pore radius of aspen wood was reduced by a factor of ten following treatment with xylanases. Presumably, this results from the opening of small cracks in the walls of the pores. Electron microscopy showed increased fibrillation in enzyme-treated pulps compared with control pulps. Noe et al. (1986) reported the characteristics of enzyme-treated pulps of birch and spruce. The Schopper-Riegler index, the amount of water retention, the breaking length, and the apparent density all increased with treatment, but the viscosity decreased by more than 30%. The wet zero-span breaking length also decreased significantly. The authors concluded that enzyme-treated pulps show enhanced beatability and better bonding as a result of increased fiber flexibility, but that the intrinsic fiber strength decreases as a result of the loss of xylan.

The effectiveness of commercial enzymes has been examined for energy savings in refining of different pulps (Bhardwaj et al. 1996). Unbleached mixed pulp (60% waste corrugated kraft cuttings and 40% unbleached softwood pulp) was treated with enzyme.

Table 10.1 Effect of enzyme treatment on beatability and strength properties of mixed pulp (60% waste corrugated kraft cuttings and 40% softwood)

Enzyme	Reduction in beating time, %	°SR	Tensile index, Nm/g	Breaking length, m	Tensile energy absorption, J/m	Burst index, kN/g
Control		28	34.05	3,473	32.25	2.26
Enzyme 2	15.0	28	34.88	3,558	28.90	2.23
Enzyme 3	15.0	28	34.79	3,549	27.50	2.21

Conditions: temperature, 50 C; pulp consistency, 4%; reaction time, 3 h; enzyme dose, 0.05% on o.d. pulp; pH, 5.0 with Enzyme 2 and 7.0 with Enzyme 3
Based on Bhardwaj et al. (1996)

Table 10.2 PFI refining of OCC pulps

	°SR					
		Cy 5%, 50 C, 1 h		Cy 5%, 50 C, 2 h		
No. of revolutions	Control	Enzyme (0.02%)	Enzyme (0.03%)	Enzyme (0.02%)	Enzyme (0.03%)	Enzyme (0.04%)
2,000	26.0	32.0	32.0	34.0	33.0	34.5
2,900	31.0	37.5	38.0	38.0	38.5	39.0
3,400	36.0	40.5	40.5	41.5	42.0	43.0
3,750	40.0	45.5	46.0	46.0	46.5	47.5

Based on Bajpai et al. (2006)

The beating time was reduced by 15% with two different enzyme samples (Table 10.1). In another case, a double-sorted old corrugated cartons (OCC) pulp sample was treated with Fibrezyme LBR before refining (Bajpai et al. 2006). The °SR of enzyme-treated pulps were higher at the same PFI revolutions. The enzyme-treated pulps required about 30% less energy to reach a °SR of 30 (Table 10.2).

Oksanen et al. (1997) and Mansfield et al. (2000) reported that the effectiveness of xylanase-aided refining varies with pulp type and that fully bleached pulps are less responsive than high Kappa pulps. Release papers, which are used as backings to hold adhesive labels, are extremely dense and are made by extensively refining a chemical pulp. Mill trials showed that treatment with a commercial cellulase reduced the refining energy required by 7.5% (Freiermuth et al. 1994). The success of this cellulase application, which has been implemented in some mills, may be due to a greater tolerance for the losses in fiber strength associated with cellulase treatments. Other product grades in this category include high-density papers used in the food industry, as well as condenser papers and glassine – the refining of all of these is enhanced by cellulase treatments (Yamaguchi and Yaguchi 1996).

Laboratory trials with Pergalase A40H on condenser, glassine, and thin papers showed about a 20% reduction in refining energy. Mill trials on glassine paper showed an energy saving of between 15 and 20%, while the opacity remained the same. Studies on thin paper showed that, even when pulp with a freeness of 40 ml higher was used, the formation improved and there was an energy saving of 10% (Yamaguchi and Yaguchi 1996).

Lecourt et al. (2010) examined the effect of three different commercial cellulose treatments on softwood bleached kraft pulp before the refining step. The refining experiments were conducted with a disc refiner to simulate industrial conditions usually present in paper mills, and the impact on energy saving, fiber characteristics, and paper properties were explored. An energy saving of 20% was obtained with two cellulase treatments to reach a given drainage index or breaking length. The water retention value (WRV) was increased by the cellulase treatment, but tear index losses were also observed. It is concluded that if a paper quality with a high tensile resistance but lowered tear strength is acceptable, cellulase treatment could save 20% of the electrical energy required for refining.

In high-yield pulps, the use of oxidative enzymes has also been evaluated. After treatment of an alkaline peroxide pulp derived from poplar, with Manganese peroxidase, 25% less PFI-refining was required to develop the equivalent pulp freeness (Petit-Conil et al. 1998). In contrast, the treatment of a high Kappa kraft pulp with the laccase-mediator system reduced the refinability by increasing the handsheet bulk (Wong et al. 1999b). A comprehensive study compared the effects that different monocomponent enzymes from a cellulolytic system have on the secondary refining of mechanical pulps (Pere et al. 1994, 1996). Cellobiohydrolase I (CBH) was capable of reducing the energy consumption during laboratory refining to develop freeness, while CBH II, different endoglucanases, xylanase, and mannanase had little effect. A subsequent trial confirmed the effects of CBH I, by demonstrating a 10% saving in the energy required for the secondary refining of primary rejects. The authors suggested that the improved refining properties were due to the ability of CBH I to decrease the crystallinity of the cellulose. In contrast, when mechanical pulps were treated with a complete cellulolytic system, the resultant fibers were more difficult to refine (Viikari et al. 1998). It appears that the fiber components, which were more resistant to cellulase treatments were also more resistant to refining. After an interstage treatment of mechanical pulp with a proteinase preparation, there were no obvious energy savings during secondary refining. However, energy savings were achieved when destructured wood chips were treated with proteinase or laccase before primary refining (Mansfield et al. 1999, Mansfield 2002). It is unclear how much of this energy saving was due to a greater efficiency in fiber separation, rather than fiber development.

The treatment of recycled fibers with cellulases reduced the refining energy required to achieve a specific freeness. At equivalent levels of refining, the cellulase treatment of recycled pulps yielded increases in freeness, but led to reductions in average fiber length (Eriksson et al. 1998). One trial revealed that the freeness of the refined stock could be increased to allow greater incorporation of the recycled fibers into a corrugating medium furnish (Moran 1996). Others, using recycled kraft fibers and old corrugated container pulps demonstrated savings in refining energy (Cabrera et al. 1996).

Mohlin and Pettersson (2001) investigated the effect of cellulase treatment. The trial was conducted on the EuroFEX paper machine. The bleached softwood market pulp was treated with a commercial cellulase (Celluclast from Novozymes), prior to refining. The potential for energy reduction was substantial and the pulps showed

Table 10.3 Effect of enzyme treatment on power consumption during manufacturing of ESKP high strength – *Process-scale trial results*

Particulars	Stock DDR, kWh/ton	Machine DDR, kWh/ton	Steam, Ton/Ton paper
Control (no enzyme)	80.12	50.97	3.18
Trial (with enzyme)	62.34	43.57	2.55
Savings	17.79	7.40	0.63

Net savings in refining power: 25.19 kWh/ton

Conditions; Temperature, 40–45 C; pH, 6.8–7.5; Enzyme dose, 145 ml/TP; Dosing point, pit pulper

Based on data from Bajpai et al. (2005, 2006)

improved formation and retained their sheet strength properties. Treatment with one unit of commercial enzyme reduced the energy required to reach a specific WRV-level by about 45–65 kWh/t (40–70%). The enzyme slightly reduced the pulp viscosity and had a significant effect on the fiber strength (its zero-span tensile index). In enzyme-treated pulps, there was a reduction in fiber length during refining which resulted in less fiber flocculation. Enzyme treatment produced a sheet which was superior in many ways to that made of untreated pulps. However, these benefits were not observed in laboratory testing. A study by Kallioinen et al. (2003) showed that enzyme-aided refining is economical and competitive in improving the energy economy of mechanical pulping.

Researchers at TCIRD, India (Bajpai et al. 2004; Bajpai et al. 2005, Bajpai et al. 2006) conducted extensive laboratory and process-scale studies with a neutral cellulase/hemicellulase enzymatic complex. They used FibreZyme LBR (from Dyadic International), which is derived from a *Chrysosporium* strain (US Patent No. 5,811,381, US Patent No. 6,015,707) for reducing the energy requirement in the refining/beating of different pulps – hardwood kraft pulp, 100% LF-3 bamboo pulp, OCC, and a mixed pulp containing NDLKC and LF-3 bamboo pulp (Bajpai et al. 2005). In the laboratory studies, the energy requirement reduced by 18–55% with different pulps. Process-scale trials in a mill producing packaging grade paper during the manufacture of high-strength ESKP (100% long fraction bamboo pulp) and normal ESKP (60% unbleached bamboo long fraction Kraft pulp and 40% NDLKC) showed reduction in refining energy and steam consumption. In case of high-strength ESKP, the refining energy reduced by 25 kWh/TP due to reduction in DDR (stock & machine) load (Bajpai et al. 2006). There was also reduction in steam consumption of ~0.6 ton/ton paper. The strength properties were not affected by enzyme treatment; in fact, mill was able to produce high-strength paper having low Gurley porosity without sacrificing other strength properties (Table 10.3). In case of normal ESKP, stock DDR was bypassed. Saving in power due to reduction in DDR load was 54 kWh/TP (Table 10.4). Saving in steam consumption was observed to be ~0.25 ton/ton paper and the strength properties were comparable.

Another process-scale trial in a mill producing coated papers, again using the same enzyme, showed a reduction in refining energy of about 70 kWh/TP in softwood pulps and 30 kWh/TP in hardwood pulps. A reduction in steam consumption of around 0.5 T/T of paper was observed. The use of the enzyme eliminated the

Table 10.4 Effect of enzyme treatment on power consumption during manufacturing of ESKP Normal – *Process-scale trial results*

Particulars	Stock DDR, kWh/ton	Machine DDR, kWh/ton	Steam, Ton/Ton paper
Control (no enzyme)	71.67	35.42	3.15
Trial (with enzyme)	0.00 (Bypassed)	52.79	2.90
Savings	71.67	−17.38	0.25

Net savings in refining power: 54.29 kWh/ton
Conditions; Temperature, 40–55 C; pH, 6.8–8.0; Enzyme dose 110 ml/TP; Dosing point, pit pulper and Tridyne pulper
Based on data from Bajpai et al. (2005, 2006)

Table 10.5 Effect of enzyme treatment on power and steam consumption during coating base manufacture – *Process-scale trial results*

	Power consumption, kWh/T pulp		
Particulars	Softwood	Hardwood	Steam, T/T paper
Control	200	150	2.57
Trial	130	120	2.07
Savings	70	30	0.50

Conditions; Temperature, 40–45 C; pH, 6.8–7.0; Enzyme dose 100 g/TP (in both the streets)
Based on data from Bajpai et al. (2005, 2006)

Table 10.6 Effect of enzyme treatment on power consumption during manufacturing of high gsm base papers (super coated art board 122 gsm and art paper 102 gsm) – *Process-scale trial results*

Condition	Normal (control)	Trial
Before refining	16–18	16–18
After refining (1 Conical, 1 TDR & 1 DDR)	23–25	25–28
After refining (1 Conical & 1 DDR)	–	23–25

Conditions; Temperature, 40–45 C; pH, 6.8–7.0; RT, 1.5 h; Stock consistency, 4%; Enzyme dose, 200 g/TP (dilution 50:50); Dosing point, mixing chest
Based on data from Bajpai et al. (2005, 2006)

debottlenecking of refining in the softwood street and increased production by 12% (Table 10.5). The strength properties were not affected. Process-scale trials in other mills producing writing and printing paper also showed encouraging results. In a mill producing heavy gsm base papers, a trial conducted with Biorefine L led to the bypass of a triple-disc refiner of 180 kWh (Table 10.6). The strength and other properties were within the specified limits and comparable to those without a trial run. This enzyme is being used regularly in mills throughout India, China, Indonesia, and North America.

In one of the North American mill using 100% OCC, use of enzyme resulted in significant increase in mullen (32%) and decrease in tear by around 1.8% in both MD and CD directions. To compensate for decreasing tear, the refiner was turned down by 20% and tear came back into specification. Drainage was increased and the machine was able to speed up by 50 fpm. A second evaluation was run on 84# grade and again similar results were achieved. In addition, the mill got very positive feedback from the converter regarding the first evaluation production. The mill is running the enzyme full time without having to use any virgin fiber and achieving better quality than before.

In one European mill producing towelling tissue using a mix of long fiber and short fiber, use of enzyme resulted in reduced refiner energy by over 30%; reduced long fiber use by over 20% (the cost of short fiber was significantly lower than long fiber); reduced starch use by 50% without loss of dry strength; increased retention by 2% most likely due to reduction in fines due to less refining; increased sretch by 20% and reduced steam consumption due to improved dewatering of the sheet (Gill 2008). In another mill in Europe producing towelling tissue, use of enzyme reduced Refiner 1 energy by over 60% and Refiner 2 energy by about 25% and reduced long fiber usage by 10%. In another mill producing towelling tissue, use of enzyme shut down one refiner without loss of strength; reduced long fiber use by over 15%; completely removed starch and improved machine runnability. In another mill in Europe producing packaging grade paper from old corrugated cardboard (OCC) and mixed grades of recovered paper, use of enzyme improved sheet dewatering giving 10% increase in production; increased sheet strength properties by 9–13%; reduced steam consumption by 10% implying better press dewatering and increase in sheet post press solids. The disc filter capacity also increased by 15% with better quality filtrate.

Liberty Paper, Becker, MN, USA, producing 500tpd of linerboard from OCC, switched to a fungal biorefining enzyme from Dyadic International and the output of its paper machine jumped 15tpd and saved energy in the refining stages (Thomas and Murdoch 2006). The mill operates a twin former paper machine with an air-pad headbox and twin LNP presses. Paper is then dried and passed to a single-stage calender. Enzyme dose to the pulp stock was optimized in trials at 100 g/t of final paper with a 2-h dwell. The result was better stock drainage and consequently an increase to paper machine speed and reduced steam demand through the driers. There was also a reduction in use of cationic starch and size.

Recently, Dyadic International has launched high-performance enzyme Fibrezyme® G200 for pulp and paper industry using new C1 production platform (Murdoch 2011). This enzyme reduces production costs while also improving fiber-to-fiber bonding and pulp-refining properties. This enzyme is produced using a new variant of Dyadic's patented and proprietary C1 platform technology which Dyadic refers to as the "white strain." Developed through sustained research efforts at Dyadic Netherlands, the white strain produces significantly less background proteins. These changes allow for more efficient and economical industrial scale production of highly targeted enzymes and proteins at greater purity levels.

Fiber modification enzymes are typically applied at the pulping stage of the papermaking process. Fiber type, process temperature, pH, pulp consistency, contact time between the fiber and enzyme, and key process and paper quality parameters are taken into consideration before and during the treatment. As yet fiber modification enzymes are found to be most effective on chemically bleached fiber and few applications have been successful in other types of raw materials. No doubt this will change as other types of cellulases are produced.

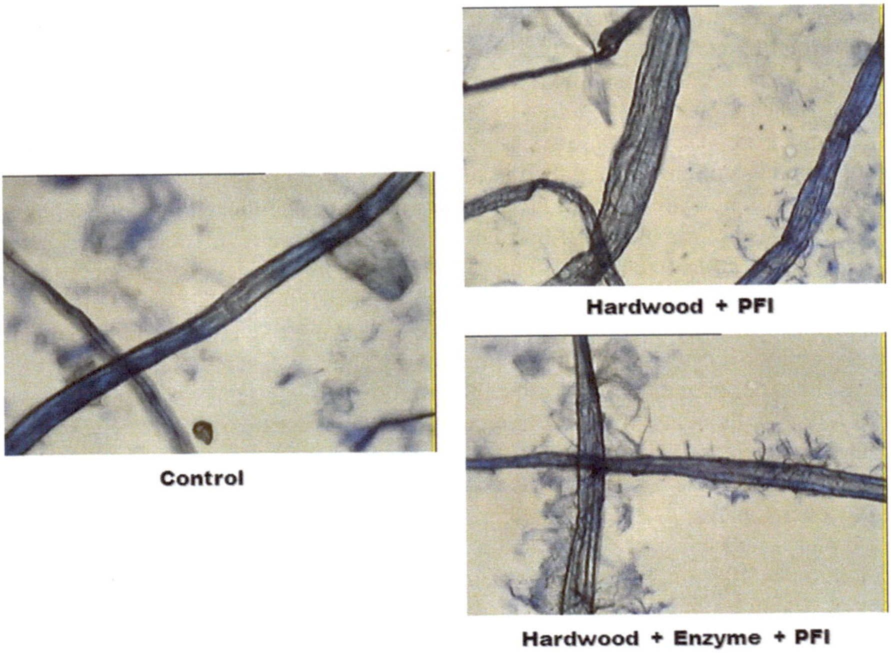

Fig. 10.1 Biorefining of hardwood fibers. Michalopoulos et al. (2005); Reproduced with permission

10.2.1 Enzyme Actions

Mixtures of cellulase and hemicellulase enzymes mainly function by partial hydrolysis of the fines, perforation, and brushing of long fibers. By hydrolysing fines, the enzyme increases drainage at the paper machine, reduces the vacuum requirement, reduces the steam load, and increases the paper machine speed. A reduction in the number of fines allows an improvement in the sheet strength due to an increase in the percentage of long fibers. Cellulases in the enzyme mixture prefer attaching to fines, rather than long fibers. This protects the long fibers from severe hydrolysis conditions. In a similar way, the xylanases collide randomly with the fines and long fibers in the pulp chest. The other main action of these enzymes is the perforation of the fibers by xylanase action. This improves the fibers' hydration (swelling) and promotes the internal fibrillation and delamination of the fiber, which improves its properties. Brushing of long fibers is another effect. The long fibers are eventually collided with by cellulases, which damage the bonds on the exposed cellulose chains. This partial depolymerization of cellulose chains on the fiber surface causes a weakening (but not a complete cutting) of external microfibers, which allows the fiber to be refined with less energy or to be more easily defibrillated. This defibrillation also facilitates fiber rehydration, internal fibrillation, and fiber reswelling.

Figures 10.1 and 10.2 show photomicrographs of enzyme-treated hardwood and softwood fibers. As can be seen use of enzyme leads to improved fibrillation.

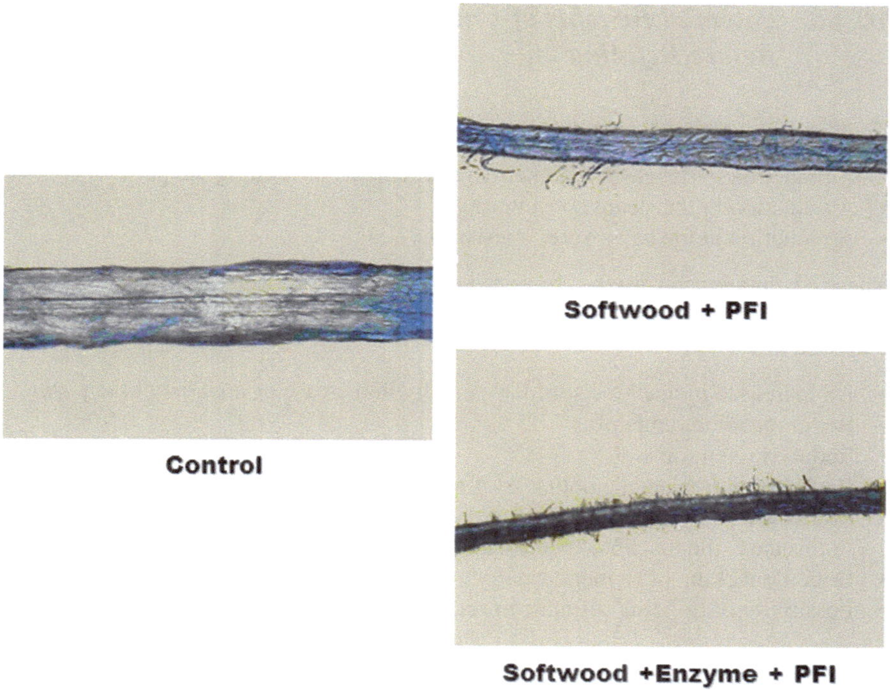

Fig. 10.2 Biorefining of softwood fibers. Michalopoulos et al. (2005). Reproduced with permission

10.2.2 Effects of Enzyme

As the enzyme promotes fiber swelling and makes fiber more flexible, the pulp gets excess refining (higher °SR) in the beginning at the same power input as that for control (without enzyme-treated pulp). Higher °SR pulp contains more fines, which results in poor drainability on the wire (more water remains with the pulp) and requires more steam to dry the paper sheet (on the dryer). Once the above effects are observed, the power input to the refiners (refining energy) is reduced so that the °SR of the pulp remains within the limit. The enzyme produces better fibrillation so those paper properties that depend on fibril content turn out better. These properties are tensile strength, bursting strength, and tensile energy adsorption. Improvement in the BOD to COD ratio in machine waste water is also expected as one component of the enzyme (endoglucanase) hydrolyses fines/fibrils and cellulosic debris in paper machine backwater to low molecular weight saccharides that are easily biodegradable. As explained above, the enzyme helps to reduce the fines/fibrils and cellulose debris in the white water (machine back water) loop, meaning the recycled water is cleaner with a minimum of fines.

10.2.3 *Potential Benefits of Enzymatic Treatment Before Refining*

The directly visible advantages are

- A reduction in the electrical energy requirement for refining the pulp
- A reduction in the steam consumption
- A reduction in the back water consistency

These advantages can also be converted into the following benefits (depending upon the situation and the requirements), but not necessarily all the benefits will be achieved

- An increased in machine speed, especially in the case of high gsm base paper
- Better machine runnability
- Reduced retention aid
- Better formation and smoothness of paper (it may be possible to reduce the head box consistency without affecting the capacity, due to improved drainability. In this case, the machine speed/steam consumption may not be reduced).
- Debottlenecking of refiner capacity to increase the production
- Possibility of utilizing difficult-to-refine pulps

Other benefits are

- Possibility of reducing toxic biocides which create problem in ETP and also denature enzyme
- Ease in operation of backwater clarification/filtration
- Possibility of reduction in pitch problem due to better dispersion
- Better biodegradability of machine effluent (due to hydrolysis of fine fiber fibrils by the enzyme
- Ease in operation of ETP (due to fewer fibrils and smaller amount of biocides)
- Reduction in greenhouse gas emissions associated with the generation of steam and power
- Ease in broke repulping – better dispersion, which may also reduce the addition of chemicals for dispersion.

10.3 Enzymes Improving Drainage

Recycled fibers have lower strength and higher drainage resistance than virgin fibers. These differences limit the paper quality and the speed at which paper machines can operate. The mechanical properties of fibers, as well as their ability to swell, are diminished after they are exposed to pulping and drying conditions imposed during the papermaking cycle. Freeness reduction during beating is much faster for secondary fibers. For equivalent beating times, a sheet containing recycled fibers is less dense and usually more absorptive than virgin fiber stock. The fines that are created when secondary fibers are beaten consist largely of microfribils that

Table 10.7 Effect of enzyme treatment on the drainability of OCC

Enzyme dose, % o.d. pulp	Reaction time, min	Drainage time for 800 ml, s	Improvement in drainage, %
0	–	31.5	–
0.1	30	27.8	11.7
0.1	45	26.5	15.9
0.1	60	26.1	17.1
0.1	120	25.1	20.3
0.1	180	23.5	25.4
0.2	30	24.8	21.3
0.2	60	22.8	27.6
0.2	180	21.5	31.7

Conditions: pH 5.0; temperature, 50 C; pulp consistency 5%; initial CSF of pulp 490 ml based on Bhardwaj et al. (1995)

were strongly coupled to each other when they were originally dried on the paper machine. When liberated during refining, they increase the specific surface area of suspension more than the swelling potential. They start to behave as fillers, with a small effect on strength but a large effect on the drainage properties. In general, the greater the degree of refining of the virgin fibers, the lower is the recovery potential of sheet properties that are a direct function of fiber bonding such as burst strength and tensile strength. Folding endurance of recycled paper is also considerably lower than for sheets made from virgin stock. Sheet density decreases each time the fibers are recycled. The strength losses may be the result of loss in binding potential, either in the strength of the interfiber bonding or in their number.

The potential of improving the drainage rates of recycled fibers by cellulase mixtures was discovered in the late 1980s (Fuentes and Robert 1986). Researchers from La Cellulose du Pin were the first to show that a mixture of cellulase and hemicellulase enzymes increases the freeness of pulp. Improved drainage and faster machine speeds, resulting from increased freeness, yields significant savings in energy and thus in overall cost. The endoglucanase activity is a prerequisite for drainage improvement of recycled pulps.

Several commercial enzymes are available which improve the drainage of secondary fibers. A commercial cellulase enzyme preparation (Pergalase A-40) based on *Trichoderma* has been used in several mills to improve drainage (Pommier et al. 1990; Eriksson et al. 1997; 1998). These types of enzymes are applied after refining/beating of the pulp, mainly to improve the dewatering. Recently, a cellulase enzyme with endoglucanase activity (FiberCare® D) developed by Novozymes has been reported to substantially increase the runnability of recycled furnishes and reduces the steam consumption in drying of paper on treating the pulp with enzyme after refining (Shaikh and Luo 2009).

The effectiveness of several commercial carbohydrate-modifying enzymes was examined for improving the drainage of secondary fibers (Bhardwaj et al. 1995). Drainage improvement over the control was substantial with Pergalase A-40 (a mixture after refining (Table 10.7). The drainage improvement was 11.7% (with 0.1% enzyme) and 21.3% (with 0.2% enzyme) at a reaction time of 30 min for low freeness pulp. An increase of the reaction time to 180 min improved the drainage by 25.4% (with 0.1% enzyme) and 31.7% (with 0.2% enzyme). The pulp retained most of the required strength properties when treated with Pergalase either at 0.1% enzyme addition and a reaction time of 45 min or at 0.2% enzyme addition and a reaction time of 30 min. Increase of reaction time beyond 30 min with 0.2% enzyme resulted in deterioration of strength properties. Pergalase treatment on pulps of different initial freeness showed that the lower the initial freeness, the higher the gain. When the pulp was treated with enzyme, the freeness increased without any loss of the mechanical properties in the paper, and when mechanical refining preceded the enzymatic treatment, better physical properties were obtained at freeness similar to the control one. In other words, better physical properties can be obtained at an identical drainability.

The effects of a commercial cellulase (Celluclast™), a commercial xylanase containing cellulase activity (Pulpzyme HA™), and a cellulase free xylanase from *Aureobasidium pullulans* were compared by Jeffries (1992). The effects of the enzymes were found to be significant. In the case of the *A. pullulans* xylanase acting on chemical fibers, the freeness was better than observed with Celluclast™ when both were used at the same dosage level.

Drainability of mechanical pulp can also be enhanced by the addition of hemicellulases (Karsila et al. 1990). Xylanase improves the freeness of deinked recycled pulp while having no detrimental effect on fiber tensile strength properties. By comparison, the tear indices of recycled pulps treated with cellulases decreased. These findings suggested that xylanases might be much more effective than cellulases or crude xylanase/cellulase mixtures. Xylanases, however, remove hemicellulases, that promote interfiber bonding. This effect can also lead to poor paper properties.

Enzymes and several chemical additives including modified polyacrylamides and modified starches to improve the drainage and strength of secondary fibers containing corrugated kraft cutting and corrugated boxes have been examined (Bhardwaj et al. 1997). It was found that the effect of enzyme treatment was limited to the improvement in drainage by 39.6% over control without any appreciable change in pulp strength properties. However, the treatment of pulp with various chemical additives resulted in substantial improvement of drainage as well as pulp strength properties. Best results were obtained using an anionic polyacrylamide.

The interaction of various types of cellulases and polymers for enhancing the freeness of a laboratory and mill furnish has also been investigated (Sarkar et al. 1995). Results showed that both enzyme and polymer are required to significantly enhance the freeness of pulp suspension. Handsheets prepared after enzyme and polymer treatment showed insignificant losses in tensile and burst strength. Enzyme treatment followed by polymer addition could provide a new biological chemical method for enhancing the freeness of recycled fiber. Although an independent treatment of pulp suspension, either with enzyme or polymer can improve the freeness of pulp stock, a combination of lower dosages of enzyme and polymer will significantly increase the freeness. Treatment of recycled fiber with polymer alone can produce large flocs. By using enzyme with lower levels of polymer, a potentially more uniform sheet can be produced.

The effects of purified cellulases and hemicellulases on the properties of recycled kraft pulps has been studied by Oksanen et al. (1995). ECF-bleached kraft pulp was recycled by subsequent drying slashing and refining steps. The recycled pulps were treated with purified *Trichoderma reesei* hemicellulases and cellulases, and their combinations. Changes in fiber properties caused by the enzymatic treatments were characterized by measuring the WRV, °SR value, and handsheet strength properties. The strength properties and WRV lost upon drying, could be recovered almost to the initial level by refining between the cycles. The tensile index increased as a function of recycles while tear index decreased slightly. However, extensive refining resulted in deteriorated drainage properties, i.e., increase in °SR value. Of the single enzymes, endoglucanases were most effective in improving the drainage, whereas cellobiohydrolases had practically no effect. The pulp strength was, however, negatively affected even with rather low endoglucanase dosages. Xylanase and mannanase treatments improved the change only slightly. Although, the drainage could be improved by enzymatic treatments, none of the enzymes could improve the swelling of the recycled fibers.

Stork et al. (1995) examined all possible enzymes of the cellulase system and relevant hemicellulases for their capacity to upgrade recycled pulps. The results have shown that presence of endoglucanase activity is a prerequisite for improvement of drainage of recycled fibers by enzymatic means. When cellobiohydrolase and xylanase activity were present, they acted synergistically with the endoglucanase to improve its effects. Mannanases were not helpful. The effect of cellulases on chemical pulp fibers and fibers containing lignin was quite different. Chemical pulp fibers were severely damaged and disintegrated completely on prolonged incubation, whereas strength properties of recycled mechanical pulp fibers were affected only to a small extent.

Stork and Puls (1995) carried more detailed investigation on endoglucanase treatment of different primary and secondary fiber sources. It was found that strength properties of different classes of primary and secondary pulps were not improved by an endoglucanase treatment. However, this treatment prevented further improvements of the breaking length by beating. Results for tear strength were found to be more complex. Endoglucanase treated and beaten kraft pulps were reduced in tear strength, whereas TMP and groundwood were not. Treated CTMP followed by beating gave superior pulp strength properties compared to control pulps.

Pala et al. (1998) investigated refining, refining with an enzymatic treatment, an enzymatic treatment plus refining, and an enzymatic treatment alone, of recycled fibers. The physical and mechanical properties were measured. The most suitable method for upgrading recycled pulps was by refining with an enzymatic treatment. Refining increased the burst and tensile resistance and the enzymatic treatment produced better drainage results under certain conditions.

Fiber which has been recycled more than once has lower papermaking qualities than virgin or once-recycled fiber. By using an enzyme blend with recycled fiber some lost freeness can be restored. Pergalase is a blend of enzymes which improves the freeness of the fiber but does not reduce the fiber strength. The enzyme is effective at an optimum pH of 5.5–6 but remains active at pH 4.5–7. The optimum temperature is between 50 and 60°C. Enzymes need time to be effective. A 15-min retention time is adequate, providing there is good mixing. Trial results from three mills show that machine speeds were increased when using Pergalase. The benefits of such an enzyme-enhanced drainage program have been shown on grades including tube stock, gypsum linerboard, and corrugating medium (Moran 1996).

The endoglucanases (EG I and EG II) of *Trichoderma reesei* have been reported to significantly improve pulp drainage at low dosage levels, and EG II was found to be more effective at a given level of carbohydrate solubilization (Oksanen et al. 2000). Combining hemicellulases with the endoglucanase treatments increased the positive effects of the endoglucanases on pulp drainage. However, as a result of the endoglucanase treatments (high dosage) a slight loss in strength was observed. EG alone has been found to be more detrimental to strength properties as compared to EG + Xyl at a given level of cellulose hydrolysis. Although the drainage properties of the pulps could be improved by selected enzymes, the water retention capacity of the dried hornified fibers could not be recovered by any of the enzymes tested. It was reported that endoglucanases enhance dewatering by hydrolyzing the amorphous hydrophilic cellulose which is the main constituent of the fines formed during refining. CBH I did not influence pulp drainage at any level of refining (Oksanen et al. 2000). All enzyme treatments except CBH 1 alone reduced the °SR value to some extent. CBH as well as xylanase did not markedly affect the strength properties.

Stork et al. (1995) used isolated cellobiohydrolases and endoglucanases of *Penicillium pinophilum* to treat recycled pulps and measured the effects on the WRV. They found that the action of endoglucanases was necessary for an improvement in the drainage of recovered paper. The effect did not appear to be due to a selective hydrolysis of the fines fraction but was a consequence of the hydrolysis of amorphous cellulose on the surface of the fibers. Depending on the origin and history of primary and secondary fibers, the endoglucanase treatment decreased the strength properties to differing degrees. It has been reported by Pere et al. (1995) that EGs dramatically decreased pulp viscosity. EG appears to attack cellulose at sites where even a low level of hydrolysis reduces pulp viscosity, resulting in a marked deterioration of strength properties.

The effectiveness of a cellulase enzyme having predominantly endoglucanase (EG) activity was evaluated in the laboratory and paper mills for improving the freeness and drainability of different types of recycled pulps (Shaikh and Luo 2009). On treating the refined pulps with enzyme, the improvement in Canadian standard freeness (CSF) was observed by 13.1% in ONP, 19.3% in old corrugated container (OCC), and 40.5% in mixed waste (MW), as shown in Table 10.8. Using OCC and addition of enzyme at different levels, the CSF value increased with the increase in enzyme dose. However, there was no appreciable change in the tensile and compression strengths of the handsheets prepared from the enzyme-treated pulps (Table 10.9). In fact, the EG activity of cellulase partially hydrolyzes the amorphous and low molecular components of cellulose present in the form of fine fribils and colloids. This helps in dewatering of the pulp, as these components constitute very small fraction of the pulp but have very high specific surface area and hold maximum water (Bajpai et al. 2006).

Table 10.8 Effect of enzyme treatment on CSF of different types of pulp

Pulp grade	CSF, ml	
	Before enzyme treatment	After enzyme treatment[a]
OCC	419	500 (19.3)
MW	304	427 (40.5)
ONP	168	190 (13.1)

[a]Values in parenthesis show percent improvement
Based on Shaikh and Luo (2009)

Table 10.9 Effect of enzyme dose on CSF and strength properties of OCC

Enzyme dose (g/TP)	CSF (ml)	Compression strength (Nm/g)	Tensile strength (Nm/g)
Nil	300	22.0	45.0
50	353	22.5	45.5
100	368	23.0	45.5
200	388	21.5	43.0
500	422	22.0	43.0

Based on Shaikh and Luo (2009)

Table 10.10 Effect of enzyme treatment on the requirement for cationic polyacrylamide for drainage control of OCC

10 s filtrate weight, g	Cationic Polyacrylamide requirement (kg/TP)	
	No enzyme treatment	After enzyme treatment
470	0.0	–
540	0.13	0.0
600	0.60	0.35
650	0.84	0.58
700	1.10	0.80

Based on Shaikh and Luo (2009)

Functional additives are generally introduced to improve retention and drainage on the paper machine and quality of the final product. The high specific surface area of the colloidal fraction of the furnish readily adsorbs a great amount of the additives without a perceptible benefit. The unproductive consumption of additives often necessitates compensation with higher additive dosing levels. Preferential EG enzyme activity upon the colloidal fraction is expected to modify the overall response of the furnish to functional additives. The impact of enzymatic treatment on the drainage of OCC pulp, conditioned with various levels of cationic polyacrylamide (CPA) was evaluated (Shaikh and Luo 2009). It is shown in Table 10.10 that the requirement of CPA decreased to a great extent for the same drainage rate when pulp was treated with enzyme.

The improvement in drainage of the pulp by enzyme treatment was demonstrated in the mill trials to observe the change in machine runnability (Shaikh and Luo 2009). Figure 10.3 shows the increase in machine speed for the production of 200 g/m^2 liner on enzyme dosing at a North American mill. Due to enhanced dewaterability, a 6–7% reduction in steam consumption within in the dryer section was also observed. It was also possible to increase the ring crush by two points. A similar trial

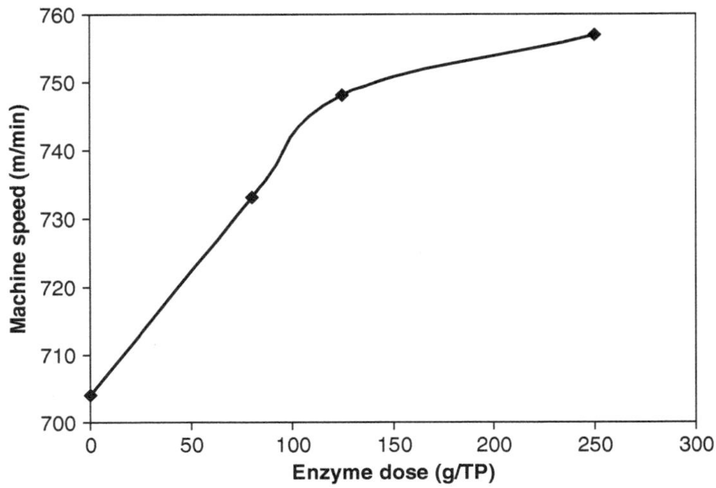

Fig. 10.3 Effect of enzyme dose on machine speed using OCC and MW pulps to produce 200-gsm liners at a North American mill. Based on Shaikh and Luo (2009)

Table 10.11 Effect of cellulase and pectinase enzymes[a] on drainage of deinked pulp

Treatment	CSF[a] (ml)	Brightness (%GE)
No enzyme	330	80.7
Cellulase (1 l/TP)	520 (190)	81.6
Pectinase (1 l/TP)	400 (70)	82.3
Cellulase (1 l/TP)+Pectinase (1 l/TP)	460 (130)	82.3

[a]Values in parenthesis indicate increase in freeness (CSF value) over control at pH 5.5

Based on Olsen et al. (2000)

with enzyme treatment (100 g/TP) before refining in an European mill, producing towel and tissue, showed an increase in machine speed from 1,650 to 1,750 m/min for tissue and from 1,600 to 1,750 m/min for towel productions. The specific refining energy also was reduced by 12.5%, probably due to the presence of some celobiohydrolase (CBH) activity in the enzyme product, which helps in pulp refining (Bajpai et al. 2006). Shaikh and Luo (2009) have also reported steam savings on enzyme treatment in the mill trials at Asian mills producing various basis weight papers from OCC pulp.

For improving the drainability (freeness) of secondary fiber, treatment with cellulase alone generally leads to loss in pulp brightness. Furthermore, when used in combination with a drainage aid polymer, the loss in brightness is even more pronounced. A method using a mixture of cellulase and pectinase enzymes has been described to accomplish the goal of simultaneously increasing freeness (drainage) without loss of brightness and physical properties (Olsen et al. 2000). As seen in Table 10.11, treatment with cellulase alone gave very high improvement in CSF value (190 ml) but not the brightness, and treatment with pectinase alone though resulted in greater enhancement of brightness (1.6 points) but very little improvement in CSF value (only 70 ml). However, treatment with the combination of both the enzymes resulted in appreciable improvement in CSF (130 ml) while 1.6 points enhancement in brightness (same as that with pectinase).

10.3.1 Enzyme Action

Several theories have been proposed to explain the freeness increase occurring after enzymatic treatment (Kantelinen and Jokinen 1997; Jackson et al. 1993; Mansfield et al. 1997; Mansfield and Wong 1999). The enzymatic attack may involve a peeling mechanism, which removes fibrils and fiber bundles that naturally have a high affinity for water, and leaves the fibers less hydrophilic and easier to drain. Alternatively it has been suggested that enzymes act preferentially on fines which have a propensity to block up interstices in the fiber network. The increase in drainage has also been attributed to the cleaving of amorphous cellulose on the surface of fines. One of the explanations suggested that due to the high specific surface area of the fines, the attack of cellulases was specific towards this fraction (i.e. fines) of the pulp. It has been reported that the fiber surface is stripped through the enzymatic hydrolysis of subsequent layers or fibrils (Jackson et al. 1993; Mansfield et al. 1997). Jackson et al. (1993) suggested that enzymes can either flocculate or hydrolyze fines and remove fibrils from the surface of large fines. The enzyme-aided flocculation occurs when a low enzyme dosage is used. In this case, fines and small fiber particles aggregate with each other or with the larger fibers, decreasing the amount of small particles in pulp and consequently improving pulp drainage. On the other hand, at higher enzyme concentration, flocculation becomes less significant, and hydrolysis of fines begins to predominate. Enzyme specifity also plays very important role. As shown by Pere et al. (1995), endo and exoglucanases affect paper properties differently. Endoglucanase can lower pulp viscosity and thus dramatically reduce the pulp strength. The differences in enzyme activity are attributed to different modes of action. Endoglucanase are more active on amorphous cellulose and randomly attack the inner part of the cellulosic chain, whereas exoglucanases can hydrolyze both crystalline and amorphous cellulose by removing cellobiose from the terminal part of cellulose chains. Although considered more detrimental to fibers, endoglucanase action is probably the main determinant of drainage improvement (Jackson et al. 1993; Stork et al. 1995).

Using mixtures of cellulases can be disadvantageous for certain pulp properties. By applying purified enzymes on specific regions of the cellulose fibers, the desired part of the pulp could be modified for a particular enzyme application. When applying cellulase enzyme mixtures, identification of the key component, responsible for the required effect on pulp and paper properties, is very difficult. In secondary fibers, the fines and fibrils, which cause low rate of drainage, decisively consist of amorphous cellulose. Since, the amorphous cellulose is more accessible than crystalline cellulose, it is not necessary to use the whole cellulose complex for the hydrolysis. That is, applying specific cellulose component may be effective enough.

Table 10.12 Benefits of improving drainage

Increased paper machine speed
Reduced steam consumption
Improved formation and appearance
Lower headbox consistency
Content of poorly draining and inexpensive recycled fibers can be increased

10.3.2 Benefits of Improving Drainage

Treatment with a mixture of cellulase and hemicellulase enzymes at low concnetration increases the drainage of pulp. The increase in drainage can enhance the caccapacity of a secondary fiber preparation plant, increase machine speed or pulp dilution in the head box, and ultimately produce paper of better quality (Table 10.12). In addition to an increase in drainage, regular use of enzymes under optimum conditions may produce beneficial secondary effects such as greater reliability of the paper machine. Less substandard paper is found to be produced partly due to lower frequency of breaks. In some cases, enzymes can be also used in conjunction with the normal retention/drainage agents to give a significant increase of freeness to the pulp. The key to a successful enzyme application is the careful selection of the right enzymes for a mill's specific furnish, process conditions, and water chemistry.

10.4 Enzymes for Vessel-Picking Problems

Significant increase in issues relating to vessel picking are faced due to increased use of tropical hardwoods. The vessel elements of tropical hardwoods are hard, large, and tubular shaped and do not fibrillate during normal beating. As a result, they stick up out of the surface of the paper. During printing, the vessels are torn out, leaving voids. This characteristic reduces the value of tropical hardwood pulps. Different HW species have different quantities and sizes of vessel elements ranging from less than 10% to greater than 50% by volume. These vessels are easily picked from the sheet by tacky inks and deteriorate offset print quality. Problems related to vessel elements include:

- Dusting on the paper machine
- Coating picks or streaks
- Linting or picking in the printing operations
- Print spots or voids (ink refusal)

Traditional methods to deal with vessel-picking problems include the use of high-intensity refining, fractionation by hydrocyclone, dry strength additives, and surface sizes. Refining to the point of eliminating vessel picking is not practical for most mills as it is expensive and generally has a negative impact on drainage and optical

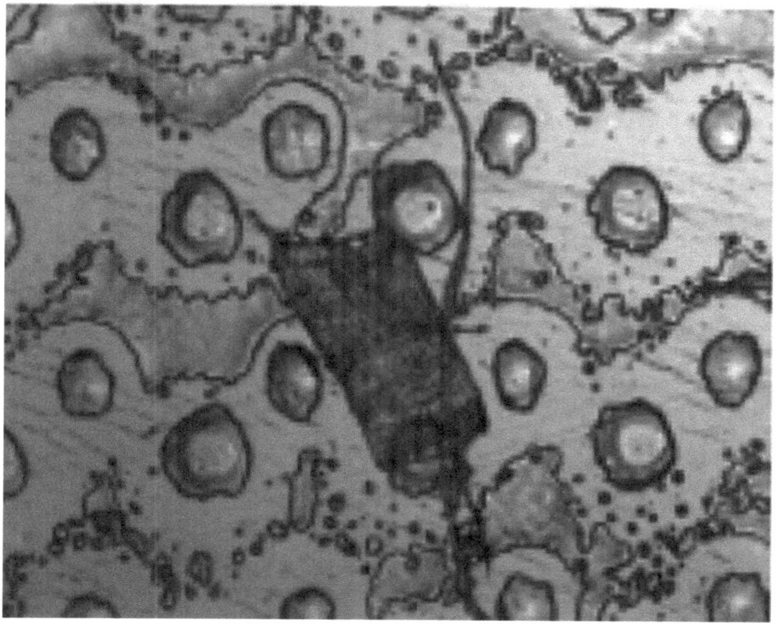

Fig. 10.4 Untreated vessel pick. Covarrubias (2009, 2010); Reproduced with permission

properties. Concentration of the vessel elements by hydrocyclone followed by subsequent high consistency refining of the vessel rich stream is probably the most effective option, but is very capital intensive for mills that do not already have the required equipment. Dry strength additives and surface sizes can help but high dosages are usually required to have much impact. In the recent years several products using proprietary enzyme blends have been developed by Buckman Laboratories that are effective for reducing vessel picking and has secured an ongoing application at a mill in Indonesia. The acacia pulp was treated with 0.5–1.0 kg/ton of enzyme and achieving a 20% reduction in refining energy while more importantly increasing IGT values. IGT is a picking test in which a paper strip is printed with a tacky ink at ever-increasing speeds. The speed just prior to the sample beginning to pick determines the pick resistance. The same instrument can also give a pick count. A laboratory test method has been developed by Buckman Laboratories for evaluating vessel picking and determining the effectiveness of the enzyme treatments to reduce vessel picking. Figures 10.4 and 10.5 show the difference in vessel element structure of treated vs untreated vessels picked from laboratory handsheets made from refined pulp (Covarrubias 2009, 2010). In the handsheets made from the enzyme-treated pulp, the picks are significantly fewer and tend to be only fragment vessel elements rather then the intact vessel elements seen from the untreated pulp handsheets. Positive results in reducing vessel picking with enzyme treatment for a number of different pulp species have been obtained (Table 10.13). As there is wide

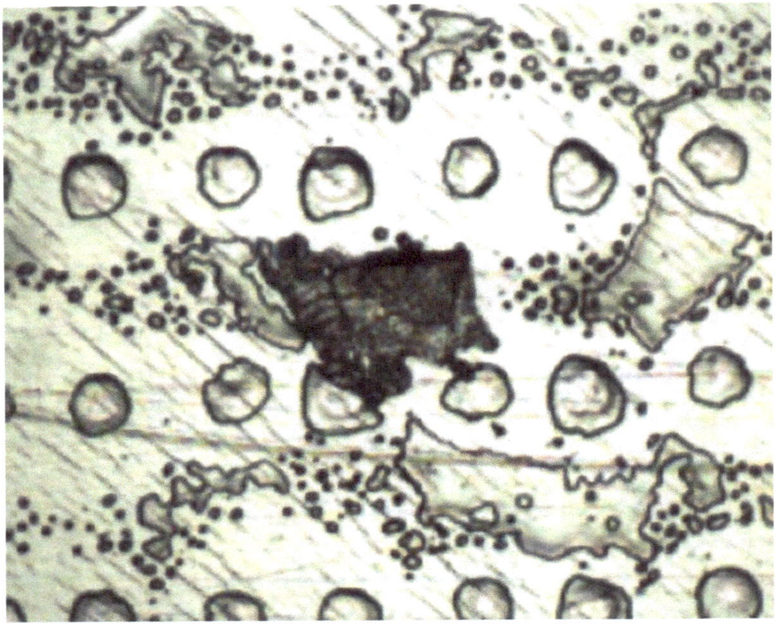

Fig. 10.5 Treated vessel pick. Covarrubias (2009, 2010); Reproduced with permission

Table 10.13 Effect of enzymes on vessel pick reduction

Hardwoods	Vessel pick reduction (%)
Brazilian Eucalyptus	70
Canadian Aspen	50
German Birch	55
Mixed Tropical Hardwood	65
Austrailian Eucalyptus	60

Based on Covarrubias (2009)

Table 10.14 Reduction in vessel element picking by fiber modification enzymes in mill trial

Vessel element (Relative area)	
Reference	1.0
Enzyme 1	0.67
Enzyme 1 + Enzyme 2	0.58

Based on Covarrubias (2009, 2010)

variety of HW pulps being used by the pulp and paper industry, a number of enzyme-based products for vessel-pick reduction have been developed by Buckman Laboratories.

In one of the mills producing writing and printing paper from HW Kraft Eucalyptus, use of fiber modification enzymes, reduced vessel element picking during printing by 30–40% (Table 10.14). The enzyme treatment was employed on brown stock prior to bleaching.

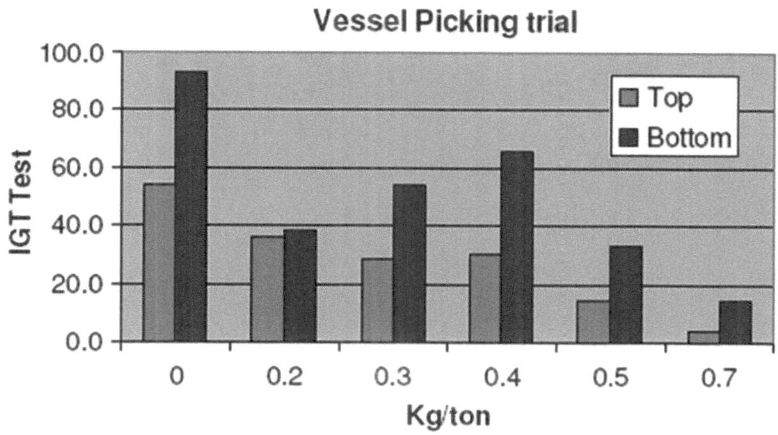

Fig. 10.6 Effect of enzyme on IGT. Gill (2008); Reproduced with permission

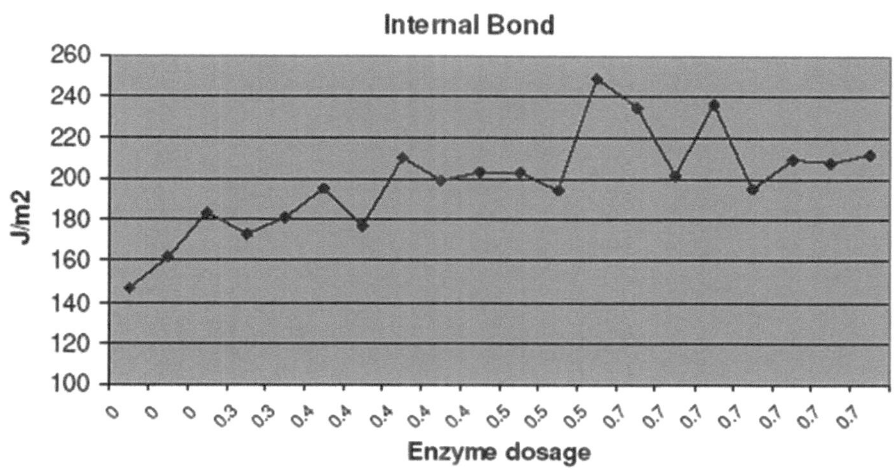

Fig. 10.7 Effect of enzyme on internal bond. Gill (2008); Reproduced with permission

In another mill trial, use of fiber modification enzymes reduced IGT counts by more than 50%; increased internal Bond; reduced vessel picking reduction even though refiner load was reduced from 70 to 55 kWh/t; reduced porosity (Gill 2008). There was some drop in tear but other strength properties remained unaffected (Figs. 10.6–10.9).

Honshu Paper Company has patented a process for the use of commercial cellulases to enhance the flexibility of hardwood vessels (Uchimoto et al. 1988). Enzyme treatment reduces vessel picking by 85% along with the improved drainability, smoothness, and tensile strength. Thus, enzymes can help in saving the beating energy for vessels.

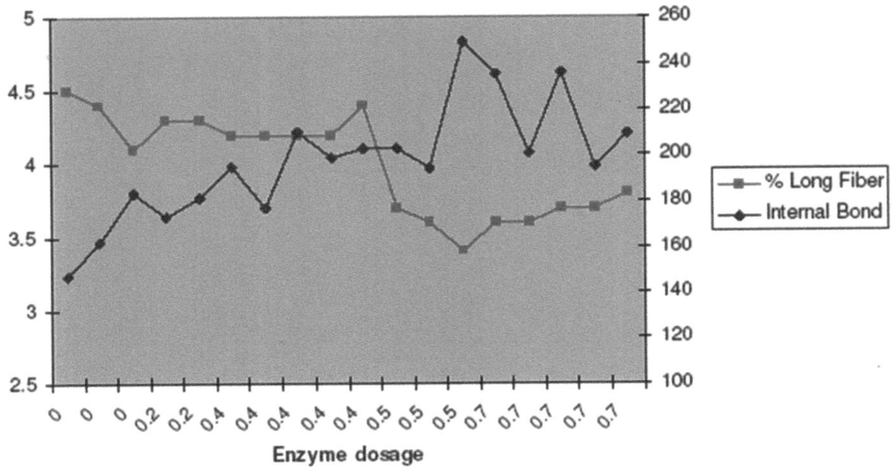

Fig. 10.8 Effect of enzyme on long fiber and internal bond. Gill (2008); Reproduced with permission

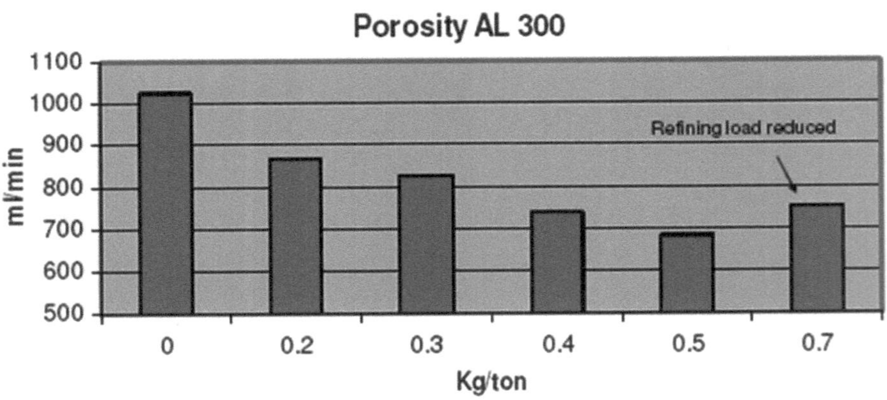

Fig. 10.9 Effect of enzyme on porosity. Gill (2008); Reproduced with permission

10.5 Conclusions

The use of fiber modification enzymes in papermaking is an emerging and exciting technology offering unique, innovative solutions for process improvements and developing sheet quality while delivering significant ROI for the paper maker.

References

Bajpai P, Bajpai PK (2001) Status of biotechnology in pulp and paper industry. InPAPER International 5(4):29

Bajpai PK, Bajpai P, Mishra OP, Mishra SP, Kumar S, Vardhan R (2004) Enzyme's role in improving papermaking, 9th international conference on biotechnology in the pulp & paper industry, Durban, Sappi Forest Products, South Africa, p 1.10, 10–14 October 2004

Bajpai P, Bajpai PK, Mishra SP, Mishra OP, Kumar S (2005) Enzymatic refining of pulp – case studies. Proceedings paperex 2005 (7th International Conference on Pulp Paper and Conversion Industry), New Delhi, Dec 3–5, 2005, pp 143–159

Bajpai P, Mishra SP, Mishra OP, Kumar S, Bajpai PK (2006) Use of enzymes for reduction in refining energy – Laboratory and process scale studies. Tappi J 5(11):25–32

Barnoud F, Comtat J, Joseleau JP, Mora F, Ruel K (1986) Interest in the Enzymatic Hydrolysis of Xylans for modifying the structure of pulp fibres, proceedings 3rd international conference on biotechnology in the pulp and paper industry, Stockholm, Sweden, p 70

Bhardwaj NK, Bajpai P, Bajpai PK (1995) Use of enzymes to improve drainability of secondary fibres. Appita 48(5):378–380

Bhardwaj NK, Bajpai P, Bajpai PK (1996) Use of enzymes in modification of fibers for improved beatability. J Biotechnol 51:21–26

Bhardwaj NK, Bajpai P, Bajpai PK (1997) Enhancement of strength and drainage of secondary fibers. Appita 50(3):230–232

Bolaski W, Gallatin A, Gallatin JC (1959) Enzymatic conversion of cellulosic fibres. US Patent No. 3,041,246

Cabrera CF, Sarkar JM, Didwania HP, Espinoza E, Benavides JC (1996) Paper mill evaluation of a cellulolytic enzyme and polymers for improving the properties of waste paper pulp, papermaker conference, Tappi Press, Philadelphia, US, p 481, 24–27 March 1996

Comtat J, Mora F, Noe P (1984) French patent No. 2,557,894

Covarrubias R (2006) Fibre modification through biotechnology. Pulp Pap 80(6)

Covarrubias RM (2007) Enzymatic strength development in OCC. 2007 TAPPI 8th Research forum on Recycling, Niagara Falls, ON, Canada, 23–26 Sept 2007, p 36

Covarrubias RM (2009) Internal Buckman R&D Newsletter

Covarrubias RM (2010) An Enzymatic Solution to Vessel Picking; PAPTAC 2010

Diehm RA (1942) Process of Manufacturing Paper, US Pat No. 2,280,307

Eriksson LA, Heitmann JA Jr. and Venditti RA (1998). Freeness Improvement of Recycled Fibres Using Enzymes With Refining, Enzymes Applications in Fiber Processing. In: Eriksson KEL (ed) ACS Symposium Series 687, American Chemical Society, Washington DC, USA, 1998, p 340

Eriksson LA, Heitmann JA, Venditti RA (1997) Drainage and strength properties of OCC and ONP using Enzymes with Refining, Recycling Symposium, Tappi Press, Chicago, IL, USA, 14–16 April 1997, p 423

Freiermuth B, Garrett M, Jokinen O (1994) The use of enzymes in the production of release papers, paper technology. Paper Industry Technical Association 35(3):21

Fuentes JL, Robert M (1986) French Patent 2,604,198

Garcia O, Torres AL, Colom JF, Pastor FIJ, Diaz P, Vidal T (2002) Effect of cellulase-assisted refining on the properties of dried and never-dried eucalyptus pulp. Cellulose 9(2):115

Gill R (2008) Advances in use of fibre modification enzymes in paper making. Conference Aticelca XXXIX Congresso Annuale, Fabriano, Italy, 29–30 May 2008

Jackson LS, Heitmann JA, Joyce TW (1993) Enzymatic modifications of secondary fiber. Tappi J 76(3):147–154

Jeffries TW (1992). Emerging Technologies for Materials and Chemicals from Biomass, ACS Symposium Series No.476 (Rowell RM, Schultz TP, Narayan R (eds) 313

Kallioinen A, Pere J, Siika-Aho M, Lehtila A, Malkki H, Syri S, Thun R (2003) Biotechnical methods for improvement of energy economy in mechanical pulping, Research Notes 2183. VTT, Espoo, Finland, p 96

Kantelinen A, Jokinen O (1997) The mechanism of cellulase/hemicellulase treatment for improved drainage. Biol Sci Sympos XXX:267–269
Karsila S, Kruss I, Puuppo O (1990) European Patent 351655 A 9001249004
Lecourt M, Meyer V, Sigoillot J-C, Petit-Conil M (2010) Energy reduction of refining cellulases. Holzforschung 64(4):441–446
Loosvelt I (2008) Enzymatic products for the pulp and paper industry. Progress 08. 16th International papermaking conference. Efficiency in papermaking and converting processes, Krakow, Poland, 23–26 Sept 2008, p 16
Loosvelt I (2009) Current applications of fibre modification enzymes in the paper industry and future possibilities. Fibre engineering, Gothenburg, Sweden, 24–26 Mar 2009, p 39
Mansfield SD (2002) Laccase impregnation during mechanical pulp processing – improved refining efficiency and sheet strength. Appita J 55(1):49
Mansfield SD, Wong KKY (1999) Improving the physical properties of linerboard via cellulolytic treatment of the recycled paper component. Prog Pap Recycl 9:20–29
Mansfield SD, Jong E, Stephens RS, Saddler JN (1997) Physical characterization of enzymatically modified kraft pulp fibers. J Biotechnol 57:205–216
Mansfield SD, Wong KKY, Richardson JD (1999) Improvements in mechanical pulp processing with proteinase treatments, 53rd annual appita conference, Rotorua, New Zealand, 2, 19–23 April 1999a, p 375
Mansfield SD, Wong KKY, Dickson AR (2000) Variation in the response of three different pinus radiata kraft pulps to xylanase treatments. Wood and Fiber Science 32(1):105
Michalopoulos DL, Ghosh D, Murdoch B (2005) Enhancement of Wood Pulps by Cellulase Treatment. 2005 TAPPI Enginnering Pulping and Environment Conference Philadelphia
Mohlin U-B, Pettersson B (2001) Improved Papermaking by Cellulase Treatment Before Refining. In: Vahala P, Lantto R (eds) 8th international conference on biotechnology in the pulp and paper industry, VTT Biotechnology, Helsinki, Finland, 4–8 June 2001, p 78
Mora F, Comtat J, Barnoud F, Pla F, Noe PJ (1986) Action of Xylanases on chemical pulp fibres, 1. investigation on cell-wall modifications. J Wood Chem Tech 6(2):147
Moran BR (1996) Enzyme treatment improves refining efficiency, recycled fibre freeness'. Pulp Pap 70(9):119
Murdoch B (2011) Dyadic International launches high performance enzyme for pulp and paper industry using new C1 production platform. Personal communication
Noe P, Chevalier J, Mora F, Comtat J (1986) Action of Xylanases on chemical pulp fibres, part ii – enzymatic beating. J Wood Chem Tech 6:167
Oksanen T, Buchert J, Pere J, Viikari L (1995) In: Proceedings 6th international conference biotechnology pulp & paper industry: recent advances in applied and fundamental research, Vienna, Austria, p 177
Oksanen T, Pere J, Buchert J, Viikari L (1997) The effect of trichoderma reesei cellulases and hemicellulases on the paper technical properties of never-dried bleached kraft pulp. Cellulose 4(4):329
Oksanen T, Pere J, Paavilainen L, Buchert J, Viikari L (2000) Treatment of recycled kraft pulps with Trichoderma reesei hemicellulases and cellulases. J Biotechnol 78:39–48
Olsen WL, Zhu H, Hubbe MA (2000) Method of improving pulp freeness using cellulase and pectinase enzymes. US Patent 6,066,233
Pala H, Mota M, Gama FM (1998) Effects of enzymatic treatment and refining on the properties of recycled pulp. associacao portuguesa dos tecnicos das industrias de celulose e papel, 478
Pastor FIJ, Garcia O, Sanchez M, Vidal T, Torres AL, Diaz P, Colom JF (2002) Cellulase modification of refining properties of dried and never-dried eucalyptus pulp, cost workshop – biotechnology in the pulp and paper industry, Cost E23 Action, Centre Technique du Papier, Poster Session, Grenoble, France, 9, 28–29 November 2002, p 10
Pere J, Siika-aho M, Viikari L (1994) Process for preparing mechanical pulp. Can Patent 2,157,513
Pere J, Siika-aho M, Buchert J, Viikari L (1995) Effects of purified trichoderma reesei cellulases on the fiber properties of kraft pulp. Tappi J 78:71–78

Pere J, Liukkonen S, Siika-aho M, Gullichsen J, Viikari L (1996) Use of purified enzymes in mechanical pulping, pulping conference, Tappi Press, Nashville, TN, USA, Book 2, 27–31 October 1996, p 693

Petit-Conil M, Sigoillot JC, Herpoel I, Kurek B, Ruel K, Moukha S, Joseleau JP, Penninckx M, Asther M, Gazza G, de Choudens C (1998) Treatment of a market poplar high-yield pulp with manganese peroxidases. 7th international conference on biotechnology in the pulp and paper industry, CPPA, Vancouver, BC, Canada, Volume A, 16–19 June 1998, p A143

Pommier JC, Goma G, Fuentes JL, Rousset C, Jokinen O (1990) Using enzymes to improve the process and the product quality in the recycled paper industry, Part 2: Industrial applications. Tappi J 73(12):197–202

Sarkar JM, Cosper DR, Hartig EJ (1995) Applying enzymes and polymers to enhance the freeness of recycled fiber. Tappi J 78(2):89

Scartazzini R, Reinhardt B, Traser G (1995) The use of enzymes as dewatering and refining additives in the paper industry. Allg Pap-Rundsch 119(40):987, 5 October 1995

Shaikh H, Luo J (2009) Identification, validation and application of a cellulase specifically to improve the runnability of recycled furnishes. Proceedings 9th international technical conference on pulp, paper and allied industry (Paperex 2009), New Delhi, India, 4–6 Dec 2009, pp 277–283

Stork G, Puls J (1995) In: Proceedings 6th international conference biotechnology pulp & paper industry: recent advances in applied and fundamental research, Vienna, Austria, p145

Stork G, Pereira H, Wood TM, Dusterhoft EM, Toft A, Puls J (1995) Upgrading recycled pulps using enzymatic treatment. Tappi J 78(2):79

Thomas N, Murdoch B (2006) Speed, production get bio-boost at liberty paper: a bio-refining enzyme helps the company run faster and cleaner. Pap 360° 1(5):17

Torres AL, Garcia O, Colom JF, Pastor FIJ, Diaz P, Roncero MB, Vidal T (1999) Cellulases-assisted refining of fibres and handsheet properties. Crossing the millennium frontier, emerging technical and scientific challenges, 27th EUCEPA Conference, Grenoble, France, 11–14 Oct 1999, p 47

Uchimoto I, Endo K, Yamagishi Y (1988) Improvement of deciduous tree pulp. Japanese Patent 135(597/88):1988

Viikari L, Pere J, Suurnakki A, Oksanen T, Buchert J (1998) Carbohydrates from Trichoderma Reesei and Other Microorganisms – Structures, Biochemistry, Genetics and Applications'. Royal Society of Chemistry, Cambridge, UK, p 245

Wong KKY, Kibblewhite RP, Signal FA (1999a) Effect of xylanase and dosage on the refining properties of unbleached softwood kraft pulp. J Wood Chem Tech 19(3):203

Wong KKY, Anderson KB, Kibblewhite RP (1999b) Effects of the laccase-mediator system on the handsheet properties of two high kappa kraft pulps. Enzyme Microbial Technol 25:125

Yamaguchi H, Yaguchi T (1996) Fibre beating with enzyme pretreatment. 50th appita annual general conference, technical association of australia and newzealand paper industry, Auckland, NZ, Volume 1, 6–10 May 1996, p 91

Yerkes WD Jr (1968) Process for the digestion of cellulosic material by enzymatic action of *Trametes Suaveolens*. US Patent 3,460,089

Chapter 11
Removal of Shives

11.1 Introduction

For fully bleached pulps, good cleanness appears more important than high brightness. A lack of cleanliness can lead to a number of problems such as dark dots, fish eyes, gravure speckle, preparations, and uneven coating in the final product. The cleanliness level at a particular brightness is the result of the competition for bleaching chemicals between dissolved organic substances pulp fibers and wood particles. There are three main sources of dirt contamination: the wood, process problems, and external origins, for example plastics. More than half of the impurities found in pulp originate from wood (Robitaille 1991). The cleanliness of bleached pulp is affected by coarse particles derived from the cellular tissue of the tree and by extraneous particles. The former group consists of shives, knots, and bark specks. The latter group consists of pitch, fungus hyphae, rust strains, lime, and sand (Annergren and Lindblad 1996). Shive is a particle or fiber bundle large enough, or in enough quantity, to produce a paper and board quality or productivity problems. Normally, this particle has a thickness, or third dimension, that separates if from being benign to being problematic. For example:

- Shives in bleached pulps show up as dirt in paper or board.
- Shives in unbleached pulps reduce print quality, reduce end strength, decrease runnability, and present visual defects.
- Shives in mechanical pulps cause paper machine breaks, offset printer linting, pick outs, coater scratches, visual defects and reduce print quality.

In short, the presence of shives in pulp is problematic to meeting customer expectations and maintaining optimum production costs. In any pulping operation, a practical balance is struck between theoretical desirability and economical feasibility.

The concentration of shives decreases significantly during most of the bleaching operation (Axegard 1980; Axegard and Teder 1976; Germgard and Sjogren 1985). Typically, bleaching to make market pulp removes 95–99% of the shives

P. Bajpai, *Biotechnology for Pulp and Paper Processing*,

DOI 10.1007/978-1-4614-1409-4_11,

Table 11.1 Methods used for improving pulp cleanliness

Method 1. Increase brightness	
	Increase in chemical use and bleaching costs
Method 2. Decrease number of stages	
	Increase in chemical use and bleaching costs
Method 3. Decrease temperature	
	No effect on chemical use if retention time is sufficient; Increase In utility costs
Method 4. Increase (CD) chemicals/decrease D1 and D2	
	Small increase in chemical use; increase AOX
Method 5. Add (CD) chemicals together rather than sequentially	
	Increase in chemical use; increase AOX

Based on Annergren and Lindblad (1996) and Bruun et al. (1993)

(Axegard 1990). Shives themselves do not consume significant amount of bleaching chemicals. However, the difficulties in removing shives by bleaching can increase bleaching chemical use significantly. The shive count of bleached pulp depends on the shive count for the pulp as it enters the bleach plant and the efficiency of shive removal by the bleaching chemicals. The efficiency of shive bleaching depends on the concentration of bleaching chemical overtime, which depends on the initial chemical charge and the temperature. An improved degree of shive removal is obtained by maintaining a high concentration of bleaching chemical for as long as possible. This can be accomplished by, for example, increasing the retention time, which is often impractical, decreasing the rate of reaction, or increasing the concentration of the bleaching chemicals. The methods that are used to do this are listed in Table 11.1 (Annergren and Lindblad 1996; Bruun et al. 1993). Bleaching in short bleaching sequences provides much better cleanliness than bleaching in long sequences. Progress in mill techniques such as whole tree utilization and closing up of the screening department including recycling the screenings aggravate cleanliness problems. Improved bleaching techniques with more extensive use of chlorine dioxide allow prolonged cooking without too much carbohydrate degradation during cooking and bleaching. The resulting pulp has fewer and probably less shives (Annergren and Lindblad 1996).

One of the technologies for improving the bleachability of kraft pulp involves the use of hemicellulase enzymes (Bajpai 2004, 2009; Bajpai and Bajpai 1992, 1997; Grant 1994, 1995; Hancock and Esteghlalian 2007; Lavielle 1993; Senior and Hamilton 1992; Skerker et al. 1991; Tolan 1992a, b; Viikari et al. 1993, 1994). This process, sometimes termed bleach boosting, has been successfully used in pulp mills throughout the world (Lavielle 1993; Viikari et al. 1994). The main enzyme needed to enhance the delignification of kraft pulp is reported to be endo-β-xylanase but enrichment of xylanase with other hemicellulolytic enzymes has been shown to improve the effect of enzymatic treatment (Viikari et al. 1994). These enzymes modify the xylan portion of the hemicellulose in such a way as to increase the efficiency of the subsequent bleaching stages. The effects of enzyme treatment on pulp brightness and delignification have been reported widely (Pedersen et al. 1992; Senior and Hamilton 1992; Skerker et al. 1991; Tolan 1992a, b). Typically, enzyme-treated pulp can be bleached to a target brightness with 15% less oxidative bleaching chemicals than untreated pulp, with no difference in pulp strength (Bajpai and Bajpai 1997; Viikari et al. 1994). These benefits have propelled xylanase enzyme treatment into regular use in many mills. Xylanase enzymes have been also used to increase the efficiency of shive removal by bleaching and encouraging results have been obtained (Tolan et al. 1994).

11.2 Application of Enzymes for Shive Removal

Researchers from Iogen Corporation, Canada used a novel enzyme treatment to improve the efficiency of shive removal by bleaching. They used three xylanase enzymes from different microbial sources (Tolan et al. 1994). One of the enzymes was a product of Iogen Corporation, which was designed to enhance the removal of shives by bleaching and sold under the trade name "Shivex." Other two enzymes were commercial bleach boosting xylanase preparations. The ability of the xylanase enzymes to enhance the shive removal was found to vary greatly. Lack of correlation of shive removal with xylanase dosage was found showing that different shive removal factor (Sf) result from treatment with different enzymes all at the same xylanase dose (Table 11.2). Shivex had the highest degree of enhancement of Sf of

Table 11.2 Effect of different xylanase enzymes on shive removal factor and bleach boosting

Pulp treatment		Brightness (Elrepho)
Shive removal factor (Sf)		
Untreated	25.0	85.5
	35.3	86.6
	63.6	87.7
Xylanase 2	26.2	86.2
	32.3	87.0
	38.9	88.0
Xylanase 3	45.0	86.5
	62.5	87.1
Shivex	37.0	85.2
	46.6	86.3
	77.2	87.2
Kappa factor		
Untreated	0.20	83.5
	0.22	85.4
	0.24	86.5
	0.26	87.7
Xylanase 2	0.19	85.2
	0.21	86.3
	0.23	87.2
Xylanase 3	0.205	86.2
	0.225	87.0
	0.235	88.0
Shivex	0.20	86.4
	0.22	87.1

Based on Tolan et al. (1994)

Table 11.3 Effect of Shivex on shive counts and shive factor in different bleaching stages at varying kappa factor

Bleaching stage	Pulp treatment	D100 kappa factor	Shives	Log Sf (% of initial)
D100	Untreated	0.15	34.0	0.44
		0.17	34.0	0.44
		0.23	32.5	0.50
		0.26	27.5	0.56
	Shivex treated	0.15	35.0	0.44
		0.17	31.0	0.50
		0.23	27.5	0.56
		0.26	23.0	0.62
E	Untreated	0.15	16.7	0.76
		0.17	15.7	0.80
		0.23	10.0	1.00
		0.26	8.0	1.10
	Shivex treated	0.15	14.0	0.83
		0.17	13.3	0.86
		0.23	8.1	1.10
		0.26	6.0	1.13
D100E-D	Untreated	0.15	0.90	1.82
		0.22	0.86	1.91
		0.23	0.70	1.88
		0.24	0.60	2.00
		0.26	0.40	2.03
	Shivex treated	0.15	1.5	2.07
		0.17	1.1	2.07
		0.23	1.3	2.16
		0.24	1.0	2.25
		0.26	0.9	2.50

Based on Tolan et al. (1994)

the three enzymes. Xylanase 2 decreased Sf at a given brightness while xylanase 3 enhanced Sf less than Shivex. The ability of enzymes to enhance shive removal was not related to the degree of bleach boosting of the enzymes (Table 11.2). Shivex showed the highest enhancement of shive removal in spite of the fact that the bleach boosting performances of the three enzymes are similar. The best performance of the Shivex was found at pH 5.2–7.8, and temperature, 40–62°C and more than 1 h treatment. The conditions for shive removal by Shivex differed slightly from those for bleaching enhancement. Shivex did not have any measurable effect on the shive count or particle size in the enzyme treatment itself. The effects of Shivex became evident after subsequent bleaching. Treatment with Shivex decreased the shive count after each of the stages in D100-E-D bleaching sequence (Table 11.3). After the D100 stage for a given chemical dose, the shive count was 10% lower for enzyme-treated pulp than for untreated pulp. This enzyme effect was found to differ

Table 11.4 Effect of Shivex on shive removal factors (Sf)[a]

Bleaching stage	Untreated	Enzyme-treated	Shive benefits
D100	3.1	3.5	1.13
E	3.2	3.5	1.09
D	10	12.5	1.25
Total	100	155	1.55

Based on Tolan et al. (1994)
[a]Kappa factor 0.24

from that observed in bleach boosting by xylanase enzymes, in which there was a little change in the brightness or kappa number of chlorinated pulp. After the extraction stage, the enzyme-treated pulp had 20% fewer shives than the untreated pulp. The benefit of enzyme treatment was greatest after the D1 stage to the extent that the shive count for treated pulp was 50% lower than for untreated pulp. Sf for the bleaching sequence increased from 100 to 155, showing about 55% increase in shive reduction (Tables 11.3 and 11.4). At a given bleached brightness, Shivex treatment was found to result in lower shive count. Enzyme treatment, therefore, helps to remove shives from the pulp beyond the associated gain in the brightness.

11.3 Mechanism of Shive Removal with Xylanase Enzymes

Enzymatic shive removal and bleach boosting are affected through different mechanisms (Tolan et al. 1994). It appears that shive removal is accomplished by an enzyme other than xylanase or that some property of xylanase other than its ability to hydrolyze xylan is important in shive removal. Most bleach boosting enzymes are the mixtures of 6–20 different proteins many of which have not been identified and only one or two of which are xylanase. Preliminary investigation of several enzymes that digest various portions of pulp including α-arabinosidase, mannanase and cellulase failed to identify a single enzyme that is responsible for enhanced shive bleaching. In terms of properties of the xylanase protein, the ability to bind on to shives or to penetrate in to shives could conceivably vary among the enzymes and cause different efficiencies of shive removal. However, there has been no direct evidence of this. The enzymes used for shive removal had molecular weights of 20–22,000 and showed similar binding tendencies on the substrate. Axegard (1990) has identified two mechanisms for shive removal by chlorine dioxide: peeling of fibers from the surface and fragmentation of the particles. The rate controlling step of both these reactions is the diffusion of chemical within the particles. Tolan et al. (1994) have suggested that enzyme acts on the surface of the shives to remove diffusion barriers and thereby increasing the efficiency of shive bleaching.

11.4 Benefits with Enzymes

Mills have following options for using enzyme treatment to decrease shives (Tolan et al. 1994).

Option 1: Maintain the same brightness, decrease shives and bleaching chemicals.
Option 2: Increase brightness and decrease shives.
Option 3: Maintain the same shive factor, decrease brightness and bleaching chemicals.

The choice at a given mill will depend on the mills operating objectives and constraints.

Option 1 is accomplished by running enzyme treatment and adjusting the control of the bleach plant to maintain the same bleached brightness as without enzyme treatment. This option saves bleaching chemicals and increases the shive factor.

Option 2 is accomplished by running enzyme treatment with no changes in the beaching chemical use. The result is higher brightness and the increase in shive factor. This option gives the maximum increase in shive factor.

Option 3 would be of interest to mills that are bleaching to a higher than necessary target brightness to maintain a low shive count. In this case, the shive factor is unchanged but the brightness target is decreased which allows a decrease in kappa factor.

Benefits of shive reduction would often be difficult to observe directly in the short term in an operating mill because mills necessarily run with Sf that allows a large margin of safety and, therefore, usually have few shives in the bleached pulp. On the contrary, treatment with Shivex would allow a mill to take benefits in chemical savings without undue risk of increasing shive counts. Shivex can be used in conjunction with other oxidizing chemicals to decrease the use of chlorine compounds without increasing the tendency to form shives. Most effective means to decrease kappa factor while increasing Sf is to combine Shivex treatment with peroxide reinforcement of the extraction stage (Table 11.5).

Table 11.5 Shive removal in different bleaching sequences

Bleaching sequence	Kappa factor	Shive removal factor
Control (D100ED)	0.26	115
3 kg/tons peroxide (D100(EP)D)	0.235	100
Shivex (SD100ED)	0.235	155
6 kg/tons peroxide [D100(EP)D]	0.20	90
Shivex plus 3 kg/tons peroxide [SD100(EP)D]	0.20	145

Based on Tolan et al. (1994)

11.5 Conclusions

Enzymes can be efficiently used to increase the efficiency of shive removal by bleaching. By treating brownstock with enzymes, mills can increase the degree of shive removal in the subsequent bleaching. Depending upon the shive level in the incoming brownstock and the desired shive level of the bleached pulp, this allows a mill to decrease its actual shive counts or to increase its margin of safety against shives. The increase in shive removal is accompanied by an increased efficiency in the bleaching of pulp. Therefore, mills can decrease chlorine use in bleach plant without compromising on shive counts. The enzyme activity responsible for the enhanced shive removal is not known. The degree of shive removal by the enzyme is not directly related to the enzyme's xylanase activity or bleach boosting effectiveness.

References

Annergren GE, Lindblad PO (1996) Shives/brightness: a problem of bleaching optimization. Tappi J 59(11):95–98

Axegard P (1980) Effect of prebleaching with chlorine and chlorine dioxide on the cleanliness of softwood kraft pulp. Svensk Papperstidn 83(10):284

Axegard P (1990) Bleaching shives and dirt – an overview. In: Paper presented at 1990 bleach plant operations short course held 17–22 June 1990, Hilton Head, SC, pp 27–34

Axegard P, Teder A (1976) Model experiments in bleaching of shives and bark particles. In: 1976 International pulp bleaching conference, 2–6 May 1976, pp 35–43

Bajpai P (2004) Biological bleaching of chemical pulps. In: Stewart GG, Russell I (eds) Critical reviews in biotechnology, vol 24. CRC Press, Boca Raton, pp 1–58

Bajpai P (2009) Xylanases. In: Schaechter M, Lederberg J (eds) Encyclopedia of microbiology, vol 4, 3rd edn. Academic, San Diego, pp 600–612

Bajpai P, Bajpai PK (1992) Biobleaching of kraft pulp. Process Biochem 27:319–325

Bajpai P, Bajpai PK (1997) Realities and trends in enzymatic prebleaching of kraft pulp. Adv Biochem Eng/Biotechnol 57:1–31

Bruun H, Henricson K, Kronlof T (1993) Particles in bleached softwood sulphate pulp and their reduction. Pap Puu 10(3):126

Germgard U, Sjogren B (1985) Ozone prebleaching of a modified cooked and oxygen bleached softwood kraft pulp. Svensk Papp 88(15):R127

Grant R (1994) Enzymes' future looks bright, as range improves and expands. Pulp Paper Int 36(8):20

Grant R (1995) Enzymes help to increase pulp and paper production. Pulp Paper Int 37(8):26

Hancock B, Esteghlalian AR (2007) New technology tools discover enzyme products to enhance pulp bleaching. In: 2007 Engineering, pulping and environmental conference, Jacksonville, FL, 21–24 Oct 2007, 4pp

Lavielle P (1993) Xylanase prebleaching. Asia Pac Papermaker 3(5):29

Pedersen LS, Choma PP, Holm HC, Kihlgren P, Munk N, Nissen AM (1992) Enzymatic bleach boosting of kraft pulp: laboratory and mill scale experiences. In: Paper presented at 1992 pulping conference, Boston, MA, Book 1, 1–5 Nov 1992, pp 31–37

Robitaille MA (1991) Policing dirt and plastic contamination Bleach plant operations short course notes. Tappi Press, Atlanta, p 39

Senior DJ, Hamilton J (1992) Use of xylanases to decrease the formation of AOX in kraft pulp bleaching. J Pulp Paper Sci 18(5):J165–J169
Skerker PS, Farrell RL, Chang HM (1991) Chlorine free bleaching with Cartazyme HS treatment. In: Proceedings of the international pulp bleaching conference, Stockholm, vol 2, pp 93–105
Tolan JS (1992a) The use of enzymes to decrease the chlorine requirements in pulp bleaching. In: 78th CPPA annual meeting, preprints, Montreal, pp A163–A168
Tolan JS (1992b) The use of enzymes to enhance pulp bleaching. In: Proceedings of Tappi pulping conference, Boston, MA, 6–10 Nov 1992, pp 13–17
Tolan JS, Guenette M, Thebault L, Winstanley C (1994) The use of a novel enzyme treatment to improve the efficiency of shive removal by bleaching. Pulp Pap Canada 95(12):T488
Viikari L, Tenkanen M, Buchert J, Ratto M, Bailey M, Siika-aho M, Linko M (1993) Hemicellulases for industrial applications. In: Saddler JN (ed) Bioconversion of forest and agricultural plant residues. CAB International, Wallingford
Viikari L, Kantelinen A, Sundquist J, Linko M (1994) Xylanases in bleaching: from an idea to the industry. FEMS Microbiol Rev 13:335–351

Chapter 12
Production of Dissolving-Grade Pulp

12.1 Introduction

Dissolving pulp is a high-grade cellulose pulp, with low contents of hemicellulose, lignin, and resin. This pulp has special properties, such as a high level of brightness and uniform molecular weight distribution. It is used to make products that include rayon and acetate textile fibers, cellophane, photographic film, and various chemical additives (Hinck et al. 1985) (Table 12.1). To a large extent, use of dissolving wood pulp depends on its purity (cellulose content), which depends mainly on the production process. To obtain products of high quality, these so-called "special" pulps must fulfill certain requirements, such as high cellulose content, low hemicellulose content, a uniform molecular weight distribution, and high cellulose reactivity. Most, if not all, of the commercial dissolving pulps accomplish these demands to a certain extent. Nevertheless, achieving high cellulose accessibility as well as solvent and reagent reactivity is not an easy task due to the compact and complex structure presented by the cellulose.

About 77% of all dissolving pulp is used in the manufacture of cellulosic fibers (rayon and acetate). These include (a) viscose rayon staple and filament yarn used mainly for textiles, tire cords, and various industrial products and (b) acetate staple and filament yarn used for textiles and acetate fiber (tow) for cigarette filters. Two basic processes are used to produce dissolving pulp (Hiett 1985). The sulfite process produces (sulfite) pulp with a cellulose content up to 92%. It can use ammonium, calcium, magnesium, or sodium as a base. The prehydrolysis sulfate process produces (sulfate) pulp with a cellulose content up to 96%. Special alkaline purification treatments can yield even higher cellulose levels – up to 96% for the sulfite process and up to 98% for the sulfate process (Hiett 1985). The 90–92% cellulose-content sulfite pulps are used mostly to make viscose rayon for textiles and cellophane. The 96–98% cellulose-content sulfate pulps are used to make rayon yarn for industrial products such as tire cord, rayon staple for high-quality fabrics, and various acetate and other specialty products (Hiett 1985). Most producers worldwide make dissolving pulp using softwood fiber and the acid–sulfite and sulfate process. The acid–sulfite process

P. Bajpai, *Biotechnology for Pulp and Paper Processing*,
DOI 10.1007/978-1-4614-1409-4_12,

Table 12.1 Derivatives and end-use products from dissolving pulp

Derivative	Total pulp use	End-use product (%)
Cellulosic fibers Viscose rayon staple		
Regular	Apparel fabric	42
High-wet modulus	Special fabric for apparel, furnishings	3
Viscose rayon filament yarn		
Regular tenacity	Apparel	10
High tenacity	Tire cord and belting, industrial uses	7
Acetate staple and tow	Cigarette filters	8
Acetate filament yarn	Apparel, furnishings	7
Others from viscose rayon		
Cellophane	Packaging	7
Sponges, sausage casings		1
Acetate plastics	Photographic films, sheets, moldings	1
Cellulose nitrates	Lacquers, film, explosives	8
Other cellulosic organic compounds	Additives in food, cosmetics	8
Special paper pulps	Filter, photographic papers	2

Based on data from Hiett (1985); Hinck et al. (1985)

is the most common that covers approximately 65% of the total dissolving pulp production and the prehydrolysis kraft process, which covers approximately 25% (Sixta 2006). Benefits of acid–sulfite process include high recovery rates of the inorganic cooking chemicals and the totally chlorine-free (TCF) bleaching. One disadvantage is that it results in pulps with a broad molecular weight distribution of cellulose (Sixta et al. 2004). Aside from the conventional pulping process, organosolv pulping has been investigated for the production of dissolving pulps because it offers several advantages and seems to be a more environmentally friendly alternative to conventional processes (Kordsachia et al. 2004; Sixta et al. 2004; Fink et al. 2004; Vila et al. 2004). Organosolv pulping can be applied in small plants, different wood species can be utilized, and one of the most important benefits is the absence of sulfur-containing chemicals (Vila et al. 2004). In addition, during the production of viscose, they exhibit superior strength than those obtained from sulfite pulps (Sixta et al. 2004). However, the main drawback is that this process is based on the use of organic solvents, and the expense of solvent recovery is large. Furthermore, it has been reported that some of the organosolv pulping reagents, Milox (peroxyformic acid) and Acetosolv (acetic acid), are not suitable for the production of viscose because of the low reactivity exhibited by these pulps, which is reflected in their structural inhomogeneity and the formation of condensation products (Sixta et al. 2004). Comparisons of the pulps produced by these different methods have been presented (Fink et al. 2004; Sixta et al. 2004).

To manufacture dissolving-grade pulps, removing hemicelluloses from the wood fiber is crucial. For instance, hemicelluloses can affect the filterability of viscose, the xanthation of cellulose, and the strength of the end product during the production of viscose (Croon et al. 1968; Christov and Prior 1993). Dissolving-grade pulp suitable for cellulose ester manufacture for fiber and film applications should contain about 97–98.5 weight percent cellulose, not more than about 3 weight percent xylans, and not more than about 0.5 weight percent mannans. Hemicelluloses are removed during the cooking of wood and the subsequent bleaching (Gamerith and Strutzenberger 1992). In sulfite pulping, the acidic conditions used are responsible for removing most of the hemicellulose while in sulfate process usually a prehydrolysis step is required to remove hemicelluloses. However, part of the hemicelluloses, mainly xylan, remain in the pulp after bleaching thus causing problems at the latter alkalization and spinning stages of viscose process. Therefore, a more complete removal of residual xylan from dissolving pulp could facilitate the following steps in rayon manufacture and improve the final quality of the product.

An alternative method to remove hemicelluloses would be by treating pulps with enzymes which react only with the hemicellulose portion of the pulp (Jeffries 1992). In last two decades, xylan degrading enzymes have been explored, for selective removal of pentosans in preparing dissolving-grade pulp (Bajpai 1997; Bernier et al. 1983; Jeffries and Lins 1990; Jeffries 1992; Paice and Jurasek 1984; Roberts et al. 1990; Senior et al. 1988; Jurasek and Paice 1988; Biely 1985; Buchert et al. 1994; Myburgh et al. 1991). Most research effort has been focused on xylanase treatment of kraft pulps as the kraft process is the dominant form of chemical pulping. The application of xylanases to sulfite pulps has received scant attention (Christov and Prior 1993, 1994; Christov et al. 1995). Enzymes have been used to enhance the cellulose accessibility and reactivity of commercial dissolving-grade pulps (Kvarnlof 2005; Ibarra et al. 2010).

12.2 Enzymes Used in the Production of Dissolving Pulp

Hemicellulose-degrading enzymes, mainly xylanases have been used for selective removal of pentosans in preparing dissolving-grade pulp. These enzymes are produced by many bacteria and fungi (Viikari et al. 1993, 1994). In the earlier studies (Viikari et al. 1986, 1987), crude culture filtrates of hemicellulases produced by different microorganisms were used and most of the reports published are based on results with unpurified enzymes. The culture filtrates used, however, contained xylanases as the main activity and the enzyme preparations were generally dosed according to the xylanase activity. Thus, when small doses were used, the amounts of other activities, mainly other hemicellulolytic and some cellulolytic enzymes were very low. Even with the unpurified enzymes, identification of the sugars released in the enzymatic treatments confirmed that xylanase was the major activity (Viikari et al. 1987, 1990). Some difference in the amount and composition of liberated sugars were observed when hemicellulase-rich culture filtrates (same xylanase dose) from different microorganisms were used (Viikari et al. 1994). The increase in reducing sugars can be influenced by the α-xylosidase activity in the enzyme preparation used, but it can reflect the presence of other hemicellulases or cellulase activities in the enzyme preparation used. The production of dissolving pulp requires the use of enzyme preparation that has little activity against amorphous cellulose. Cellulases have been considered to be detrimental in pulp treatments, if present in hemicellulase preparations (Viikari et al. 1994). This is especially true, when both endoglucanase and cellobiohydrolase are present in crude enzyme preparations. Due to the synergistic action of cellulolytic enzymes, a rapid depolymerization of cellulose occurs.

12.3 Application of Enzymes in Production of Dissolving Pulp

Xylanases from different microorganisms have been explored for selective removal of pentosans in preparing dissolving-grade pulp. Some xylanases can remove xylan from pulps without affecting other components. The purified cellulose can then be used in making dissolving pulps. Paice and Jurasek (1984) used xylanase enzyme from *Schizophyllum commune* for removal of hemicellulose from delignified mechanical aspen pulp. Hydrolysis of the pulp with crude enzyme gave no significant relative decrease in hemicellulose but poor yield (Table 12.2). On the other hand, treatment with xylanase rich fraction at two different concentrations gave better yields of pulp and the hemicellulose content was lower than the control (Table 12.2). The hemicellulose content decreased from 20.4 to 13.4% in 1 h hydrolysis. After 24 h hydrolysis, the hemicellulose content was further reduced to 9.1%. However, all treatments resulted in lower viscosity pulps, indicating cellulose hydrolysis. The higher specificity with xylanase was confirmed by the HPLC analysis of the sugars produced from the pulps during hydrolysis. The pentose sugars xylobiose and xylose predominated in both xylanase treatments while glucose was found to be the major sugar from the total enzyme hydrolysis. Treatment of low yield chemical pulp (low pentosan) with xylan rich fraction did not result in significant further removal of pentosan (Table 12.3). A target value for dissolving pulp

Table 12.2 Effect of xylanase enzyme from *Schizophyllum commune* on removal of hemicellulose from delignified mechanical aspen pulp

Treatment	Time (h)	Xylanase (U/mL)	Yield (%)	18% NaOH solubility (%)	Pentosans (%)	Viscosity (mPa.s)
Buffer control	1	0	99	23.4	20.4	16.8
Crude total	1	300	67	22.0	17.7	8.4
enzyme	24	300	18	28.0	n.d	n.d
Xylanase fraction	1	300	87	22.3	16.9	10.4
	24	300	74	20.2	12.7	9.2
Xylanase fraction	1	1,600	71	18.2	13.4	8.7
	24	1,600	65	12.9	9.1	6.4

Based on data from Paice and Jurasek (1984)

Table 12.3 Effect of xylanase enzyme from *S. commune* on pentosan content and viscosity of chemical pulp

		Pentosan content (%)		Viscosity (mPa.s)	
Enzyme	Time (h)	Partially bleached	Fully bleached	Partially bleached	Fully bleached
Total enzyme	0	7.8	4.6	13.8	13.6
Total enzyme	1	7.5	4.4	11.6	9.7
Total enzyme	4	7.3	4.3	10.7	8.6
Xylanase fraction	4	7.2	4.0	11.8	9.2
Total enzyme	24	7.0	4.0	10.4	8.5
Xylanase fraction	24	6.4	3.5	9.9	8.4

Based on data from Paice and Jurasek (1984)

Table 12.4 Effect of xylanase enzyme from *Escherichia coli* on pentosan removal from dissolving pulp

Treatment	Pentosans (%)	Viscosity (mPa.s)	Yield (%)
Control	4.6	11.4	100
Xylanase from *E. coli*	3.7	12.0	97

Based on data from Bernier et al. (1983)
Bleaching sequence – CED
Bleached pulp at 3% consistency was treated with 300 U/mL enzyme for 24 h at 40°C, pH 5.0

Table 12.5 Effect of successive xylanase treatments[a] from *Saccharomonospora virdis* for selective removal of xylan from bleached birchwood kraft pulp

	Xylan solubilization	
Treatment cycles	Xylose (mg equivalent)	Percent of initial xylan (100%)
1	360	12.5
2	486	16.9
3	576	20.0
4	576	20.0

Based on data from Roberts et al. (1990)
[a]Enzyme treatment for 24 h

(2%) could not be obtained. Viscosities of the treated pulps decreased markedly as a result of traces of cellulase present in the enzymatic preparation. The α-cellulose content also dropped. These studies revealed that xylanase preparation of high purity must be used.

Bernier et al. (1983) undertook to construct a specific xylanase producing organism by genetic engineering. A *Bacillus subtilis* strain producing both cellulase and xylanase was chosen as a source of the enzyme. The xylanase gene was transferred to an *Escherichia coli* strain which normally does not produce either cellulase or xylanase. As a result, a new strain of *E. coli* was obtained which was an exclusive xylanase producer. Experiments with xylanase produced from this new microorganism showed that a partial xylan removal from dissolving pulp can be obtained without sacrificing pulp viscosity (Table 12.4).

Roberts et al. (1990) used enzyme preparation from thermophilic actinomycete *Saccharomonospora virdis* for selective removal of xylan from bleached birchwood kraft pulp. Successive treatments with the enzyme resulted in the removal of 20% of the total xylan content, as estimated by reducing sugar determinations (Table 12.5).

Senior et al. (1988) treated bleached and unbleached kraft pulp with xylanase from *Trichoderma harzianum* and tested for the liberation of oligosaccharides of glucose and xylose. Only xylose oligosaccharides were released from both pulps, with xylobiose detected as the predominant sugar. The greater solubilization of the bleached pulp after a 24 h treatment probably reflected the increased accessibility of the residual xylan, which resulted from the prior bleaching and extraction steps. The subsequent 24 h enzyme treatments of the unbleached pulp resulted in the progressive solubilization of the xylan component with about 25% of the original xylan released after three

Table 12.6 Effect of xylanase from *Trichoderma harzianum* on xylan content of unbleached and bleached kraft pulps

Pulp/enzyme treatment	Xylan remaining in pulp (%)	Xylan removed (%)
Unbleached kraft		
Control	17.0	0
First 24 h hydrolysis	15.0	11.8
Second 24 h hydrolysis	14.4	15.3
Third 24 h hydrolysis	12.7	25.3
Bleached kraft		
Control	21.0	0
First 24 h hydrolysis	11.4	54.3

Based on data from Senior et al. (1988)

24 h treatments. More than 54% of the original xylan present in the bleached pulp was released after one 24 h treatment (Table 12.6). These results were similar to those reported by Chauvet et al. (1987). They also found from analysis of neutral sugar products that only xylan-derived sugars were solubilized. The accessibility of the substrate to the enzyme appears to be a major problem as prolonged incubation was not as effective as a series of short treatments during which the pulp was water extracted, consequently providing better access to the remaining xylan.

Jeffries and Lins (1990) investigated the use of *Aureobasidium pullulans* xylanase to selectively remove xylan from aspen kraft and thermomechanical pulps (TMP). *A. pullulans* secretes a highly active, specific, extracellular xylanase that exhibits little or no glucanase activity. About 33% removal of xylan was achieved. Dilute alkali treatment was found to have major effect on the susceptibility of TMP to xylanase. The enzyme was found to be stable under the conditions employed and only small amounts were necessary to remove significant amounts of sugar from TMP treated with dilute alkali.

Christov and Prior (1994) used enzymes from *A. pullulans* in removing pentosans from bleached sulfite dissolving-grade pulp. The release of reducing sugars at varying dose levels of enzyme was seen. The amount of reducing sugars in the hydrolysate increased with the increase of incubation time and enzyme loadings. The best result was obtained with 1,500 U xylanase/g for 24 h (about 10 mg reducing sugars/g pulp). Low enzyme dosages (15 U/g) and buffer alone failed to release reducing sugars from dissolving pulp. The main degradation product detected in the enzyme hydrolysates was xylose (Table 12.7). Xylose accumulated in the incubation mixture gradually with time. The xylosidase activity of the enzyme preparation was sufficient enough (ratio of xylanase to xylosidase activity 20:1) to hydrolyze the major part of xyloligomers, produced by the xylanase action to xylose. The pentosan content of enzyme-treated samples was reduced simultaneously with the progress of hydrolysis time and enzyme charges (Table 12.7). Buffer alone slightly decreased pentosans in pulp. The most pronounced effect was observed after treatment with 1,500 U/g for 24 h (about 31% reduction).

Table 12.7 Effect of xylanase enzyme from *Aureobasidium pullulans* on pentosans from bleached sulfite dissolving-grade pulp

Reaction time (h)	Xylanase dose (U/g)	Pentosans (%w/w)	Xylose (mg/mL)
0		2.62	0.00
1	0	2.46	
	15	2.36	
	150	2.30	0.011
	450	2.18	
	1,500	2.12	
3	0	2.44	
	15	2.27	
	150	2.19	0.015
	450	2.13	
	1,500	2.08	
5	0	2.44	
	15	2.26	
	150	2.16	0.021
	450	2.07	
	1,500	2.01	
24	0	2.43	
	15	2.24	
	150	2.11	0.045
	450	1.95	
	1,500	1.81	

Based on data from Christov and Prior (1994)

These researchers also investigated the removal of pentosans from unbleached sulfite pulps with crude enzyme of *A. pullulans* (Christov and Prior 1993). The crude enzyme preparation contained 61 U/mL xylanase activity and 3 U/mL xylosidase activity at a protein concentration of 0.25 mg/mL. Increasing dosages and times of treatment resulted in greater amounts of reducing sugars being released. A maximum of 12.8 mg reducing sugars/g pulp was reached when 1,500 U/g xylanase activity was applied for 24 h. The proportional release of reducing sugars with time and enzyme loadings indicated that no inhibition of enzyme hydrolysis by chemicals remaining in pulp after delignification apparently occurred. Enzyme treatment, when carried out for shorter periods (1 and 5 h) was more successful at 50°C than at 40°C. On the other hand, from 30 to 70% more reducing sugars were released at 40°C that at 50°C when treated for longer periods (10–24 h). Residual xylanase activity in the pulp slurry after treatment at 50°C was from 5 to 30% lower than that at 40°C. It decreased also with the progress of hydrolysis time and in some instances less than 50% activity remained after 24 h. Statistical analysis of results from pulp treatments showed that the treatment type X (enzyme treatment) and XE (enzyme treatment and alkaline extraction) and the effect of enzyme charges of pulp were both highly significant. However, there was no significant difference between mean effect at 40 and 50°C. The pentosan content significantly decreased with the duration of treatment and enzyme loading (Table 12.8). From 5 to 28% of pentosans was removed by

Table 12.8 Effect of xylanase enzyme from *A. pullulans* on properties of unbleached sulfite pulps

Reaction time (h)	Enzyme (U/g)	Pentosans (%w/v)		Viscosity (mPa.s)	
		Enzyme	Enzyme alkali	Enzyme	Enzyme alkali
1	0	4.1	3.8	14.4	14.6
	1.5	4.1	3.8	14.4	14.7
	15	4.0	3.7	14.6	14.8
	45	4.0	3.5	14.5	14.7
	150	3.8	3.4	14.4	14.6
	450	3.7	3.3	14.1	14.4
5	0	4.1	3.7	14.4	14.6
	1.5	4.0	3.7	14.4	14.8
	15	3.9	3.5	14.6	14.9
	45	3.7	3.4	14.5	14.7
	150	3.6	3.3	14.3	14.5
	450	3.6	3.3	14.0	14.4
10	0	4.1	3.7	14.4	14.6
	1.5	3.9	3.5	14.4	14.8
	15	3.8	3.4	14.6	14.9
	45	3.8	3.4	14.5	14.8
	150	3.5	3.3	14.2	14.5
	450	3.3	3.1	13.8	14.2
24	0	4.1	3.7	14.4	14.6
	1.5	3.8	3.4	14.6	14.8
	15	3.6	3.2	14.8	15.0
	45	3.7	3.1	14.7	14.9
	150	3.3	2.9	14.2	14.4
	450	3.1	2.8	13.8	14.2

Based on data from Christov and Prior (1994)

enzyme treatments only, but after alkaline extraction, a further 6–13% was released. By treatment with 450 U/g for 24 h followed by alkaline extraction, 35% of the pentosan was removed. Kappa numbers of xylanase treated pulps were not significantly influenced by the hydrolysis time but increased with the enzyme dosages, especially if 45 U/g or more had been used. The kappa number of pulp treated with 450 U/g for 1, 5, or 10 h was 8% higher than the untreated control (4.6) and 25% higher than the blank when treated only with buffer. However, after alkaline extraction, kappa numbers were significantly lowered by 15–30% resulting in pulps with kappa numbers at least one unit lower than the control of 4.6. Viscosity of unbleached untreated pulps (14.4 mPa.S) was hardly affected or slightly improved (Table 12.8) by relatively low enzyme dosages (up to 45 U/g), whereas a drop of 0.2 and 0.6 mPa.S was observed with pulps treated, respectively, by 150 and 450 U/g for longer periods (10 and 24 h). However, enzyme–alkali-treated pulps had slightly improved viscosities compared to pulps obtained after enzyme treatment alone. In all enzymatic hydrolysates, xylose was detected as the main degradation product and its concentration increased with the length of treatment and xylanase dosages. Xylobiose and

Table 12.9 Effect of xylanase enzyme from *A. pullulans* on properties of sulfite pulp

Pulp treatment	Kappa number (% w/w)	α-cellulose (% w/w)	Pentosan	Brightness (% ISO)
Control	6.7	89.9	3.9	56.0
Xylanase treated	6.5	90.7	3.4	56.5

Based on data from Christov et al. (1995)

Table 12.10 Bleaching of sulfite pulp with *A. pullulans* xylanase and reduced amount of active chlorine in OD1EOD2H sequence

Pulp treatment	% Total active chlorine reduction	α-cellulose content	Pentosan (%w/w)	Brightness (%ISO)
Control	0	91.6	2.4	87.5
Xylanase treated	0	92.4	1.9	88.5
	13	92.2	1.7	88.4
	23	92.2	1.8	89.6
	37	92.1	1.8	88.3

Based on data from Christov et al. (1995)

xylotriose also were detected after prolonged incubation (10 and 24 h) with enzyme loadings of 45 U/g or more. In the alkaline extracts, the major component found was xylobiose, accompanied by other longer chain oligosaccharides. However, monosugar analysis confirmed that xylose was the main hydrolysis product of alkaline extracts. Scanning electron micrographs indicated improved fibrillation and flexibility of the fiber structure by enzyme treatment. No cut fragments typical of cellulose degradation were seen. Untreated sulfite pulp fibers were observed to be narrow, separated with relatively uniform length and stiff with moderate fibrillation.

Christov et al. (1995) used xylanases from *A. pullulans* in biobleaching of sulfite pulps for dissolving pulp production. The pentosan content was lowered by 13% and α-cellulose enriched by 1 point, respectively after 1 h enzyme treatment with 15 IU xylanase/g pulp. At the same time, the impact of xylanase treatment on brightness and kappa number was insignificant (Table 12.9). Bleaching of sulfite pulp with *A. pullulans* xylanases and chemicals in sequence OD1E0D2H resulted in production of a dissolving pulp having the highest α-cellulose content (92.1–92.4%) as related to control (91.6%) (Table 12.10). This is indicative for the high selectivity of xylanolytic enzymes of *A. pullulans* in hydrolyzing only the hemicelluloses mostly xylan in the pulp as well as for the lack of cellulase activities in the crude enzyme preparation. The enzyme pretreatment of pulp contributed about 72% to the total release of reducing sugars during bleaching with the latter being approximately four times greater than that of control. Pentosans of xylanase pretreated dissolving pulp were reduced by 25–29% when pulp was bleached with 13–37% less active chlorine than the control of enzymatically untreated dissolving pulp (Table 12.10). Therefore, the amount of pentosans removed from unbleached pulp by the xylanase treatment (13%, Table 12.9) was doubled after bleaching (Table 12.10). Brightness was also improved over the control by 1–2 points.

Table 12.11 Properties of pulp before and after treatment with *A. pullulans* hemicellulases and alkali

Pulp treatment	Reducing sugars	Pentosans (%)	Kappa number	Brightness (%)
X	1.8 (0.0)	3.8 (3.9)	6.0 (6.1)	57.8 (57.4)
E	1.1	3.4	4.2	57.9
XE	3.2 (1.3)	3.0 (3.4)	4.3 (4.3)	60.7 (60.0)
XEX	4.3 (1.3)	2.9 (3.4)	4.3 (4.2)	63.9 (63.0)
XEXE	5.1 (2.1)	2.5 (2.9)	3.8 (3.7)	66.9 (65.7)
EX	2.7 (1.1)	3.2 (3.4)	4.4 (4.4)	60.6 (59.9)
EXE	3.7 (2.1)	2.8 (2.9)	3.8 (3.8)	65.6 (64.9)
EXEX	4.6 (2.1)	2.6 (2.9)	3.8 (3.7)	66.4 (65.5)
Unbleached pulp (blank)	–	3.9	6.5	5.6

Based on data from Christov and Prior (1996)
X-Enzyme treatment (15 U xylanase/g pulp; pH 4.5; 1 h; 55°C; 9% pulp consistency)
E-Alkaline extraction (1.5 g NaOH/100 g pulp; 1 h; 100°C; 9% pulp consistency)
Data in parentheses represent results of control determinations in which the enzyme treatment step (X) was carried out using only buffer and no enzymes

Christov et al. (1996) reported that pretreatment of sulfite pulp with *A. pullulans* xylanases could improve alkali solubility and brightness, important parameters of dissolving pulp for producing viscose rayon. On the other hand, biobleaching with the white-rot fungus *C. subvermispora* could enhance significantly the brightness, however, affecting the cellulose content of dissolving pulp. A combined fungus–enzyme pretreatment of sulfite pulp leads to production of dissolving pulp with 63% less active chlorine and a brightness of over 93% ISO. A TCF bleaching for dissolving pulp production would be feasible once the fungal treatment conditions are optimized in terms of preventing the nonselective degradation of cellulose by *C. subvermispora.*

Biobleaching of acid bisulfite pulp with *A. pullulans* xylanases and reduced amounts of chlorine dioxide (up to 50% as active chlorine) produced dissolving pulp with improved properties such as alkali solubility (S_{10} and S_{18}) and brightness over the conventionally ECF-bleached control (Christov and Prior 1996).

Christov and Prior (1996) also studied repeated consecutive *A. pullulans* hemicellulases and alkali–oxygen treatments in a TCF bleaching sequence for their effects on pulp characteristics related to hemicellulose and lignin content of the pulp. The treatment sequence of hemicelluloses and alkali produced different effects on sulfite pulps. The combination of treatments, where the enzymes were applied first, followed by extraction (XE), solubilized more pentosans from pulp than the combination beginning with alkali (EX) (Table 12.11). Pentosans were further reduced using the treatment sequence XEXE by increasing the xylanase loadings from 15 to 120 U/g pulp. The amounts of reducing sugars released by the treatment sequence XO and XOXO were slightly less than those released by the XE and XEXE combinations. The triple treated xylanase and oxygen treatments of sulfite pulps resulted in significant modifications of pulp properties. Pentosan content of pulp was reduced twofolds, the K number decreased by 60%, brightness was enhanced by 18 points, and α-cellulose content was enriched by 3 points.

Monocarbohydrate analysis indicated that treated pulp contained less xylan (48%) and glucomannan (15%), than the untreated reference of dissolving pulp.

Extensive research has been undertaken into the use of xylanolytic enzymes in bleaching of dissolving pulp and a process developed for bleaching of dissolving pulp using an enzymatic pretreatment with xylanase enzyme (Bajpai and Bajpai 2001; Bajpai et al. 2005). This process involves using mild prehydrolysis stage to retain relatively higher pentosans in the unbleached pulp. The higher pentosan content results in higher unbleached pulp yield at the same kappa number. Results of bleaching of enzyme-treated pulp in a CEHED sequence indicated reductions in bleach chemical requirement, together with improved brightness of the final bleached pulp, desired levels of pentosans and a considerable increase in bleached pulp yield. The rayon yield and reactivity were also higher for the enzyme-treated pulp.

Jackson et al. (1998) reported that alkali extraction of dissolving pulps made from bleached hardwood kraft fiber and recycled paper rich in hardwood fiber, followed by xylanase treatment and a second alkali extraction, provided a reduction in the hemicellulose content of the pulp, acceptable viscosity, and alkali solubility.

Kopcke et al. (2010a) studied the feasibility of converting different kraft pulps from wood and nonwood pulps into commercial dissolving-grade pulps. Several sequences of treatments were examined based on enzymatic treatments and alkali extraction steps and the process parameters were optimized in order to improve the cellulose reactivity, hemicellulose content, pulp viscosity, and molecular weight distribution. Commercial dried elemental chlorine free (ECF)-bleached kraft pulps from birch and eucalypt were investigated, together with several dried ECF-bleached soda/anthraquinone pulps from flax, hemp, sisal, abaca, and jute. It was found that birch, eucalypt, and sisal pulps were suitable for use as dissolving-grade pulps. The xylanase treatment and the alkali extraction stages reduced the amount of hemicelluloses, reaching values to those of a commercial dissolving-grade pulp. Moreover, it was observed that the endoglucanase treatment was the main contributor to the enhancement of the cellulose reactivity. On the other hand, the pulp viscosity exhibited values slightly lower than those of a commercial dissolving pulp, but suitable for the production of viscose. The pulps also contained only cellulose I, while the molecular weight distribution was more uniform than that of a commercial dissolving-grade pulp.

Enzymatic treatments, especially using cellulases, hold great potential for increasing cellulose reactivity in dissolving pulps (Rahkamo et al. 1988, 1996, 1998; Cao and Tan 2002, 2006; Henriksson et al. 2005; Engström et al. 2006). Cellulases are enzymes that hydrolyze the 1,4-α-D-glucosidic bonds of the cellulose chain. For hydrolysis to take place, cellulases must bind to the surface of the substrate (i.e., cellulose) (Zhang and Lynd 2004). There are three major groups of cellulases: endoglucanases (EC. 3.2.1.4), cellobiohydrolases or exoglucanases (EC. 3.2.1.91), and glucosidases (EC. 3.2.1.21) as illustrated in Fig. 12.1. These enzymes can act on the cellulose chain either alone or in combination. When they act together, synergy is often generated, resulting in the efficient degradation of the cellulose structure (Fig. 12.2). Endoglucanases (EG) are enzymes that randomly cleave the amorphous sites of cellulose, which creates shorter chains and, therefore, new chain ends.

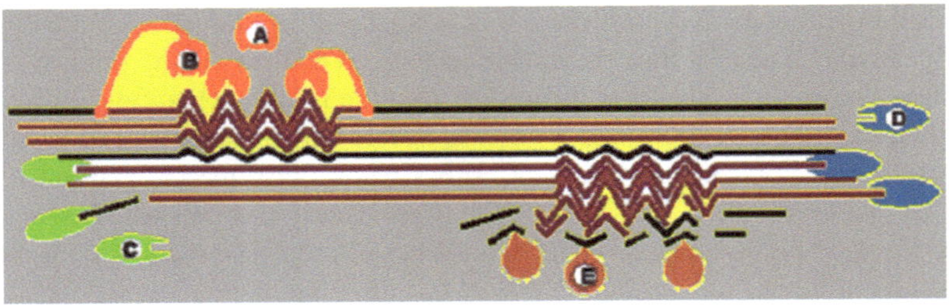

Fig. 12.1 Types of cellulases: (**a**) Endoglucanases without cellulose-binding domain (**b**) endoglucanases with cellulose-binding domain; (**c**, **d**) cellobiohydrolases (**e**) glucosidases. Based on Köpcke (2010b)

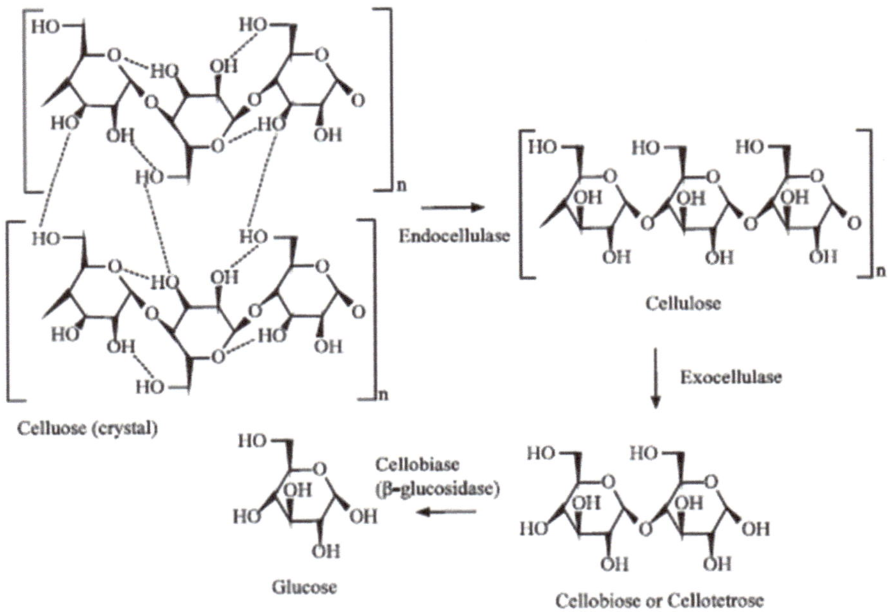

Fig. 12.2 Mode of action of various components of cellulose. Based on Wood and McCrae (1979)

Cellobiohydrolases or exoglucanases (CBH) attack the reducing and nonreducing ends of the cellulose chains, generating mainly cellobiose units. This type of cellulose can also act on microcrystalline cellulose by a peeling mechanism. Glucosidases act on cellobiose, which is two glucose units linked by a 1,4-α-D-glucosidic bond, generating glucose units (Lynd et al. 2002). It has been suggested that there are three primary parameters that affect the degree of enzymatic hydrolysis: the crystallinity, the specific surface area, and the degree of polymerization of the cellulose (Mansfield et al. 1999; Eremeeva et al. 2001). Most cellulases consist of two domains. The first is a catalytic domain, which is responsible for the hydrolysis of the cellulose chain. The catalytic domain of endoglucanases is cleft-shaped Exoglucanases, however, have a tunnel-shaped catalytic domain structure. The second is a cellulose-binding domain (CBD), which helps the enzyme bind to the cellulose chain and bring the catalytic domain close to the substrate. An interdomain linker serves as a connection between the two domains (Rabinovich et al. 2002a, b). The abilities of these enzymes to degrade their substrates are directly related to their size and structure, and their mode of action is at the substrate surfaces, which is the main region where the modifications occur (Mansfield et al. 1999; Zhang and Lynd 2004). Endoglucanases have been demonstrated to significantly enhance the cellulose reactivity of dissolving pulps (Henriksson et al. 2005; Cao and Tan 2006; Engström et al. 2006). Reported that Endoglucanase treatment decreased the viscosity and chain length and increased the reactivity of a pulp made from eucalyptus and acacia. The endoglucanases were more efficient at hydrolyzing the pulp than were cellobiohydrolases at the same protein dosage. The use of cellulases in a two-step process, where the alkali insoluble material was recovered, treated with enzymes, and recombined with the first extraction, resulted in a dissolving pulp that was more soluble in alkali than pulp treated directly with enzymes followed by alkali. Although when compared at the same level of hydrolysis, the fibers from the two-step process were actually slightly lower in solubility.

Ibarra et al. (2010) studied the behavior of different endoglucanases on the accessibility and reactivity of hardwood and softwood dissolving-grade pulps. It was found that endoglucanase with a CBD and an inverting hydrolysis mechanism was the most effective of the different assayed enzymes in increasing the reactivity of both pulps. At the same time, the viscosity decreased, which was more marked for softwood dissolving pulp. In both pulps, a narrower molecular weight distribution with a considerable reduction in the amount of long-chain cellulose molecules was observed and this was more pronounced for softwood dissolving pulps. On the other hand, endoglucanase without a CBD and a retaining hydrolysis mechanism exhibited an inefficient action on the studied properties. When never-dried dissolving pulps were used, it was found that the effects of the different endoglucanase treatments were more pronounced.

Henriksson et al. (2005) and Engström et al. (2006) reported that monocomponent endoglucanase treatment increased the reactivity of softwood sulfite dissolving pulp. The reactivity of the pulp as determined with the Fock method increased drastically with relatively low amounts of enzyme, and the yield loss and decrease of viscosity were moderate.

Kvarnlöf (2005) undertook research study to improve the reactivity of the dissolving pulp in order to reduce the chemical demand in the viscose process, and thus reduce the cost and indirectly the environmental impact. The work showed that it is possible to enhance the pulp reactivity and to use less carbon disulfide in the production of viscose, while maintaining a good quality viscose dope, by two entirely different pretreatment methods, one chemical and one enzymatic. The chemical method used pressurized oxygen after the mercerisation step, which increased the reactivity of the alkali cellulose. The viscose dopes produced from the pressurized oxygen treated alkali cellulose had lower filter clogging values, Kw, compared to conventionally produced viscoses. The temperature and the oxygen treatment time of the alkali cellulose were, however, crucial for the viscose quality. The best performing enzyme of several tested was a cellulase of the monocomponent endoglucanase preparation Carezyme®. This enzymatic treatment was optimized with respect to viscose dope preparation. The study showed that the enzyme treatment could be carried out under industrially interesting conditions with respect to temperature, enzyme dose, and reaction time. A recirculation study of the enzyme showed that it was possible to reuse the spent press water from the enzymatic treatment step several times, and thus lower the production cost. Some of the viscose process stages were modified to properly fit the enzymatically treated dissolving pulp and a comparison between viscose made from enzyme-treated pulp and viscose made from conventional pulp, showed that the enzyme-treated samples had a lower filter clogging value, Kw. This indirectly indicates that the enzyme pretreatment could reduce the carbon disulfide charge in the viscose manufacturing process.

12.4 Conclusions

Attempts to remove hemicellulose, for production of dissolving pulps with very low hemicellulose contents, have shown that complete enzymatic hydrolysis of hemicellulose within the pulp is difficult to achieve. The complete removal of residual hemicellulose seems thus unachievable, due to the modification of the substrate or to structural barrier. It appears that pentosans in the bleached pulp are well shielded by other pulp components and therefore not susceptible to enzymatic attack. Even with very high enzyme loadings and prolonged incubation periods, xylan hydrolysis is limited. The wood species and the method of their pulping, the accessibility of pentosans and their quantity in pulp, the penetration capabilities and substrate specificity of the enzymes, the inhibitory action of bleaching chemicals, and the linkage of xylan to lignin and cellulose by covalent and hydrogen bonds, respectively, may be the factors contributing to the difficulties in removing xylan from the bleached pulp (Gamerith and Strutzenberger 1992; Minor 1986; Vikkari et al. 1990). Xylanase treatment of unbleached pulp appears to be more effective because of the presence at this stage of more hemicellulose accessible for enzymatic degradation. Alkaline extraction in conjunction with enzyme treatment leads to some improvement of the pulp characteristics. Enzymatic treatments hold great potential for increasing cellulose reactivity in dissolving pulps. A monocomponent endoglucanase with a CBD significantly improves the cellulose reactivity.

Little is known about the mechanisms producing the effects of xylanase-cellulase interactions and the role that other essential enzymes may play thereby. There is a need to evaluate the effectiveness of accessory enzymes alone and in combination for production of dissolving pulps.

References

Bajpai P (1997) Microbial xylanolytic enzyme system: properties and applications. Adv Microbiol 43:141–194

Bajpai P, Bajpai PK (2001) Development of a process for the production of dissolving kraft pulp using xylanase enzyme. Appita 54(4):381–384

Bajpai P, Bajpai PK, Varadhan R (2005) Production of dissolving grade pulp with hemicellulase enzyme. In: Proceedings international pulp bleaching conference, Stockholm, Sweden, pp 303–305

Bernier R Jr, Driguez H, Desrochers M (1983) Molecular cloning of a *Bacillus subtilis* xylanase gene in *Escherichia coli*. Gene 26:59–65

Biely P (1985) Microbial xylanolytic systems. Trends Biotechnol 3:286–290

Buchert J, Tenkanen M, Kantelinen A, Viikari L (1994) Application of xylanases in the pulp and paper industry. Bioresource Technol 50:65–72

Cao Y, Tan H (2002) Effects of cellulase on the modification of cellulose. Carbohydr Res 337(14):1291–1296

Cao Y, Tan H (2006) Improvement of alkali solubility of cellulose with enzymatic treatment. Appl Microbiol Biotechnol 70(2):176–182

Chauvet JM, Comtat J, Noe P (1987) Assistance in bleaching of never-dried pulps by the use of xylanases, consequences on pulp properties. In: Symposium on wood and pulping chemistry, Paris, pp 325–327

Christov LP, Prior BA (1993) Xylan removal from dissolving pulp using enzymes of *Aureobasidium pullulans*. Biotechnol Lett 15:1269–1274

Christov LP, Prior BA (1994) Enzymatic prebleaching of sulphite pulps. Appl Microbiol Biotechnol 42:492–498

Christov LP, Akhtar M, Prior BA (1995) Biobleaching in dissolving pulp production. In: Proceedings of the sixth international conference on biotechnology in the pulp and paper industry: advances in applied and fundamental research. Vienna, Austria, pp 625–628

Christov LP, Prior BA (1996) Repeated treatments with *Aureobasidium pullulans* hemicellulases and alkali enhance biobleaching of sulphite pulps. Enz Microbial Technol 18(4):244–250 (1996)

Christov LP, Akhtar M, Prior BA (1996) Impact of xylanase and fungal pretreatment on alkali solubility and brightness of dissolving pulp. Holzforschung 50:579–582

Croon I, Jonsén H, Olofsson HG (1968) Hemicellulose in pulp, viscose and yarn. Svensk Papperstidn 71(2):40–45

Eremeeva T, Bikova T, Eismonte M, Viesturs U, Treimanis A (2001) Fractionation and molecular characteristics of cellulose during enzymatic hydrolysis. Cellulose 8(1):69–79

Engström AC, Ek M, Henriksson G (2006) Improved accessibility and reactivity of dissolving pulp for the viscose process: pretreatment with monocomponent endoglucanase. Biomacromolecules 7(6):2027–2031

Fink HP, Weigel P, Ganster J, Rihm R, Puls J, Sixta H, Parajo JC (2004) Evaluation of new organosolv dissolving pulps. Part II: structure and N; O processability of the pulps. Cellulose 11(1):85–98

Gamerith G, Strutzenberger H (1992) In: Visser J (ed) Xylans and xylanases. Elsevier, Amsterdam, pp 339–348

Henriksson G, Christiernin M, Agnemo R (2005) Monocomponent endoglucanase treatment increases the reactivity of softwood sulphite dissolving pulp. J Ind Microbiol Biotechnol 32(5):211–214
Hiett LA (1985) Dissolving cellulose: its present position and prospects for future development. Tappi J 68(2):42–48
Hinck JF, Casebier RL, Hamilton JK (1985) Dissolving pulp manufacture. In: Ingruber OV, Kocurec MJ, Wong A (eds) Pulp and paper manufacture, vol 4. Joint Text-book Committee of the Paper Industry TAPPI, Atlanta, pp 213–243
Ibarra D, Kopcke V, Ek M (2010) Behavior of different monocomponent endoglucanases on the accessibility and reactivity of dissolving-grade pulps for viscose process. Enzyme Microb Technol 47(7):355–362
Jackson LS, Heitmann JA Jr, Joyce TW (1998) Production of dissolving pulp from recovered paper using enzymes. Tappi J 81:171
Jeffries TW, Lins CW (1990) Effects of Aureobasidium pullulans xylanase on aspen thermomechanical and kraft pulps. In: Kirk TK, Chang H-M (eds) Biotechnology in pulp and paper manufacture. Butterworth-Heinemann, Boston, pp 191–202
Jeffries TW (1992) Enzymatic treatments of pulps. ACS Symp Ser 476:313–329
Jurasek L, Paice MG (1988) Biological treatments of pulps. Biomass 15:103–108
Kopcke V, Ibarra D, Larsson PT, Ek M (2010a) Feasibility study on converting paper-grade pulps to dissolving-grade pulps. In: 11th European workshop on lignocellulosics and pulp, Hamburg, Germany, 16–19 Aug 2010, pp 149–152
Köpcke V (2010b) Conversion of wood and non-wood papergrade pulps to dissolving-grade pulps. Doctoral Thesis, Royal Institute of Technology, Stockholm
Kvarnlöf N (2005) Novel chemical and enzymatic approaches to activate dissolving pulps. University of Karlstad, Karlstad, p 100
Kordsachia O, Rosskopf S, Patt R (2004) Production of spruce dissolvingpulp with the prehydrolysis-alkaline sulfite process (PH-ASA). Lenzinger Berichte 83:24–34
Lynd LR, Weimer PJ, van Zyl WH, Pretorius IS (2002) Microbial cellulose utilization: fundamentals and biotechnology. Microbiol Mol Biol Rev 66(3):506–577
Mansfield SD, Mooney C, Saddler JN (1999) Substrate and enzyme characteristics that limit cellulose hydrolysis. Biotechnol Prog 15(5):804–816
Minor JL (1986) Chemical linkage of polysaccharides to residual lignin in loblolly pine kraft pulps. J Wood Chem Technol 6(2):185–201
Myburgh J, Prior BA, Kilian SG (1991) The temperature and pH properties of the extracellular hemicellulose-degrading enzymes of *Aureobasidium pullulans* NRRL Y 2311-1. Process Biochem 26:343–348
Paice MG, Jurasek L (1984) Removing hemicellulose from pulps by specific enzymic hydrolysis. J Wood Chem Technol 4:187–198
Rabinovich ML, Melnick MS, Bolobova AV (2002a) The structure and mechanism of action of cellulolytic enzymes. Biochemistry 67(8):850–871
Rabinovich ML, Melnick MS, Bolobova AV (2002b) Microbial cellulases (review). Appl Biochem Microbiol 38(4):305–321
Rahkamo L, Siika-Aho M, Vehviläinen M, Dolk M, Viikari L, Nousiainen P, Buchert J (1996) Modification of hardwood dissolving pulp with purified *Trichoderma reesei* cellulases. Cellulose 3(1):153–163
Rahkamo L, Viikari L, Buchert J, Paakkari T, Suortti T (1988) Enzymatic and alkaline treatments of hardwood dissolving pulp. Cellulose 5(2):79–88
Rahkamo L, Siika-Aho M, Viikari L, Leppänen T, Buchert J (1998) Effects of cellulases and hemicellulases on the alkaline solubility of dissolving pulps. Holzforschung 52(6):630–634
Roberts JC, McCarthy AJ, Flynn NJ, Broda P (1990) Modification of paper properties by pretreatment with *Saccharomonospora viridis* xylanase. Enz Microbial Technol 12:210–213
Senior DJ, Mayers PR, Miller D, Sutcliffe R, Tan L, Saddler JN (1988) Selective solubilization of xylan in pulp using a purified xylanase from *Trichoderma harzianum*. Biotechnol Lett 10(12):907–912

Sixta H (2006) Chemical pulping. Handbook of pulp, Wiley-VCH Verlag GMbH & Co, KGaA, Weinheim, pp 3–19
Sixta H, Harms H, Dapia S, Parajo JC, Puls J, Saake B, Fink HP, Röder T (2004) Evaluation of new organosolv dissolving pulps. Part I: preparation, analytical characterization and viscose processability. Cellulose 11(1):73–83
Viikari L, Ranua M, Kantelinen A, Linko M, Sundquist J (1986) Bleaching with enzymes In: Proceedings of third international conference, Stockholm, p 67
Viikari L, Ranua M, Kantelinen A, Linko M, Sundquist J (1987) Application of enzymes in bleaching. In: Proceedings of fourth international symposium on wood and pulping chemistry, Paris, vol 1, p 151
Vikkari L, Kantelinen A, Poutanen K, Ranua M (1990) Characterization of pulps treated with hemicellulolytic enzymes prior to bleaching. In: Kirk TK, Chang HM (eds) Biotechnology in pulp and paper manufacture. Butterworth-Heinemann, Boston, pp 145–151
Viikari L, Tenkanen M, Buchert J, Ratto M, Bailey M, Siika-aho M, Linko M (1993) Hemicellulases for industrial applications. In: Saddler JN (ed) Bioconversion of forest and agricultural plant residues. CAB International, Wallingford, pp 131–182
Viikari L, Kantelinen A, Sundquist J, Linko M (1994) Xylanases in bleaching: from an idea to the industry. FEMS Microbiol Rev 13:335–350
Vila C, Santos V, Parajo JC (2004) Dissolving pulp from TCF bleached acetosolv beech pulp. J Chem Technol Biotechnol 79:1098–1104
Wood TM, McCrae SI (1979) Synergism between enzymes involved in the solubilization of native cellulose. Adv Chem Ser 181:181–209
Zhang YHP, Lynd LR (2004) Toward an aggregated understanding of enzymatic hydrolysis of cellulose: noncomplexed cellulase system. Biotechnol Bioeng 88(7):797–824

Chapter 13
Biological Treatment of Pulp and Paper Mill Effluents

13.1 Introduction

The pulp and paper industry is an important economic activity in several countries around the world although being considered as one of the largest polluters. The pulp and paper making industry is very water intensive and ranks third in terms of water consumption and ranks fifth among the major industries in its contribution to water pollution problems in the USA. The consumption of water varies depending on the type of paper being produced.

Pulping, bleaching, and paper making operations are the three major wastewater sources of the industry. Of the different wastewaters generated by pulp and paper industry, bleach plant effluents are considered to be the most polluting. Bleach effluents are colored and toxic in nature. They contain chlorinated and nonchlorinated products of lignin and extractives of wood. Because of the color, productivity of aquatic ecosystems gets affected when these effluents are discharged into the water bodies. Color also affects the downstream uses (municipal and industrial) of water. It makes the water treatment difficult and costly. Color reduces the visual appeal and recreational value of the water. It retards sunlight transmission, thus, reducing the productivity of the aquatic community by interfering with the photosynthesis. Color imparting substances complex with metal ions, such as iron, or copper, and form tar-like residues. These residues may have direct inhibitory effects on some of the lower organisms in the food chain.

The chlorinated organic compounds and the lignin derivatives of the bleach effluents are recalcitrant and get bioaccumulated along the food chains in the aquatic ecosystems. The low molecular weight fraction of bleach effluent contains potentially problematic (toxic) compounds. These have the ability to penetrate cell membranes and tendency to bioaccumulate. Low molecular weight chlorinated organic compounds significantly affect the biology of aquatic ecosystems. Disappearance of benthic invertebrates, high incidence of fish diseases, and mutagenic effects on the aquatic fauna are some of the consequences of the disposal of bleach effluents into surface waters.

P. Bajpai, *Biotechnology for Pulp and Paper Processing*,
DOI 10.1007/978-1-4614-1409-4_13, © Springer Science+Business Media, LLC 2012

Table 13.1 The effect of various technologies on effluent parameters

					Polychlorinated organics		Acute toxicity
	BOD	COD	TSS	AOX	PCDD/Fs	PCPs	LC_{50}
Modified cooking	+	+	φ	+	φ	+	?
Oxygen delignification	++	++	φ	++	+	++	+
Modified bleaching with small amounts of chlorine	φ	φ	φ	++	+++	+++	++
External treatment							
1°/2° clarification	φ	φ	+++	φ	+	φ	φ
2° biological treatment	+++	+	+	++	+	++	+++

Based on Annergren (1990), Annergren et al. (1990)
+ Weak positive effect
++ Positive effect
+++ Strong positive effect
φ Little or no effect

Increasing awareness of environmental consequences of bleach effluent has led to stringent environmental regulations. Most of the nations have imposed limits on adsorbable organic halides (AOX) of the effluents. In some nations, limits have also been set on individual chlorinated organic compounds of bleach effluents. In response to environmental concerns and environmental regulations, pulp and paper industry has reacted by making process modifications based on existing and new proven technologies. For a bleached kraft mill, a number of alternative technologies are available (Table 13.1).

From these, it is possible to select a combination that can meet the present or future effluent discharge limits. Initially, the effluent requirements varied from country to country for reasons of differing national priorities and this has led to diverse range of technological responses. However, as a result of the concern regarding dioxins and polychlorinated organic materials, and as a result of more stringent regulations, the industry would tend to evaluate and avail the multitude of wide ranging options available.

13.2 Bleaching and Environmental Impact

In kraft pulping, more than 90% of wood lignin gets solubilized during cooking process. The remaining lignin is mainly responsible for the brown color of the kraft pulp and unbleached paper. The basic aim of bleaching is to remove the residual lignin from the pulp as selectively as possible, without degrading the pulp carbohydrate, especially cellulose, which would lead to decrease in strength of the pulp. Pulp bleaching is carried out in a series of steps employing bleach chemicals such as Cl_2, ClO_2, O_3, O_2, H_2O_2, etc. During bleaching, the wood components, mainly lignin, get degraded, heavily modified, chlorinated, and finally, dissolved in the effluent (Dence and Reeve 1990). As a result, the effluent from the bleaching process is dark

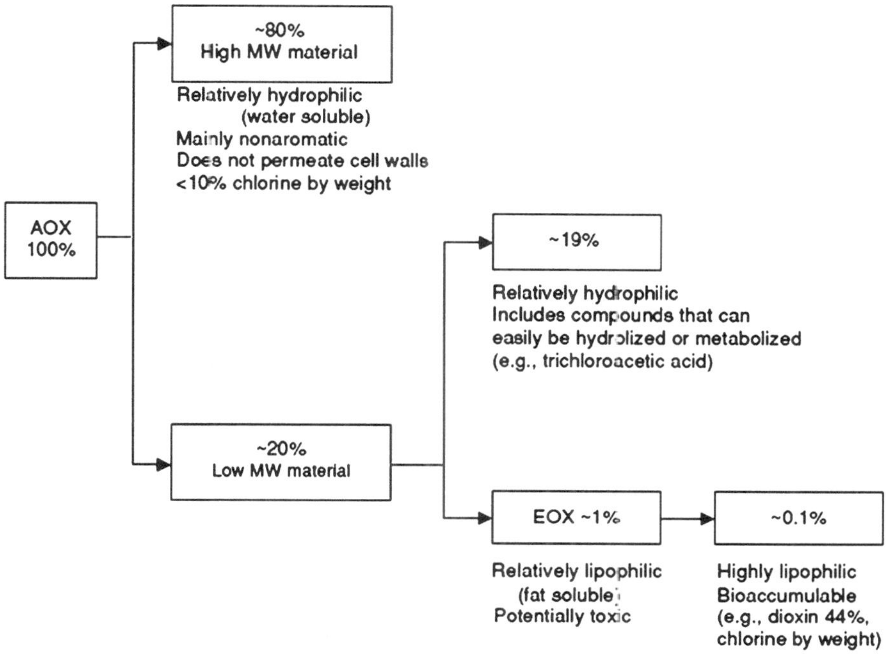

Fig. 13.1 The character of AOX in the effluent from conventionally pulped and bleached kraft pulp. Based on Bajpai and Bajpai (1996) and Gergov et al. (1988)

brown in color due to the presence of chromophoric polymeric lignin derivatives. The amount of chlorinated organics produced, during the pulp bleaching, varies with wood species, kappa number of the pulp, bleaching sequence, and conditions employed. The pollution loads from a hardwood kraft pulp bleach plant are, generally, lower than those from a softwood pulp bleaching plant. Of the total chlorine used in the bleach plant, about 90% form the common salt, and 10% or less gets bound to the organic material removed from the pulp. This organically bound chlorine is also termed as AOX. A physicochemical classification of this chlorinated organic material, present in spent liquors from conventionally pulped and bleached softwood kraft pulp is shown in Fig. 13.1 (Axegard et al. 1993; Gergov et al. 1988; Lindstrom and Mohamed 1988). About 80% or more of the organically bound chlorine corresponds to high molecular weight ($M_W > 1{,}000$) chlorinated lignin material, commonly referred to as chlorolignin. About 20% of the organically bound chlorine correspond to relatively low molecular weight material. This fraction is expected to contain potentially problematic compounds (toxic to aquatic organisms) due to their ability to penetrate cell membranes or their tendency to bioaccumulate in the fatty tissues of higher organisms. Some of the major components of this low molecular weight fraction have been found to consist of relatively water soluble substances, such as chlorinated acetic acids or chlorinated acetone (Gergov et al. 1988; Lindstrom and Mohamed 1988) which are easily broken down before or during biotreatment

Table 13.2 Chlorinated organic compounds in bleach plant effluents

Types of chlorinated compounds	Variations (numbers)
Acids, esters, aldehydes, furans, pyrenes	77
Phenols and phenol ethers	52
Aldehydes and ketones	66
Hydrocarbons	75
Alcohols	25
Dioxins and furans	20
Miscellaneous	15
Total	330

Based on McKague and Carlberg (1996)

and are, therefore, of little environmental significance. The fraction of AOX which is extractable by a nonpolar organic and referred to as EOX, account for about 1–3% of the TOCl. This fraction contains lipophilic neutral organic compounds primarily of low molecular weight and, therefore, of greater environmental significance than the remaining 99% of the AOX material. About 456 different compounds have been identified in the effluents from conventional bleach plants. About 330 of those are chlorinated organic compounds, which include chlorinated phenolics, dioxins, hydrocarbons, and resin acids (Table 13.2) (Mckague and Carlberg 1996). The most common chlorinated phenolics in bleached kraft pulp mill effluents are tri- and tetra chloroguaiacols (Fig. 13.2) (Liebergott et al. 1990).

Bleach kraft mill effluent is a complex mixture of chlorinated and nonchlorinated products of lignin and/or extractives of wood that imparts dark color to the effluent. Bleached kraft mill effluent may have a noticeable effect on the biological quality of the receiving water. Disappearance of benthic invertebrates, such as mussels, and high incidence of fish diseases are some of the effects (Sundelin 1988; Sodergren et al. 1993). Bleached kraft and bleached sulphite mill effluents have been demonstrated to impair the functions of liver, enzyme systems, and metabolic cycles in the exposed fish. Further more, such exposures have been demonstrated to increase the incidence of spinal deformities and reduce gonad development.

Major part of the organically bound chlorine (80%) is believed to be heterogeneous material of relatively high molecular weight compounds. These compounds, apparently, contribute little to the effluent BOD and acute toxicity. Their major contribution is towards color, COD, and chronic toxicity. Ecological/natural processes, such as sedimentation, biodegradation, and bioaccumulation, are apparently correlated to the molecular size and hydrophobicity of the compounds. Highly polar and high molecular mass constituents are responsible for the toxicity of the bleach effluents during early life stages of marine animals and plants (Higachi et al. 1992). Chlorocymenes and chlorocymenenes, of the bleach effluent, have been reported to bioaccumulate in fish and mussels (Suntio et al. 1988). Chlorinated dioxins, which are present in very low concentrations in the bleach plant effluent (usually in ppt levels), account to a ten billionth of the total AOX discharged. About 210 dioxins, belonging to the two families, namely, PCDDs and PCDFs, have been reported in the bleach effluents. 2,3,7,8-TCDF and 2,3,7,8-TCDD are

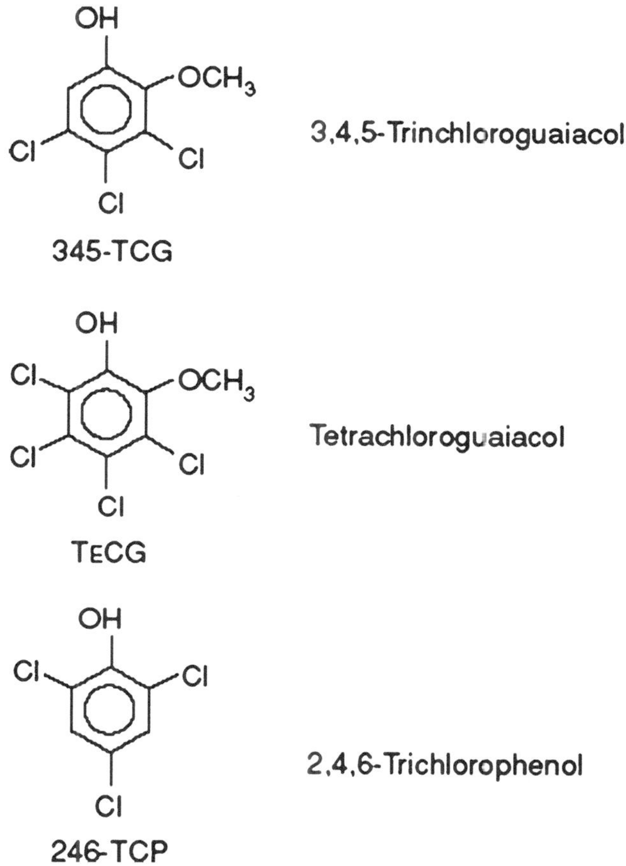

Fig. 13.2 Specific compounds discharged from bleached pulp mills. Based on Gavrilescu (2006); Liebergott et al. (1990)

important from the toxicological point of view. These two chemicals are known to be highly toxic, carcinogenic, and bioaccumulable. Dioxins are almost insoluble in water. They tend to enter the food chains and accumulate in high concentrations in predators, such as fish eating birds (McCubbin 1989; McCubbin et al. 1990). Adverse effects of dioxins have been observed in almost all the species tested. According to an environmental protection agency (EPA) report (Anon 1994), human beings lie somewhere in the middle of the sensitivity range (from extremely responsive to extremely resistant) for dioxins. Even in trace amounts, dioxins may cause a wide range of adverse health conditions, such as disruption of regulatory hormones, reproductive and immune system disorders, and abnormal fetal development (Bajpai and Bajpai 1996, 1997; Bajpai et al. 1999). The structures of the most toxic forms of dioxin and furan molecules are shown in Fig. 13.3 (Rappe and Wagman 1995).

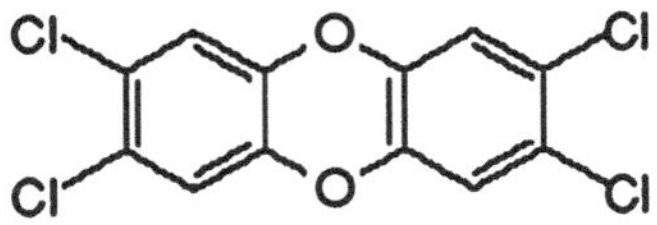

2,3,7,8,-TCDD

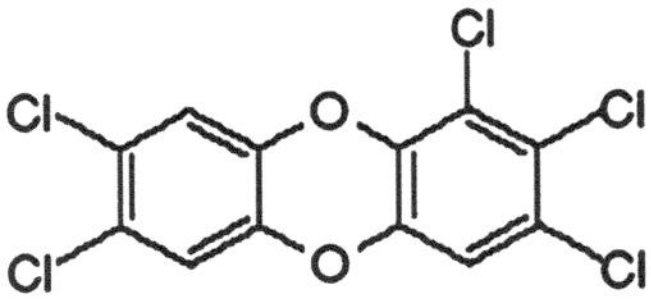

1,2,3,7,8,-pentaCDD

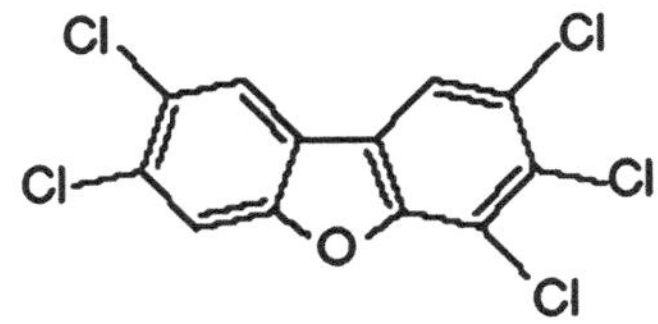

2,3,4,7,8,-pentaCDF

Fig. 13.3 Most toxic isomers of polychlorinated dioxins and furans. Based on Gavrilescu (2006) and Rappe and Wagman (1995)

13.3 Biotechnological Methods for Treatment of Pulp and Paper Mill Effluents

Several methods have been attempted for decolorization and detoxification of pulp and paper mill effluents. These include physicochemical and biotechnological methods (Bajpai and Bajpai 1994, 1996, 1997; Bajpai et al. 1999; Bajpai 2001; Thompson et al. 2001; Pokhrel and Viraraghavan 2004; Ali and Sreekrishnan 2001). The problems underlying the physicochemical treatments are those associated with cost and reliability. Coagulation and precipitation produce a voluminous sludge, which is very difficult to dewater. Usually, an extreme pH range is used for optimum treatment and the pH needs to be readjusted to neutral before discharge. Oxidation using ozone and hydrogen peroxide are costly and oxidation using chlorine species generates secondary pollutants such as chlorinated organics. The membrane techniques require pretreatment and a large capital investment. Membrane fouling is also a problem with the membrane technique. Biotechnological methods have the potential to eliminate/reduce the problems associated with physicochemical methods. These methods are described below:

13.3.1 Enzymatic Treatment

Some of the enzymes seem to have the potential to decolorize and detoxify effluents from pulp and paper mill effluents. The use of microbial- or enzyme-based treatment offers some distinct advantages over physical and chemical decolorization and AOX precipitation methods, in that only catalytic and not stoichiometric amounts of the reagents are needed, and the low organic concentrations and large volumes typical of bleaching effluents are therefore less of a problem. Also, both complete microbial systems and isolated enzymes have been shown to reduce the acute toxicity by polymerizing and thereby rendering less soluble many of the low molecular mass nonchlorinated and polychlorinated phenolics (Bollag et al. 1988; Klibanev and Morris 1981; Ruggiero et al. 1989). In 1988, a review on the use of enzymes for wastewater treatment in the pulp and paper industry examined the new possibilities of using enzymes like laccase, peroxidase, and ligninase for this effect (Hakulinen 1988). Forss et al. (1987) examined the use of laccase for effluent treatment. They aerated pulp bleaching wastewater in the presence of laccase for 1 h at pH 4.8 and subsequently flocculated with aluminum sulphate. High removal efficiencies were obtained for chlorinated phenols, guaiacols, vanilins, and catechols. Roy-Arcand and Archibald (1991a) studied direct dechlorination of chlorophenolic compounds in pulp and paper mill effluent by laccases from *Trametes versicolor* and found that all the major laccases, secreted by *T. versicolor*, could partially dechlorinate a variety of chlorophenolics. These researchers also studied effects of horseradish peroxidase (HRP) and *Phanerochaete chrysosporium* peroxidase on the mixture of five chlorophenolics (pentachlorophenol, tetrachloroguaicol, 4,5,6-trichloroguaicol, 4,5-dichloroguaicol, 2,4,6-trichlorophenol). Both peroxidase enzymes were found to degrade the majority of substrates except pentachlorophenol whereas the *P. chrysosporium* peroxidase was superior to both HRP and laccase in degrading pentachlorophenol, and it was inferior to HRP in degrading the other four phenolics

Field (1986) patented a method for the biological treatment of wastewaters containing nondegradable phenolic compounds and degradable nonphenolic compounds. It consisted of an oxidative treatment to reduce or eliminate toxicity of the phenolic compounds followed by an anaerobic purification. This oxidative pretreatment could be performed with laccase enzymes and it was claimed to reduce chemical oxygen demand (COD) by 1,000-fold. Call (1991) patented a process on the use of laccase for wastewater treatment. He claimed that wastewater from delignification and bleaching could be treated with laccases in the presence of nonaromatic oxidants and reductants and aromatic compounds. Almost complete polymerization of the lignins is obtained which is 20–50% above the values attainable with the addition of laccase alone. About 70–90% lignin is developed into insoluble form, which is removed by flocculation and filtration.

Milstein et al. (1988a, b) described the removal of chlorophenols and chlorolignins from bleaching effluents by combined chemical and biological treatments. The organic matter from spent bleaching effluents of the chlorination, extraction, or a mixture of both stages was precipitated as a water insoluble complex with polyethylene imide. The color, COD, and AOX were reduced by 92, 65, and 84%, respectively, for the chlorination effluent and by 76, 70, and 73% for the extraction effluent. No significant reduction in BOD of treated effluent was detected but fish toxicity was greatly reduced. Enzyme treatment results in coprecipitation of the bulk mono- and dichlorophenols with the liquors of the chlorination and extraction bleaching stages. Lyr (1963) reported that laccase of *T. versicolor* partially dechlorinates PCP and Hammel and Tardone (1988) reported that peroxidase from *P. chrysosporium* can partially dechlorinate PCP and 2,4,6-trichlorophenol.

Paice and Jurasek (1984) studied the ability of HRP to catalyze color removal from bleach plant effluents. The color removal from effluents at neutral pH by low levels of hydrogen peroxide was enhanced by the addition of peroxidase. No precipitation occurred during the decolorization process. The catalysis with peroxidase (20 mg/L) was observed over a wide range of peroxide concentrations (0.1–800 mM) but the largest effect was between 1 and 100 mM. The pH optimum for catalysis was around 5.0. Compared with mycelial color removal by *Coriolus versicolor*, the rate of color removal by peroxide plus peroxidase was initially faster (for the first 4 h) but the extent of color removal after 45 h was higher with the fungal treatment. Further addition of peroxidase to the enzyme-treated effluents did not produce additional catalysis. Thus, the peroxide/peroxidase system did not fully represent the metabolic route used by the fungus. One working hypothesis has been proposed to explain the behavior of enzymes in the decolorization process (Paice and Jurasek 1984). Glucose is used by the cell to produce peroxidase which is one of the extracellular enzymes often found in white rot fungi. It seems that this enzyme oxidizes the chromophores and so removes the color from bleaching wastewater.

Though the use of enzyme-based treatments offers some distinct advantages over physical and chemical methods in that only catalytic amounts of reagents are needed, biochemical instability and difficulty in reusing of the enzyme are its disadvantages. Immobilization of the enzymes is required for biochemical stability and reuse of the enzymes. Carbon immobilized laccase was used by Davis and Burns (1992) to decolorize extraction stage effluent at the rate of 115 PCU/enzyme U/h. The removal rate was found to increase with the increasing effluent concentration. Dezotti et al. (1995) developed a simple immobilization method where activated silica was used as a support and used it for enzymatic color removal from extraction stage effluent by lignin peroxidase (LiP) from *Chrysonilia sitophila* and by commercial HRP. Immobilized HRP gave 73% decolorization and LiP gave 65 and 12% reductions in COD and color, respectively. Immobilized enzymes were found to retain activity even after 5 days of contact with the kraft mill effluent. Ferrer et al. (1991) reported that immobilized lignin peroxidase decolorized kraft effluent. Novel lignin peroxidases called pulpases produced by *P. chrysosporium* mutant strain SC 26 and described in two patents assigned to the Repligen Corporation have been claimed to decolorize bleaching effluents (Farrel 1987a, b).

Karimi et al. (2010) investigated the efficiency of advanced oxidation processes (AOPs), enzymatic treatment, and combined enzymatic/AOP sequences for the color remediation of soda and chemimechanical pulp and paper mill effluent. The results indicated that under all circumstances, the AOP using ultraviolet irradiation (photo-Fenton) was more efficient in the degradation of effluent components in comparison with the dark reaction. It was found that both versatile peroxidase (VP) from Bjerkandera adusta and laccase from *T. versicolor*, as pure enzymes, decolorize the deep brown effluent to a clear light-yellow solution. In addition, it was found that in the laccase treatment, the decolorization rates of both effluents were enhanced in the presence of 2, 2′-azinobis (3-ethylbenzthiazoline-6-sulfonate), while in the case of VP, Mn(+2) decreased the efficiency of the decolorization treatment. The concomitant use of enzymes and AOPs imposes a considerable effect on the color remediation of effluent samples.

13.3.2 *Bacterial Treatment*

Bacterial treatment includes aerobic treatment, anaerobic treatment, and combination of both treatments. Combinations of anaerobic and aerobic treatment processes are found to be efficient in the removal of soluble biodegradable organic pollutants (Pokhrel and Viraraghavan 2004).

13.3.2.1 Aerobic Treatment

The most common aerobic biological methods used in the treatment of pulp mill effluents are aerated lagoons or stabilization basins (ASB), activated sludge treatment (AST) processes, and sequencing batch reactors (SBR). Rotating biological contactors (RBCs) and trickling filters are rarely used.

Aerated Lagoon Treatment (ASB)

The aerated lagoon is a low rate aerobic biological process and is the oldest and simplest type of aerobic biological treatment system to construct and operate. The system consists of an aeration system for supplying dissolved oxygen. The wastewater to be treated is continuously fed into the aeration lagoon where biooxidation of the organic matter occurs, then directly flows out to the receiving environment. In some cases, a settling pond following the aeration basin is installed to remove the biological and other solids from the treated wastewater. A large portion of the sludge produced settles in the lagoon or settling pond, where it subsequently undergoes autooxidation or endogenous respiration, which not only reduces the sludge production but releases and reuses the nutrients from the sludge. Since the ASBs do not recycle the biomass, they normally require a long HRT (volume/volumetric flow) of 5–10 days, and microorganisms concentration in the lagoon is too (<200 mg/L) low.

ASB have long been widely employed in the treatment of kraft mill effluent (Tomar and Allen 1991; McCubbin 1983). ASBs have also been used for other types of pulp mill effluents including TMP and CTMP for the removal of BOD and toxicity chlorophenols, low molecular weight AOX, resin and fatty acids of pulp mill effluents (Liu et al. 1996; Johnson and Chatterjee 1995; Saunamaki et al. 1991; Tomar and Allen 1991 and Jokela et al. 1993). To produce a reliable high quality effluent, this method generally uses a long HRT of 5–7 days, which achieves high BOD removal (85–95%) and effluent detoxification. Extensive experience in applying ASBs in the treatment of pulp mill effluent is available. In both Canada and USA (Wilson and Holloran 1992; Turk 1988), most of the early constructed secondary treatment systems in pulp and paper mills, where available land space is not limited, are aerated lagoons. In developing countries, lagoons are the major process for the treatment of pulp mill effluent. Removals of AOX from bleached kraft mill effluents are quite variable among systems, ranging from 15 to 60% with an average of 30% (Wilson and Holloran 1992). The complex and variable properties of AOX compounds from different processes account for this range in the removal rate.

Experiments on recirculation of biomass in aerated lagoons have indicated that a fourfold increase in lagoon biomass could increase removal efficiency from 50 to 60% (Boman et al. 1988). Significant work has been done to determine the mechanism of AOX removal in aerated lagoon (Yin et al. 1989a; Bryant et al. 1987, 1988). It has been postulated that AOX removal occurs by biosorption of organohalides to biomass and anaerobic dehalogenation and degradation in the benthyl layer, of the lagoon with biosorption providing the transport mechanism (Bryant et al. 1987, 1988; Amy et al. 1988). Both high and low molecular weight chlorolignins are reported to adsorb to aerobic biomass but aerobic dehalogenation has not been reported.

Conversely, it has been suggested that the majority of AOX removal in an aerated lagoon is due to aeration enhanced hydrolytic splitting of chlorine from the organic substrate (Yin et al. 1989a). It is said that MLSS levels in an aerated lagoon are too small to allow significant biosorption to sludge. Removal of resin and fatty acids (RFAs) in CTMP effluent is generally >95% and degradation seems to take place by three mechanisms: biooxidation by microorganisms, adsorption onto sludge, and air oxidation. Biooxidation is the main removal mechanism. Adsorption onto sludge is the primary mechanism. However, when the treatment time is very short, air oxidation plays a minor role (Liu et al. 1996). Analysis of relative removals of different M_w fractions in three North American ASBs was reported (Bryant and Barkley 1990). Low molecular weight AOX was removed more effectively (43–63%) than high molecular weight AOX (4–31%). Effluent AOX removal from mills using hardwood and softwood furnishes was comparable but furnish changeovers reduced the removal performance. In a separate lab scale ASB study, degradation of hardwood derived TOCl was greater (44–52%) than for softwood derived TOCl (44%) (Yin 1989).

Fulthorpe and Allen (1995) studied the ability of three bacterial species to reduce AOX in bleached kraft mill effluents. *Ancylobacter aquaticus* A7 exhibited the broadest substrate range but could only affect significant AOX reduction in softwood effluents. *Methylobacterium* CP13 exhibited a limited range but was capable of removing significant amounts of AOX from both hardwood and softwood effluents. By contrast, *Pseudomonas* sp. Pl exhibited a limited substrate range and poor to negligible reductions in AOX levels from both effluent types. Mixed inocula of all the three species combined and inocula of sludge from mill treatment systems removed as much AOX from softwood effluents as did pure populations of *Methylobacterium* CP13. Rogers et al. (1975) treated the bleached kraft mill effluent in a bench scale aerated lagoon for 29, 58, and 99 h showed that toxicity, BOD, and resin acids were most consistently reduced during the 99 h treatment. Leach et al. (1978) reported the biodegradation of seven compounds representing the major categories of toxicants in a laboratory scale batch aerated lagoon. Resin acids which are the major source of acute toxicity were readily biodegradable but only part (less than 30%) of the load of chlorophenolic compounds was removed. Deardorff et al. (1994) reported that the efficiency of AOX removal through biotreatment of combined bleach plant effluent increases with increasing chlorine dioxide substitution.

Table 13.3 Reported activated sludge removal efficiencies for chlorophenols

Compound	Reduction range (%)
Dichlorophenols	22–63
Trichlorophenols	−11–57
Tetrachlorophenols	22–67
Pentachlorophenols	25
Dichloroguaiacols	0–89
Trichloroguaiacols	−400–81
Tetrachloroguaiacols	80
Dichlorocatechols	−115–45
Trichlorocatechols	−30–41
Tetrachlorocatechols	13–57
Monochlorovanillins	92–100
Dichlorovanillins	81–96

Based on Wilson and Holloran (1992), Boman et al. (1988), Bryant et al. (1987), Gergov et al. (1988), Voss (1983), Lindstrom and Mohamed (1988)

Biological treatment in an aerated lagoon reduced the concentration of polychlorinated phenolic compounds by 97%. Jokela et al. (1993) reported that aerobic lagoon systems removed 58–60% of the organochlorine compounds from the water phase whereas the full-scale activated sludge plants removed 19–55%. Both biotreatments removed all sizes and classes of organochlorine molecules and slightly changed the relative size distribution of the compounds remaining in the water phase towards the large molecular weights. Eriksson and Kolar (1985) have shown that high molecular weight fraction in bleach plant effluents cannot be degraded in an aerated lagoon. In another study, it has been shown that chloroform is stripped during the biological treatment and COD, AOX, and high molecular weight material are reduced to a lesser extent (SSVL-85 Project 4, Final report).

Reduction of individual chlorinated organics across aerated basins has been reported by various authors (Boman et al. 1988; Saunamaki et al. 1991; Lindstrom and Mohamed 1988). Individual removal efficiencies for various chlorophenols provided in Table 13.3 range from 30 to 89%. Information obtained from Paprican has indicated removal efficiency up to 100% for chlorovanillins (Wilson and Holloran 1992).

ASBs provide distinct advantages over high rate systems such as ASTs, including little or no nutrient addition required, expect at initial start up, lower net settleable solids generation, lower energy consumptions due to avoidance of sludge handling and reduced aeration requirement, and better toxicity removal. Lagoons are generally able to detoxify pulp mill effluents because of the long HRT, thus ASBs have been universally accepted by pulp and paper mills, where land space is not limited. One other major merit of ASBs is much lower capital and operating costs than AST processes. However, since the HRT in an ASB is long, the required land space for constructing an aeration basin is large, which can be a major disadvantage. This has led to the introduction of AST systems, which require much less land.

Activated Sludge Treatment (AST)

AST is a high rate biological process adapted largely from sanitary waste treatment. In contrast to an ASB, in an AST process there is sludge settler following the aeration basin. The function of the settler is to separate the sludge from the treated effluent so that it can be recycled to the aeration basin and bacterial concentration in the aeration basin can be maintained at a high level (2,000–5,000 mg/L). The high biomass concentration increases the rate of treatment, so the required HRT for treating the same effluent is much shorter than in an ASB and aeration basin size required is also greatly reduced. Two major AST processes used in paper mills are air and pure oxygen AST systems.

AST has been used by the pulp and paper industry when the available land space is small and/or a low treated effluent suspended solids concentration is required. ASTs have been adopted initially in the paper mills in the USA. They have also emerged into the Canadian paper mills and are also common in other countries. A number of full-scale AST systems are operated in the USA and in Canada for the treatment of various pulp mill effluents, including those from kraft, paper board, deinking, TMP and CTMP, sulphite and newsprint mill operations (Buckley 1992; Paice 1995; Johnson and Chatterjee 1995). ASTs are generally reported to remove much higher quantities of AOX than aerated lagoons. Removal efficiencies ranging from 14 to 65% have been reported.

Melcer et al. (1995) carried out pilot-scale investigation of AST of bleached kraft effluent at a Northern Ontario mill site over a 8-month period. The AS system was operated at a 1-day HRT, 25–30 days SRT and 30°C. Treated effluents were found to pass all acute and chronic toxicity tests as measured by Rainbrow trout LC_{50}, Microtox, and *Ceriodaphnia* LC_{50} and IC_{25} tests. A high level of effluent quality was achieved with low concentrations of AOX (4–13 mg/L), total chlorophenolics (0.3–0.32 mg/L), toxicity equivalents (TEQ-PCP) (0.4–5 mg/L), total resin and fatty acids (0–4 mg/L), BOD (4–12 mg/L), and soluble COD (142–274 mg/L) being recorded over the whole period of investigation. An 8- to 17-fold reduction in hepatic MFO enzyme activity was measured in the treated effluents over the influent wastewaters.

Valenzuela et al. (1997) studied the degradation of chlorophenols by *Alcaligenes eutrophus* TMP 134 in bleached kraft mill effluent. After 6 days of incubation, 2,4-dichlorophenoxyacetate (400 ppm) or 2,4,6 trichlorophenol (40–100 ppm) was extensively degraded (70–100%). In short-term batch incubations, indigenous microorganisms were unable to degrade such compounds. Degradation of 2,4,6-trichlorophenol by strain JMP 134 was significantly lower at 200–400 ppm of compound. This strain was also able to degrade 2,4-dichlorophenoxyacetate, 2,4,6-trichlorophenol, 4-chlorophenol, and 2,4,5-trichlorophenol, when bleached kraft mill effluent was amended with mixtures of these compounds. On the other hand, the chlorophenol concentration and the indigenous microorganisms inhibited the growth and survival of the strain in short-term incubations. In long-term (>1 month) incubations, strain JMP 134 was unable to maintain a large, stable population, but an extensive 2,4,6-trichlorophenol degradation was still observed.

When combined effluents of a kraft pulp mill were treated in a lab scale activated sludge system, the average TOC and AOX removal efficiencies were found to be 83 and 21%, respectively (Ataberk and Gokcay 1997). The highest AOX removal occurred at larger SRTs. Mass balance on the system revealed that the principal AOX removal mechanism was metabolization at long SRT. About 90% of the AOX removed was metabolized. As SRT was lowered, AOX removal efficiency decreased.

When the bleaching effluents from chlorination and extraction stage were treated in an activated sludge process, the AOX reduction was found to be 30–40% in 8 days. About 70–80% of the total AOX reduction was achieved in about 4 days (Mortha et al. 1991). The presence of high molecular weight material in the bleached kraft effluent was found to improve the removal of chlorophenolic compounds. Growth experiments using microorganisms from a lab scale activated sludge reactor showed that high molecular weight material had a significant role in soluble COD and chlorophenol removal (Bullock et al. 1994). Large decreases in the soluble COD and increases in the biomass were observed with the addition of high molecular weight materials to the low molecular weight fraction. The addition of mono- and dichlorinated phenolic compounds at concentrations up to 10 mg/L was found to have no effect on the metabolism or growth of the microorganisms in the activated sludge. While 6-chlorovanillin (6-CV), 2,4-dichlorophenol (2,4-DCP), and 4,5 dichloroguaiacol (4,5-DCG) were found to be stable in uninoculated controls and inoculated low molecular weight effluent over a 160-h period, these compounds decreased significantly, when low molecular weight with 3 times the original concentration of high molecular weight material was inoculated with microorganisms. The removal rates of these compounds increased in the order: 6-CV > 4,5-DCP > 2,4-DCP. Gergov et al. (1988) investigated pollutant removal efficiencies in mill scale biological treatment systems. About 48–65% AOX was removed in the activated sludge process.

The combined effects of oxygen delignification, ClO_2 substitution, and biological treatment on pollutants levels in bleach plant effluents were examined. Biological treatments did not reduce color but reduced COD, BOD, AOX, and toxicity (Graves et al. 1993). ClO_2 substitution reduced the discharge of all five pollutants with a large reduction in AOX. Oxygen delignification reduced discharges of the five pollutants, and effluents from the sequence with oxygen delignification were easier to treat by aerobic methods. Treatment of bleaching effluent in sequential activated sludge and nitrification systems revealed that dechlorination of bleaching effluent took place in both the systems (Altnbas and Eroglu 1997). In the activated sludge system, released inorganic chloride was 4.5–7 mg/L at TOC loading rate of 0.03–0.07 mg/mg VSS/day, respectively; but it was decreased from 10 to 3 mg/L at TOC loading rate of 0.006–0.06 mg/mg VSS/d, respectively. Removal efficiencies for individual chlorinated organics range from 18 to 100% and are presented in Table 13.4.

Liu et al. (1996) demonstrated that AOX removal mechanism includes biodegradation, adsorption to biomass, and air oxidation. Among these three, biodegradation is the major mechanism. Apart from achieving high AOX removals in ASTs, high performance COD, BOD, and TSS removals were recorded (Goronzy et al. 1996).

Table 13.4 Reported activated sludge removal efficiencies for chlorophenols

Compound	Reduction range (%)
Dichlorophenols	78
Trichlorophenols	51–69
Tetrachlorophenols	86–100
Pentachlorophenols	50–80
Dichloroguaiacols	67–97
Trichloroguaiacols	18–97
Tetrachloroguaiacols	59–99
Dichlorocatechols	37
Trichlorocatechols	63–95
Tetrachlorocatechols	59–90
Monochlorovanillins	94
Dichlorovanillins	100

Based on Wilson and Holloran (1992), Gergov et al. (1988), Saunamaki (1989), Rempel et al. (1990), Mcleay (1987)

AOX removal efficiency was correlated to SRT and HRT (Rempel et al. 1990) in pilot-scale tests of air and oxygen activated sludge systems. The maximum reported AOX removal efficiencies (>40%) were achieved for SRTs greater than 20 days and HRTs greater than 15 h. In a separate report on Finish activated sludge systems, the highest AOX removals (45%) in mill scale units were reported for SRTs greater than 50 days (Salkinoja-Salonen 1990). Varying the HRTs and SRTs indicated that HRT had more of an effect on treatment performance than SRT. Longer HRTs led to improved BOD, COD, toxicity, and AOX removal, whereas longer SRTs were not shown to significantly affect performance (Barr et al. 1996). Paice et al. (1996) investigated effluents from CMP/newsprint operation that was treated in two parallel laboratory scale activated sludge systems. Removal of BOD and resin fatty acids in excess of 90% was achieved with an HRT of 24 h. Anoxic conditioning of the sludge (Liu et al. 1997) and hydrolysis pretreatment of bleachery effluents (Zheng and Allen 1997) have been demonstrated to enhance AOX removal by about 8 and 20–30%, respectively, in AST. As the temperature of mill effluent is high (60°C), research is underway to use thermophilic (50–60°C) bacteria in ASTs (Barr et al. 1996; Rempel et al. 1990; NCASI 1990 and Puhakka 1994).

Tiku et al. (2010) studied the capability of three bacteria, *Pseudomonas aeruginosa* (DSMZ 03504), *P. aeruginosa* (DSMZ 03505), and *Bacillus megaterium* (MTCC 6544) to reduce the BOD and COD level of pulp and paper mill effluents up to permissible levels, i.e., 30 and 250 mg/L, respectively, within a retention time of 24 h in batch cultures. A concomitant reduction in total dissolved solids (TDS), AOX, and color (76%) was also observed. This is the first report on the use of bacterial cultures for the holistic bioremediation of pulp mill effluent.

Rai et al. (2007) examined three lignin-degrading bacterial strains, identified as *Paenibacillus* sp., *Aneurinibacillus aneurinilyticus,* and *Bacillus* sp. for the treatment of pulp and paper mill effluent. The results of this study revealed that all three bacterial strains effectively reduced color (39–61%), lignin (28–53%), biochemical oxygen demand (BOD) (65–82%), COD (52–78%), and total phenol (64–77%) within 6 days of incubation. However, the highest reduction in color (61%), lignin (53%), BOD (82%), and COD (78%) was recorded by *Bacillus* sp., while maximum reduction in total phenol (77%) was recorded with *Paenibacillus* sp. treatment. Significant reduction in color and lignin content by these bacterial strains was observed after 2 days of incubation, indicating that bacterium initially utilized growth supportive substrates and subsequently chromophoric compounds thereby reducing lignin content and color in the effluent. The total ion chromatograph (TIC) of compounds present in the ethyl acetate extract of control and bacterial treated samples revealed the formation of several lignin-related aromatic compounds. The compounds identified in extracts of treated samples by *Paenibacillus* sp. were t-cinnamic acid and ferulic acid, while 3-hydroxy-4-methoxyphenol, vanillic acid by *A. aneurinilyticus* and gallic acid and ferulic acid by *Bacillus* sp., respectively, indicating the degradation of lignin present in the effluent. The identified compounds obtained after different bacterial treatments were found to be strain specific. Among these identified compounds, ferulic acid, vanillic acid, and vanillin could have immense value for their use in preservatives and in the food flavor industry.

Mishra and Thakur (2010) isolated four different bacterial strains from pulp and paper mill sludge in which one alkalotolerant isolate (LP1) having higher capability to remove color and lignin was identified as *Bacillus* sp. by 16S RNA sequencing. Optimization of process parameters for decolorization was initially performed to select growth factors which were further substantiated by Taguchi approach in which seven factors, % carbon, % black liquor, duration, pH, temperature, stirring, and inoculum size, at two levels, applying L-8 orthogonal array were taken. Maximum color was removed at pH 8, temperature 35°C, stirring 200 rpm, sucrose (2.5%), 48 h, 5% (w/v) inoculum size, and 10% black liquor. After optimization twofold increase in color and lignin removal from 25 to 69 and 28 to 53%, respectively, indicated significance of Taguchi approach in decolorization and delignification of lignin in pulp and paper mill effluent. Enzymes involved in the process of decolorization of effluent were found to be xylanase (54 U/mL) and manganese peroxidase (28 U/mL). Treated effluent was also evaluated for toxicity by Comet assay using *Saccharomyces cerevisiae* MTCC 36 as model organism, which indicated 58% reduction after treatment by bacterium.

Chandra et al. (2008) isolated eight aerobic bacterial strains from pulp paper mill waste and screened for tolerance of kraft lignin (KL) using the nutrient enrichment technique in mineral salt media (MSM) agar plate (15 g/L) amended with different concentrations of KL along with 1% glucose and 0.5% peptone (w/v) as additional carbon and nitrogen sources. The strains ITRC S6 and ITRC S8 were found to have the most potential for tolerance of the highest concentration of KL. These organisms were characterized by biochemical tests and further 16S rRNA gene (rDNA) sequencing, which showed 96.5 and 95% sequence similarity of ITRC S(6) and ITRC S(8) and confirmed them as *Paenibacillus* sp. and *Bacillus* sp., respectively. KL decolorization was routinely monitored with a spectrophotometer and further confirmed by HPLC analysis. Among eight strains, ITRC S(6) and ITRC S(8) were found to degrade 500 mg/L of KL up to 47.97 and 65.58%, respectively, within

144 h of incubation in the presence of 1% glucose and 0.5% (w/v) peptone as a supplementary source of carbon and nitrogen. In the absence of glucose and peptone, these bacteria were unable to utilize KL.

Monje et al. (2010) evaluated the aerobic and anaerobic biodegradability and toxicity to *Vibrium fischeri* of generated L-stage and total bleaching sequence effluents. The highest levels of aerobic and anaerobic degradation of the generated effluents were achieved for treatments with laccase plus violuric acid, with 80% of aerobic degradation and 68% of anaerobic biodegradation. *V. fischeri* toxicity was remarkably reduced for all the effluents after aerobic degradation.

Sequencing Batch Reactors (SBR)

The SBR process is a fill and draw cyclic batch activated sludge process. The operation of each cycle normally consists of four sequential steps: fill reaction, settle, withdraw, and idle. During the fill period, wastewater is fed to SBR under anoxic conditions (without aeration) and biosorption takes place. After completion of fill, the aerobic reaction starts with aeration. Following reaction, the biomass is allowed to settle in quiescent conditions in the reactor. Finally, about one-third of the SBR of the clarified treated effluent is withdrawn. For multi-SBR systems without sufficient wastewater an idle period may be necessary. The next cycle starts again at the fill stage. Sludge wasting occurs at the end of the settle period or during the idle period. Essentially, the SBR's batch stage can be compared to the unit operations in an AST, with the react stage corresponding to the aeration basin and the settle draw stages corresponding to the secondary clarifier and sludge recycle.

SBRs have the following advantages compared to conventional ASTs: lower operating costs as there is no aeration for 30–40% of the total time, no sludge settler or recycling pumps are required. Control of filamentous bulking due to the anoxic fill, ability to tolerate peak flow and shock loads, and denitrification during the anoxic fill and settle stages are the main advantages. In addition, the control and operation of a SBR are flexible.

SBRs were initially used for the treatment of small and medium size municipal wastewaters. Before 1980s, the application of SBR processes was limited mainly due to the lack of automatic control devices. But now with the rapid development of modern automatic control devices and computer technology, operation of an SBR can be easily accomplished through automatic control devices. As such, the application of SBRs for the treatment of various effluents has rapidly increased. SBRs have been used for the treatment of pulp mill effluents and, in North America, there are several full-scale SBR systems treating various pulp mill effluents including kraft, TMP, high yield sulphite, deinking, and fine paper mill effluents. SBRs generally produce smaller quantity of effluent as ASTs. One of the major problem is the lack of experience for both design and operation of SBR systems for the treatment of such large quantities of effluents.

Other Aerobic Treatment Systems

Other aerobic biological processes include rotary disc contractors and trickling filters. Mathys et al. (1993, 1997) studied the treatment of CTMP mill wastewater in laboratory scale RBC. Application of these two processes for the treatment of pulp mill effluents are limited (Lunan et al. 1995; Mathys et al. 1997).

13.3.2.2 Anaerobic Treatment

The major anaerobic processes currently used for the treatment of pulp mill effluents include anaerobic lagoons, anaerobic contract process, upflow anaerobic sludge blanket (UASB), anaerobic fluidized bed, and anaerobic filter.

Anaerobic Lagoon

The anaerobic lagoon is the oldest low rate anaerobic treatment process. It generally consists of large flow through basin where the SRT equals HRT. To achieve a high treatment efficiency, the HRT is generally long (from 10 to 30 days), requiring large land areas, which is the major limitation of the system.

Anaerobic Contact Process

The anaerobic contact process was developed in 1950s and was first high rate anaerobic treatment system (Lee 1993). Separation of the sludge from the settling tank is the critical factor for maintaining high biomass concentration and for operating the contact process. It is an outgrowth of anaerobic lagoon and is similar to activated sludge process, consisting of fully mixed anaerobic reactor and sludge settling tank. A portion of the sludge is returned to the contact reactor to maintain high biomass concentration (3,000–10,000 mg/L) in the reactor. Due to the recycling of sludge, the SRT can be controlled to be much longer than the HRT. Separation of the sludge from the settling tank is the critical factor for maintaining high biomass concentration and for operating the contact process. This system is suitable for treating effluents containing a high concentration of suspended solids. It can be operated at an organic loading from 1 to 2 kg BOD/m^3/day.

Upflow Anaerobic Sludge Blanket Reactor (UASB)

The UASB reactor was developed in the Netherlands in the 1970s (Lettinga 1980). This reactor operates entirely as a suspended growth system and consequently does not contain any packing material. It contains a gas–liquid–solid separation device for the separation of biogas, treated effluent, and suspended solids at the top surface of the reactor to minimize the loss of biomass. Wastewater to be treated is distributed into the bottom of the reactor and flows upward in the reactor. A dense granular sludge formation in the reactor is the critical factor in process performance, since it ensures proper settling characteristics of sludge. The SRT value is extremely high for well adapted systems, and generally this process seems to have the potential to treat more dilute and colder effluents than contact process. Loading rates generally range from 3.5 to 5.0 kg BOD/m^3/day and can be up to 8 kg BOD/m^3/day (Lee et al. 1995). At present, most of the full-scale high rate anaerobic systems in use in pulp and paper industry are UASB reactors.

Fluidized Bed Reactor

The effluent is distributed into the bottom of the reactor and flows upwards through a fluidized bed of microorganisms attached on a carrier. A certain amount of water usually has to be re-circulated in order to keep the bed fluidized. The SRT value may be extremely high, comparable to the UASB reactor. Loading rates are in the range of 17–41 kg BOD/m^3/day (Lee et al. 1995). However, operating costs of this reactor are elevated since recycling of effluent inside the reactor consumes a large amount of power.

Anaerobic Filter

The anaerobic filter, also known as fixed bed or fixed film, contains a packing material, usually plastic material, with a large specific area. The microorganisms grow on surface of the material and in the void space between surfaces. The effluent may pass the bed upflow or downflow. The loading rates range from 4 to 15 kg BOD/m^3/day.

The application of anaerobic treatment system at pulp and paper mills is experiencing a notable increase in the last years. Anaerobic technologies are already in use for many types of forest industry effluents. The upflow anaerobic sludge bed (UASB) reactor and the contact process are the most widely applied anaerobic systems. Most of the existing anaerobic full-scale plants are treating noninhibitory forest industry wastewater rich in readily biodegradable organic matter (carbohydrates and organic acids) such as recycling wastewater, thermomechanical pulping (TMP) effluents. Full-scale application of anaerobic systems for chemical, semichemical and chemithermomechanical bleaching and debarking liquors is still limited.

Thermomechanical pulping wastewaters are known to be highly biodegradable during anaerobic digestion and not toxic to methanogenic bacteria. This makes them highly suitable for anaerobic wastewater treatment (Sierra-Alvarez et al. 1990, 1991; Jurgensen et al. 1985). In mesophilic anaerobic process, loading rates up to 12–31 kg COD/m^3/d with about 60–70% COD removal efficiency have been obtained (Sierra-Alvarez et al. 1990, 1991; Rintala and Vuoriranta 1988). In thermophillic anaerobic process conditions up to 65–75%, COD removal was obtained at 55°C at loading rate of 14–22 kg COD/m^3/day in UASB reactors (Rintala and Vuoriranta 1988; Rintala and Lepisto 1992).

About 60% COD removal was maintained at 50% in the UASB reactor at a loading rates as high as 80 kg COD/m^3/day, which corresponds with HRT of 55 min. Kortekaas et al. (1998) studied anaerobic treatment of wastewaters from thermomechanical pulping of hemp. Hemp stem wood and hemp bark thermomechanical pulping wastewaters were treated in a laboratory scale UASB reactor. For both the types of wastewaters, maximum COD removal of 72% were obtained at loading rates of 13–16 g COD/L/day providing 59–63% recovery of the influent COD as methane. The reactors continued to provide excellent COD removal efficiencies of 63–66% up to loading rate of 27 g COD/L/day, being the highest loading rate tested. Batch toxicity assays revealed the absence of methanogenic inhibition by hemp TMP wastewaters, coinciding with the high acetolastic activity of the reactor sludge of approximately 1 g COD/g VSS/day. Due to the relatively low molecular weight of hemp TMP lignin, its removal which was measured as UV 280 during anaerobic treatment was markedly high and averaged 45 and 31% for the hemp stem wood and the hemp bark TMP UASB reactors, respectively. Subsequent batch aerobic posttreatment led to considerable increase of color levels and polymerization of the residual lignin to molecular weight in excess of 34 kD.

The application of anaerobic treatment for degradation and dechlorination of kraft bleach plant effluent has been studied. The COD removals in the anaerobic treatment of bleaching effluents have ranged from 28 to 50% (Lafond and Ferguson 1991; Raizer-Neto et al. 1991; Rintala and Lepisto 1992). Removal of AOX was improved when easily degradable co-substrate (methanol or ethanol) was used to supplement the influent (Parker et al. 1993a). Many chlorophenolic compounds, chlorinated guaiacols – catechols and chlorovanillins were removed at greater than 95% efficiency (Parker et al. 1993b). Fitzsimonas et al. (1990) investigated anaerobic dechlorination/degradation of chlorinated organic compounds of different molecular masses in bleach plant effluents. A decrease in organically bound chlorine measured as adsorbable organic halogen was found with all molecular mass fraction. The rate and extent of dechlorination and degradation of soluble AOX decreased with increasing molecular mass (Table 13.5). As high molecular weight chlorolignins are not amenable to anaerobic microorganisms, dechlorination of high molecular weight compounds may be due to combination of energy metabolism, growth, adsorption, and hydrolysis.

Black liquor and bleach effluent from an agroresidue based pulp and paper mill were treated anaerobically to reduce their high COD and AOX contents (Ali and Sreekrishnan. 2007). Addition of 1% w/v glucose yielded 80% methane from black liquor with concomitant reduction of COD by 71%, while bleach effluent generated 76% methane and produced 73 and 66% reductions in AOX and COD, respectively. In the absence of glucose, black liquor and bleach effluent produced only 33 and 27% methane with COD reductions of 43 and 31%, respectively.

NSSC pulping is the most widely used semichemical pulping process. Chemical recovery in semichemical pulping is not practiced in all the mills and thus there is a need to treat the spent liquor. Hall et al. (1986) and Wilson et al. (1987) demonstrated anaerobic treatability of NSSC spent liquor together with other pulping and paper mill wastewater streams. The methanogenic inhibition by NSSC spent liquor was apparently the effect of the tannins present in these wastewaters (Habets et al. 1985).

Table 13.5 Reduction of COD and AOX in the continuous reactor by anaerobic treatment

Fraction	Sampling pt	Total COD (% reduction)	AOX (% reduction)
I (M_w>20,000)	1		
	2	8	0
	3	62	5
	4	56	14
II (6,000<M_w<20,000)	1		
	2	11	8
	3	67	23
	4	69	34
III (2,000<M_w<6,000)	1		
	2	3	9
	3	80	46
	4	84	58
IV (M_w<2,000)	1		
	2	14	31
	3	85	66
	4	87	66

Based on Fitzsimonas et al. (1990)

Formation of H_2S in the anaerobic treatment of NSSC spent liquor has been reported but not related to methanogenic toxicity. Apparently, the evaporator condensates from the NSSC production are amenable to anaerobic treatment because of their high volatile fatty acid mainly acetate (Pertulla et al. 1991).

Unstable operations have been encountered in anaerobic treatment of pulp mill effluents, in particular with CTMP and NSSC wastewaters. The exact reason for these operation problems is still unclear although it is believed that they may be associated with the toxicants in these effluents, particularly wood extractives (RFA). Because of the unstable operation problems, application of anaerobic treatment technology in the paper industry sector is still limited. Research is underway to develop treatment systems that combine aerobic technology or ultrafiltration process. The sequential treatment of bleached kraft effluent in anaerobic fluidized bed and aerobic trickling filter was found to be effective in degrading the chlorinated, high and low molecular material (Haggblom and Salkinoja-Salonen 1991). The treatment significantly reduced the COD, BOD, and AOX of the wastewater. COD and BOD reduction was greatest in the aerobic process whereas dechlorination was significant in the anaerobic process. With the combined aerobic and anaerobic treatment, over 65% reduction of AOX and over 75% reduction of chlorinated phenolics was observed. The similar COD/AOX ratio of the wastewater before and after treatment indicates that the chlorinated material was as biodegradable as the nonchlorinated.

Dorica and Elliott (1994) studied the treatability of bleached kraft effluent using anaerobic and aerobic plus aerobic processes. BOD reduction in the anaerobic stage varied between 31 and 53% with hardwood effluent. Similarly the AOX removal from the hardwood effluents were higher 65 and 71%, for single- and the two-stage

Table 13.6 Removal of pollutants by anaerobic–aerobic treatment of bleaching effluent

Parameter	Reduction (%)
Chemical oxygen demand (mg O_2/L)	61
Biochemical oxygen demand (mg O_2/L)	78
Adsorbable organic halogens (mg Cl/L)	68
Chlorophenolic compound	
2,3,4,6 tetrachlorophenol	71
2,4,6 trichlorophenol	91
2,4 dichlorophenol	77
Tetrachloroguaiacols	84
3,4,5 trichloroguaiacols	78
4,5,6 trichloroguaiacols	78
4,5 dichloroguaiacols	76
Trichlorosyringol	64

Based on Haggblom and Salkinoja-Salonen (1991)

treatment, respectively, than that for softwood effluents (34 and 40%). Chlorate was removed easily from both softwood and hardwood effluents (99 and 96%, respectively) with little difference in efficiency between the single- and two-stage anaerobic systems. At organic loadings between 0.4 and 1.0 kg COD/m^3/day, the biogas yields in the reactors were 0.16–0.37 L/g BOD in the feed. Biogas yield decreased with increasing BOD load for both the softwood and hardwood effluents. Anaerobic plus aerobic treatment removed more than 92% of BOD and chlorate. AOX removal was 72–78% with hardwood effluents and 35–43% with softwood effluents. Most of the AOX was found to be removed from hardwood effluents during feed preparation and storage. Parallel control treatment tests in nonbiological reactors confirmed the presence of chemical mechanisms during the treatment of hardwood effluent at 55°C. The AOX removal that could be attributed to the anaerobic biomass ranged between 0 and 12%. The Enso-Fenox process was capable of removing 64–94% of the chlorophenol load, toxicity, mutagenicity, and chloroform in the bleaching effluent (Hakulinen 1982).

The sequential treatment of bleached Kraft effluent in an anaerobic fluidized bed and aerobic trickling filter was found to be effective in degrading the chlorinated high and low molecular weight material (Haggblom and Salkinoja-Salonen 1991). The treatment significantly reduced the COD, BOD, and the AOX of the wastewater. COD and BOD reduction was greatest in the aerobic process whereas dechlorination was significant in the anaerobic process. With the combined aerobic and anaerobic treatment, over 65% reduction of AOX and over 75% reduction of chlorinated phenolic compounds was observed (Table 13.6). The COD/AOX ratio of the wastewater was similar before and after treatment indicating that the chlorinated material was as biodegradable as the nonchlorinated. Microbes capable of mineralizing pentachlorophenol constituted approximately 3% of the total heterotrophic microbial population in the aerobic trickling filter. Two aerobic polychlorophenol degrading *Rhodococcus* strains were able to degrade polychlorinated phenols, guaiacols, and syringols in the bleaching effluent.

Singh and Thakur (2006) and Singh (2007) studied sequential anaerobic and aerobic treatment in two steps bioreactor for removal of color in the pulp and paper mill effluent. In anaerobic treatment, color (70%), lignin (25%), COD (42%), AOX (15%), and phenol (39%) were reduced in 15 days. The anaerobically treated effluent was separately applied in bioreactor in presence of fungal strain, *Paecilomyces* sp., and bacterial strain, *Microbrevis luteum*. Data of study indicated reduction in color (95%), AOX (67%), lignin (86%), COD (88%), and phenol (63%) by *Paecilomyces* sp. whereas *M. luteum* showed removal in color (76%), lignin (69%), COD (75%), AOX (82%), and phenol (93%) by day third when 7 days anaerobically treated effluent was further treated by aerobic microorganisms. Change in pH of the effluent and increase in biomass of microorganisms substantiated results of the study, which was concomitant to the treatment method.

Pudumjee Pulp and Paper Mills in Maharashtra, India, which is having a 30-tpd bagasse pulping capacity and a paper manufacturing capacity of 50 tpd, is running a full-scale anaerobic–aerobic plant for treatment of black liquor. The process is known as Pudumjee-An-OPUR-P. The anaerobic treatment scheme includes two digesters each of 6,200 m^3 capacity to treat not only the existing effluent coming from the 30 tpd pulping operations but also to treat increased flow coming from an enhanced 50 tpd production capacity. The anaerobic pretreatment of black liquor has reduced COD and BOD by 70 and 90%, respectively (Deshpande et al. 1991). The biogas produced is used as a fuel in boilers along with LSHS oil. The anaerobic pretreatment of black liquor has reduced organic loading at aerobic treatment plant thereby reducing the electrical energy and chemical nutrient consumption.

Swedish MoDo Paper's Domsjo Sulfitfabrik is using anaerobic effluent treatment at its sulphite pulp mill and produces all the energy required at the mill (Olofsson 1996). It also fulfills 90% of the heating requirements of the inner town of Ornskoldvik. Two bioreactors at the mill transform effluent into biogas and slime. The anaerobic unit is used to 70% capacity. A reduction of 99% has been achieved for BOD_7 and the figure for COD is 80%. There are plans to use the slime produced as a fertilizer.

A process based on UF and anaerobic and aerobic biological treatments has been proposed (Ek and Eriksson 1987; Ek and Kolar 1989; Eriksson 1990). The UF was used to separate the high molecular weight mass, which is relatively resistant to biological degradation. Anaerobic microorganisms were believed to be able to more efficiently remove highly chlorinated substances than aerobic microorganisms. The remaining chlorine atoms were removed by aerobic microorganisms. The combined treatments typically removed 80% of the AOX, COD, and chlorinated phenolics and completely removed chlorate (Table 13.7).

Anaerobic processes were previously regarded as being too sensitive to inhibitory compounds (Lettinga et al. 1990; Rinzema and Lettinga 1988). But now the advances in the identification of inhibitory compounds/substances in paper mill effluents as well as increasing insight over the biodegradative capacity and toxicity tolerance of anaerobic microorganisms have helped to demonstrate that anaerobic treatment of various inhibitory wastewaters is feasible.

Table 13.7 Removal of pollutants with ultrafiltration plus anaerobic/aerobic system and the aerated lagoon technique

Parameter	UF plus anaerobic/ aerobic predicted reductions (%)	Aerated lagoon estimated reductions (%)
BOD	95	40–55
COD	70–85	15–30
AOX	70–85	20–30
Color	50	0
Toxicity	100	Variable
Chlorinated phenols	>90	0–30
Chlorate	>99	Variable

Based on Eriksson (1990), Ek and Eriksson (1987), Ek and Kolar (1989)

The capacity of anaerobic treatment to reduce organic load depends on the presence of considerable amounts of persistent organic matter and toxic substances. Most important toxicants are sulphate and sulphite (Pichon et al. 1988), wood resin compounds (Sierra-Alvarez and Lettinga 1990; McCarthy et al. 1990), chlorinated phenolics (Sierra-Alvarez and Lettinga 1991), and tannins (Field and Lettinga 1991). These compounds are highly toxic to methanogenic bacteria at very low concentration. In addition, a number of low molecular weight derivatives have also been identified as methanogenic inhibitors (Sierra-Alvarez and Lettinga 1991).

In CTMP wastewaters, resins and volatile terpenes may account for up to 10% of the wastewater COD (1,000 mg/L) (Welander and Anderson 1985). The solids present in the CTMP effluent were found to contribute to 80–90% of the acetoclastic inhibition (Richardson et al. 1991). The apparent inhibition by resin acids was overcome by diluting anaerobic reactor influent with water or aerobically treated CTMP effluent which contained less than 10% of the resin acids present in the untreated wastewater (Habets and de Vegt. 1991; MacLean et al. 1990). Similarly, the inhibition by resin acids was overcome by diluting the anaerobic reactor influent with water and by aerating the wastewater to oxidize sulphite to sulphate prior to anaerobic treatment (Eeckhaut et al. 1986).

The chlorinated organic compounds formed in the chlorination and alkaline extraction stages are generally considered responsible for a major portion of the methanogenic toxicity in bleaching effluents (Rintala et al. 1992; Yu and Welander 1988; Ferguson et al. 1990). Anaerobic technologies can be successfully applied for reducing the organic load in the inhibitory wastewaters if dilution of the influent concentration to subtoxic levels is feasible (Ferguson and Dalentoft 1991; Lafond and Ferguson 1991). Dilution will prevent methanogenic inhibition and favor possible microbial adaptation to the inhibitory compounds. In practice, considerable dilution might be feasible with other noninhibitory waste streams such as kraft condensates (Edeline et al. 1988) and sulphite evaporator condensates (Sarner 1988) prior to anaerobic treatment, and has been shown to be an efficient measure for reducing the methanogenic toxicity.

Tannic compounds present at fairly high concentrations contribute 30–50% of the COD of the debarking wastewaters and inhibit methanogenesis (Field et al. 1988, 1991). Dilution of wastewater or polymerization of toxic tannins to high molecular weight compounds by autooxidation at high pH as the only treatment (Field et al. 1991) was shown to enable anaerobic treatment of debarking effluents.

13.3.3 Fungal Treatment

Fungi have been harnessed and utilized by humans for thousands of years for many diverse applications. In response to demand for innovative technologies to degrade recalcitrant materials, fungi have been used and found to have nonspecific ability to degrade many of the recalcitrant chemicals, including PCB's PCP, DDT, and several other polycyclic hydrocarbons (Bumpus and Aust 1995). Work with fungi-based biological processes has shown that certain fungi are capable of degrading complex xenobiotic chemicals (organochlorines) and sorb heavy metals from aqueous solutions (Kapoor and Viraraghavan 1995).

Fundamental research on biological treatment of pulp mill wastewaters especially bleach effluents has been considered as one of the important fields of study during the last 3 decades. The research indicates that white rot fungi (*P. chrysosporium* and *T. versicolor*) are the known microbes capable of degrading and decolorizing bleach plant effluents. White rot fungi have been evaluated in trickling filters, fluidized bed reactors, and airlift reactors at bench scale and found technologically feasible (Pellinen et al. 1988a, b; Prouty 1990). Only mycelia color removal (MyCoR) process which uses *P. chrysosporium* to metabolize lignin color bodies has crossed the bench scale and has been evaluated at pilot scale level (Campbell et al. 1982; Jaklin-Farcher et al. 1992) and found to be very efficient in destroying organochlorines. However, no reactors/process studied so far have been found economically feasible because of the following reasons:

1. Energy required for lignins/chloro-lignin degradation by white rot fungi has to be derived from an easily metabolizable, low molecular mass sugars.
2. Process is not self-sustaining from the angle of growth of white rot fungi used.

Factors affecting fungal treatment of pulp mill effluents/bleach effluents include concentration of nutrients and dissolved oxygen, pH, and temperature. Fungi, like other living organisms, require certain essential minerals for their growth. The essential mineral nutrients required can be divided into two categories, viz., macronutrients required at 10^{-3} M or more, and micronutrients required at 10^{-6} M or less. Fungal decolorization involves a series of complex reactions many of which are catalyzed by enzymes. The addition of mineral solution presumably activates the specific enzymes necessary for normal metabolism, growth, and decolorization. The fungus can tolerate a wide range of pH and temperature during decolorization compared to the growth stage. Decolorization is maximal under high oxygen concentration and the fungus requires a carbon source such as glucose or cellulose.

A small addition of nitrogen is required to sustain decolorization because nitrogen is lost from the system by the extracellular enzyme secreted by the fungus.

To identify the potential fungal strains for the treatment of bleach effluents, many researchers have screened cultures obtained from a variety of sources. Most of the papers dealt with effluent supplemented with nutrients. Fukuzumi et al. (1977) were probably the first to study the use of white rot fungi for effluent treatment. The fungi were grown in Erlenmeyer flasks in a liquid medium containing nutrients, vitamins, and spent liquor from the first alkali extraction stage of pulp bleaching. Among the fungi selected from 29 species of tropical fungi and 10 species of Japanese isolates, *Tinctoporia* sp. showed the best results for decolorization of the extraction stage effluents. *Phlebia brevispora, Phlebia subserialis, Poria cinerascens,* and *T. versicolour* were tested by Eaton et al. (1982) and found to reduce the effluents color efficiently. In another study (Livernoche et al. 1983), 15 strains of white rot fungi were screened for their ability to decolorize bleaching effluents. Five fungal strains – *T. versicolor, P. chrysosporium, Pleurotus ostreatus, Polyporus versicolour,* and one unidentified strain – showed decolorizing activity. *T. versicolor* was found to be most efficient in shaken cultures. Galeno and Agasin (1990) evaluated several white rot fungi collected in the south of Chile for their ability to decolorize bleaching effluents and found *Ramaira* sp. strain 158 to have the highest potential. Over 90% of the color (initial color 14,500 color units) was removed after 140 h under air with a similar rate and extent of decolorization as *P. chrysosporium* did under oxygen.

The addition of an easily metabolizable nutrient such as glucose or cellulose is required for obtaining the maximum decolorization efficiency with most of the white rot fungal cultures. However, this would increase the operational cost of the process. Moreover, if the added nutrients are not completely consumed during the decolorization stage, they could increase the BOD and COD of the effluents after fungal treatment. Esposito et al. (1991) and Lee et al. (1994) examined fungi that showed efficient decolorization of the extraction stage effluents without any addition of nutrients. Through a screening of 51 ligninolytic strains of fungi, the *Lentinus edodes* strain was shown to remove 73% of the color in 5 days without any additional carbon source. Under these conditions, *L. edodes* was more efficient than the known *P. chrysosporium* strains (Esposito et al. 1991). Lee et al. (1994) screened fungi having high decolorization activity. The fungus KS-62 showed 70 and 80% reduction of the color after 7 and 10 days of incubations, respectively. To obtain a reasonable basis for evaluation of an industrial fungal treatment, Lee et al. (1995) performed treatment of the extraction stage effluent with the immobilized mycelium of the fungus KS-62. This fungus showed 70% color removal (initial color 6,600 PCU) without any nutrient within 1 day of incubation with 4 times effluent replacement; however, the color removal started to decrease at the fifth replacement with the fresh extraction stage effluent. The decolorization activity of the fungus was restored by one replacement of extraction stage effluent containing 0.5 of glucose and the high decolorization was continuously observed for four replacements in the absence of glucose. With the fungus KS-62, such decolorization activity was reportedly obtained for 29 days of total treatment period. Through screening of 100 strains at low glucose concentration, *Rhizopus oryzae* – a zygomycete and *Ceriporiopsis subvermispora* – a wood degrading white rot fungi were shown to remove 95 and 88% of the color, respectively. Even in the absence of carbohydrates, significant amount of color reductions was achieved (Nagarathnamma et al. 1999a, b). Glucose has been found to be the most effective cosubstrate for decolorization by most of the white rot fungi (Nagarathnamma et al. 1999a, b; Bajpai et al. 1993; Mehna et al. 1995; Fukuzumi 1980; Prasad and Joyce 1991; Bergbauer et al. 1991; Pallerla and Chambers 1995). Belsare and Prasad (1988) showed that the decolorization efficiency of *Schizophyllum commune* could be rated in the following order: sucrose (60%), glucose (58%), cellulose (35%), and pulp (20%). With the fungus *Tinctoporia*, ethanol was also found to be very effective cosubstrate for decolorization of waste liquor (Fukuzumi 1980). Ramaswammy (1987) observed that addition of 1% bagasse pith as a supplementary carbon source resulted in 80% color reduction in 7 days with *S. commune*. Eaton et al. (1982) compared the suitabilities of three primary sludges and combined sludge with that of cellulose powder for use as a carbon source for *P. chrysosporium* cultures. Archibald et al. (1990) reported that *T. versicolor* removed color efficiently in the presence of inexpensive sugar refining or brewery waste. With *R. oryzae* (Nagarathnamma et al. 1999a, b), maximum decolorization of the order of 92% was obtained with addition of glucose in 24 h. Ninety percent color reduction was measured with mycrocrystalline cellulose and lactose, 89% was measured with sucrose, and 88% was measured with CMC and Xylose. Starch and ethyl alcohol showed about 87 and 84% color reduction, respectively.

P. chrysosporium has been the most studied white rot fungus for waste treatment. Eaton et al. (1980) studied extensively, for the first time, the application of this fungus for the treatment of bleaching effluents. Their report indicated that 60% decolorization of extraction stage effluent (initial color 3,500 PCU) could be accomplished with *P. chrysosporium* in shake flasks. The same mycelium could be recycled up to 60 days or 6 successive batches. Mittar et al. (1992) also showed that under shaking conditions, the 7-day-old growth of the culture at 20% (v/v) inoculum concentrations resulted in maximum decolorization (70%) of the effluent along with more than 50% reduction in BOD and COD.

Sundman et al. (1981) studied the reactions of the chromophoric material of extraction stage effluent during the fungal treatment without agitation. The results of these studies showed no preference towards degradation of lower molecular weight polymeric material over high molecular weight material. They noticed that the yield of high molecular weight material decreased to half during the fungal treatment. As the color also decreased by 80%, they concluded that chromophores were destroyed. Further, they noticed that the fungal attack led to a decrease in the content of phenolic hydroxyl groups and to an increase in oxygen content.

Joyce and Pellinen (1990) have explored ways to use white rot fungus to decolorize and detoxify pulp and paper mill effluents. They proposed a process termed FPL NCSU MyCoR using *P. chrysosporium* for decolorization of pulp mill effluents. It resulted from the cooperative research between the U.S. Forest Products Laboratory and North Carolina State University. A fixed film MyCoR reactor is charged with growth nutrients which can include primary sludge as the carbon source and is inoculated with the fungus. The sludge will provide some of the required mineral nutrients and trace elements as well as carbon. Nitrogen rich secondary sludge can be also used to supply the nitrogen required for growth. After the mycelium has grown over the reactor surface, it depletes the available nitrogen and becomes ligninolytic (pre-growth stage 2–4 days). The reactor is then ready for use. Operations for over 60 days have been achieved in bench reactors in a batch mode. This process converts to 70% of the organic chlorides to inorganic chlorides in 48 h while decolorizing the effluent and reducing both COD and BOD by about half.

Huynh et al. (1985) used the MyCoR process for the treatment of chlorinated low molecular mass phenols of the extraction stage effluent. It was found that most of the chlorinated phenols and low molecular mass components of the effluent were removed during the fungal treatment. Pellinen et al. (1988a, b) have reported that the MyCoR process can be considerably improved in terms of COD removal by simply using less glucose as the carbon source for the fungi – *P. chrysosporium*. However, the decolorization was reported to be faster at high glucose concentration. Yin et al. (1989b) studied the kinetics of decolorization of extraction stage effluent with *P. chrysosporium* in an RBC under improved conditions. The kinetic model developed for 1 and 2 days retention times showed a characteristic pattern. The overall decolorization process can be divided into three stages, viz., a rapid color reduction in the first hour of contact between the effluent and the fungus followed by a zero-order reaction and then a first-order reaction. The color removal rate on the second day of the 2-day batch treatment was less than that on the first day. The decolorization in a continuous flow reactor achieved approximately the same daily color removal rate, but the fungus had a larger working life than when in the batch reactor, thereby removing more color over the fungal life time. Pellinen et al. (1988a, b) studied decolorization of high molecular mass chlorolignin in first extraction stage effluent with white rot fungus – *P. chrysosporium* immobilized on RBC. The AOX decreased almost by 50% during one day treatment. Correlation studies suggested that decolorization and degradation of chlorolignin (as COD decrease) are metabolically connected, although these processes have different rates.

The combined treatment of extraction stage effluent with white rot fungi and bacteria has been also reported. Yin et al. (1990) studied a sequential biological treatment using *P. chrysosporium* and bacteria to reduce AOX, color, and COD in conventional softwood kraft pulp bleaching effluent. In six variations of the white rot fungus/bacterial systems studied, only the degree of fungal treatment was varied. In three of the six variations, ultrafiltration was also used to concentrate high molecular mass chlorolignins and to reduce effluent volume (and thus cost) prior to fungal treatment. The best sequence, using ultrafiltration/white rot fungus/bacteria, removed 71% TOCl, 50% COD, and 65% color in the effluent. Fungal treatment enhances the ability of bacteria to degrade and dechlorinate chlorinated organics in the effluent.

The degradation of model compounds – chlorophenols, and chloroguaiacols in pure water solution by fungal treatment using an RBC has been studied by Guo et al. (1990). It was found that at concentration of 30 mg/L, 80–85% of chlorophenols and chloroguaiacols could be degraded after 3–4 h treatment.

Prouty (1990) proposed an aerated reactor in order to eliminate some of the problems associated with the RBC process. The fungal life in the aerated reactor was longer and the color removal rate was significantly higher than those of the RBC process in an air atmosphere. A preliminary economic evaluation of the RBC process indicated that the rate of decolorization and the life span of the fungus are the most critical factors (Joyce and Pellinen 1990). Yin et al. (1989b) and Yin (1989) suggested that treatment of the extraction stage effluent by ultrafiltration before RBC treatment would be economically attractive. Their study also suggested that combining ultrafiltration and the MyCoR system could maximize the efficiency of the MyCoR process and reduce the treatment cost, thereby making the process more economically feasible for industrial use.

Although the MyCoR process was efficient in removing color and AOX from bleaching effluents, it also had certain limitations. The biggest problem was the relatively short active life of the reactor. Therefore, several other bioreactors such as packed bed and fixed bed reactors were studied (Lankinen et al. 1991; Messner et al. 1990; Cammarota and Santanna 1992). The use of trickling filter-type bioreactor, in which the fungus immobilized on porous carrier material, was adopted in the MyCOPOR system (Messner et al. 1990). For extraction stage effluent with an initial color between 2,600 and 3,700 PCU, the mean rate of color reduction was 60% during consecutive 12 h run. The mean AOX reduction value at a color reduction of 50–70% in 12 h was 45–55%. Cammarota and Santanna (1992) developed a continuous packed bed bioreactor in which *P. chrysosporium* was immobilized on polyurethane foam particles. The bioreactor operation at a hydraulic retention time of 5·8 days was able to promote 70% decolorization. In comparison with the MyCoR process, the fungal biomass could be maintained in this process for at least 66 days without any appreciable loss of activity.

To apply the MyCOPOR process on an industrial scale, relatively big reactors (diameter, 70 and 100 mm; volume 4–16 L) were prepared and filled with polyurethane-foam cubes (1 cm^3) as carrier material. Long-term experiments were successfully carried out and it was decided to build a small pilot reactor at a large paper mill in Austria (Jaklin-Farcher et al. 1992). However, many aspects related to the operating conditions must be further investigated and improved. A disadvantage of these treatment processes is that *P. chrysosporium* required high concentrations of oxygen and energy sources such as glucose or cellulose as well as various basal nutrients, mineral solution, and tween 80 (Messner et al. 1990). Kang et al. (1996) developed a submerged biofilter system in which mycelia of *P. chrysosporium* were attached to media (net ring type) and used to dispose wastewater from a pulp mill. Maximum reduction of BOD, COD, and lignin concentrations were 94, 91, and 90%, respectively, in 12 h of hydraulic retention time.

Fukui et al. (1992) determined the toxicity by the microtox bacterial assay of E_p (alkaline extraction with hydrogen peroxide) effluent and ultrafiltration fractionated E_p effluent before and after fungal treatment. The overall toxicity of unfractionated effluent was reduced; however, fungal degradation of higher molecular weight fractions led to an increase in toxicity because of the generation of lower molecular weight compounds after enzyme cleavage.

Matsumoto et al. (1985) demonstrated that RBC treatment of extraction stage effluent was effective for the removal of organically bound chlorine as well as color. Removal of AOX was determined to be 62, 43, and 45% per day for the low molecular weight fraction of extraction stage effluent, high molecular weight fraction of the same, and unfractionated extraction stage effluent, respectively. After further optimization, 49% of the high molecular weight AOX was transformed to inorganic chloride in 1 day and 62% in 2 days. The chloride concentration increased simultaneously with decreasing AOX including decolorization.

Singhal et al. (2005) studied treatment of pulp and paper mill effluent by *P. chlysosporium* at two different pH, 5.5 and 8.5. At both the pH, color, COD, lignin content, and total phenols of the effluent significantly declined after bioremediation. However, greater decolorization and reduction in COD, lignin content, and total phenols were observed at pH 5.5. Such bioremediated effluent of pulp and paper mill could gainfully be utilized for crop irrigation.

Egyptian researchers applied the fungus, *P. chrysosporium* DSMZ 1,556, to the microbiological processing of mill effluents (Abdel-Fattah et al. 2001). Experiments were conducted to compare the decolorization of paper mill effluents using this fungus under free cell, repeated batch, and coimmobilization systems. Immobilization and coimmobilization of the fungus was accomplished using alginate and activated charcoal. A twofold increase in color reduction was achieved using fungus that was immobilized in alginate compared with alginate used alone as bioadsorbent. Similarly, a further 40% increase in decolorization was found to occur with the cells coimmobilized with alginate and charcoal compared with alginate and charcoal used alone. The results are ascribed to the ability of the immobilization and the protective barrier formed by the adsorbent to provide greater control over the remediation process.

Another white rot fungus – *C. versicolour* removed 60% of the color of combined bleach kraft effluents within 6 days in the presence of sucrose (Livernoche et al. 1983). Decolorization of effluent was more efficient when the concentration of sucrose and inoculum was high. When the fungus was immobilized in calcium alginate gel, it removed 80% color from the same effluent in 3 days in the presence of sucrose. The decolorization process affected not only the dissolved chromophores but also the suspended solids. The solids after centrifugation of the zero time samples were dark brown while the solids after 4-day incubation were light brown. The beads with the immobilized mycelium remained light colored throughout the experiments with no indication of accumulation of the effluent chromophores. Recycled beads were found to remove color efficiently and repeatedly in the presence of air but not under anaerobic conditions. Biological reactors of the airlift type using calcium alginate beads to immobilize the fungus *C. versicolour* have been used to study the continuous decolorization of kraft mill effluents (Royer et al. 1985). The effluent used contained only sucrose and no other nutrient source. An empirical kinetic model was proposed to describe the decolorization process caused by this fungus, but it did not shed any light on the chemical mechanism involved in the decolorization.

Bergbauer et al. (1991) showed that *C. versicolour* efficiently degraded chlorolignins from bleaching effluents. More than 50% of the chlorolignins were degraded in 9-day incubation period, resulting in a 39% reduction in AOX and 84% decrease in effluent color. In a 3-L laboratory fermenter, with 0.8% glucose and 12 mM ammonium sulphate, about 88% color reduction was achieved in 3 days. Simultaneously, the concentration of AOX dropped from 40 to 21.9 mg/L, a 45% reduction in 2 days.

Direct use of suspended mycelium of the fungus *C. versicolour* may not be feasible because of the problem of viscosity, oxygen transfer, and recycling of the fungus. The fungus was therefore grown in the form of pellets, thus eliminating the problems with biomass recycling and making it possible to use a larger amount (Royer et al. 1985). Rate of decolorization with fungal pellets was almost 10 times as high in batch culture as in continuous culture under similar conditions. The capacity for decolorization decreased markedly with increase in lignin loading (Royer et al. 1985).

Bajpai et al. (1993) reported 93% color removal and 35% COD reduction, from first extraction stage effluent (7,000 PCU) with mycelial pellets of *C. versicolour* in 48 h in batch reactor, whereas, in a continuous reactor, the same level of color and COD reduction was obtained in 38 h. No loss in decolorization ability of mycelial pellets was obtained when the reactor was operated continuously for more than 30 days. Mehna et al. (1995) also used *C. versicolor* for decolorization of effluents from a pulp mill using agriresidues. With an effluent of 18,500 color units, the color reduction of 88–92% with COD reduction of 69–72% was obtained. Royer et al. (1991) described the use of pellets of *C. versicolour* to decolorize ultrafiltered kraft liquor in nonsterile conditions with a negligible loss of activity. The rate of decolorization was observed to be linearly related to the liquor concentration and was lower than that obtained in the MyCoR process. This could be due to lower temperature used in this work and to the use of pellets with relatively large diameters which could limit the microbial activity as compared to the free mycelium used in the MyCoR process. An effective decolorization of effluent having 400–500 color U/L can be obtained in presence of a simple carbon source such as glucose. In the repeated batch culture, the pellets exhibited a loss of activity dependent on the initial color concentration. Simple carbohydrates were found to be essential for effective decolorization with this fungus and a medium composed of inexpensive industrial by-products provided excellent growth and decolorization (Archibald et al. 1990).

Pallerla and Chambers (1996) have shown that immobilization of *T. versicolor* in urethane prepolymers leads to significant reductions in color and chlorinated organic levels in the treatment of kraft bleach effluents. Color reduction ranging from 72 to 80% and AOX reduction ranging from 52 to 59% are possible from a continuous bioreactor at a residence time of 24 h. The highest color removal rate of 1,920 PCU per day was achieved at an initial color concentration of 2,700 PCU. The decolorization process was linearly dependent on the concentration of glucose cosubstrate up to a level of 0.8% by weight. The biocatalyst remained intact and stable after an extended 32-day operation.

Treatment of extraction stage effluent with ozone and the fungus *C. versicolour* has also been tried (Roy-Arcand and Archibald 1991b). Both ozone treatment and biological treatment effectively destroyed effluent chromophores but the fungal process resulted in greater degradation as expressed by COD removal. Monoaromatic chlorophenolics and toxicity were removed partially by ozone and completely by *C. versicolour*. Molecular weight distributions showed roughly equal degradation of all sizes of molecules in both the treatments. The combination of a brief ozone treatment with a subsequent fungal treatment revealed a synergism between the two decolorization mechanisms on extraction stage effluent. Effluent was pretreated with ozone (110–160 mg/L) or *C. versicolour* (24 h with 2–5 g/L wet weight fungal biomass). The pretreatment was followed by 5-day incubation with *C. versicolour*. It was noted that partial color removal by ozone pretreatment allowed more effective removal by the fungus than that by fungal pretreatment. After 20 h, 46–53% decolorization was observed for ozone pretreated effluents, compared to 29% for fungal treatment alone. The contribution of ozone seemed to be most important in the first 24 h following the pretreatment. Ozone pretreatment also produced a small improvement in the bioavailability of effluent organics to the fungus. A partial replacement of chlorine by ozone in the bleach plant or a brief ozone pretreatment of extraction stage effluent should considerably reduce the low molecular mass toxic chlorophenolics. In addition, the use of ozone should also improve decolorization by subsequent fungal and possibly bacterial treatments.

A white rot fungus, *Tinctoporia borbonica,* has been reported to decolorize the kraft waste liquor to a light yellow color (Fukuzumi 1980). About 99% color reduction was achieved after 4 days of cultivation. Measurement of the culture filtrate by ultraviolet spectroscopy showed that the chlorine-oxylignin content also decreased with time and measurement of the culture filtrate plus mycelial extract after 14 days of cultivation showed the total removal of the chlorine-oxylignin content.

Another white rot fungus – *S. commune* has also been found to decolorize and degrade lignin in pulp and paper mill effluent (Belsare and Prasad 1988). The fungus was able to degrade lignin in the presence of an easily metabolizable carbon source. The addition of carbon and nitrogen not only improved the decolorizing efficiency of the fungus but also resulted in reduction of the BOD and COD of the effluent. Sucrose was found to be best carbon source for the degradation of the lignin. A 2-day incubation period was sufficient for lignin degradation by this fungus. Under optimum conditions, this fungus reduced the color of the effluent by 90% and also reduced BOD and COD by 70 and 72% during a 2-day incubation.

Duran et al. (1991) reported that preradiation of the effluent, followed by fungal culture filtrate treatment resulted in efficient decolorization. Moreover, when an effluent preirradiated in the presence of ZnO was treated with *L. edodes* (Esposito et al. 1991), a marked enhancement of the decolorization at 48 h was obtained (Duran et al. 1994). They proposed that the combined photo-biological decolorization procedure appears to be an efficient decontamination method with potential for industrial effluent treatment.

White rot fungus *C. subvermispora* has been found to decolorize, dechlorinate, and detoxify the pulp mill effluents at low cosubstrate concentration (Nagarathnamma

Table 13.8 Effect of treatment with *C. subvermispora* CZ-3 on chlorophenols and chloroaldehydes in the effluent from extraction stage

Compounds	Untreated effluent (mg/L)	Treated effluent (mg/L)	Removal[a] (%)
2-Chlorophenol	14.20	8.88	36.5±1.7
4-Chlorophenol	48.60	3.20	93.4±2.4
3-Chlorocatechol	1.82	Nil	100
6-Chloroguaiacol	90.12	31.59	67.0±2.6
5-Chloroguaiacol	3,202.00	184.70	94.0±1.9
3,6-Dichloroguaiacol	50.00	1.95	96.0±1.6
3,6-Dichlorocatechol	4.92	Nil	100
4,5-Dichlorocatechol	50.27	Nil	100
3,4,5-Trichloroguaiacol	0.147	Nil	100
3,4,6-Trichlorocatechol	4.14	Nil	100
4,5,6-Trichloroguaiacol	8.89	2.39	73±2.0
Pentachlorophenol	2.11	Nil	100
Trichlorosyringaldehyde	2.38	1.29	45.9±1.8
Tetrachlorocatechol	64.35	Nil	100
2,6-Dichlorosyringaldehyde	46.42	16.50	64.5±2.2

Based on Nagarathnamma et al. (1999a, b)
[a]Results are reported as mean of three measurements, ± the standard deviation

et al. 1999a, b). The fungus removed 91% color and 45% COD in 48 h under optimum conditions. The reductions in lignin, AOX, and EOX were 62, 32, and 36%, respectively. The color removal rate was 3,185 PCU/day at an initial color concentration of 7,000 PCU. Monomeric chlorinated aromatic compounds were removed almost completely and toxicity to Zebra fish was eliminated (Table 13.8).

A zygomycete *R. oryzae* has been reported to decolorize, dechlorinate, and detoxify extraction stage effluent at low cosubstrate concentration. Optimum conditions for treatability were determined as pH 3–4.5 and temperature 25–40°C (Nagarathnamma et al. 1999a, b). Under optimum conditions, the fungus removed 92–95% color, 50% COD, 72% AOX, and 37% EOX and complete removal of monoaromatic phenolics and toxicity. Significant reduction in chlorinated aromatic compounds was observed and toxicity to zebra fish was completely eliminated (Table 13.9). The molecular weight of chlorolignins was substantially reduced after the fungal treatment. Another thermotolerant zygomycete strain *Rhizomucor pusillus* RM 7 could remove up to 71% of color and substantially reduce COD, toxicity, and AOX levels in the effluent (Christov and Steyn 1998).

Kannan (1990) reported about 80% color removal and over 40% BOD and COD reduction with fungus *Aspergillus niger* in 2 days. Tono et al. (1968) reported that *Aspergillus* sp. and *Penicillium* sp. achieved 90% decolorization in 1 week's treatment at 30°C and at pH 6.8. Later Milstein et al. (1988a, b) reported that these microorganisms removed appreciable levels of chlorophenols as well as chloroorganics from the bleach effluent. Gokcay and Taseli (1997) have reported over 50% AOX and color removal from softwood bleach effluents in less than 2 days of contact with *Penicillium* sp. Bergbauer et al. (1992) reported AOX reduction by 68%

Table 13.9 Effect of treatment with *R. oryzae* on chlorophenols and chloroaldehydes in the effluent from extraction stage

Compounds	Untreated effluent (mg/L)	Treated effluent (mg/L)	Removal (%)
2-Chlorophenol	14.20	Nil	100
4-Chlorophenol	48.60	2.90	94 ± 1.8
3-Chlorocatechol	1.82	Nil	100
6-Chloroguaiacol	90.12	Nil	100
5-Chloroguaiacol	3,202.00	Nil	100
3,6-Dichloroguaiacol	50.00	Nil	100
3,6-Dichlorocatechol	4.92	Nil	100
4,5-Dichlorocatechol	50.27	Nil	100
3,4,5-Trichloroguaiacol	0.147	Nil	100
3,4,6-Trichlorocatechol	4.14	Nil	100
4,5,6-Trichloroguaiacol	8.89	2.49	72 ± 1.9
Pentachlorophenol	2.11	Nil	100
Trichlorosyringaldehyde	2.38	Nil	100
Tetrachlorocatechol	64.35	27.02	58 ± 2.2
2,6-Dichlorosyringaldehyde	46.42	Nil	100

Based on Nagarathnamma et al. (1999a, b)

[a]Results are reported as mean of three measurements, ± the standard deviation

and color reduction by 90% in 5 days with the coelomycetous fungus *Stagonospora gigaspora*. Toxicity of the effluent was reduced significantly with this fungus. Few marine fungi have been also reported to decolorize the bleach plant effluents (Raghukumar et al. 1996, 2008). With *Trichoderma* sp. under optimal conditions, total color and COD decreased by almost 85 and 25%, respectively, after cultivation for 3 days (Prasad and Joyce 1991).

Wu et al. (2005) explored the lignin-degrading capacity of attached-growth white rot fungi. Five white rot fungi, *P. chrysosporium*, *P. ostreatus*, *L. edodes*, *T. versicolor*, and S22, grown on a porous plastic media, were individually used to treat black liquor from a pulp and paper mill. Over 71% of lignin and 48% of COD were removed from the wastewater. Several factors, including pH, concentrations of carbon, nitrogen, and trace elements in wastewater, all had significant effects on the degradation of lignin and the removal of COD. Three white rot fungi, *P. chrysosporium*, *P. ostreatus,* and S22, showed high capacity for lignin degradation at pH 9.0–11.0. The addition of 1 $g\,L^{-1}$ glucose and 0.2 $g\,L^{-1}$ ammonium tartrate was beneficial for the degradation of lignin by the white rot fungi studied.

Malaviya and Rathore (2007) reported bioremediation of pulp and paper mill effluent by an immobilized fungal consortium for the first time. They immobilized two basidiomycetous fungi (*Merulius aureus* syn. *Phlebia* sp. and an unidentified genus) and a deuteromycetous fungus (*Fusarium sambucinum* Fuckel MTCC 3,788) on nylon mesh and used the consortium for bioremediation of pulp and paper mill effluent in a continuously aerated bench-top bioreactor. The treatment resulted in the reduction of color, lignin, and COD of the effluent in the order of 78.6, 79.0, and 89.4% in 4 days. A major part of reductions in these parameters occurred within first 24 h of the treatment, which was also characterized by a steep decline in the pH of the effluent. During this period, total dissolved solids, electrical conductivity, and salinity of the effluent also registered marked decline.

Singhal and Thakur (2009) took up genotoxicity analysis along with effluent treatment to evaluate the efficiency of biological treatment process for safe disposal of treated effluent. Four fungi were isolated from sediments of pulp and paper mill in which PF4 reduced color (30%) and lignin content (24%) of the effluent on third day. The fungal strain was identified as *Emericella nidulans* var. *nidulans* (anamorph: *Aspergillus nidulans*) on the basis of rDNA ITS1 and rDNA ITS2 region sequences. The process of decolorization was optimized by Taguchi approach. The optimum conditions were temperature (30–35°C), rpm (125), dextrose (0.25%), tryptone (0.1%), inoculum size (7.5%), pH (5), and duration (24 h). Decolorization of effluent improved by 31% with reduction in color (66.66%) and lignin (37%) after treatment by fungi in shake flask. Variation in pH from 6 to 5 had most significant effect on decolorization (71%) while variation in temperature from 30 to 35°C had no effect on the process. Treated effluent was further evaluated for genotoxicity by alkaline single cell gel electrophoresis (SCGE) assay using *S. cerevisiae* MTCC 36 as model organism, indicated 60% reduction.

Chuphal et al. (2005) applied *Paecilomyces* sp. and *Pseudomonas syringae* pv *myricae* (CSA105) for treatment of pulp and paper mill effluent in a two-step and three-step fixed film sequential bioreactor containing sand and gravel at the bottom of the reactor for immobilization of microbial cells. The microbes exhibited significant reduction in color (88.5%), lignin (79.5%), COD (87.2%), and phenol (87.7%) in two-step aerobic sequential bioreactor, and color (87.7%), lignin (76.5%), COD (83.9%), and phenol (87.2%) in three-step anaerobic–aerobic sequential bioreactor.

Selvam et al. (2002) used white rot fungi *Fomes lividus* and *T. versicolor*, isolated from the Western Ghats region of Tamil Nadu, India, to treat pulp and paper industry effluents on a laboratory scale and in a pilot scale. On the laboratory scale a maximum decolorization of 63.9% was achieved by *T. versicolor* on the fourth day. Inorganic chloride at a concentration of 765 mg/L, which corresponded to 227% of that in the untreated effluent, was liberated by *F. lividus* on the tenth day. The COD was also reduced to 1,984 mg/L (59.3%) by each of the two fungi. On the pilot scale, a maximum decolorization of 68% was obtained with the 6-day incubation by *T. versicolor*, inorganic chloride 475 mg/L (103%) was liberated on the seventh day by *T. versicolor*, and the COD was reduced to 1,984 mg/L corresponding to 59.32% by *F. lividus*. These results suggested that *F. lividus* seems to be another candidate efficient for dechlorination of wastewater.

Ragunathan and Swaminathan (2004) studied the ability of *Pleurotus* spp. – such as *P. sajor-caju*, *P. platypus*, and *P. citrinopileatus* – to treat pulp and paper mill effluent on a laboratory and pilot scale. On the laboratory scale treatment, *P. sajor-caju* decolorized the effluent by 66.7% on day 6 of incubation. Inorganic chloride liberated by *P. sajor-caju* was 230.9% (814.0 mg/dL) and the COD was reduced by 61.3% (1,302.0 mg/dL) on day 10 of treatment. In the pilot scale treatment, maximum decolorization was obtained by *P. sajor-caju* (60.1%) on day 6 of the incubation. Inorganic chloride content was increased by 524.0 mg/dL (113.0%) and the

COD was reduced by 1,442.0 mg/dL (57.2%) by *P. sajor-caju* on day 7 of incubation. These results revealed that the treatment of pulp and paper mill effluent by *P. sajor-caju* proved as better candidate for the purpose than *P. platypus* and *P. citrinopileatus*.

Belém et al. (2008) used *Pleurotus sajor-caju* and *P. ostreatus* to promote degradation of organic matter and remove color from kraft pulp mill effluent by an activated sludge process. Absorbance reduction of 57 and 76% was observed after 14 days of treatment of final effluent with glucose by *P. sajor-caju*, at 400 and 460 nm, respectively. Lower values of absorbance reduction were observed in final effluent with additives and inoculated with the same species (22–29%). Treatment with *P. ostreatus* was more efficient in the effluent with additives, 38.9–43.9% of reduction. Higher growth rate of *P. sajor-caju* was observed in the effluent with glucose. Biological treatment resulted in 65–67% reduction of COD after 14 days revealing no differences for each effluent composition and inoculated species. Profiles of composition of organic compounds obtained by GC-MS showed no significant differences between the two effluents treated with *P. sajor-caju* or *P. ostreatus*, but longer incubation time reflected higher reduction of organic compounds.

Pendroza et al. (2007) carried out lab experiments with the fungus *T. versicolor* to see the effect of using a sequential biological and photocatalytic treatment on COD, color removal, the degradation of chlorophenolic compounds in bleaching effluent, produced during paper manufacture. *T. versicolor* was cultured in an Erlenmeyer flask with wheat bran broth and 100 polyurethane foam (PDF) cubes, 1 cm^3 in size. The culture was incubated for 9 days at 25°C. Samples of fungi were placed in effluent samples. This was tested for its chemical, physical, and microbiological characteristics before *T. versicolor* was added, and after 4 days treatment with the fungus. Some samples were then treated with UV/titanium dioxide/ruthenium (Ru)-selenium (Se) chalcogenide for 20 min. After treatment with *T. versicolor* there was an 82% reduction in COD and color, and significant reductions in chlorophenols. When this was followed by photolysis with titanium dioxide/Ru-Se, COD fell by 97%, and there was a 92% reduction in color. The chlorophenols were reduced by 99%.

Shintani et al. (2002) used a newly isolated fungus *Geotrichium candidum* Dec 1, capable of decolorizing a wide range of synthetic dyes for the treatment of kraft pulp bleaching effluent. With a glucose content of 30 g/L, a color removal of 78% and a reduction in adsorbable organic halogen (AOX) concentration of 43% could be obtained after 1 week. Decolorization was not observed in the absence of added glucose. The average molecular weight of colored substances was reduced from 5,600 to below 3,000. It would appear that *G. candidum* Dec 1 has a different mechanism to that of peroxidase, manganese peroxidase, and laccase in the decolorization of bleaching effluents. Color removal is believed to proceed via color adsorption to the cells followed by decomposition of the adsorbed materials.

A study was undertaken by van Driessel and Christov (2001) to evaluate the bioremediating abilities of *C. versicolor* (a white rot fungus) and *R. pusillus* (a mucoralean fungus) applied to plant effluents with special emphasis on the color removal from a bleaching plant effluent and the mechanisms involved. Effluent decolorization was studied in a RBC reactor. The decolorization by both fungi was

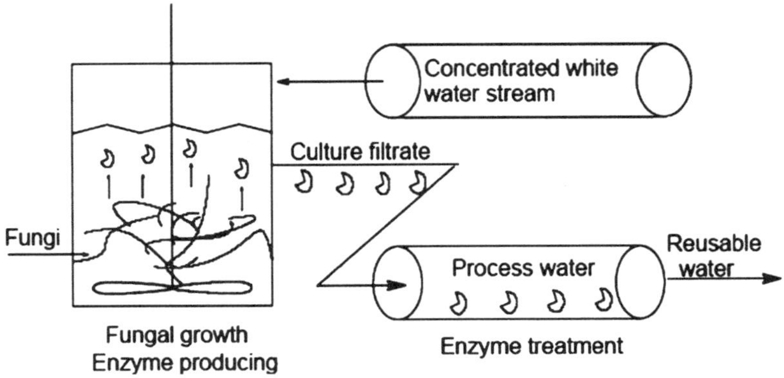

Fig. 13.4 The principle of combined fungal and enzyme treatment system. Based on Zhang (2001)

directly proportional to initial color intensities and its extent was not adversely affected by color intensity, except at the lowest level tested. Decolorization of 53–73% was found to be attainable using a hydraulic retention time of 23 h. With *R. pusillus*, 55% of adsorbable organic halogen (AOX) were removed compared to 40% by *C. versicolor*. Fungal treatment with both *R. pusillus* and *C. versicolor* rendered the effluent virtually nontoxic and the addition of glucose to the decolorization media stimulated color removal by *C. versicolor*, but not with *R. pusillus*. Ligninolytic enzymes (manganese peroxidase and laccase) were only detected in effluent treated by *C. versicolor*. There are definite differences in the decolorizing mechanisms between the *C. versicolor* fungus (adsorption plus biodegradation) and the mucoralean fungus (adsorption).

Sequential treatment using fungal process followed by photo-catalytical treatment on COD, color removal, degradation of chlorophenolic compounds in bleach effluent has been studied by Pendroza et al. (2007). The overall reduction was 97% in COD, 92% in color, and 99% in chlorophenols.

The potential of using a combined fungal and enzyme system as an internal treatment was investigated for the control of the build-up of detrimental dissolved and colloidal substances present in a thermo-mechanical pulp (TMP)/newsprint mill process water (Zhang 2001). The experimental data obtained concerning the composition of a typical TMP process white water provided a better understanding of the influence that the dissolved colloidal substances (DCS) components in the recycling white water system had on the paper properties and also helped in the design of an appropriate treatment technology. The white rot fungus, *T. versicolor*, was shown to be able to grow on unsupplemented process waters, while effectively removing white water organics. The fungus also produced a large spectrum of enzymes during its growth on the different white water streams and the fungal enzyme treatments resulted in a significant degradation or modification of various DCS components. A combined fungal and enzyme system could possibly be used as an internal treatment "kidney" to remove detrimental organic substances present in a TMP/newsprint mill with a closed water system. A concept flow diagram for the combined fungal and enzymatic treatment system is shown in Fig. 13.4.

Different studies have shown that a combined fungal enzyme system showed promise as one way of decreasing the detrimental substances present within a closed water system, while treatments with different enzyme preparations revealed that fungal laccases play an important part in the removal of white water extractives. Potentials of laccase enzymes to modify model extractive compounds found in thermo-mechanical pulp (TMP)/newsprint process waters were investigated by Zhang et al. (2001). The model compounds used were representative of the fatty acids, resin acids, and triglycerides found in mill process waters. It was found that these compounds were significantly degraded or modified by the laccase treatments.

Sorce Inc. technology uses a combination of fungi and facultative bacteria to degrade the highly cross-linked structure of lignin. This technology was started up on a Southeastern Kraft Pulp Mill with a wastewater flow of 15 million gallons/day. The starting color value of the wastewater averaged 1,880 Pt-Co units. After nutrient conditioning and application of the microorganisms to the mill's lagoon, color reduced more than 50% with the anticipation of more than 80% color removal (Sorce Inc 2003). This technology is called fungal/bacterial sequencing or biogeochemical cycling. Biogeochemical cycling uses specific microorganisms and manipulates their life cycles so that their degrading activities can be predicted and harnessed in a beneficial way. In this case, white rot fungi and its ability to secrete highly oxidative enzymes is used to fragment the lignin structure into smaller compounds. These fragments are mineralized into CO_2 and water with facultative anaerobic bacteria. Mineralization is accelerated with the use of co-metabolites. The latent phase of the fungal growth is a time of limited food, nutrients, or adverse environmental conditions that results in a decrease in the microbial population. The life cycle manipulation is the basis of the white rot/facultative bacteria sequencing technology. In other words, keeping the fungi in a latent phase at the same time keeping facultative bacteria in a prolonged exponential growth phase. Under these conditions, the white rot fungi secretes a tremendous amount of enzymes catalyzing the lignin degradation reaction that become the prime food source for the bacteria ultimately mineralizing the degraded lignin fragments and reducing color, toxicity, BOD, COD, etc., of the water to make it suitable for certain applications. This technology seems to be quite effective but requires large land area – tens of hectares.

A novel two-stage process using chemical hydrogenation as a first-stage treatment, followed by biological oxidation showed promise in substantially reducing the color of pulp mill effluents. In a pilot plant study using two 20 L reactors in series, the addition of sodium borohydride to the first reactor, for a residence time of 1 day, resulted in a 97% reduction in color. Subsequent biological oxidation in the second reactor reduced BOD (99%), COD (92%), and TSS (97%) (Ghoreishi and Haghighi 2007).

Tables 13.10 and 13.11 show the comparison of the results for color and AOX reduction by a few white rot fungi.

Table 13.10 Comparison of systems used for the treatment of bleaching effluents with different fungi in batch process

Evaluation method	Operation mode	Residence time (day)	Max. color reduction (%)	Max. AOX reduction (%)	Reference
Lentinus edodes					
Mycelial pellets	Batch	5	73	–	Esposito et al. (1991)
Pleurotus sajor-caju					
Mycelial pellets	Batch	6	66.7	–	Ragunathan and Swaminathan (2004)
Geotrichium candidum Dec 1					
Mycelial pellets		7	78	43	Shintani et al. (2002)
Phanerochaete chrysosporium					
Mycelium immobilized on rotating disc	Batch	1	90	–	Yin et al. (1989a)
Tinctoporia borbonica					
Mycelial pellets	Batch	4	99	–	Fukuzumi (1980)
Schizophyllum commune					
Mycelial pellets	Batch	2	90		Belsare and Prasad (1988)
Trametes versicolour					
Mycelial pellets immobilized in Ca-alginate	Batch	3	80	–	Livernoche et al. (1983)
Mycelial pellets	Batch	2	61	–	Royer et al. (1985)
Free cells	Batch	3	88	45	Bergbauer et al. (1991)
Mycelial pellets	Batch	5	80	–	Archibald et al. (1990)
Mycelial pellets	Batch	3	88	–	Mehna et al. (1995)
Rhizopus oryzae					
Mycelial pellets	Batch	2	91	–	Nagarathnamma et al. (1999a, b)
Ceriporiopsis subvermispora					
Mycelial pellets	Batch	1	95	–	Nagarathnamma et al. (1999a, b)

Table 13.11 Comparison of systems used for the treatment of bleaching effluents with different fungi in continuous process

Fungus	Operation mode	Residence time (day)	Maximum color reduction (%)	Maximum AOX reduction (%)	Reference
Rhizomucor pusillus					
Mycelium	Continuous *RBC reactor*	0.95	73	55	van Driessel and Christov (2001)
Phanerochaete chrysosporium					
Mycelium immobilized on porous material	Continuous	0:5	60	55	Messner et al. (1990)
Mycelium immobilized on polyurethane foam	Continuous	5–8	70	–	Cammarota and Santanna (1992)
Mycelium immobilized on net ring type	Continuous	0.5	91	–	Kang et al. (1996)
Trametes versicolour					
Mycelial pellets immobilized in Ca-alginate	Continuous	0.7	45	–	Royer et al. (1983)
Mycelial pellets	Continuous	0.6–1.2	50	–	Royer et al. (1985)
Mycelial pellets	Continuous	1	78	42	Pallerla and Chambers (1995)
Mycelial pellets immobilized in Ca-alginate beads	Continuous	1	80	40	Pallerla and Chambers (1996)
Mycelial pellets	Continuous	1.6	93	–	Bajpai et al. (1993)

13.3.4 Ligninolytic Enzymes and Their Role in Decolorization of Bleaching Effluents

The enzymes lignin peroxidase, manganese peroxidase, and laccase have been implicated in the decolorization of bleaching effluents (Momohara et al. 1989; Esposito et al. 1991) but their roles were not critically examined until 1991. The results of Momohora et al. (1989) indirectly indicated that decolorization of extraction stage effluent by *P. chrysosporium* was not catalyzed by lignin peroxidase. Lackner et al. (1991) concluded for the first time that MnP plays the major role in the initial breakdown and decolorization of high molecular weight chlorolignin in bleaching effluents with *P. chrysosporium* in vivo, by demonstrating the following:

1. *P. chrysosporium* degraded high molecular weight chlorolignin in bleaching effluents even though a direct contact between ligninolytic enzymes and chlorolignins was prevented by a dialysis tubing.
2. Manganese peroxidase effectively catalyzed the depolymerization of chlorolignin in the presence of Mn (11) and H_2O_2.

These researchers also investigated the biochemical mechanism of chlorolignin degradation in the MYCOPOR reactor and found that the amount of mycelium bound manganese peroxidase correlated with decolorization rates. This explains the fact that bleaching effluents can be degraded during continuous operation of the MYCOPOR reactor for months even though the enzymes are washed out. Mycelium-bound manganese peroxidase could generate Mn (iii) which can freely diffuse into the effluent and depolymerize the chlorolignins trickling through the reactor. Michel et al. (1991) also investigated the role of ligninolytic enzymes of *P. chrysosporium* in decolorizing bleaching effluents. They concluded that manganese peroxidase plays an important role in effluent decolorization. Moreover, Lee et al. (1994) demonstrated high levels of manganese peroxidase but no lignin peroxidase activity during extraction stage effluent treatment with the fungus KS-62 which showed excellent decolorization without any additional nutrients. Because significant reduction was observed for the decolorization of a catalase added culture, they suggested that manganese peroxidase plays an important role in the decolorization of extraction stage effluent by this fungus. The role of manganese peroxidase in decolorization of bleach plant effluent has been also confirmed by Jaspers et al. (1994). Archibald and Roy (1992) reported that laccase and not peroxidase plays the primary role in effluent decolorization by *T. versicolor*. Archibald and Roy (1992) later demonstrated that *T. versicolor* laccase, in the presence of phenolic substrate, was able to generate Mn (iii) chelates similar to those produced by manganese peroxidase and which were shown by Lackner et al. (1991) to be responsible for the oxidation of bleaching effluent.

Manzaners et al. (1995) evaluated the enzymatic activities when the effluents from alkaline cooking of cereal straw were treated with *T. versicolor*. They reported that the production of laccase activity was much higher than that obtained under the same conditions in synthetic growth media and that there was a clear relationship between the effluent concentration in the medium and laccase activity. In the decolorization medium, manganese peroxidase activity was detected when $MnSO_4$ was added to these media, although no lignin peroxidase activity was detected in any of the conditions assayed. Lankinen et al. (1991) treated softwood pulp bleaching effluents with carrier immobilized *Phlebia radiata* and noticed the production of large amounts of lignin peroxidase (the most characteristic lignin peroxidase isozymes in effluent media were lignin peroxidase 2 and lignin peroxidase 3) during AOX decrease and color removal.

13.4 Conclusions and Future Perspectives

Several methods have been attempted for decolorization and detoxification of bleached kraft effluents. These include physicochemical and biotechnological methods. The problems underlying the physicochemical treatments are those associated with cost and reliability. Biotechnological methods have the potential to eliminate/reduce the problems associated with physicochemical methods. These methods may be bacterial treatment (aerobic as well anaerobic), fungal treatment, or enzymatic treatment. The bacterial processes are not very effective due to the limitation that they cannot degrade the high molecular weight chlorolignin compounds and enzymatic processes are not cost effective. Among the biological methods tried so far, fungal treatment technology using white rot fungi appears to be the most promising in this regard. One of the drawbacks associated with the fungal treatment has been the requirement of easily metabolizable cosubstrate like glucose for the growth and development of ligninolytic activity. To make the fungal treatment method economically feasible, there is a need to reduce the requirement of cosubstrate or identify a cheaper cosubstrate. Hence, efforts should be made to identify the strains that show good decolorization with less or no cosubstrate and can utilize industrial waste as a cosubstrate. Efforts should be also made to utilize the spent fungal biomass for preparing the culture medium required in the synthesis of active fungal biomass. If succeeded, the cost of treatment may be further reduced. As lignin degrading system of white rot fungus has a high oxygen requirement, use of oxygen instead of air as fluidizing media should be explored. Increasing the oxygen concentration in the culture atmosphere is expected to have a dual effect: it would lead to an increased titer of the lignin degrading system and to increased stability of the existing system. A quantitative study of extracellular enzymes is also required in order to gain insight into the possible enzymatic mechanism involved in the degradation of lignin derived compounds present in the effluents.

Use of white rot fungi can serve as a pretreatment method to bacterial treatment and to enhance the bacterial ability to remove organic chlorine and to degrade the relatively higher molecular weight chlorolignins. This process can be used as an alternative to internal process modifications (modified cooking, oxygen bleaching, high level chlorine dioxide substitution, etc.) and conventional biological treatment.

Since the majority of AOX and color is in high molecular weight chlorolignins, the priority of research should concentrate on the fate of high molecular weight chlorolignins in biological treatment or in the natural environment. Since bacteria degrade significantly only those chloroorganics with molecular weights lower than 800–1,000 Da, research is needed to decrease the chlorolignin molecular weight or to remove high molecular weight chlorolignins before biological treatment is applied in order to enhance the biotreatability of bleaching effluents.

References

Abdel-Fattah YR, Hoda HH, El-Kassas HY, Sabry SA (2001) Bioprocess development of paper mill effluent's decolorization by *Phanerochaete chrysosporium* DSMZ 1556. Fresenius Environ Bull 10(10):761–765

Ali M, Sreekrishnan TR (2001) Aquatic toxicity from pulp and paper mill effluents: a review. Adv Environ Res 5:175–196

Ali M, Sreekrishnan TR (2007) Anaerobic treatment of agricultural residue based pulp and paper mill effluents for AOX and COD reduction. Process Biochem 36(1–2):25–29

Altnbas U, Eroglu V (1997) Treatment of bleaching effluent in sequential activated sludge and nitrification systems. Fresenius Environ Bull 6:103–108

Amy GL, Bryant CW, Alleman BC, Barrley WA (1988) Biosorption of organic halides in a kraft mill generated lagoon. J Water Pollut Cant Fed 60(8):1445–1458

Annergren GE (1990) Environmental harmonization of high quality bleached kraft pulp production – a high tech development. 24th EuCepa Conference, Stockholm, Sweden

Annergren GE, Osterberg F, Lindblad P (1990) Tappi pulping conference, Tappi, Toronto

Anon (1994) EPA report thrusts dioxins back into the spot light. ENDS Ref. No. 236, pp 21–24

Archibald FS, Roy B (1992) The role of fungus-fiber contact in biobleaching of kraft brownstock by *Trametes versicolor*. Appl Environ Microbiol 58:1496–1499

Archibald F, Paice MG, Jurasek L (1990) Decolourization of kraft bleaching effluent chromophores by *Coriolus (Trametes) versicolor*. Enzyme Microb Technol 12:846–853

Ataberk S, Gokcay CF (1997) Removal of chlorinated organics from pulping effluents by activated sludge process. Fersenius Environ Bull 6:147–153

Axegard P, Berry RM, Gellerstedt G, Lindblad PO, Luthie CE, Popke I, Voss RH, Wrist PE (1993) The effects of recent changes in bleached softwood kraft mill technology on organochlorine emissions: an international perspective. In: Jameel H (ed) Bleaching, vol 2. Tappi Press, Atlanta, pp 759–770

Bajpai P (2001) Microbial degradation of pollutants in pulp mill effluents. In: Neidleman S, Laskin A (eds) Advances in applied microbiology, vol 48. Academic, New York, pp 79–134

Bajpai P, Bajpai PK (1994) Biological colour removal of pulp and paper mill wastewaters. J Biotechnol 33:211–220

Bajpai P, Bajpai PK (1996) Organochlorine compounds in bleach plant effluents – genesis and control. PIRA International, Leatherhead

Bajpai P, Bajpai PK (1997) Reduction of organochlorine compounds in bleach plant effluents in biotechnol. In: Eriksson K-E (ed) Pulp and paper industry (special edition) for advances in biochemical engineering and biotechnology, vol 57. Springer, Berlin, pp 213–259

Bajpai P, Mehna A, Bajpai PK (1993) Decolourization of kraft bleach effluent with white rot fungus *Trametes versicolor*. Process Biochem 28:377–384

Bajpai P, Bajpai PK, Kondo R (1999) Biotechnology for environmental protection in pulp and paper industry. Springer, Germany, pp 239–261 (Chapter 11)

Barr TA, Taylor T, Duff S (1996) Effect of HRT, SRT and temperature on the performance of activated sludge reactors treating bleached mill effluent. Water Res 30(4):799–802

Belém A, Panteleitchouk AV, Duarte AC, Rocha-Santos TAP, Freitas AC (2008) Treatment of the effluent from a kraft bleach plant with white rot fungi *Pleurotus sajor caju* and *Pleurotus ostreatu*. Global NEST J 10(3):426–431

Belsare DK, Prasad DY (1988) Decolourization of effluent from the bagasse based pulp mills by white-rot fungus *Schizophyllum commune*. Appl Microbiol Biotechnol 28:301–304

Bergbauer M, Eggert C, Kraepelin G (1991) Degradation of chlorinated lignin compounds in a bleach effluent by the white-rot fungus *Trametes versicolor* Appl Microbiol Biotechnol 35(1):105–109

Bergbauer M, Eggert C, Kalnowski G (1992) Biotreatment of pulp mill bleachery effluent with the Coelomycetous fungus *Stagonospora gigaspora*. Biotechnol Lett 14(4):317–322

Bollag JM, Shottleworth KL, Anderson DH (1988) Laccase-mediated detoxification of phenolic compounds. Appl Environ Microbiol 54:3086–3091

Boman B, Frostell B, Ek M, Eriksson KE (1988) Some aspects on biological treatment of bleached pulp effluents. Nordic Pulp Paper Res J 1:13–18

Bryant CW, Barkley WA (1990) The capabilities of conventional treatment systems for removal of chlorinated organic compounds from pulp and paper wastewater. Pacific Paper EXPO Technical Conference Proc. Program 7: Environment, Vancouver, BC

Bryant CV, Amy GL, Allemen BC (1987) Organic halide and organic carbon distribution and removal in a pulp and paper wastewater lagoon. J Water Pollut Control Fed 59(10):890–896

Bryant CW, Amy GL, Neil R, Ahmed S (1988) Partitioning of organic chlorine between bulk water and benthal interstitial water through a kraft mill aerated lagoon. Water Sci Technol 20(1): 73–79

Buckley DB (1992) A review of pulp and paper industry experience with biological treatment process bacterial augmentation. Tappi Environmental Conference. Tappi Press, Atlanta, pp 750–810

Bullock JM, Bicho PA, Saddler JN (1994) The effect of high molecular weight organics in bleached kraft mill effluent on the biological removal of chlorinated phenolics. Proc. of 1994 Environmental Conference, pp 371–378

Bumpus JA, Aust SD (1995) Biodegradation of environmental pollutants by the white-rot fungus – *P. chrysosporium*. Bio Essays 6(4):166–170

Call HP (1991) Laccases in delignification, bleaching and wastewater treatment. Patent No. DE 4137761

Cammarota MC, Santanna GL Jr (1992) Decolourization of kraft bleach plant E_1 stage effluent in a fungal bioreactor. Environ Technol 13:65–71

Campbell AG, Gerrard ED and Joyce TW (1982) The MyCoR process for colour removal from bleach plant effluent: bench-scale studies. In Proc. of the Tappi Research and Development Conference, North Carolina, Tappi Press, Atlanta, pp 209–214

Chandra R, Singh S, Krishna Reddy MM, Patel DK, Purohit HJ, Kapley A (2008) Isolation and characterization of bacterial strains *Paenibacillus* sp. and *Bacillus* sp. for kraft lignin decolorization from pulp paper mill waste. J Gen Appl Microbiol 54(6):399–407

Christov LP, Steyn MG (1998) Modifying the quality of a bleach effluent using Mucoralean and white-rot fungi. In Proc. of 7th International Conference on Biotechnology in the pulp and Paper Industry. Vancouver, Canada, pp C203–C206

Chuphal Y, Kumar V, Thakur IS (2005) Biodegradation and decolorization of pulp and paper mill effluent by anaerobic and aerobic microorganisms in a sequential bioreactor. World J Microbiol Biotechnol 21(8–9):1439–1445

Davis S, Burns RG (1992) Covalent immobilisation of laccase on activated carbon for phenolic effluent treatment. Appl Microbiol Biotechnol 37:474–479

Deardorff TL, Willhelm RR, Nonni AJ, Renard JJ, Phillips RB (1994) Formation of polychlorinated phenolic compounds during high chlorine dioxide substitution and bleaching. Tappi J 77(8):163–173

Dence CW, Reeve DW (1990) The technology of chemical pulp bleaching. In: Dence CW, Reeve DW (eds) Pulp bleaching – principles and practice. Tappi Press, Atlanta, pp 91–183
Deshpande SH, Khanolkar VD, Pudumjee KD (1991) Anaerobic-aerobic treatment of pulp mill effluents – a viable technological option. Proc. International Workshop on Small Scale Chemical Recovery, High Yield Pulping and Effluent Treatment, 16–20 Sept, New Delhi, India, pp 201–213
Dezotti M, Innocentini-Mei LH, Duran N (1995) Silica immobilised enzyme catalysed removal of chlorolignins from kraft effluent. J Biotechnol 43:161–167
Dorica J, Elliott A (1994) Contribution of non-biological mechanisms of AOX reduction attained in anaerobic treatment of peroxide bleached TMP mill effluent. In Proc. Tappi Intl. Environmental Conference, pp 157–163
Duran N, Dezotti M, Rodriguez J (1991) Biomass photochemistry – XV: photobleaching and biobleaching of kraft effluent. J Photochem Photobiol 62:269–279
Duran N, Esposito E, Innocentini-Mei LH, Canhos VP (1994) A new alternative process for kraft E_1 effluent treatment, a combination of photochemical and biological methods. Biodegradation 5(1):13–19
Eaton DC, Chang HM, Kirk TK (1980) Fungal decolourization of kraft bleach plant effluents. Tappi J 63(10):103–106
Eaton DC, Chang HM, Joyce TW, Jeffries TW, Kirk TK (1982) Method obtains fungal reduction of the color of extraction-stage kraft bleach effluents. Tappi J 65(6):89–92
Edeline F, Lambert G, Fatliccioni H (1988) Anaerobic treatment of mixed hardwood kraft pulp cooking condendates with first alkaline stage effluents. In: Grassi G (ed) Energy from Biomass – 4. Proc. 3rd Contractors. Meeting. pp 463–475
Eeckhaut M, Alaerts G, Pipyn P (1986) Anaerobic treatment of paper mill effluents using polyurethane foam carriers reactor (PCR) technology. In PIRA Paper & Board Division Seminar. Cost effective treatment of paper mill effluents using anaerobic technologies, 14–15 Jan, Leatherhead, UK
Ek M, Eriksson KE (1987) External treatment of bleach plant effluent. 4th Int. Symp. on Wood and Pulping Chemistry, Paris
Ek M, Kolar MC (1989) Reduction of AOX in bleach plant effluents by a combination of ultrafiltration and biological methods. Proc. of 4th Int. Biotech. Conf. in Pulp and Paper industry, Raleigh, North Carolina, 16–19 May, pp 271–278
Eriksson KE (1990) Biotechnology in the pulp and paper industry. Water Sci Technol 24:79–101
Eriksson KE, Kolar MC (1985) Studies on microbial and chemical conversions of chlorolignins. Environ Sci Technol 19(12):1219–1224
Esposito E, Canhos VP, Duran N (1991) Screening of lignin degrading fungi for removal of colour from kraft mill wastewater with no additional extra carbon source. Biotechnol Lett 13(8):571–576
Farrel RL (1987a) Use of rLDM ™ 1–6 and othe ligninolytic enzymes. WO 87/00564
Farrel RL (1987b) Industrial applications of lignin transforming enzymes. Philos Trans R Soc London A 321:549–553
Ferguson JF, Dalentoft E (1991) Investigation of anaerobic removal and degradation of organic chlorine from kraft bleaching wastewaters. Water Sci Technol 24:241–250
Ferguson JF, Luonsi A, Ritter D (1990) Sequential anaerobic/aerobic biological treatment of bleaching wastewaters. In Proc. Tappi. Environmental Conference, Atlanta, GA, pp 333–338
Ferrer I, Dezotti M, Duran N (1991) Decolorization of kraft effluent by free and immobilized lignin peroxidase and horseradish peroxidase. Biotechnol Lett 13:577–582
Field JA (1986) Method for biological treatment of waste waters containing nondegradable phenolic compounds and degradable nonphenolic compounds. EP Patent 1986; EP 238148
Field JA, Lettinga G (1991) Treatment and detoxification of aqueous spruce bark extracts by *Aspergillus niger*. Water Sci Technol 24:127–137
Field JA, Leyendeckers MJH, Sierra-Alvarez R, lettinga G, Habets LHA (1988) The methanogenic toxicity of bark tannins and the anaerobic biodegradability of water soluble bark matter. Water Sci Technol 20:219–240

Field JA, Leyendeckers MJH, Sierra-Alvarez R, Lettinga G (1991) Continuous anaerobic treatment of auto-oxidized bark extracts in laboratory-scale columns. Biotechnol Bioeng 37:247–255

Fitzsimonas R, Ek M, Eriksson K-EL (1990) Anaerobic dechlorination/degradation of chlorinated organic compounds of different molecular masses in bleach plant effluents. Environ Sci Technol 24:1744–1748

Forss K, Jokinen K, Savolainen M, Williamson H (1987) Utilization of enzymes for effluent treatment in the pulp and paper industry. In: Proc 4th International Symposium on Wood and Pulping Chemistry, vol 1, Paris, France, pp 179–183

Fukui H, Presnell TL, Joyce TW, Chang HM (1992) Dechlorination and detoxification of kraft E^P effluent by *Phanerochaete chrysosporium*. In: Kuwahara M, Shimada M (eds) Biotechnology in Pulp and Paper manufacture. Uni Publishers, Tokyo, pp 75–80

Fukuzumi T (1980) Microbial decolourization and defoaming of pulping waste liquor. In: Kirk TK, Chang HM, Higuchi T (eds) Lignin biodegradation: microbiology, chemistry and potential applications, vol 2. CRC Press, Boca Raton, pp 161–171

Fukuzumi T, Nishida A, Aoshima K, Minami K (1977) Decolourization of kraft waste liquor with white-rot fungi – I: screening of the fungi and culturing condition for decolourization of kraft waste liquor. Mokuzai Gakkaishi 23(6):290–298

Fulthorpe RR, Allen DG (1995) A comparison of organochlorine removal from bleached kraft pulp and paper mill effluents by dehalogenating *Pseudomonas* Ancylcbacter and Methylobacterium strains. Appl Microbiol Biotechonol 42:782–789

Galeno G, Agasin E (1990) Screening of white-rot fungi for efficient decolourization of bleach pulp effluents. Biotechnol Lett 12(11):869–872

Gavrilescu D (2006) Environmental consequences of pulp and paper manufacture 1. Bleached kraft pulp mills. Environ Eng Manage J 5(1):37–49

Gergov M, Priha M, Talka E, Valtilla O, Kangas A, Kukkonen K (1988) Chlorinated organic compounds in effluent treatment at kraft mills. Tappi J 71(12):175–184

Ghoreishi SM, Haghighi MR (2007) Chromophores removal in pulp and paper mill effluent via hydrogenation-biological batch reactors. Chem Eng J 127(1–3):59–70

Gokcay CF, Taseli BK (1997) Biological treatability of pulping effluents by using a *Penicillium* sp. Fresenus Environ Bull 6:220–226

Goronzy M, Demoulin G, Jager A, Srebotnik E, Messener K (1996) The use of the cyclic activated treatment technology for wastewater treatment in the pulp and paper industry. In Proc. Sixth Intl. Conference Biotechnology in Pulp and Paper Industry. Vienna, Austria, pp 239–245

Graves JW, Joyce TW, Jameel H (1993) Effect of chlorine dioxide substitution, oxygen delignification and biological treatment on bleach plant effluent. Tappi J 76(7):153–159

Guo HY, Chang HM, Joyce TW, Glasser JH (1990) Degradation of chlorinated phenols and guaiacols by the white rot fungus *Phanerochaete chrysosporium*. In: Kirk TK, Chang HM (eds) Biotechnology in pulp and paper manufacture. Butterworth-Heinemman, Stoneham, pp 223–230

Habets LHA, de Vegt AL (1991) Anaerobic treatment of bleached TMP and CTMP effluent in Biopaq UASB system. Water Sci Technol 24:331–345

Habets LHA, Tielboard MH, Ferguson AMD, Prong CF, Chmelauskas AJ (1985) Onsite high rate UASB anaerobic demonstration plant treatment of NSSC waste water. Water Sci Technol 20:87–97

Haggblom M, Salkinoja-Salonen M (1991) Biodegradability of chlorinated organic compounds in pulp bleaching effluents. Water Sci Technol 24(3/4):161–170

Hakulinen R (1982) The Enso-Fenox process for the treatment of Kraft pulp bleaching effluent and other waste waters of the forest industry. Paperi Ja Puu Paper Och Tra 5:341–354

Hakulinen R (1988) The use of enzymes in the waste water treatment of pulp and paper industry-a new possibility. Water Sci Technol 20(1):251–262

Hall ER, Robson RD, Prong CF, Chmelauskas AJ (1986) Evaluation of anaerobic treatment for NSSC wastewater. In Proc. Tappi Environmental Conference, Atlanta, GA, pp 207–217

Hammel KE, Tardone PJ (1988) The oxidative 4-dechlorination of polychlorinated phenols is catalyzed by extracellular fungal lignin peroxidases. Biochemistry 27:6563–6568

Higachi R, Cherr G, Shenker J, Macdonald J, Crosby D (1992) A polar high molecular mass constituent of bleached kraft mill effluent is toxic to marine organisms. Environ Sci Technol 26(12):2413–2420

Huynh VB, Chang HM, Joyce TW, Kirk TK (1985) Dechlorination of chloroorganics by white-rot fungus. Tappi J 68(7):98–102

Jaklin-Farcher S, Szeker E, Stifter U, Messner K (1992) Scale up of the MYCOPOR reactor. In: Kuwahara M, Shimada M (eds) Biotechnology in pulp and paper industry. Tokyo, Japan, pp 81–85

Jaspers CJ, Jimenez G, Penninck MJ (1994) Evidence for the role of manganese peroxidase in the decolourization of kraft pulp bleach plant effluent by *Phanerochaete chrysosporium*: effect of initial culture conditions on enzyme production. J Biotechnol 37:229–236

Johnson T, Chatterjee A (1995) Activated sludge and surface aerators treat combined CTMP and kraft effluent. Pulp Paper Can 96(8):26–29

Jokela JK, Laine M, Ek M, Salkinoja-Salonen M (1993) Effect of biological treatment on halogenated organics in bleached kraft pulp mill effluents studied by molecular weight distribution analysis. Environ Sci Technol 27(3):547–552

Joyce TW, Pellinen J (1990) White rot fungi for the treatment of pulp and paper industry waste water. Tappi Env. Conference, Seattle, 9–11 Apr

Jurgensen SJ, Benjamin MM, Ferguson JF (1985) Treatability of thermomechanical pulping process effluents with anaerobic biological reactor. In Proc. Tappi Environmental Conference. Tappi Press, Atlanta, pp 83–92

Kang CH, On HK, Won CH (1996) Studies on the treatment of paper mill wastewater by *Phanerochaete chrysosporium*. In: Srebotnik E, Messener K (eds) Biotechnology in pulp and paper industry. Facultas-Universitatsverlag, Vienna, pp 263–266

Kannan K (1990) Decolourization of pulp and paper mill effluent by *Aspergillus niger*. World J Microbiol Biotechnol 6(2):114–116

Kapoor A, Viraraghavan T (1995) Fungal biosorption – an alternative treatment option for heavy metal bearing wastewaters – a review. Bioresour Technol 53:195–206

Karimi S, Abdulkhani A, Karimi A, Ghazali AB, Ahmadun FR (2010) The effect of combination enzymatic and advanced oxidation process treatments on the colour of pulp and paper mill effluent. Environ Technol 31(4):347–356

Klibanev AM, Morris ED (1981) Horseradish peroxidase for the removal of carcinogenic aromatic amines from water. Enzyme Microb Technol 3:119–122

Kortekaas S, Wijngaarde RR, Klomp JW, Lettinga G, Field JA (1998) Anaerobic treatment of hemp thermomechanical pulping wastewater. Water Res 32(11):3362–3370

Lackner R, Srebotnik E, Messner K (1991) Oxidative degradation of high molecular weight chlorolignin by manganese peroxidase of *Phanerochaete chrysosporium*. Biophys Res Commun 178(3):1092–1098

Lafond RA, Ferguson JF (1991) Anaerobic and aerobic biological treatment processes for removal of chlorinated organics from kraft bleaching wastes. In Proc. Tappi Environmental Conference. Tappi press, Atlanta, GA, USA pp 797–812

Lankinen VP, Inkeroinen MM, Pellinen J, Al H (1991) The onset of lignin modifying enzymes; decrease of AOX and color removal by white rot fungi grown on bleach plant effluents. Water Sci Technol 24(3/4):189–198

Leach JM, Mueller JC, Walden CC (1978) Biological detoxication of pulp mill effluents. Process Biochem 13(1):18–26

Lee JW (1993) Anaerobic treatment of pulp and paper mill wastewaters. In: Springer AM (ed) Industrial environmental control, pulp and paper industry. Tappi Press, Atlanta, pp 405–446

Lee SH, Kondo R, Sakai K (1994) Treatment of kraft bleaching effluents by lignin degrading fungi – III: treatment by newly found fungus KS-62 without additional nutrients. Mokuzai Gakkaishi 40(6):612–619

Lee SH, Kondo R, Sakai K, Sonomoto K (1995) Treatment of kraft bleaching effluents by lignin – degrading fungi V. Successive treatments with immobilized mycelium of the fungus KS-62. Mokuzai Gakkaishi 41(1):63–68

Lettinga G (1980) Use of upflow anaerobic sludge blanket wastewater treatment system: a technology review. Biotech Bioeng 22(6):699–734
Lettinga G, Field JA, Sierra-Alvarez R, Vanlier JB, Rintala J (1990) Future perspective for the anaerobic treatment of forest industry wastewaters. Water Sci Technol 24:91–102
Liebergott N, van Lierop B, Kovacs T, Nolin A (1990) Comparison of the order of addition of chlorine and chlorine dioxide in the chlorination stage. Tappi J 73:207–213
Lindstrom K, Mohamed M (1988) Selective removal of chlorinated organics from kraft mill effluents in aerated lagoons. Nordic Pulp Paper Res J 3:26–33
Liu HW, Lo SN, Lavallee HC (1996) Mechanisms of removing resin and fatty acids in CTMP effluent during aerobic biological treatment. Tappi J 79(5):145–154
Liu HW, Liss SN, Allen D (1997) Influence of Anoxic conditioning of sludge on enhanced AOX removal in aerobic biological treatment systems. Water Sci Technol 35(2):77–82
Livernoche D, Jurasek L, Desrochers M, Dorica J, Veliky IA (1983) Removal of colour from kraft mill waste waters with cultures of white-rot fungi and immobilised mycelium of *Coriolous versicolor*. Biotechnol Bioeng 25:2055–2065
Lunan WE, Harden C, Krupa K (1995) Pilot trials of trickling filter for treatment of waste water from a newsprint mill. In Proc. Tappi Environmental Conference, Atlanta, GA, USA pp 241–247
Lyr VH (1963) Enzymic detoxification of chlorinated phenols. Phytopathology 47:73–83
MacLean B, de Vegt A, Droste RL (1990) Role of resin acids in the anaerobic toxicity of Chemithermomechanical pulp waste water. Water Res 24:1401–1405
Malaviya P, Rathore VS (2007) Bioremediation of pulp and paper mill effluent by a novel fungal consortium isolated from polluted soil. Bioresour Technol 98(18):3647–3651
Manzaners P, Fajardo S, Martin C (1995) Production of ligninolytic activities when treating paper pulp effluents by *T. versicolor*. J Biotechnol 43:125–1430
Mathys RG, Branion RMR, Lo KV (1993) CTMP waste water treatment using rotating biological contactor. Proc. 79th CPPA Annual Proc., pp 370–374
Mathys RG, Branion RMR, Lo KV, Anderson KB, Leyen P, Louie D (1997) CTMP waste water treatment using rotating biological contactor. Water Qual Res J Can 32:771–774
Matsumoto Y, Yin CF, Chang HM, Joyce TW, Kirk TK (1985) Degradation of chlorinated lignin and chlorinated organics by a white rot fungus. Proc. of 3rd ISWPC, Vancouver, pp 45–53
McCarthy PJ, Kennedy KJ, Droste RL (1990) Role of resin acids in the anaerobic toxicity of chemithermomechanical pulp wastewater. Water Res 24:1401–1405
McCubbin N (1983) The basic technology of the pulp and paper industry and its environment protection practices, Report EPS 6-EP-83-1, Environment Canada, Ottawa
McCubbin N (1989) Dioxin '89: a bit like the adventures in Alice and wonderland. Pulp Paper Can 90(11):17–19
McCubbin N, Sprague JB, Bonsor N (1990) Kraft mill effluents in Ontario. Pulp Paper Can 91(3):T112–T114
Mckague AB, Carlberg G (1996) Pulp bleaching and the environment. In: Dence CW, Reeve DW (eds) Pulp bleaching – principles and practice. Tappi Press, Atlanta, pp 746–820
McLeay DJ (1987) Aquatic toxicity of pulp paper mill effluent: a review. Environment Canada Report EPS 4/PF/1 Beauregard Press Ltd, Ottawa Ontario, Canada
Mehna A, Bajpai P, Bajpai PK (1995) Studies on decolourization of effluent from a small pulp mill utilizing agriresidues with *Trametes versicolor*. Enzyme Microb Technol 17(1):18–22
Melcer H, Steel P, Mckinley A, Cook CR (1995) The removal of toxic contaminants from bleached kraft mill waste water with enhanced activated sludge treatment. Proc. Tappi International Environmental Conference, pp 795–807
Messner K, Ertler G, Jaklin-Farcher S, Boskovsky P, Regensperger V, Blaha A (1990) Treatment of bleach plant effluents by Mycor system. In: Kirk TK, Chang HM (eds) Biotechnology in pulp and paper manufacture. Butterworth-Heinemann, Newton, pp 245–251
Michel FC, Dass SB, Grulke EA, Reddy CA (1991) Role of manganese peroxidases and lignin peroxidases of *Phanerochaete chrysosporium* in the decolourization of kraft bleach plant effluent. Appl Environ Microbiol 57(8):2368–2370

Milstein O, Haars A, Majcherczyk A, Trojanowski J, Tautz D, Zanker H, Huttermann A (1988a) Removal of chlorophenols and chlorolignins from bleaching effluents by combined chemical and biological treatment. Water Sci Technol 20(1):161–170
Milstein O, Trojanowski J, Majcherczyk A, Tautz D, Zanker H, Huettermann A (1988b) Removal of chlorophenols and chlorolignins from bleaching effluent by combined chemical and biological treament. Water Sci Technol 20(1):161–170
Mishra M, Thakur IS (2010) Isolation and characterization of alkalotolerant bacteria and optimization of process parameters for decolorization and detoxification of pulp and paper mill effluent by Taguchi approach. Biodegradation 21(6):967–978
Mittar D, Khanna PK, Marwaha SS, Kennedy JF (1992) Biobleaching of pulp and paper mill effluents by *Phanerochaete chrysosporium*. J Chem Technol Biotechnol 53:81–92
Momohara I, Mateumoto Y, Ishizu A, Chang HM (1989) Mechanism of kraft pulp bleaching mill effluent by *Phanerochaete chrysosporium*: characteristics of colour and its change during decolourization. Mokuzai Gakkaishi 35(12):1110–1115
Monje PG, González-García S, Moldes D, Vidal T, Romero J, Moreira MT, Feijoo G (2010) Biodegradability of kraft mill TCF biobleaching effluents: application of enzymatic laccase-mediator system. Water Res 44(7):2211–2220
Mortha G, Mckay LR, Cadel F, Rouger J (1991) AOX reduction in an activated sludge treatment of kraft bleaching effluent. 6th ISWPC. 1–2
Nagarathnamma R, Bajpai P, Bajpai PK (1999a) Studies on decolourization, degradation and detoxification of chlorinated lignin compounds in kraft bleaching effluents by *Ceriporiopsis subvermispora*. Process Biochem 34:939–948
Nagarathnamma R, Bajpai P, Bajpai PK (1999b) Decolourization of extraction stage effluent from chlorine bleaching of kraft pulp by *Rhizopus oryzae*. Appl Environ Microbiol 65(3):1078–1082
NCASI (1990) Field verification of the NCASI predictive model for organic compound removal by biological wastewater treatment processes. NCASI Technical Bulletin No. 582
Olofsson A (1996) Domsjo heats up Ornskoldsvik with biogas. Svensk Papperstidin 99(11): 33–34
Paice MG (1995). Activated sludge treatment of mechanical pulp mill effluents containing sulfite. Proc. of CPPA Environmental Conference, pp 81–86
Paice MG, Jurasek L (1984) Peroxidase catalyzed color removal from bleach plant effluent. Biotechnol Bioeng 26:477–480
Paice M, Elloitt A, Young F, Dorica J (1996) Activated sludge treatment of mechanical pulp mill effluents containing sulfites. Pulp Paper Can 97(9):88–92
Pallerla S, Chambers RP (1995) Continuous decolourization and AOX reduction of bleach plant effluents by free and immobilized *Trametes versicolor*. J Environ Sci Health A30(2):432–437
Pallerla S, Chambers RP (1996) New urethane prepolymer immobilized fungal bioreactor for decolourization and dechlorination of kraft bleach effluents. Tappi J 79(5):156–161
Parker WJ, Eric R, Farguhar GJ (1993a) Assessment of design and operating parameters for high rate anaerobic fermentation of segregated kraft mill bleach plant effluents. Water Environ Res 65(3):264–270
Parker WJ, Eric R, Farguhar GJ (1993b) Removal of chlorophenolics and toxicity during high rate anaerobic treatment of kraft mill bleach plant effluents. Environ Sci Technol 27(9):1783–1789
Pellinen J, Joyce TW, Chang HM (1988a) Dechlorination of high molecular weight chlorolignin by white rot fungus *Phanerochaete chrysosporium*. Tappi J 71(9):191–194
Pellinen J, Yin CF, Joyce TW, Chang HM (1988b) Treatment of chlorine bleaching effluent using a white-rot fungus. J Biotechnol 8(1):67–76
Pendroza AM, Mosqueda R, Alonso-Vante N, Rodriguez-Vazquez R (2007) Sequential treatment via Trametes versicolor and UV/titanium dioxide/RuxSey to reduce contaminants in waste water resulting from the bleaching process during paper production. Chemosphere 67(4):793–801
Pertulla M, Konrusdottin M, Pere J, Kristjansson JK, Viikari L (1991) Removal of acetate from NSSC sulphite pulp mill condensates using thermophilic bacteria. Water Res 25:599–604

Pichon M, Rouger J, Junet E (1988) Anaerobic treatment of sulphur containing effluents. Water Sci Technol 20:133–141

Pokhrel D, Viraraghavan T (2004) Treatment of pulp and paper mill wastewater – a review. Sci Total Environ 333(1–3):37–58

Prasad DY, Joyce TW (1991) Colour removal from kraft bleach plant effluents by *Trichoderma* sp. Tappi J 74(1):165–169

Prouty AL (1990) Bench scale development and evaluation of a fungal bioreactor for colour removal from bleach effluents. Appl Microbiol Biotechnol 32:490–493

Puhakka JA (1994) Aerobic and anaerobic biotransformation and treatment of chlorinated pulp bleach waste constituents. Water Sci Technol 29:73–80

Raghukumar C, Chandramohan D, Michael FC Jr, Reddy CA (1996) Degradation of lignin and decolourization of paper mill bleach plant effluent by marine fungi. Biotechnol Lett 18(1):105–106

Raghukumar C, D'Souza-Ticlo D, Verma AK (2008) Treatment of colored effluents with lignin-degrading enzymes: an emerging role of marine-derived fungi. Crit Rev Microbiol 34:189–206

Ragunathan R, Swaminathan K (2004) Biological treatment of a pulp and paper industry effluent by *Pleurotus* spp. World J Microbiol Biotechnol 20(4):389–393

Rai A, Krishna Reddy MM, Chandra R (2007) Decolourisation and treatment of pulp and paper mill effluent by lignin-degrading *Bacillus* sp. J Chem Technol Biotechnol 82(4):399–406

Raizer-Neto E, Pichon M, Beniamin MM (1991) Decreasing chlorinated organics in bleaching effluents in an anaerobic fixed bed reactor. In: Kirk TK, Chang HM (eds) Biotechnology in pulp and paper manufacture. Butterworth-Heinmann, Stoneham, pp 271–278

Ramaswammy V (1987) Biotechnology application in waste utilization and pollution abatement. IPPTA convention issue. Indian Pulp and Paper Association, Saharanpur India

Rappe C, Wagman N (1995) Trace analysis of PCDDs and PCDFs in unbleached and bleached pulp samples. Organohalogen Comp 23:377–382

Rempel W, Turk O, Sikes JEG (1990) Side by side activated sludge pilot plant investigations focussing on organochlorines. CPPA Spring Conference Jasper, Alberta

Richardson DA, Andras E, Kennedy KJ (1991) Anaerobic toxicity of fines in Chemithermomechanical pulp wastewaters: a batch assay-reactor study comparison. Water Sci Technol 24:103–112

Rintala J, Lepisto S (1992) Anaerobic treatment of thermomechanical pulping wastewater at 35–70°C. Water Res 26:1297–1305

Rintala J, Vuoriranta P (1988) Anaerobic-aerobic treatment of pulping effluents. Tappi J 71: 201–207

Rintala J, Lepisto R, Chand S (1992) Toxicity of kraft bleaching effluents on thermophilic and mesophilic VFA methanation. Bioresour Technol 42:17–26

Rinzema A, Lettinga G (1988) Anaerobic treatment of sulfate containing wastewater. Biotreatment Syst 3:65–109

Rogers IH, Davis JC, Kruzynski GM, Mahood HW, Servizi JA, Gordon JW (1975) Fish toxicants in kraft effluents. Tappi J 58(7):136–140

Roy-Arcand L, Archibald FS (1991a) Direct dechlorination of chlorophenolic compounds by laccases from Trametes versicolor. Enzyme Microb Technol 13:194–203

Roy-Arcand L, Archibald FS (1991b) Comparison and combination of ozone and fungal treatments of a kraft bleaching effluents. Tappi J 74(9):211–218

Royer G, Livernoche D, Desrochers M, Jurasek L, Rouleau D, Mayer RC (1983) Decolourization of kraft mill effluent: kinetics of a continuous process using immobilized *Coriolus versicolor*. Biotechnol Lett 5:321–326

Royer G, Desrochers M, Jurasek L, Rouleau D, Mayer RC (1985) Batch and continuous decolourization of bleached kraft effluents by white-rot fungus. J Chem Technol Biotechnol 35B:14–22

Royer G, Yerushalmi L, Rouleau D, Desrochers M (1991) Continuous decolourization of bleached kraft effluents by *Coriolus versicolor* in the form of pellets. J Indl Microbiol 7:269–278

Ruggiero P, Sarkar JM, Bollag JM (1989) Detoxification of 2,4-dichlorophenol by a laccase immobilized on soil or clay. Soil Sci 147:361–370

Salkinoja-Salonen M (1990) Biodegradability and ecological considerations of organochlorine from bleached kraft pulp mill effluent. 6th Colloquium on pulp and paper mill effluents, Toranto

Sarner E (1988) Anaerobic treatment of a mixture of condensate and caustic extraction liquor from dissolving pulp mill. Water Sci Technol 20(1):279–281

Saunamaki R (1989) Biological waste water treatment in the finnish pulp and paper industry. Paperi Ja Puu 2:158–164

Saunamaki R, Jokinen K, Jarvinen R, Savolainen M (1991) Factors affecting the removal and discharge of organic chlorine compound at activated sludge treatment plants. Water Sci Technol 24(3/4):295–308

Selvam K, Swaminathan K, Song MH, Chae KS (2002) Biological treatment of a pulp and paper industry effluent by *Fomes lividus* and *Trametes versicolor*. World J Microbiol Biotechnol 18(6):523–526

Shintani N, Sugano Y, Shoda M (2002) Decolorization of kraft pulp bleaching effluent by a newly isolated fungus, Geotrichum candidum Dec 1. J Wood Sci 48(5):402–408

Sierra-Alvarez R, Lettinga G (1990) The methanogenic toxicity of wood resin constituents. Biol Wastes 33:211–226

Sierra-Alvarez R, Lettinga G (1991) The methanogenic toxicity of wastewater lignins and lignin related compounds. J Chem Technol Biotechnol 50:443–455

Sierra-Alvarez R, Kato M, Lettinga G (1990) The anaerobic biodegradability of paper mill wastewater constituents. Environ Technol Lett 11:891–898

Sierra-Alvarez R, Kortekaas S, van Eckort M, Harbrecht J, Lettinga G (1991) The anaerobic biodegradability and methanogenic toxicity of pulping wastewaters. Water Sci Technol 24:113–125

Singh P (2007) Sequential anaerobic and aerobic treatment of pulp and paper mill effluent in pilot scale bioreactor. J Environ Biol 28(1):77–82

Singh P, Thakur IS (2006) Colour removal of anaerobically treated pulp and paper mill effluent by microorganisms in two steps bioreactor. Bioresour Technol 97(2):218–223

Singhal A, Thakur J (2009) Decolourization and detoxification of pulp and paper mill effluent by Emericella nidulans var. nidulans. Hazard Mater 171(1–3):619–625

Singhal V, Kumar A, Rai JP (2005) Bioremediation of pulp and paper mill effluent with Phanerochaete chrysosporium. J Environ Biol 26(3):525–529

Sodergren A, Adolfsson-Erici M, Bengtsson BE, Jonsson P, Lagergren S, Rahm R, Wulff F (1993) Environmental impact of bleached pulp mill effluent. In: Sodergren A (ed) Bleached pulp mill effluent composition, fate and effects in Baltic Sea. Environmental Protection Agency Report 4047. Arlow, Sweden, pp 26–46

Sorce Inc (2003) Fungal/bacterium sequencing program for lignin degradation and colour removal from wastewater at a southeastern kraft pulp mill. Personal communication with Mike Bowling

Sundelin B (1988) Effects of sulphate pulp mill effluents on soft bottom organisms – a microcosm study. Water Sci Technol 20(2):175–177

Sundman G, Kirk TK, Chang HM (1981) Fungal decolourization of kraft bleach plant effluent: fate of the chromopheric material. Tappi J 64(9):145–148

Suntio LR, Shiu WY, Mackay DA (1988) A review of the nature and properties of chemicals present in pulp mill effluents. Chemosphere 17(7):1249–1299

Thompson G, Swain J, Kay M, Forster CF (2001) The treatment of pulp and paper mill effluent: a review. Bioresour Technol 77(3):275–286

Tiku DK, Kumar A, Chaturvedi R, Makhijani SD, Manoharan A, Kumar R (2010) Holistic bioremediation of pulp mill effluents using autochthonous bacteria. Int Biodeter Biodegr 64(3): 173–183

Tomar P, Allen DG (1991) Removal of organochlorines from kraft mill effluents by an aerated lagoon. Water Pollut Res J Can 26(1):101–104

Tono T, Tani Y, Ono KJ (1968) Microbial treatment of agricultural industrial waste. Part 1: adsorptions of lignins and clarification of lignin containing liquor by mold. Ferment Technol 46:569–574

Turk S (1988) Option for treatment of CTMP effluents. Report WTC Bio-7-1988. Environment Canada, Ottawa

Valenzuela J, Bumann U, Cespedes R, Padilla L, Gonzalez B (1997) Degradation of chlorophenols by *Alcaligenes eutrophus* TMP 134 (p JP4) in bleached kraft mill effluent. Appl Environ Microbiol 63(1):227–232

van Driessel B, Christov L (2001) Decolorization of bleach plant effluent by Mucoralean and white-rot fungi in a rotating biological contactor reactor. J Biosci Bioeng 92(3):271–276

Voss RH (1983) Chlorinated neutral organics in biologically treated bleached kraft mill effluents. Environ Sci Technol 17(9):530–537

Welander T, Anderson PE (1985) Anaerobic treatment of wastewater from the production of chemithermomechanical pulp. Water Sci Technol 17:103–107

Wilson DG, Holloran MF (1992) Decrease of AOX with various external effluent treatments. Pulp Paper Can 93(12):T372–T378

Wilson RW, Murphy KL, Frenelte EG (1987) Aerobic and anaerobic pretreatment of NSSC and CTMP effluent. Pulp Paper Can 88:T4–T8

Wu J, Xiao Y-Z, Yu H-Q (2005) Degradation of lignin in pulp mill wastewaters by white rot fungi on biofilm. Bioresour Technol 96:1357–1363

Yin CF (1989) Characterization, bacterial and fungal degradation, dechlorination and decolorization of chlorolignins in bleaching effluents. Ph.D. thesis North Carolina State University, Raleigh, NC

Yin CF, Joyce TW, Chang HM (1989a) Bacterial degradation and dechlorination of bleaching effluent – effect of wood species and O_2 bleaching. Tappi Intl. Symposium on Wood and Pulping Chemistry. Raleigh, NC, Tappi Press, Atlanta, pp 753–758

Yin CF, Joyce TW, Chang HM (1989b) Kinetics of bleach plant effluent decolourization by *Phanerochaete chrysosporium*. J Biotechnol 10:67–76

Yin CF, Joyce TW, Chang HM (1990) Dechlorination of conventional softwood bleaching effluent by sequential biological treatment. In: Kirk TK, Chang HM (eds) Biotechnology in pulp and paper manufacture. Butterworth-Heinemann, Newton, pp 231–234

Yu P, Welander T (1988) Anaerobic toxicity of kraft bleach effluent. In: Tilche A, Rozzi A (eds). Proc. of 5th International Symposium on Anaerobic Digestion Bologna, Itlay, pp 865–867

Zhang X (2001) The potential of using a combined fungal and enzyme treatment system to remove detrimental dissolved and colloidal substances from TMP/newsprint mill process waters. UMI Dissertation Services, University of British Columbia, Vancover, Canada, 157pp

Zhang X, Stebbing DW, Beatson RP, Mansfield SD, Saddler JN (2001) Laccase catalyzed modification of lipophilic extractives found in TMP/newsprint mill process waters. 8th International Conference on Biotechnology in the Pulp and paper Industry, Helsinki, Finland, 4–8 June 2001, edited by Vahala P, Lantto R, p 80 [Espoo, Finland: VTT Biotechnology]

Zheng Y, Allen DG (1997) Effect of prehydrolysis of D-stage filtrate on the biotreatability of chlorinated organic compounds in bleached kraft effluent. Water Res 31:1595–1600

Chapter 14
Slime Control

14.1 Introduction

Significant advances have taken place in the pulp and paper industry in recent years. The dissemination of knowledge, the development of new and improved raw materials, the stronger emphasis on process development, and the increased use of automation have combined to produce a stronger industry having a broader range of products in terms of quality level. Despite these advances, problems still remain that increase production costs and reduce profit margins. The control of slime is perhaps the most troublesome of these problems (Johnsrud 1997; Bajpai 1999; Bajpai and Bajpai 2001).

Paper mills provide the natural conditions of nutrient supply, temperature, and moisture for the breeding of slime-forming microorganisms which can cause spoilage of raw materials and additives (pulp, mechanical pulp, recycled paper, starch, filler and pigment dispersions) (Safade 1988; Bennett 1985; Bjorklund 2000, 2002a; Blanco et al. 1996, 2002; Brown and Gilbert 1993; Chaudhary et al. 1997; King 1990; Kulkarni et al. 2003). Intense growth of spore-forming bacteria can cause problems with the hygienic quality of the end products in manufacturing of food-quality packaging paper and board. Volatile, malodorous metabolic products of microbes may enter the end products. According to Salzburger (1996) the most important economic loss caused by microbes in paper machines arises from the growth of biofilms, i.e., slime layers on the machine surfaces that are mostly made of stainless steel. Unless microbial growth is controlled, these tiny organisms can bring the large, modern, computer-controlled, hi-tech paper machines – up to 9 m wide and 200 m long – to a standstill (Salzburger 1996). The reuse of white water inoculates previously sterile parts of the system, spreading and magnifying the problem. If the growth of these microorganisms is unchecked for too long, the deposits they cause may dislodge and be carried into the paper, lowering product quality. The slime may also cause breaks in the paper at the wet end, interfere with the free flow of stock or cause excessive downtime for cleaning (Jokinen 1999; Blankenburg and Schulte 1997). This chapter presents cause and nature of the slime problems in pulp and

P. Bajpai, *Biotechnology for Pulp and Paper Processing*,

DOI 10.1007/978-1-4614-1409-4_14, © Springer Science+Business Media, LLC 2012

paper industry, and the use of microbicides, enzymes, bacteriophages, competing organisms, biological equilibrium, biodispersants, and the use of biocontrol agents in combination with biocides to counteract the slime.

14.2 Slime Problems in the Mills

Slime is the generic name for deposits of microbial origin in a paper mill (Bendt 1971; Bennett 1985; Safade 1988). Problematic slimes in the paper and board machines are mixed deposits with thick microbial biofilms as major components. Table 14.1 present the primary characteristics of biofilm and general paper machine deposits.

Paper-machine biofilms are usually composed of bacteria, EPS produced by the bacteria, wood fibers, and miscellaneous papermaking additives from the process (Latorre et al. 1991; Nurmiaho-lassila et al. 1990). Generally, EPS are composed of polysaccharides (Glucose, galactose, mannose, glucosamine, galactosamine, L-rhamnose, fucose) but may also contain proteins, nucleic acids, and polymeric lipophilic compounds. The percentage of various sugars in EPS in different paper grade furnish is presented in Fig. 14.1 (Grant 1998).

Verhoef et al. (2002) have purified EPS from bacteria *Brevundimonas vesicularis,* isolated from a paper mill. Chemical, mass spectrometry, and NMR experiments showed that *B. vesicularis* sp. produces a linear exopolysaccharide without nonsugar substituents containing a tetrasaccharide-repeating unit with the following structure:

→4)–α–L–GlcpA(1→4)–α–D–GalpA–(1→4)–β–L–Rhap–(1→4)–β–D–Glcp(1→.

The novel EPS consists of only one distinct homologue population with a molecular wt distributed around 2,000–4,000 kDa and an intrinsic viscosity of around 0.5 dL/g. The novel EPS contains a liner backbone consisting of four sugar residues. In terms of weight and volume, EPS represent the major structural component of biofilms, being responsible for the interaction of microbes with each other as well as with interfaces (Flemming 2002; Neu et al. 2001; Hoyle et al. 1990; Corpe 1980;

Table 14.1 Primary characteristics of biofilms and general paper machine deposition

Characteristic	Range
Biofilms	
Water content	High (70–95%)
Organic content	High (70–95% dry wt.)
Bacterial content	High (10^6 cells/g wet wt.)
Biological activity	High
Inorganic content	Low
Paper machine deposition	
Microbiological	Filamentous bacteria, unicellular bacteria, fungi
Particle materials (organic and inorganic)	Calcium carbonate, clay fines, fibers, pitch, chemical deposition
Dissolved substances	Minerals, sodium, chlorides

Based on Bunnage et al. (2000)

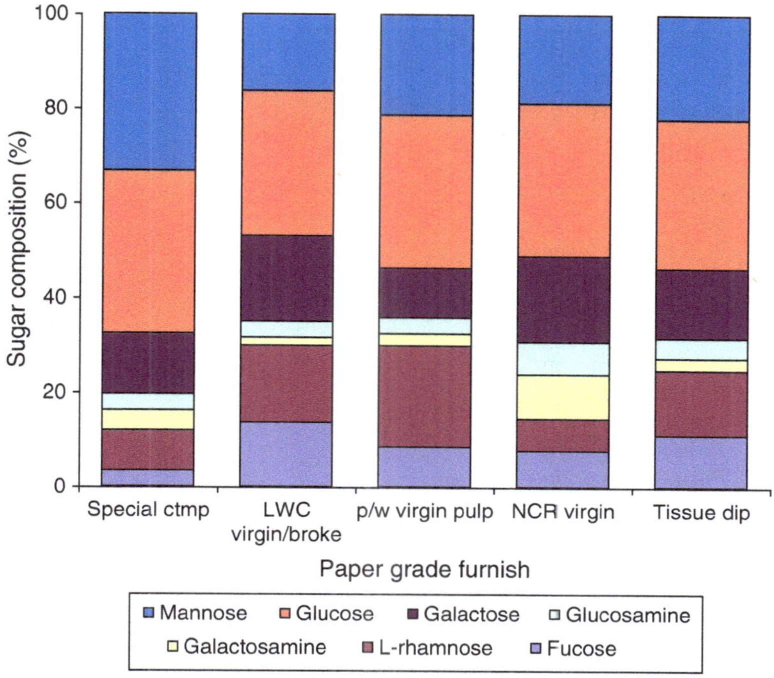

Fig. 14.1 Composition of EPS (extracellular polysaccharides) from paper machines (Grant 1998; reproduced with permission)

Costerton et al. 1978; Dudman 1977; Srinivasan et al. 1995). Slime binds in soft and viscous masses, which hook onto the sections of the paper machine where the amount of flow is not sufficiently powerful to dislodge them. These masses increase in volume until they fall off under their own weight and contaminate the pulp.

The true nature of slime and the causes of its formation are complex and many factors interact to establish the necessary conditions for slime formation. Paper mills, especially those employing increasingly closed processes and higher use of secondary fibers, have high nutrient levels as well as optimal temperature and pH ranges to support serious microbial proliferation (Oyaas 2001). Many of these microorganisms develop slimy capsular materials around the cell. This capsular material enables the cells to attach to each other and to adhere to surfaces. The slime may be homopolysaccharides or heteropolysaccharides. The homopolysaccharides are usually fructans containing a terminal glucose unit. The polymer is known as inulin or levan depending on the linkages – β (2-6) linkages occur in levan and β (2-1) linkages in inulin. Many strains of *Bacillus polymyxa* produce heteropolysaccharides that consists of D-glucose, D-mannose, D-galactose, D-fructose, glucuronic acid, and pyruvate. The proportions of various sugars in the polymer depend on the environment and the microbial strain. Levans, polymers of fructose, are generated by various forms of bacteria (Table 14.2) (Purkiss 1973; Dedonder 1966; Chaudhary et al. 1996, 1997).

Table 14.2 Levanase-producing bacteria

Bacillus mesentricus
Bacillus subtilis
Bacterium prune (Phytomonas pruni)
Bacillus megaterium
Bacillus vulgatus
Pseudomonas mors-prunosum
Bacillus polymyxa levan
Pseudomonas prunicola
Aerobacter levanicum
Corynebacterium sp.
Aerobacter acetigenum
Corynebacterium sp.
Pseudomonas aureofaciens
Pseudomonas caryophyll
Pseudomonas chlororaphis
Pseudomonas denitroflourescens
Pseudomonas syringae
Aerobacter levanicus
Serratia killensis
Aerobacter sp.

Based on Purkiss (1973), Dedonder (1966), Chaudhary et al. (1996, 1997)

Majority of bacterial extracellular polysaccharides are made of more than one type of sugar residue, and they often contain uronic acids (Chaudhary et al. 1996, 1997; Han and Clarke 1990; Hestrin et al. 1943; Tanaka and Yamamoto 1979). Many bacteria secrete organic polymers with limited solubility in water, which tend to accumulate as loose, confluent layers in the immediate neighborhood of the cell just outside the wall. Unlike the cell wall, the capsule or slime layer seems to have no important direct role in the maintenance of cellular function. Some bacteria form the slime layer only when growing at the expense of a specific substrate, which is a direct biochemical precursor of the slime substance in question. The behavior is characteristic of certain *Streptococci*, *Bacillus*, *Pseudomonads*, and *Xanthomonads*, which form copious quantities of either dextrans or levans when growing at the expense of the disaccharide sucrose. No other metabolizable sugar, including glucose and fructose themselves, can serve as a substrate for the synthesis of these polysaccharides. Consequently, dextran- and levan-forming bacteria produce these capsular materials only when growing on sucrose-containing medium. Certain kinds of capsular substances can be removed from the cells by treatment with specific hydrolytic enzymes. Such enzymatic treatment leaves the cell unharmed (Stanier et al. 1968). Effluents rich in sucrose specifically encourage the growth of dextran- and levans-producing organisms.

The slime problem is a natural phenomenon. Microorganisms such as bacteria, fungi, and yeast thrive in the presence of nourishment, moisture, and warmth. The nutrient-laden waters of the paper mill provide a hospitable environment, especially in the areas where the water temperature is around 30°C. These are the areas where microbial metabolic substances such as slime, chemical deposits, or fine debris are found. Each papermaker's experience with slime is unique. The type of slime varies from one mill to another, from one paper machine to another within the same mill and even at different areas in the same machine. An economically sound slime control program for a given mill must bear the proper relationship to the total loss caused by the slime. This loss can be gauged by considering the following features: loss of finished product, loss of production time, loss of heat, chemicals, filler, fiber and water, decreased life of equipment or clothing. Literature shows that significant problems are caused by several genera of bacteria such as *Aerobacter* particularly the species *Aerogenes*, *Cellulmonas*, *Chromobacter*, *Achromobacter*, and *Crenothrix*. Pink and red slime is a familiar presence in the mills producing printing paper (Nason et al. 1940). Sanborn (1965) noted that such slime was composed of organisms closely resembling *Micrococcus agilus*, *Serratia marcescens*, *S. piscatora*, *Ascophyta* sp., *Gliocladium roseum*, *Penicillium pinophilum*, *Fusarium* sp., *Rhodotorula*, and *Monilia*. Various organic biocides are typically added to the papermaking system in an attempt to control these microorganisms. However, the antimicrobial activity of the chemicals is nonselective. As a result, operation of sewage-treatment systems containing active sludge is often impaired. On the contrary, microorganisms can develop a resistance to a given biocide. Most papermakers try to avoid this phenomenon by alternating the use of different biocides.

Understanding the nature of slime formation in paper mills requires an understanding of the functioning of the mill. This helps in identifying possible areas where slime deposits may form and interfere with the operation of the mill. The papermaking process consists of several steps. In the paper mill, the pulp produced in a pulp mill is suspended in water to 3–4% solids. The resulting slurry is passed through a series of machine chests and refiners in which the pulp is mechanically processed and diluted to a workable consistency. These steps make up the beginning of wet end. A complex system of flow spreaders and pressurized head boxes eject the dilute slurry of fibers onto a rapidly moving paper machine wire screen. Fibers are laid onto the surface of the screen, the sheet is formed, and the water is removed. This water is known as white water. Excessive biological activity in paper mill is generally seen at this end. The stored pulp from various recycle streams may be mixed with fresh pulp. Microbial contamination may take place due to prolonged storage in the open. Microbial contamination is further aggravated in closed loop mills where white water is recycled for dilution of the pulp. Closing up of white water system contributes to the cycling of nutrients (Baker 1981; Barnes 1984; Baurich et al. 1998; Bendt 1971; Bennett 1985). Factors such as pH, temperature, and levels of organic nutrients also play a significant role in slime development (Dubout 1979). From the nutrient point of view, microorganisms are divided into two groups: those which demand relatively complex organic molecules and those which, like higher plants, are capable of synthesizing their food stuffs on the basis of simple mineral salts and carbon dioxide. These two categories of microorganisms have free field in the pulp and white water systems. If we cast a glance at the solutions, which are normally found in the water of the paper mill, it is easy to see that the number of invading organisms is considerable. The conditions normally found in the paper machine are pH 5–8, temperature 20–78°C, and an abundance of nutrients, which is an excellent environment for the growth of bacteria and fungi (Kolari et al. 2003). The starch, glues, and coatings that are used as additives in the mill are excellent food sources for many microorganisms (Oppong et al. 2000). Nutrients from pulp and white water are coated on surfaces to produce films of concentrated food, which becomes the basis for microbiological growth and activity. Changes in equipment design, use of complex chemical mixtures or additives, changes in operating practices from high to low grammage grades, bad housekeeping, and storage of pulps, recycled fiber, and sludges are the major factors that aggravate the slime problem. Surface water supply from lakes, rivers, ponds, and wells can also be a serious source of inorganic nutrients (e.g., iron, sulfur) and bacterial contamination. The deposits formed can be classified as biological (slime) and nonbiological (scale and pitch).

14.3 Microorganisms Within the Slime and Contamination Sources

The biological components of the slime of a paper mill are all unicellular members of the plant kingdom. Bacteria are the most common, but these are often accompanied by fungi and yeasts (Stanier et al. 1968; Sanborn 1965; Väisänen et al. 1989, 1991, 1994, 1998; Lutey 1972; Eveleigh and Brewer 1963; Brewer 1960) (Table 14.3).

The bacteria most commonly found in paper mills are those, which are normally present in natural water. The most common of the slime-generating bacteria belonging to this group is *Aerobacter aerogenes*, which is a nonsporing rod and has a large capacity for adapting to the oxygen supply of the environment. It is very frequently found in paper mills all over the world and gives rise to a soft and gelatinous slime, which requires the mechanical support of a network of paper fibers. Its development seems particularly large in mill waters, which have a substantial biochemical oxygen demand because of the substantial concentration of organic materials. There are several variants of this species, which are mobile and are capable, to a certain extent, of choosing the most favorable location of the paper machine for forming colonies. Other bacterial species encountered in slime are *Escherichia coli*, *Pseudomonads*, *Chlamydobacterials* (encapsulated bacteria), *Alcaligenes*, *Arthrobacter*, *Proteus*, *Bacillus*, and others. *E. coli* is essentially a microorganism of intestinal flora, which is similar to *A. aerogenes* in many respects but habitually forms a film of slime, which progressively increases in thickness with time and finally tears off in the form of small grayish scales. In general, its nutrient needs are similar to those of *Aerobacter*. *Pseudomonads* make a special contribution to slime formation. This group comprises a range of badly defined species, which are notoriously resistant to low concentrations of antislime agents. *Pseudomonads* can be identified by the fluorescent tint, which they give to slime. *Chlamydobacterials* is an extremely important group of slime-generating bacteria, but in contrast with the *A. aerogenes* and *E. coli*, they require a great deal of oxygen. They prefer water with a low BOD, and this is why

Table 14.3 Microorganisms commonly found in mill environment

Microorganisms	Type	Characteristics
Desulfovibrio (sulfate reducing)	Anaerobic bacteria (pH 3.5–10.0)	Corrosive, odorous due to H_2S
Clostridium	Anaerobic spore forming bacteria (pH 4.0–10.0)	Corrosive, putrefactive odors, decompose starches and proteins. Form heat-resistant spores
Aspergillus *Pencillium* *Trichoderma* *Mucor*	Aerobic filamentous fungi (pH 2.0–7.0)	May attack cellulose, also found in wood piles, chips, and bales of dry lap stock or waste paper. May cause mildew and musty odor
Oospora *Manilla* *Torula* *Rhodotorula* *Saccharomyces*	Aerobic, yeast-like fungi (pH 2.0–7.0)	Discoloration spotting of pulps
Mucoid types *Pseudomonas* *Bacillus mycoides* *Bacillus subtilis* *Bacillus cereus* *Bacillus megaterium*	Aerobic slime forming bacteria (pH 3.5–9.5)	Stock odors (souring)
Flavobacterium *Enterobacter* *Escherichia coli*	Aerobic slime-forming bacteria (pH 5.0–8.0)	Stock odors. Heat resistant

Based on Stanier et al. (1968), Sanborn (1965), Väisänen et al. (1989, 1991, 1994, 1998)

they are most often found in mills, which do not recycle their process water. One member of this group forms abundant white filamentous plumules. Another converts, by metabolism, the iron salts and deposits iron oxide in its external envelope. In addition, bacteria found most often in paper and board machine slime include species of *Flavobacterium*, *Clavibacter*, *Sphaerotilus*, and *Leptothrix*.

Slime/biofilm systems may provide an anaerobic zone enabling the sulfate-reducing bacteria – *Desulfotomaculum, Desulfovibrio* (Postgate 1979) – to grow and metabolize even when the bulk water contains oxygen close to saturation. The outer layer of this film will consist of heterotrophic bacteria depleting oxygen and developing an environment suitable for sulfate-reducing bacteria to grow and metabolize if sulfate and excess organic carbon are present. Sulfate reduction in biofilm systems is of concern in many water systems (Piluso 1977; Robichaud 1991). Sulfate-reducing bacteria produce hydrogen sulfide, which can deteriorate concrete or corrode metals. A general hierarchy of heterotrophic organisms, denitrifiers, iron-reducing bacteria, sulfate-reducing bacteria, and methanogens may be found in thick microbial films where excess organic material, nitrate, iron, and sulfate are present.

Sulfur bacteria make only a small contribution to the formation of deposits of sludge, one of their members, "sulfate-reducing bacteria" is very commonly encountered in the highly sulfurized water of paper machines. These bacteria develop poorly in the presence of gaseous oxygen but they are frequently found in full growth under the thick deposits of slime and those of fibers in the storage tanks of white water. They reduce the sulfate ions in hydrogen sulfide and are responsible for odors and for the black-enemy of the fourdriner wires and accessories made of copper alloy. They can also deactivate the mercurial slimicides by the formation of insoluble sulfurs. Proteolytic bacteria decompose the formulations of casein and gelatin employed for sizing. Among others, *Pseudomonas* sp. is in this category.

Common fungal species found in slime are *Aspergillus*, *Penicillium*, and *Cephalosporium* (Brewer 1960; Huster 1992a, b). In closed process water systems, where temperatures are often maintained above 30°C, the thermotolerant fungi play an important role in slime formation in paper mills (Eveleigh and Brewer 1963). Several mold species are also known to be synergistic with bacterial species, forming tough slimes that are resistant to biocides and difficult to control. These include *Cladosporium*, *Geotrichum*, and *Mucor* (Sanborn 1965). The extremely varied group of microorganisms produces an equally varied number of extracellular polysaccharides, reserve materials, and other excretory products that make up the slime. Fungi, unlike bacteria, are not really at home in process water. On the contrary, they often form the fibrous mat on which the deposits of slime will accumulate as a consequence. They are to a large extent responsible for the deterioration of balls of wet pulp in sheets in the course of stacking or of finished paper products. The colored patches of the pulp so well known to papermakers are, in general, due to colonies of Basidiomycetes and *Penicillium.* It is then essential for the slime control program to include the use of fungicidal element. *Penicillium roqueforti* deactivates the mercury compounds.

Väisänen et al. (1998) performed a thorough examination of microbial communities of a printing paper machine. They isolated 390 strains of aerobic bacteria representing atleast 34 species, thus demonstrating the large number of bacteria living in a single paper machine. The most frequent contaminants of the machine wet end were species of genera *Bacillus, Burkholderia*, *Pantoea*, *Ralstonia*, and *Thermomonas*. Chaudhary et al. (1997) reported that *Bacillus alvei* and *A. aerogenes* were the most prevalent contaminants in an Indian paper mill. Oppong et al. (2000) isolated pink-pigmented bacteria *Deinococcus grandis*, *Flectobacillus* sp., *Methylobacterium zatmanii*, *Micrococcus* sp., and *Roseomonas* sp. from slime deposits of American paper machines. Harju-Jeanty and Vaatanen (1984) showed that the filamentous bacterium *Sphaerotilus natans* can produce slimy aggregates in groundwater pulp (1%) at 28°C, but it failed to grow at above 40°C (Pellegrin et al. 1999), making it an unlikely biofilm-former in modern paper machines with operation temperatures near 50°C. Raaska et al. (2002) observed that contamination of starch-based glues with spore-forming bacteria (e.g., *Bacillus* and *Paenibacillus*) and enterobacteria (e.g., *Citrobacter* and *Enterobacter*) was the most important factor threatening the process hygiene and product safety in one machine manufacturing refined paper products for food packaging. Generally, the major contaminants of the dry paper products seem to be limited to three genera of spore-forming bacteria: *Bacillus*, *Brevibacillus* and *Paenibacillus* (Hughes-van and Kregten 1988; Pirttijärvi et al. 1996; Raaska et al. 2002; Suominen et al. 1997; Väisänen et al. 1998). Kolari et al. (2003) have reported microbiological survey of colored biofilms in six paper and board machines, including two case studies of outbreaks of colored slimes in which the causative bacteria were found. A total of 95 pink-, red-, orange-, or yellow-pigmented strains were isolated. Nearly all (99%) of the strains grew at 52°C, 72% grew at 56°C, but only 30% grew at 28°C, indicating that most of the strains were moderately thermophilic. Biofilm formation potential and biocide susceptibility of the strains were analyzed with a microtiter plate assay. In the presence of 5 ppm of methylene bisthiocyanate (MBT) or 2,2-dibromo-3 nitrilopropionamide in paper-machine water, 55 strains formed biofilm. Moreover, 39 strains increased biofilm production in the presence of biocide, suggesting that biocide concentrations inhibitory to planktonic but not to surface-attached cells may actually promote biofouling. The cells may have inactivated a portion of the biocides, as the cell density in this assay was high, corresponding to the highest cell densities occurring in the circulating waters. Four groups of colored bacteria that were isolated from several mills were identified. Pink pigmented *Deinococcus geothermalis* and red-pigmented *Meiothermus silvanus* occurred as common primary biofilm-formers in paper machines. The third group of bacteria (putative new species related to *Roseomonas*) contained strains that were not biofilm formers, but which were commonly found in slimes of neutral or alkaline machines. The fourth group contained red-pigmented biofilm-forming strains representing a novel genus of a-Proteobacteria related to *Rhodobacter*. Many colored paper-machine bacteria are species previously known from microbial mats of hot springs.

The main sources of microbiological contamination are:

1. Fresh water, especially when surface water without previous treatment is used.
2. The cellulosic raw material, particularly when secondary fibers are used.
3. The brokes, especially when sizing and coating additives are used.
4. The solutions or suspensions of additives, fillers, pigments, starches, coatings, etc.
5. The recycled water.
6. The environment in which the paper machine is place.

When the dissolved oxygen concentration is high, aerobic bacteria will be developed, which are the main producers of the slime. Alternatively, if the concentration of oxygen is decreased, the population shifts toward anaerobic species, which are responsible for the problems of odors and corrosion (Bennett 1985). If the temperature increases, the population could vary from mesophilic to thermophilic species, which form spores. The reuse of the white water also increases the concentration of filamentous microorganisms present in the system. These species, which initially enter the system with the fresh water, are developed in the process, continuously contaminating the system wherever the white water is reused. Although the fresh water is the main cause of the presence of algae in the system, the recycled pulp is the main source of bacterial and fungal contamination. The concentration of microorganisms in the recycled pulp is a thousand times higher than in the virgin fiber pulp. The starches and coating products that appear in the recycled paper are an important source of microorganisms (Robertson 1994). Also, the fillers and adhesives that are present in the recycled pulps enhance the formation of slime, as they are ideal places for the attachment of fungi and bacterial colonies. However, rosin and aluminum compounds generally reduce the growth of microorganisms.

14.4 Sites Chosen by the Microorganisms in the Paper Mill

On a theoretical level, microorganisms can develop in any location of the water systems of a paper machine but, in practice, it must be stated that they form at various points of relatively small isolated packets of proliferation (Blanco et al. 1996). The placing of these chosen sites depends on the combined effect of several factors such as: model of machine, type of microorganism, chemical environment, flow speed of water, pulp consistency, composition of the pulp, program for combating the slime. Even in the absence of heavy contamination, the type of construction of the paper machine can exert a substantial influence on the formation of deposits of slime by supplying a multiplicity of points with zero flow. By this, we understand the sectors of the machine in the pipe conduits where a certain degree of stagnation is produced, for example: the inner side of the angles of pulp pipes, the T or Y shaped joints, the orifice plates, the rotary cleaners, cross-bars, basins and pulp lines, the tanks of rich white water, the wire rolls, the fourdriner, etc. All the efforts taken at the moment with studying the machine for avoiding these dead points and making the interior of pipes smooth will contribute to reducing the problems caused by slime.

The very nature of microorganisms developing in slime plays an important part in the location of slime on the machine (Purkiss 1973; Blanco et al. 1996). Anaerobic bacteria are to be expected at points where the oxygen tension of the water is zero, under already existing deposits of slime or accumulations of fiber deposits. One will find anaerobic bacteria in most of the other locations. These bacteria form a thin gelatinous film proliferate where the content of abrasive products of the water is at its minimum, since when the consistency of the pulp has risen (1–3%), the displacement of the cellulose fibers tends to remove the fine film as soon as it forms. The pseudofilamentous bacteria require a certain turbulence and a surface suitable for their fixing. They do not like substantial quantities of electrolyte and avoid points where the alum concentration is too high. Certain microorganisms can only form coherent masses when they find a suitable mechanical base and establish their colonies around the fibers, progressively constructing hanging growths of a dirty gray color, which generally becomes too dense and tear off under their own weight. The fungi and molds often develop on the dried splashes situated above the water line of the wooden chests. If the contamination is not all that substantial, the deposits of slime tend to form very far from the point of introduction but the converse is not true when it is a question of massive contamination. Majority of mill infections are borne at a very localized principal point, a point where the bacterial proliferation is able to develop and from where the greater part of the other infections expands. In the majority of cases, the center of infection is the white water circuit. The most satisfying method for localizing the center of infection of a paper machine consists of making a quantitative microbiological examination, for example by using the plate count method. When this counting operation is carried out on a machine installation, it is normally found that the greatest concentration of microorganisms is in the buffer tank or a white water storage tank even if the deposits of slime are not apparent there. It would seem that the microorganisms reproduce in the white water in which the conditions are favorable and take the form of slime where the chemical milieu is adverse. By biological means, it is possible to predict the most favorable locations for microbial proliferation but the spores where slime forms seem to be a question relating to the hydraulics and type of machine.

The manifestations of infections most commonly encountered in a paper mill are the following – Formation of slime, Blocking of the felts, Degradation of the felt, Fermentation of rosins, Stains in the pulp, Cellulolytic action, Mold, Musty odors (Purkiss 1973).

14.4.1 Formation of Slime

In its most accentuated form, slime does not leave any doubt as to its presence. Its glutinous, blackish and fibrous mass reveals the latter, even to the least observant. On the contrary, it is in the least serious cases of infection that slime is best able to show its insidious nature. As a consequence of their surface electrical charge, the bacteria associate very rapidly with the cellulose fibers in suspension in the pulp. The effect of this association is to slow down dewatering in the wire. This variation in the rate of dewatering is often the precursor of machine contamination. The most serious infections are revealed by the appearance of slime stains in the finished product, a situation which worsens progressively until windows appear, holes and finally tears in the sheet caused by the localized adhesion of the latter to the rolls.

14.4.2 Blocking of the Felts

A microbiological examination of a warped felt always reveals a substantial bacteriological examination. The accumulation of slime in the interslices of the fabric can very quickly make the felt unusable until it is cleaned. As the proliferation of bacteria is progressive, the loss of yield will take place for a certain period before the complete working of the felt.

14.4.3 Degradation of the Felt

When the microbial population of a machine contains proteolytic elements, the degradation of the fibers making up the woolen felts is frequently apparent. This state of affairs leads to a reduction in tensile strength and shear strength. Often the rapture of a felt is attributed to "normal mechanical wear" although a closer examination of this shows that numerous points display week resistance while purely mechanical wear is slight there. Tanning can reduce this problem to a certain extent but never completely.

14.4.4 Fermentation of Rosins

Rosin preparations constitute a nutrient milieu particularly suitable for microorganisms. The contamination of rosin can normally be detected by the characteristic odor, which is released. Protein-based adhesives release a bad odor of ammonia or carbylamine, while starch normally smells like alcohol. Frequently, these bad odors are not revealed until the paper is in the hands of the Converter. Rosins are contaminated in a fashion rather similar to that of the machine but cause a problem only if insufficient attention is paid to the cleaning of the mixers or the resin formulations are left too long for a period.

14.4.5 Stains in the Pulp

As the pulp rarely contains much free water, it tends to promote the development of fungi rather than that of bacteria. Proliferations of fungi are often pigmented and appear in the form of stains in the pulp. It may also be the case that these stains appear only after a certain period of storage of the latter. The fact that mercury compounds have been added does not provide any guarantee in respect of the nonappearance of these stains, since there are certain forms of fungi, which are capable of absorbing mercury is *Penicillium roquieforti.* A great deal of research work has been done on this microorganism and it has been found that it can reduce the concentration of mercury in the sheet of pulp to such an extent that the development of other types of microorganisms becomes possible.

14.4.6 Cellulolytic Action

The effects of the action of fungi and cellulolytic bacteria are not generally felt in the mill itself. In fact, given that these microorganisms bring about, by degrading the cellulose, a reduction in the tensile strength of the fibers, the effect of their attack is not apparent in the paper until the latter is delivered into the hands of the final consumer. In the case of products such as kraft racks of several thicknesses, this problem can largely be resolved by paying suitable attention to slime control or, in limited cases, by subjecting them to a treatment to make the product nonbiodegradable.

14.4.7 Mold

It is very common to receive complaints concerning the development of mold on the finished products, and the paper produced is rarely at fault in these cases. It should also be noted that many things can be carried out on the machine to prevent the proliferation of mold on papers destined for special uses.

14.4.8 Musty Odors

Here, we have to deal with strange odors, generally disagreeable, which are associated with various phases of the production of paper and board. The most common of these odors are the following: Firstly the musty earthy smell, which is normally released from stacks and covered tubs, and finally that offensive smell of hydrogen sulfide, which is often noticed when the channels under the machine are cleared. These are then a result of the manifestations of the microbiological activity. The musty earthy odor is normally due to proliferation of fungi under the lids of the stacks, as well as that of hydrogen sulfide resulting from the action of sulfate-reducing bacteria, in conditions where there is only little or no free oxygen.

14.5 Methods for Detection of Slime

Following methods are available for detecting slime problems (Farkas et al. 1987).

14.5.1 Slime Collection Boards

A collecting board is a unit placed in the water of a paper machine in such a way as to cause slime formation. Inspection of the board at regular intervals makes it possible to reveal the formation of slimes on the machine in the early stages before they become a source of problems. Numerous authors have studied these boards and it has been found that the most effective and least costly ones were those made of plastic material, for example of Plexiglas. It is sufficient to take small rectangular pieces of Plexiglas in which two holes are pierced so they can be suspended. These pieces must be slightly curved and placed in a flow of water in such a way that an area of slight turbulence is created on the concave side. It is easy to detect the formation of slime on a collecting board. In fact an extremely fine coating gives greasy sensation. These detecting boards must be placed in any locations as are necessary and inspected at regular and frequent intervals.

14.5.2 Identification of the Contaminated Points

The stains and black areas of the paper are not necessarily caused by slime and it is thus necessary to have some method for distinguishing the marks of biological origin from the others. As most mills do not have any microbiologist, traditional methods of microbiology have relatively little interest for the identification of stains. So chemical methods are used. Two tests belonging to this last category have been developed in the last few years. These are (1) Ninhydrin test and (2) Tetrazolium chloride test.

14.5.3 Standard Plate Count Method

The generally accepted method for quantifying the level of biological activity in a system is the standard plate count. In this method, a sample from the system is serially diluted and placed in a sterile nutrient medium in a petri plate. This accommodates the normal paper mill populations, which range anywhere from less than 10,000 organisms/mL of sample to more than 100 million organisms/mL. The plates are incubated and individual organisms allowed to reproduce in the solidified nutrient until colonies of organisms, which were present in the original sample. Incubation of bacterial plates generally takes 48–72 h; fungal counts require 5–7 days.

The plate count method has several well documented, but minor, technical limitations which are overcome using slightly modified procedures or nutrients.

14.5.4 Dip Sticks

Dip sticks marketed by a number of companies contain a solid nutrient and a color changing indicator, which allows colonies to be counted. With these methods, the device is momentarily dipped into a water sample. Following incubation of 24–48 h, the number of colonies on the surface of the device is counted, or a comparison is made with a chart to gauge a color change, which has been correlated to standard plate counts. This simple method gives a rough indication of the biological activity, which was present in the sample. The rapidity of the test is a great improvement over the standard plate count method. The dip method is widely used for control of cooling tower systems, but the lack of precision and 24-h turnaround time has led paper mills to search for a yet more rapid precise simple test to alert the management of potential problems.

14.5.5 Luminescence

Adenosine triphosphate (ATP) is the molecule associated with all biological activity. A luminescence method can be rapidly used to quantify the amount of ATP in a liquid sample. The level of ATP correlates to the level of biological activity at the sample point. Since different species of bacteria and fungi function with different levels of ATP, a comparison with plate counts on a stable population must be performed before the luminescence measure has meaning. The method is most useful for study of pure laboratory cultures. Following sample preparation, the ATP method provides useful information in less than 30 min. ATP-luminescence is rapid and precise, but it is fairly complicated and expensive to perform. Few mills find the method useful as a routine control method.

14.5.6 Bio-Lert Method

The Bio-Lert method uses an indicator system with a sterile stabilized nutrient broth in a convenient vial. The liquor portion of a paper mill sample (stock, filler slurry, starch slurry, coating) is poured into the vial, filling it to the 10 cc mark. The vial is capped and the contents shaken to thoroughly mix the nutrient and indicator with the contents. The vial is then incubated at 37–42°C (98.6–107.6°F). The time required for the obvious color change from blue to pink is correlated to standard plate counts. Laboratory testing has demonstrated that the time for a sample to cause color change is dependent solely upon the biological population in the original sample. Populations in the range of 100 organisms/mL to 1 billion organisms/mL have been studied. Using the Bio-Lert vials, excessive biological populations can be detected in as little as 1 h. This allows the mill to take corrective action before the effects of potential problems are translated into lost production or spoiled raw material.

The test method has proven to be effective and stable despite wide variation in incubation temperatures or sample pH environments. The optimum incubation temperature has been found to be 37–42°C, but the test has utility even when incubation occurs in a shirt pocket. A variety of sample types respond to this method, including wood pulp fiber suspensions, filler slurries, and starch slurries. Table 14.4 compares the practicality and rapidness of the Bio-Lert method to the other methods discussed.

Because different species of microorganisms may react differently in the test, a comparison with standard plate counts should be made for each sample point to improve accuracy. The test is intended to provide an alert to high levels of biological activity. When systems containing low levels of biological activity are studied, it will take several hours for the color change to occur. If a sample produces unusually rapid color change, a biological control upset is indicated. Then, the cause of the upset can be investigated and corrective measures taken before there has been an economic impact upon the mill. The Bio-Lert method has proven its simplicity, reliability and a rapidity in a variety of mill situations. The Bio-Lert method has provided accurate, valuable information about the biological activity in pulp and paper mill systems on a timely basis, leading to raw material cost savings, continued up-time, and greater peace of mind.

Table 14.4 Comparison of biological activity test methods

Method	Time	Accuracy	Comments
Bio-Lert	1–4 h	Very good	Rapid simple procedure, especially useful as a Bio-Lert
Standard plates	48–72 h	Excellent	Time consuming procedure
Dip-stick	24 h	Yes	Results not rapid enough
ATP-luminescence	<30 min Excluding sample preparation	Good	Simple test, results not rapid enough, pulp times can interfere with test
TTC, Indicator dyes	4–48 h excluding sample preparation	Good	Results not rapid enough, sample preparation sometimes complicated
Ninhydrin spray	5 min	Fair	Rapid amino-nitrogen test (careful; do not tough spot), not quantitative

Based on Farkas et al. (1987)

14.5.7 Slime Monitor

Sugi (1999) has developed a new monitoring tool – *Slime Monitor* to establish a suitable slime control program in paper mills. This monitor allows online, real time measurement of slime deposition and makes sure that the minimum possible amount of biocide is used to the greatest possible effect.

Buckman Laboratories International Inc. has developed a dual-channel biofouling monitor to assess slime buildup in white water and cooling water circuits (personal communication). The monitor displays a numeric and graphical index of biofouling and incorporates nutrient augmentation and other technology to moderate sensitivity to slime buildup. These features provide an early warning of biofouling that can be used to take corrective action prior to damaging slime buildup in the industrial process. The features also permit the sensitivity of the device to be adjusted to better correlate monitor output with paper machine performance. The instrument was used to measure fouling in an alkaline fine paper mill following a wet end boil-out. Results showed that biofouling in the nutrient augmented channel was accelerated by several days compared with a white-water only control. The data was modeled with an exponential curve to obtain a specific fouling rate of 1.27 and 1.03/day for the accelerated and control channels respectively. ATP measurements and microscopic analysis indicated the slime buildup contained both microbial and abiotic components. The new instrument can be used to optimize microbial control programs in paper mills by providing a quantitative measure of slime accumulation

14.6 Biofilm Formation in Paper Systems

A well-known and long-standing problem in paper manufacture is the proliferation of biological slimes on machinery. Virtually every surface examined in natural, industrial and pathogenic ecosystems is colonized by biofilms consisting of adherent (sessile) bacterial populations enmeshed within a glycocalyx, matrix, which is defined as a polysaccharide containing material lying outside the cell (Lawrence et al. 2002; Mattila et al. 2002; Bryers 1990; Brading et al. 1995). Although many studies on free-living bacteria (planktonic) most commonly isolated from process waters have been reported (Hughes-van Kregten 1988; Jung and Kutzner 1978; Vaatanen and Niemela 1983), relatively few of these have been concerned with the biofilm environment. Biofilm buildup is characterized by an increased resistance to biocides, by the development of anaerobic zones, by structural heterogeneity, and by protection from predative grazing microorganisms and from desiccation (Lawrence et al. 1994). Some reports relate to the formation of biofilm in pulp and paper mills (Stoner and King 1994; Väisänen et al. 1994; Wiatr 1990, 1994; Mueller 1994). A few studies have been made on biofilms in the paper machine environment (Väisänen et al. 1994; Latorre et al. 1991; Nurmiaho-lassila, et al. 1990). Most investigators of biofilm development in engineered systems have concentrated on enumerating the bacterial population, with relatively fewer studies considering the possible importance of filamentous fungi in such systems (Nagy and Olson 1986; Eveleigh and Brewer 1964). Today, much effort is being directed toward elucidating those features that determine the resistance of bacterial biofilms toward antibacterial agents.

Although biofilm formation has been studied extensively in natural aquatic systems, wastewater treatment systems, and medical appliances, only a few studies have been made on biofilms in the paper machine environment. Work by Väisanen et al. (1994) has shown that paper machine biofilm has a complicated biological and chemical structure, which morphologically resembles those of wastewater biofilms, biofilms in industrial cooling towers and natural river ecosystems. Within the slime, microbial dominance is dynamic. The microbial composition of the slime is dependent on its original and subsequent development in the mill system. Slime formation depends both on a mechanism of inception and on the subsequent conditions for growth of the organisms. The relative importance of a specific species is difficult to assess, as many factors such as the influence of nutrients (Eveleigh and Brewer 1964; Humiston 1955; Shema 1955), the presence of specific inhibitors and stimulants in woods, and physical conditions can control both its amount and type of growth. Seasonal changes, slime treatment variations, fresh water recirculation vs. nonrecirculation, type of pulp stock, and cleanliness also cause variations in microflora (Eveleigh and Brewer 1964; Hughes-van Kregten 1988; Martin 1955). Cell concentrations in biofilms are three to four orders of magnitude higher than those of planktonic cells. Concentrations of 10^{12} cells/cm^3 are commonly found within biofilms, whereas the maximum concentration for cells in suspension is 10^8–10^9 cells/cm^3. As late as 1987, the biofilms were regarded as surface-attached microbes embedded in their EPS, growing in a selfish manner and producing unorganized slime layers on the surfaces (Stoodley et al. 2002). The current opinion is that biofilm formation is a much more complicated and also a well-controlled phenomenon. Increasing evidence exists that bacteria possess different growth modes and that cells in biofilms differ profoundly from planktonically growing cells of the same species (Kuchma and O'Toole 2000; Stoodley et al. 2002). Biofilm formation is a dynamic process that begins with bacterial cells attaching to, or "colonizing," a submerged surface and continues until a mature biofilm develops. The process of biofilm formation can be divided into five stages – Conditioning Layer; Bacterial Attachment; Biofilm Formation and EPS Production, Biofilm Maturation and Detachment (Brading et al. 1995; Bunnage et al. 2000; Costerton et al. 1995; Glazer 1991; Marshall et al. 1971; Väisänen et al. 1994; Werres 1998).

The maturation of a biofilm, resulting in the complex architecture with water channels, is influenced by a number of biological factors and by hydrodynamic features (Stoodley et al. 2002). The biological factors include cell-to-cell signaling between the biofilm bacteria, growth rates of the bacteria, extent of EPS production, motility of the biofilm bacteria as well as possible competition or cooperation between the bacteria. A mature biofilm with its complex architecture provides niches with distinct physicochemical conditions, differing e.g., in oxygen availability, in concentration of diffusible substrates and metabolic side products, in pH, and in the cell density. Consequently, cells in different regions of a biofilm can exhibit different patterns of gene expression (Costerton et al. 1999). Mixed-species biofilms can contain niches with distinct groups of bacteria having metabolic cooperation (Kuchma and O'Toole 2000). The EPS matrix of a biofilm community can also contain microzones with different charge and hydrophobicity (Wolfaardt et al. 1998). Watnick and Kolter (2000) summarized that a mixed species biofilm is a dynamic community harboring bacteria that stay and leave with purpose, compete and cooperate, share their genetic material, and fill distinct niches within the biofilm. They stated, "The natural biofilm is a complex, highly differentiated, multicultural community much like our own city."

Bacteria in a mature biofilm are far more resistant to antimicrobials (biocides and antibiotics) than freely swimming cells. Different mechanisms have been proposed to account for this increased resistance that is most likely multifactorial (Costerton et al. 1999; Czechowski and Stoodley 2002; Donlan and Costerton 2002; Dunne 2002; McBain et al. 2002; Stoodley et al. 2002; Watnick and Kolter 2000; Whiteley et al. 2001). EPS may form permeability barriers or make complexes with the antimicrobials thus interfering with the antimicrobial action. Reactive oxidants may be deactivated in the outer layers of EPS faster than they diffuse. Extracellular enzymatic activity inside the biofilm may be high enough to destroy the antimicrobials. Different microenvironments existing at the deepest biofilm layers with altered pH, pCO_2, pO_2, cation concentrations etc., may affect the activity of antimicrobials. Bacterial cells of the biofilm phenotype may have reduced susceptibility because of altered cellular permeability, metabolism, or growth rate. Das et al. (1998) reported changes in susceptibility that were greater than twofold and occurred immediately on attachment, and also in the presence of biocide concentrations which exceeded the minimum inhibitory concentrations for the planktonic cells. Grobe et al. (2002) quantified the degree of biofilm resistance by calculating a resistance factor that compared killing times for biofilm and planktonic cells in response to the same concentration of biocides. The resistance factors averaged 29 for 2,2-dibromo-3-nitrilopropionamide (DBNPA), 34 for glutaraldehyde, 120 for chlorine and 1,900 for a quaternary ammonium compound. This indicates that data on antimicrobial efficacy obtained with planktonic bacteria are not reliable indicators of biocide performance against biofilms. A part of the biofilm resistance could be attributed to incomplete penetration of chlorine, glutaraldehyde and DBNPA due to neutralizing reactions with EPS. The results also implicated presence of a tolerant subpopulation for the quaternary ammonium compound.

14.7 Control of Slime

Slime control is necessary in paper and board production to ensure high quality and trouble-free operation, without downtime caused by slime-induced paper breaks and fouling of paper machinery (Safade 1988; Bendt 1971; Bennett 1985; Blankenburg and Schulte 1997; Huster 1992a, b; Morrcs 1995; Robertson 1994; Wiatr 1990; Chaudhary et al. 1998). At the same time, paper manufacturers are demanding more of their microbial control programs to meet higher quality criteria from customers, face hard competition in the market place and comply with tougher environmental legislation. The control of slime problems in paper or board mills is carried out mainly using chemical substances called biocides. Because of environmental and legislative demands, several alternative slime control methods such as enzymes, bacteriophages, introduction of competing microorganisms, biological equilibrium and biodispersants have been developed (Gould 1992; Geller 1996).

14.7.1 Traditional Methods

Regular inspection of all areas, a planned program of frequent and effective boil-outs, and regularly scheduled wash-up reduces slime development (Lindvall 2000). A thorough survey of the particular machine of interest is necessary for determining the extent of microbiological problem. In this regard, the key machine profile points have been noted by Bendt (1971). These include machine chest, head box, white water storage tanks, save-all, and the wet-end additives. A total microbial count is performed on the key areas in the wet-end system (Kolari et al. 2003; Goldstein 1987). Information on pH, temperature, number of slime holes, sheet breaks, sheet shift, count history, flow diagram of the system and the extent of recycling also provide a good background for developing a slime control program and monitoring its effectiveness. Design of the system has a significant impact on the microbial load. Some of the inexpensive hygienic design practices established in the bioprocessing industry need to be evaluated for use in paper mill operations (Chisti 1992a, b). Simple strategies such as keeping tanks covered and limiting the number of valves and fitting in the pulping system help reduce contamination. Microbial activity is relatively easy to control in machines, which use a series of short recycle loops with save-alls for removing solids from the water (Bennett 1985). Alternatively, existing effluent treatment plant may be utilized to separate solids from water. The return of water and sludge serves as a major source of contamination unless these streams are treated. The recycled water could be treated with either oxidizing or nonoxidizing biocides. Several mills treat the sludge either at the clarifier or at the holding tank stages.

The deposit control program consists of thorough washing of the machine whenever there is a shut drawn for maintenance or for other reasons (Lindvall 2000). Hot (80–90°C) caustic dispersant boilout has been used as an interim measure for slime control (Lustenberger and Deuber 1991; Baker 1981; Bendt 1971). After circulating the hot caustic solution, the system is flushed with fresh water to remove all loosened deposits. The frequency of boilout is directly proportional to the rate of prevention of slime deposits. Deposits are brought down to tolerable levels by wash-ups and boilouts (Patterson 1986; Kolari et al. 2003). Boilout can address four major problems: formation of inorganic scales, organic deposition, microbiological slimes and removal of general wet-end additives (Goldstein 1987). Mechanical cleaning and boilout are often impracticable and can be costly because they usually involve equipment down time. In addition, the harsh chemicals used in a boilout can challenge what is often already stressed waste water treatment system.

In addition to above practices, the paper industry utilizes various biocides (Barnes 1984; Bigotte 1979; Blanco et al. 2002; Brattka 1992; Bunnage and Schenker 1995; Schenker 1996; Camp 1989; Carvalho 1978; Cloete and Brozel 1991; Cloete and Gray 1985; Eriksson et al. 1995; Farkas 1990; Giatti 1993; Goldstein 1983; Haack et al. 1997; Himpler et al. 2001; Hootmann 2002; Lindvall 1998b; Lustenberger and Deuber 1991; Mobius et al. 1986; Nelson 1982; Palcic and Teodorescu 2002; Paulus 1993; Purvis and Tomlin 1991; Rantakokko et al. 1994; Scharschmied 1975; Schirch et al. 1993; Stomps 1995; Van Haute 2000; Weir et al. 1981). One class of these includes oxidizing agents such as chlorine dioxide, hydrogen peroxide, and peracetic acid. Chlorine dioxide and hydrogen peroxide happen to be the same chemicals that are also usually used for pulp bleaching. Therefore, it is convenient for facilities employing these bleaching agents to add small amounts to the paper machine system as well. The oxidizing action either kills the bacteria and fungi outright or weakens the cell walls so that they are more susceptible to the main class of biocides. The other class includes highly toxic organic chemicals. Subclasses of toxic biocides go by such names as thiazoles, thiocynates, isothiozolines, cyanobutane, dithiocarbamate, thione, and bromo-compounds (Bunnage et al. 2000). As the names imply, many of them contain sulfur ("thio"). Grades composed of bleached pulps very often involve a combination of treatments with oxidizing biocides, supplemented by toxic organic biocides. The usual recommendation is to treat each of the incoming stream, including the freshwater, filler slurries, chemical additives and make down streams. Special attention has to be paid to the starch preparation area since starch is such a favored food for slime growth. The level of oxidizing agent has to be monitored at a sufficiently low level that there are not any problems with bleaching of dyes or decomposition of starch etc. A possible starting level to consider is a residual of 1 ppm of active oxidizing agent in the paper machine system. Hydrogen peroxide, when used as a biocide, is slower to act but longer lasting. As a consequence, it needs to be controlled at a higher level of residual activity in the system. The toxic organic biocides can be selected based on the temperature of the system and on the relative needs to control either bacteria or fungal growth. It is common to cycle the toxic biocide on and off over periods from several minutes to several hours; by this means it is possible to reach a required threshold of activity and also minimize the cost of the chemicals. Such practices of intermittent addition need to be checked to make sure that do not cause excessive savings in first pass retention or other problems. Certain biocides are known to contain anionic dispersants that interfere with retention. The residual level of oxidizing chemicals is most conveniently estimated by measuring the redox potential of the furnish with a Pt electrode relative to a standard reference electrode. The effectiveness of a biocide program is best evaluated with a combination of measurements, including petri-dish cultures of water, tests for the presence of biological deposits as surfaces, the slipperiness of wetted surfaces, and the level of smells within the facility. Biocides have become an indispensable part of paper production particularly because modern process and the water supply conditions in most plants greatly promote the growth of microorganisms. In many mills, trouble free working with largely closed water circuits and continuous production of coated papers and boards has been made possible only by the systematic use of slimicides and preservatives. Since the different types of microorganisms have varying degree of resistance to biocides, there is no single preparation, which can solve all the preservation problems occurring in the paper industry. The properties demanded of biocides, in addition to high activity and cost-effectiveness, also vary according to their specific field of application. Biocides must be used in concentrations that do not disturb the papermaking processes, even if they are added in massive doses. The products must be compatible with the many auxiliaries used in papermaking. Furthermore, biocides must have low toxicity and, in particular, low ecotoxicity, and no appreciable amount must be present in the finished product, so that the paper can also be used in food packaging. Preservatives for coating mixes, filler suspensions and sizes and active ingredients for the antimicrobial finish of paper and board have to meet stringent requirements with regard to the absence of color and odor, compatibility and physiological harmlessness in their use concentrations.

Chlorine dioxide has several strong points in its favor as a paper mill slime and odor control agent (Baker 1981; Giatti 1993; Nelson 1982, Anonymous 1990b). As a strong oxidizing agent, it provides broad-spectrum kill of microorganisms. Many mills alternately apply several types of biocides to avoid developing resistance by certain troublesome microorganisms to a single product. This possibility is eliminated with oxidizing type of chemical. It performs well over a broad pH range. This is important in mills that operate paper machine system at different pH levels, whether because of the paper grades produced or because sizing changes have raised the pH in the papermaking processes. Chlorine and some nonoxidizing biocides drop off in performance in alkaline pH environment, but chlorine dioxide does not. Effective dosages are generally low, making Chlorine dioxide programs cost competitive with other biocides. Low usage rates can result in cost reduction for effective slime control and greatly reduce the potential harmful effects to the environment from the mill effluent water. Chlorine dioxide provides a rapid kill of slime-forming microorganisms by the mechanism of interrupting protein synthesis. Processing equipment such as screens, with high water flows and short contact times, can be kept free of slime buildups by the quick kill rates of Chlorine dioxide. An improved slime control program utilizing Chlorine dioxide improves the quality of paper products by reducing defects such as slime specks, spots and holes in the sheet. Removal of these defects reduces sheet breaks and avoids subsequent production losses. Chlorine dioxide does not form considerable reaction products with other chemical substances that are associated with use of Chlorine dioxide. Chlorine dioxide does not react with ammonia and most ammonia-nitrogen compounds. It can be used for controlling odors. Malodors caused by microbiological growths, phenols, sulfides, or mercaptons are destroyed by Chlorine dioxide. The use of Chlorine dioxide has eliminated time-consuming microbiological testing. Chemical spot testing for Chlorine dioxide at various points in the system is easily done and can be used to make adjustments in the treatment program to compensate for demand changes. This permits slime control by chemical control.

In western countries, increased public consciousness regarding environmental issues has led to strict regulations on the use of biocides (including slimicides). New microbiological control agents are being developed and research is going on all the time to find the ideal ecofriendly microbial deposits control program. One option being investigated is an equilibrium blend of peroxyacetic acid and hydrogen peroxide as part of an integrated microbial control package. This approach can provide a solution to the problem of residual toxicity associated with conventional programs and synergy has been demonstrated upon using the combination approach. Hydrogen peroxide is not a powerful biocide at low temperatures or low concentrations, but it does exhibit a strong biostatic effect inhibiting growth of many microbiological species (Chiari et al. 1990; Rantakokko et al. 1994; Schirch et al. 1993). It has been used as a sterilant for aseptic packaging for milk and fruit juice containers. It has also been used to control anaerobic bacteria in the paper industry. Hydrogen peroxide is used often in conjunction with peroxyacetic acid, existing in an equilibrium mixture engineered to specific formulations to achieve the greatest effectiveness for paper industry applications. Peroxyacetic acid is an extremely powerful, fast acting biocide having broad spectrum of activity over a wide temperature range (below °C to above 100°C). Although claimed as an oxidizing biocide, the mode of activity is not simply oxidation, as the molecule penetrates the cell wall to give a greater effect than pure oxidation. There is also no known immunity to peroxyacetic acid, provided sufficiently high active levels are maintained. Peroxyacetic acid has long been used in sewage treatment and in the sugar, dairy, and brewery industries as well as for medical sterilization for renal dialysis machines. Once reacted, it breaks down to nontoxic end products — water, oxygen, and acetic acid — which itself breaks down to CO_2 and H_2O. It is nonfoaming and can reduce chemical and biological oxygen demand in effluent. The active ingredients can be easily monitored using proprietary electro-optical measuring equipment, giving parts per million concentrations within seconds. The chemistry does have some limitations. It is less successful against organisms with thicker cell walls such as filamentous bacteria and molds. Certain system chemistries — for example, high levels of carbonate filler — can mean that higher dosage rates are required to effect control. The control program has been used successfully in UK Paper New Thames Mill, Kant, England (Bhattacharjee and Farr 1977).

Glutaraldehyde, a broad-spectrum biocide, is effective against the bacteria-yeast and mold present in papermaking systems and useful in preventing the formation of slime in all the areas of the papermaking process (Purvis and Tomlin 1991). It is found to be readily biodegradable and is effective against the aerobic and anaerobic microorganisms including sulfate-reducing bacteria. It is fully compatible with commonly used wet-end additives and greatly reduces the level of microorganisms in both acidic (pH 5.3) and alkaline systems. It is found to show more than 90% reduction at 25 ppm and essentially complete kill at 50 ppm in ASTM (American Society for Microbiology) paper slimicide test. It is also able to drastically reduce the level of sulfate-reducing bacteria present in the solutions at any time point.

Over the past several years, a number of new oxidant products have entered the market place (Bruce 2003; Thomas 1999). These products consist of halogens, bromine and/or chlorine, combined with an organic or inorganic carrier. One key advantage to combining the halogen is that it can often reduce the negative impact of the oxidant while maintaining its biocidal properties. The idea of combining a halogen with another molecule so that halogen is less aggressive but still biocidal is nothing new. In the 1930s and 1940s, mixing bleach (or chlorine gas) with ammonia to make chlorammonia for microbiological control in papermaking systems was very common. Even though chloramines treatment was often touted as being more effective than chlorine, eventually the chemistry was abandoned. Some of the reasons for abandoning chloramines treatment were increased corrosion and increased microbiological activity presumably due to the ammonia, a good source of nitrogen for bacteria. In the 1960s and 1970s, bleach in combination with sulfamic acid (chlorosulfamate) was proposed as a biocide with low reactivity to process equipment and chemistries. However, the product never obtained commercial success in paper probably due to the weak biocidal activity of chlorosulfamate.

In the past several years, interest in using combined halogens has been rejuvenated (Bruce 2003). Papermakers are currently using a number of combined halogen products for microbiological control in papermaking systems. The various trade names often make it confusing to determine the number and nature of the products available. In reality, papermakers currently use only four types of chemistries, hydantoin, sulfamate, ammonia/ammonium, and isocyanurate. *Hydantoin* group consists of bromine, chlorine, or both attached to a hydantoin molecule. Halohydantoins are not very stable in liquor form so they are manufactured and sold as a solid product either as powder, granules or briquettes. Feeding halohydantoins requires the use of a powder feeder or a brominator. A brominator consists of a vessel with granules or briquettes. Water flows through the vessel, dissolves product, and is sent to the process. Powder feeders work by making a slurry and delivering that slurry to the process. Once in the process, the product completely dissolves. *Sulfamate* group consists of bromine or chlorine attached to a sulfamate molecule. Unlike the halohydantoins, halosulfamates can be manufactured as stable liquid products. Currently, only bromosulfamate is used for paper mill water treatment. Chlorosulfamate is used in some cooling tower applications. *Ammonia/ammonium* can be generated by mixing chlorine or bleach with ammonia or an ammonium salt. The two can be mixed in the process water or premixed before application to the process water. Like halohydantoins, these types of haloamines cannot be manufactured as stable liquid products.

Recently, bleach in combination with ammonium bromide has been introduced as a way to produce a new haloamine oxidant. *Isocyanurate* group consists of chlorine attached to an isocyanurate molecule. However, sodium bromide may also be present, allowing for the formation of hypobromide. Like halohydantoins, halocyanurates are produced in solid from. One halocyanurate product, which is applied using a specialized power feeder, has enjoyed commercial success in cooling tower applications and utilities, but has seen only limited application in paper.

Potential benefits of combined halogens are the following: persistence, increased efficacy in high oxidant demand systems, better slime penetration and removal, better compatibility with papermaking chemistries and with papermaking equipment.

Bromochloro dimethylhydantoin (BCDMH) has been found to be cost effective, fast acting biocide, fully compatible with the conditions found in modern paper manufacturing practices. It offers the paper manufacturer a unique solution to microbiological problems by overcoming the disadvantages of traditional nonoxidizing biocides. The product is proven to be 3 times more effective toward filamentous bacteria selectively found in paper machine fresh water and slime deposits. Knapick et al. (2003) have reported the use of brominated methylethylhydantoin in a tissue mill led to reduced costs, improved safety, greater selectibility, low biological costs and less corrosion.

A new Biocide – Spectrum® Ammonium Bromide Technology – has been launched by Hercules Pulp and Paper Division that effectively controls microorganisms in alkaline systems without the adverse side effects associated with strong oxidizing biocides (personal communication). The new biocide, which degrades into inert compounds before effluent discharge, is produced onsite by blending an ammonium bromide solution with sodium hypochlorite and mill fresh water. Dedicated blending and dosing effluent ensures safe, consistent production of the biocide.

A new biocide system for use in paper production has been developed (Hootmann 2002). It is based on the reaction of ammonium bromide and sodium hypochlorite. The two inorganic reagents combine to form an oxidizing biocide. Unlike conventional biocides, the new system does not require limited life stocks to be held. Instead, it is dosed into circuits at rates that can be varied for different chemical oxygen demand (COD) levels. The in situ system is effective against slime-forming and sulfate-reducing microorganisms as well as molds and algae. Two year mill trials confirm wet end biocide treatment improves runnability, reduces the incidence of holes and spots in finished paper and reduces hydrogen sulfide levels. The trials covered closed loop production of unbleached liner from 100% waste paper as well as mills producing high quality bleached graphic papers from chemical pulp. The in situ biocide does not affect adsorbable organic halogen (AOX) and there is no accumulation of active biocide or of byproducts.

An entirely new class of antimicrobial chemistry has been registered with EPA for use in papermaking, based on the biocidal molecule – TetrakisHydroxymethyl-Phosphonium Sulfate (THPS) (Haack et al. 1997). The most outstanding characteristic of THPS is its ability to combine broad-spectrum antimicrobial efficacy with a relatively benign human and environmental toxicity profile. THPS biocides are classified by DOT as nonhazardous, and they have very low acute toxicity in the

Table 14.5 Effect of tetrakishydroxymethylphosphonium sulfate (THPS) against *Enterobacter aerogenes* and SRB

E. aerogenes			
THPS concentration (ppm a.i.)	Surviving bacteria per mL after stated exposure time[a]		
	2 h	6 h	24 h
0 (control)	2.3×10^5	1.3×10^5	1.9×10^6
15	4.0×10^4	1.0×10^0	0
37.5	1.0×10^0	1.0×10^0	0
75	0	0	0
150	0	0	0

SRB		
THPS concentration (ppm a.i.)	Surviving SRB per mL after stated exposure time[b]	
	6 h	24 h
0 (control)	8.8×10^6	1.4×10^7
10	1.1×10^5	1.4×10^7
25	2.0×10^0	0
50	0	0
100	0	0

Based on Haack et al. (1997)

[a]The initial bacterial level was 1.8×10^6

[b]The initial SRB level was 2.0×10^6

Table 14.6 Effect of THPS on Activated sludge in the biological effluent treatment (BET) plant

THPS dose (ppm)	Oxygen demand (% O_2/min)
0 (control)	40
100	44
200	39
300	20

Based on Haack et al. (1997)

environment. In addition, THPS degrades rapidly on discharge to a molecule that is virtually nontoxic, thereby minimizing the risk of pollution and/or harm to biological effluent treatment (BET) plants. THPS biocides are fast acting and offer excellent activity against slime-forming microorganisms, sulfate-reducing bacteria, and biofilms. The nonfoaming THPS molecule can be monitored on-site with a simple and rapid titration procedure to ensure proper dosing levels. As an illustration, Table 14.5 shows effect of THPS against *Enterobacter aerogenes* and SRB. Laboratory tests indicated upto 200 ppm of THPS formation could safely be dosed to the paper mill without adversely affecting the BET plant (Haack et al. 1997) (Table 14.6). This indication was substantiated during the plant trial. The plant trial confirmed that microbial control could be achieved in the process at a dose rate of only 9.6 ppm THPS (as active ingredient). Hydrogen sulfide levels were controlled at acceptable levels. There were no detrimental effects in the BET plant.

Biocides possess different mechanisms of antimicrobial activity. Glutaraldehyde reacts with amino and thiol groups in proteins causing irreversible cross-links in the cellular constituents (Paulus 1993). In gram-negative bacteria glutaraldehyde interacts principally with outer components of the cells, notably lipoproteins (Maillard 2002). High degree of cross-linking means that the cells are unable to perform their essential functions, resulting in bactericidal effect. BCDMH (1-bromo-3-chloro-5,5-dimethylhydantoin) is not biologically active as such, but its rapid hydrolysis yields hypobromic and hypochloric acids (Kemira Chemicals Oy 2003). Glutaraldehyde and oxidizing biocides are effective also against bacterial spores (Paulus 1993). MBT chelates Fe^{3+} ions essential for the microbial growth. Isothiazolones (5-chloro-2-methyl-4-isothiazolin-3-one and 2-methyl-4-isothiazolin-3-one in a mixture) and DBNPA are electrophilic active compounds reacting with cytoplasmic constituents such as thiol groups of proteins, thus inhibiting cellular metabolism (Maillard 2002; Paulus 1993). Bronopol (2-bromo-2-nitro-propane-1,3-diol) also contains an active halogen group, but can release formaldehyde as well (Paulus 1993). Dazomet (3,5-dimethyl-1,3,5-2H-tetrahydrothiadiazine-2-tion) is rapidly hydrolyzed in water to methylene isothiocyanate (Kemira Chemicals Oy 2003), but can also release formaldehyde (Paulus 1993).

Biocides are generally toxic with low biological selectivity. Some such as DBNPA, isothiazolone mixtures, glutaraldehyde, and MBT are also sensitizing (Kemira Chemicals Oy 2003; Paulus 1993; Pirttijärvi 2000). Many of the currently used biocides such as peracetic acid, hydrogen peroxide, BCDMH, DBNPA, glutaraldehyde or isothiazolone mixtures are reactive molecules that are rapidly (bio) degraded to nontoxic molecules.

Thus, these biocides are not detrimental for the biological wastewater treatment processes and are not persistent (bioconcentrated) in the environment (Kemira Chemicals Oy 2003; Paulus 1993).

14.7.2 Use of Enzymes for Control of Slime

Enzymes are biologically produced nonliving proteins that act as a catalyst. A catalyst effects the rate of a chemical reaction, generally accelerating it, but is not altered or consumed by the reactions, so they can perform the same reaction repeatedly. Enzymes have been evaluated in both the laboratory and in paper process streams for control of microbiological slime deposits (Anonymous 1984; Anonymous 1990a; Anstey et al. 1998a, b; Fischer and Baurich 1999; Freis 1984; Galon 1997; Gould 1998; Hagelsieb et al. 1996, 1999; Hart 2001; Jaquess 1994; Kanto and Brutar 1996; Kupfer and Baurich 1999; Lindvall 1998b; Siika-aho et al. 2000; Schuetz and Wollenweber 1999; Van Haute 1997b; Benard 2010; Rivera and Jara 2007; Loosvelt and Datweiler 2007; Xu 2005; Paice and Zhang 2005; Buchert et al. 2004). Enzymes are composed of one or more polypeptide chains consisting of amino acids, with molecular weights ranging from around 10,000 to a million or more. Published molecular weights for the more common industrial enzymes – α amylases, cellulases, lipases, proteases, and xylanases range from around 20,000 to 250,000 Da. The structure and function of enzymes are determined by the sequence of amino acids in the polypeptide chain(s), which leads to one of the most important characteristic of enzymes – their specificity, i.e., catalyzing one particular reaction on a particular substrate without effecting other elements. This specificity can be pictured like a lock and key where polypeptide chains fold and arrange in such a way as to form a unique binding active site for the substrate. The chemical reaction occurs at this active site. Enzymes have been used for several years in the food, beverage and pharmaceutical industries, but have had limited use and acceptance until recently in the pulp and paper industry. The accepted theory behind slime control with enzymes is that the enzymes degrade extracellular polysaccharides by cleaving a specific bond in the EPS, thus dissolving the slime. Once the slime has been digested away, the cells are then directly exposed to the biocides, which is now effective at lower concentration. The slime, which is a fructan polymer, is broken down to monomers by carbohydrase type of enzymes. There are two types of enzymes involved in degradations of fructans — hydrolases and transferases. The hydrolytic enzymes, according to their mode of action are either endo- or exoenzymes, producing a homogeneous series of oligofructans or only fructose, respectively. The transferases, on the contrary, split off fructose dimers and by simultaneous transfructosylation give rise to difructose anhydride. Levan in liquid systems is generated by various forms of bacteria, most of which are attached to surfaces. Levanases from *Streptomyces*, *Bacillus*, and *Rhodotorula* have been purified and their molecular weights are found to be 54, 135, and 39 KDC, respectively. The enzyme of *Streptomyces* sp. no. 7-3 and *Streptomyces exfolratus* F3-2 have been found to hydrolyze levan to produce levanbiose. Levanase from *Bacillus* sp. produces levanheptose as the predominant product. It begins forming when colonizing bacteria cement to a surface and creates a biofilm by covering themselves with layers of levan polysaccharides. This creates a protective niche that covers the cell completely. It eventually grows into the large, complex slime deposits — by trapping fines, fibers, mat-forming organisms, and general debris — that create productivity problems for paper mills. The use of enzymes in the control of microbiological slime deposits, particularly EDC-I (enzymatic deposit control) has proved useful under modern mill conditions (Hatcher 1983; Grussenmeyer and Wollenweber 1992, 1993). EDC-I is an enzyme that hydrolyses and depolymerizes the fructose polysaccharide levan, which has been identified in paper mill slime (Colasurdo and Wilton 1988). It has been patented by Economics Laboratory (Hatcher 1973). This enzyme is produced during an aerobic fermentation of a common nonpathogenic, non-spore-forming soil bacterium. The product is environmentally safe and has no effect on the operation of activated sludge effluent plants. Obviously personal safety is greatly enhanced through the use of an enzyme product rather than conventional toxic biocides. The enzyme does not require any special handling, is nontoxic, does not emit dangerous fumes, and is not corrosive. Spills can be flushed away with water. Once the system has been charged with EDC-I, the enzymes will continue to work until they are inactivated by the natural forces in the system. The exact period for which the enzyme is stable depends on a number of factors peculiar to any paper machine system, including the temperature and pH profiles as well as any other additive. EDC-I has been successfully applied in systems with wide range of temperature and pH. It is used by paper mills in UK, Scandinavia, Japan, and USA. Applications have been on machines producing writing and printing, fine paper grades and paperboards.

The combination treatment of enzyme and selected biocides has a significant effect on cell survival. In all cases, the combination treatment resulted in significantly greater population reduction compared to use of the biocide alone at the same concentration (Ferris et al. 1989). The enzyme alone showed no cidal or static effects. Further, it was shown that on application of enzyme (0.10 kg/metric ton) on machines producing printing grade paper, the biocide concentration could be reduced from 0.15 to 0.02 kg/metric ton (Ferris et al. 1989). The amount of biocide was reduced to ~15% of what had been employed prior to using the enzyme. A similar level of reduction in biocide concentration was achieved in the white water system of a paper mill when an enzyme preparation was deployed (Hatcher 1983). Patterson (1986) has reported that with the use of enzyme, the concentration of biocide may be reduced by 50% and slime breaks are reduced from three per day to three per month. Similarly Colasurdo and Wilton (1988) noted that with the use of an enzyme preparation at SONOCO Products Co., Hartsville, USA, the slime breaks were virtually eliminated leading to increased productivity. The Levanase enzyme produced by *Rhodotorula* sp. has been found to reduce the needed biocide concentration by 25% without adversely affecting the paper properties (Chaudhary 1992). Despite the strong indications that exopolysaccharides are involved in adhesion, enzyme digestion has tended to show that proteases are more effective than carbohydrate-degrading enzymes in removing bacteria from surfaces.

Microorganisms have cell walls, which confer rigidity, and, if the cell wall is removed in some manner, the cell usually dies because of lysis resulting from osmotic balance. Cell walls in microorganisms are composed of substances such as cellulose, chitin, mucopeptide and β-1,3-glucan. In addition, many organisms have capsules, slime layers or other surface components, which are polysaccharides or proteins attached to the outside of the rigid layer or interdigitated with it, conferring additional rigidity and/or protection. Enzymes are known, which will attack all of these polymers. The application of lytic enzymes with gluconase and protease activity in paper machine process water to destroy microbes causing precipitates and/or forming slime and/or adversely affecting product quality, has been advocated. Enzymatic dispersion of biological slimes has been claimed for a pentosanase–hexosanase enzyme, Rhodozyme HP-150. This enzyme is particularly effective in treating industrial waters used in the operation of cooling towers to disperse slimes and slime-forming masses within such waters to prevent the deposit of such slimes on the heat exchange surfaces of cooling towers and surface associated with such units. Orndorff (1983) in his patent has described a method of killing and inhibiting the growth of microorganisms in industrial process streams using peroxidase- or lacasse-catalyzed oxidation of various phenolic compounds to generate microbial oxidation products e.g., poisonous quinines.

Enzyme based antislime – Betzdearborn's Spectrum has been developed in response to such pressures of the European biocides Directive ecolabelling, ISO 1400 and mill water system closure. It has now become available in the market and among its tool kit of products is a gluconase enzyme, which catalyzes the attack of glucose, which represents a significant proportion of the polysaccharides in the EPS on paper machines. In addition, formation of this troublesome polysaccharide coating is inhibited by new and nonenzymatic materials in Spectrum, which interfere with slime formation mechanism.

The use of a blend of enzymes has been shown to be more effective in treating microbially produced extracellular polysaccharides in cooling water and in papermaking broke water than the use of a single enzyme. The composite enzyme system tested consisted of cellulase, α-amylase and protease. Some manufacturers and service companies assert that the enzymes available are too species-specific and can be used only on a smaller scale or as a supplement to conventional biocidal methods.

14.7.3 Biological Equilibrium

A process has been invented, which works wholly or largely without biocides on the principal of natural biological equilibrium (Oberkofler 1987, 1989, 1992; Oberkofler and Braunsperger 1994; Braunsperger et al. 1996). The reduction of chemical deposits and efficient system cleaning is achieved with the aid of controlled bacterial activity. This process uses nontoxic modified lignosulfonates with specific components (chelating agent different from those commonly used in papermaking). The objective of the lignosulfonate treatment is to maintain the organisms and the suspended matter in the white water in a colloidal state thereby preventing the bacteria from agglomerating and forming deposits. The bacteria, which carry a net negative charge, adhere to the modified lignosulfonate. As the bacteria digest the organic substances within the colloids the material becomes more flocculant. The lignosulfonate, which is water soluble, is drained off on the wire press, whereas bacteria enter the finished paper. The process allows biocide-free operation and is applicable to all paper grades. Counter-indications are open water systems, unbleached kraft pulp, use of recycled fibers and an increase in organic matter.

Bimogard, produced by BIM Kemi AB, has been employed since 1991, as a biocide free slime control agent (Anonymous 1986; Bjorklund 1999, 2000, 2001a, b, 2002a, b; Gavelin 1996). It is based on modified lignins and surfactants. It works by first cleaning surfaces and thus reducing the adhesion of bacteria and by reducing microorganism activity by inhibiting the production of extracellular polysaccharides and reducing bacterial growth and spore formation. The Bimogard system, which resembles the Biochem Process (Morros 1995) is claimed to be the first completely biocide-free slime control system in Europe. Several Bimogard systems are in operation on paper machines for newsprint, lightweight coated fine paper, tissue, greaseproof paper and paperboard. Reports from the mills indicate that the system has brought about significant improvements in efficiency. For example, the hole frequency drops, the time between wash-ups increases, cleaning is easier and there are fewer wet-end breaks and therefore lower broke production. This slime control has been successfully used in low as well as neutral to high pH conditions, and at

Table 14.7 Effect of Bimogard on EPS after introduction to a mill previously using biocides

EPS (mg/g)				
Sample	Before introduction of Bimogard	2 Weeks after introduction of Bimogard	4 Weeks after introduction of Bimogard	16 Weeks after introduction of Bimogard
Deposit from section under forming roll	13.0	6.1	5.5	0.0
Back water	6.3	3.1	1.2	0.0
Process water from wire pit	3.8	1.3	0.0	0.0
Fresh water	0.0	0.0	0.0	0.0

Based on Bjorklund (2001b)

small, old as well as large, modern, fast moving paper machines. It is also used as intermittent board machines in pulp mills. The cost for using Bimogard for slime control is the same or, most often, less than using biocides. The bacterial count is usually 10^5–10^7/mL in mills using Bimogard. Most of these bacteria are harmless and die in the drying section without causing slime problems. Unlike biocides, it does not trigger bacterial defense mechanisms. Table 14.7 shows the effect of Bimogard on the amount of EPS after introduction to a mill previously using biocides. It can be clearly seen that the amount of EPS decreases after introduction of biocide-free slime control in paper mill using 100% DIP.

A new biochem method for closed water systems from PETROMONTAN uses a multifractionated and modified lignosulfate called Petrodis (Anonymous 1986). It reduces the need for biocides when added to the system and leads to better functioning of the paper machine because of reduced chemical scaling and more efficient biological control, reduction of toxic substances in the system and in the effluent. It is safer for workers, more efficient, and reduces costs for the control of slime and scale.

14.7.4 Biodispersants

Biodispersants offer an ecologically attractive method of controlling slime in the pulp and paper industry by eliminating or reducing the needs for biocides (Anstey et al. 1998a; Blankenburg and Schulte 1999; Gould 1998, 2001; Pauly 2001; Robertson and Taylor 1994; Saner 1998; Stenqvist 1992; Van Haute 1997a, b, 1999; Weisshuhn et al. 2000; Wright 1997). Biodispersants dissolve and disperse deposits, preventing the biofilm from reestablishing itself. Conventional wisdom has held that dispersants act as biopenetrants opening the biofilms and allowing biocides to penetrate the exopolysaccharides. There are additional claims that dispersants facilitate penetration of biocides into the cell or that they trigger sloughing of biofilms. Biodispersant technology is based on nonionic polymers, which are nontoxic, nonfoaming, colorless, and free of organic solvents. Because of their nonionic character, they will neither increase the system anionicity nor interfere with other papermaking chemicals. Biodispersants have no pH limitations and are suitable for use in both acidic and alkaline papermaking. The nonionic products, which have seen large-scale commercial use, fall into four general types, i.e., alcohol ethoxylates, alkylphenol ethoxylates, polyoxyethylene esters, and polyoxyethylene–polyoxypropylene derivatives. The latter are mixed polymers with hydrophobic groups derived from propylene oxide, further reacted with ethylene oxide until the desired properties are attained. Ordinarily they are of rather high molecular mass, often with much more than the usual 8–15 molecules of ethylene oxide characteristics of the other nonionics.

Buckman Laboratories have developed neoteric programs, which are specifically designed for use in preventing and slowing the development of deposits in paper machine systems (Van Haute 1999; Hart 2001; Koopmans and de Vreese 2002). These programs have been successfully being employed in acidic, neutral and alkaline conditions and in the manufacture of all types of paper, paperboard, tissue and toweling (Van Haute 1999, 2000). Neoteric programs consist of dispersants, enzymes, and potentiators added to the wet end of the paper machine. Neoteric programs are designed to enhance the effectiveness of traditional biocides, reducing and in some cases eliminating the use of biocides in wet end. Neoteric dispersants are carefully selected group of chemicals that prevent agglomeration and deposition on machine surfaces of common deposit causing components in the paper machine furnish. Neoteric enzymes are biodegradable and nonbiocidal, have relatively low level of toxicity to test animals and are FDA allowed. These enzymes include formulation of amylase enzymes and protease enzymes, which catalyze the attack on starches and proteins. Potentiators are nonbiocidal products that enhance biocide performance. These products allow the use of less organic biocide while still achieving the same effect as the original dosage of biocide. Simply, potentiators lower the effective concentration of biocide required for a given application. These products, when used alone, do not reduce or slow microbial growth or metabolism. Buckman says that atleast 50 paper machines in Europe use Neoterics in their short circulation loop, with Busperse and Buzyme.

For closed paper machine systems, Nalco recommends lignosulfonate-based biodispersants plus thorough machinery and system cleansing; elimination of static areas and rough surfaces; fresh water and additive contamination controls; good retention and prevention of anaerobic bacteria in storage tanks through oxygenization.

Although the name biodispersant suggests the activity on a biological entity, laboratory testing has shown that many commercially available biodispersants, in the absence of a microbicide have little or no ability to prevent or remove an existing biofilm. Indeed, the majority of commercial biodispersant applications include the use of a microbicide. Properly applied biodispersant technology can increase the overall efficiency of a given microbicide program.

Buckman Laboratories have developed an enzyme-based product line of biodispersants for boilouts and other application (Wolfanger 2001). The company's patented enzyme stabilization process promotes greater shelf life and reduced temperature instability concerns. It also has the potential to eliminate significant safety and health hazards in paper machine operations that are present during the use of caustic for boilout. In addition, it is now possible to release nontoxic cleaning solutions without additional treatment before discharge. Enzyme based boilouts are successful because they target and remove the specific compounds that hold deposits together including starch, slime, pitch, adhesives, latex and other synthetic binders. The type of enzyme used and need for an additional dispersant will depend on the type and amount of deposit present in the system. Starch slurry systems typically contain deposits that are microbiological and/or starch protein based. Boilouts using a product containing a stabilized protease enzyme are effective in these systems. For the cooked starch system, an amylase product is used to remove deposits comprised mainly of cooked starch. In both cases, a pre-boilout system flush is desirable. This removes slurry or cooked starch from the system and allows the enzyme to work specifically on deposits. An ideal starch system boilout would utilize 0.2–0.5% of the boilout product; and temperature of 120–150°F with a minimum recirulation time of 1 h. Starch-based coating systems can be effectively cleaned via a boilout using the amylase product mentioned above for cooked starch systems. The parameters for this boilout are similar with the enzyme based boilout product being used at a 0.1–0.5% concentration. It is, however, desirable to use a neutral dispersant/penetrant capable of removing latex, and other binders that may be present in the coating formulation in addition to the enzyme based product. In addition, a tighter temperature range, 130–150°F, and a longer recirculation time, up to 2 h, will provide performance benefits for a starch-based coating system boilout. Enzyme boilouts for coating systems that do not contain starch are not recommended. Enzyme boilouts have been successfully applied, however, in systems that regularly process both starch-based and non-starch-based coatings. Enzyme-based chemistries are proven successful for all types of mills: from acid to alkaline papermaking, recycled to virgin, board to fine paper to tissue, including all variety of additives (and deposits associated with these additives) linked with these various furnishes. Given the specific mode of action of enzymes, how is this possible? If the majority of the machine system deposits are microbiological slime, this makes sense. A protease will break down these deposits. But in a system that has deposits containing various fillers, fiber, and hydrolyzed additives, how will the protease have an impact? The answer is that most deposits in the paper machine system, regardless of whether or not they are microbiological slime based, will have microbiological growth in, on, and around the deposit. When the enzyme breaks down this microbiological matrix, the deposit sloughs off and is carried away in the boilout solution. This phenomenon is based on the unique ability of bioengineered enzymatics to target and attack specific deposits by consuming the organisms that compose them, making them simple and easy to rinse away. If necessary, there are pH neutral additives that can be added to aid in penetrating, breaking down, and removing deposits that are nonmicrobiological in nature. These additives help the protease-based enzyme product to penetrate into and behind a deposit where microbiological growth is helping to anchor and hold a deposit together. The result is a cleaning effect that is equal to or better than caustic boilouts with none of the problems. As with starch and coating boilouts, the operational parameters associated with an enzyme boilout for the paper machine are very similar to traditional caustic boilouts. Temperature between 120 and 150°F, recirculation time of 1–2 h and the proper additive selection and concentration will assure for optimum results. The similarities, however, end here. It is with the paper machine boilout that the biggest difference can be found when comparing traditional to enzyme boilouts. Enzymatic boilouts are pH neutral. This means that the boilout can be dumped directly to the sewer, without taking the time to neutralize the solution, and without the potential for upsets at the waste treatment facility. A neutral boilout also removes the safety concerns associated with working with and around the caustic solution. Maintenance crews can access areas sooner, or work in areas considered unsafe during a caustic boilout. The neutral boilout eliminates the need for excessive rinsing to purge caustic from the system. Each of these factors contributes to a reduction in the downtime necessary for a boilout and maintenance outage, and this means cost savings.

Paper companies strive toward worker health and safety, environmental stewardship, and manufacturing a product of the highest quality. Neutral boilouts demonstrate a major step forward in application technology to aid the papermaker in meeting these demanding goals. Through the further development of stabilized enzyme products, the future is sure to bring new and varied used of enzyme technology to further benefit the paper industry.

14.7.5 Use of Competing Microorganisms

During the last decade, some nonconventional approaches have been adopted in solving slime problems (Lindvall 1998a). The inoculation of non-slime-forming organisms to outcompete the slime formers has been advocated. Oberkofler (1993) has patented a slime control process, which is based on the introduction of a consortium of bacteria commercially available in freeze-dried or liquid form. The bacteria are pregrown prior to inoculation of the circuit water, and the amount added is calculated on the basis of TOC present. It is also suggested that additives may be introduced together with the bacteria, thus favoring their proliferation. These additives include tensides to prevent the adhesion of bacteria, lignosulfonate to increase the nutrients, and enzymes to catalyze the breakdown of organic substances. Aeration of the circulating water with O_2 or air or the addition of oxygen-releasing compounds such as hydrogen peroxide is also suggested. The invention is not restricted to bacteria. Fungi alone or the blend of fungi and bacteria can be used. The patent outlines an indiscriminate use of bacteria and fungi, of which the mixed culture of freeze-dried bacteria used contains many of the genera associated not only with slime formation in pulp and paper mills but also such genera undesirable for paper and board intended to come into contact with food stuffs. Plant scale trials run continuously for more than 9 months have nevertheless shown that the addition of the selected microorganisms e.g., bacteria to the circulating water reduced the buildup of slime on solid surfaces and in the liquid phase.

Table 14.8 Modes of action of microbicides, biodispersants, enzymes, and biofilm inhibitors

Microbiocides
Reduce/control microbial populations
Biodispersants
Loosen wet-end deposits and support the effect of microbicides
Enzymes
Cleave specific bonds in the EPS
Biofilm inhibitors
Prevent the formation of a concentrated EPS layer around cells, thus preventing biofilm growth

14.7.6 Biofilm Inhibitors

Biofilm inhibitors are an effective environment friendly alternative treatment for short loops. A family of molecules discovered to be very efficacious at inhibitory deposit formation is the sulfosuccinates (Davis et al. 1999; Scharpf 1998; Schenker et al. 1998). A specifically designed sulfosuccinate molecule has been found to act at an earlier stage than biodispersants and enzymes in the inhibition of biofilm deposit formation. The mechanism of biofilm inhibitors differs from that of biodispersants and enzymes in that bacterial attachment and biofilm formation is prevented by hindering the creation of a concentrated extracellular EPS layer around a bacterial cell (Table 14.8). Therefore, biofilm inhibitors prevent biofilm formation at an earlier stage than biodispersants and enzymes.

Furthermore, in all the technical validation trials that have been run to date, no adverse effects were observed on the sheet properties or operating parameters. A conventional microbicide strategy is still recommended for incoming furnish to maintain a consistent microbiological population level and as treatment for mill fresh water.

14.7.7 Use of Bacteriophages

Bacteriophages are viruses that inject and destroy bacteria and are completely harmless to humans or the lower animals including birds and fish. The term "Bacteriophage" means literally "bacteria eater." These useful microorganisms are noted for their highly selective lytic action and high activity in the presence of host cells. However, the development of industrial application has been slow. Viruses lack an independent metabolism. They multiply only inside living cells, using the metabolic machinery of the host cells. A bacteriophage can proliferate only by coming into contact with its specific host bacterium. When this occurs, the phage lyses the bacterium completely. Some of the bacteriophages that attack *E. coli* (ψ, λ and T-4 phage) are well known for their extensive contribution to development in molecular biology.

Table 14.9 Colony count of slime-forming bacteria (S-1) following application of a synthetic biocide MBT and combined application of MBT and the corresponding bacteriophage (PS-1)

MBT addition (ppm)	Bacteriophage addition (10^5 PFU/mL)	Colony count for slime-forming bacteria (10^5 CFU/mL) Initial	After 24 h
10	–	87.0	110.0
10	10.0	100.0	2.4

Based on Araki and Hosomi (1990)
PFU Plaque-forming unit; *CFU* Colony-forming unit

In general, the bacteriophage has an icosahedral head of 50–100 nm in diameter, a long tail of about 100 nm in length (with or without a sheath around it) and six tail fibers at the end of its tail. The bacteriophage comprises only protein with nucleic acid as gene. The bacteriophage selectively attacks its particular host bacterium. The gene of the bacteriophage is inserted into the host bacterium, where it is multiplied. The host cells also begin to biosynthesize the specific protein that constitutes the bacteriophage. A large number of bacteriophages are formed in the host cell, and when the phages are released from the host, the cycle begins once again. The bacteriophage formed in the host cell has the same properties and morphology as the original. The reproduced phages thus can repeatedly lyse the same host bacteria. The number of bacteriophages formed after one lytic cycle is called "burst size" and the time required for the cycle is called latent period. These values are reported for evaluating the activity of bacteriophage.

The use of bacteriophages in paper mill white water systems was reported by Vaatanen and Harju-Jeanty (1986). Their concept was based on the idea of isolating harmful bacterial streams of process waters and thereafter searching for virulent, lytic bacteriophage for these bacteria. They studied bacteriophages lytic for the bacteria *Enterobacter* and *Klebsiella* in model systems and found that phage activity against *Enterobacter agglomerans* prevented its growth for more than 19 h. When *Klebsiella pneumoniae* was dosed with phage at 3 h intervals, bacterial growth was no more curtailed than when the culture was injected only at the beginning of the test. They concluded that further studies to determine the efficacy of bacterial control in process waters must take into consideration the expected development of bacterial resistance to phage attack and the variation in bacterial strains among paper mills. The main advantage of using phages in combating bacteria is their bactericidal (killing) nature, selectivity and nontoxicity to man and the environment. The main drawback is that the phages have to be isolated for each harmful bacterium, the type of which may vary between paper mills. Japanese researchers have reported similar investigations using *Pseudomonads* sp. (S-1) and its corresponding bacteriophage (PS-1) (Araki and Hosomi 1990). In vitro experiments showed that growth of slime-forming bacteria was greatly retarded in the presence of its corresponding bacteriophage. Results with bacteriophage and conventional biocides are presented in Table 14.9. After addition of 10 ppm of methylenebisthiocynate (MBT) to the test solution, the colony count of the slime-forming bacteria increased from 87×10^5 CFU/mL to 110×10^5 CFU/mL after 24 h at 28°C indicating that MBT was effective in keeping the activity of the slime-forming bacteria to a low level. However, the simultaneous addition of MBT and bacteriophage PS-1 reduced the colony count of the slime-forming bacteria from 100×10^5 to 2.4×10^5 CFU/mL.

Results showed that addition of bacteriophage to mill white water was an effective technique for slime control. Simultaneous addition of bacteriophage with conventional biocides also was found to be effective. Practical application of this technique on a commercial scale awaits completion of fundamental studies in several key areas. Unlike conventional biocides, bacteriophages will not impair the activity of the sludge used in waste treatment systems.

References

Anonymous (1984) Slime control findings. Paper 201(10):12

Anonymous (1986) Biochem-method reduces use of biocides. Wochenbl Papierfabr 114(11–12):464–465

Anonymous (1990a) Slime control using an enzyme product; NOPCO EDC 1. Papeterie 140:32

Anonymous (1990b) Chlorine dioxide simplifies wet end chemistry. Pap Technol 31:28–29

Anstey MR, King VM, Dykstra GM (1998a) The practical side of newer deposit control technologies. In: 84th annual meeting technical section. Preprints A, Montreal, QC, pp A299–A306

Anstey MR, Rouleau C, King VM, Dykstra GM (1998b) Practical aspects of newest technologies used for deposit control. In: Conference on echnologique estivale, Quebec, Canada, pp 87–91

Araki M, Hosomi M (1990) Using bacteriophage for slime control in the paper mill. Tappi J 73(8):155–158

Bajpai P (1999) Application of enzymes in the pulp and paper industry. Biotechnol Prog 15(2): 147–157

Bajpai P, Bajpai PK (2001) Status of biotechnology in pulp and paper industry. Pap Int 5(4): 29–35

Baker ER (1981) Using chlorine dioxide for slime control in alkaline paper machine systems. Tappi J 64:91–93

Barnes RW (1984) Biocide update: current practices for cost effective mill slime control. Pulp Pap 58(6):113–115

Baurich C, Fischer K, Scheen J (1998) Laboratory study of slime control in paper machine process water. Wochenbl Papierfabr 126(10):446–450

Benard D (2010) More production by using enzymes. Wochenbl Papierfabr 138(10):838–839

Bendt HT (1971) Slime control: a better way. Pulp Pap 45:129–133

Bennett C (1985) Control of microbial problems and corrosion in closed systems. Paper Technol Ind 26:331–335

Bhattacharjee S, Farr R (1977) A low residual toxicity microbiological control programme. Tappi J 80(12):43–46

Bigotte B (1979) Slimicides: a long term solution. Papeterie 4:103, 106–108 (in French)

Bjorklund M (1999) Biocide-free slime control – a practical reality in pulp and paper mills. In: 6th International conference on new available technologies, Stockholm, Sweden, pp 243–246

Bjorklund M (2000) Alternatives to conventional biocides in the pulp and paper industry. IPPTA 12(4):1–4

Bjorklund M (2001a) Slime control without biocide: a practical reality. Skogindustri 55(3):22–24

Bjorklund M (2001b) Biocide-free slime control in pulp and paper mills. TAPPSA J 6:17–18

Bjorklund M (2002a) Process water cleaning and slime control in tissue and food grade mills. In: 7th International conference on new available technologies, Stockholm, Sweden, pp 216–218

Bjorklund M (2002b) Methods for slime control meets the environmental demands. Nord Papp Massa 2:54

Blanco MA, Negro C, Gaspar I, Tijero J (1996) Slime problems in the paper and board industry. Appl Microbiol Biotechnol 46:203–208

Blanco A, Negro C, Monte C, Tijero J (2002) Overview of two major deposit problems in recycling: slime and stickies. Part 1: slime problems in recycling. Prog Pap Recycl 11(2):14–25

Blankenburg I, Schulte J (1997) An ecological method for slime and deposit control. Pulp Pap Int 39(6):67, 69

Blankenburg I, Schulte J (1999) An ecological method for slime and deposit control. IPPTA 11(1):51–56

Brading MG, Jass J, Lappin-Scott HM (1995) Dynamics of bacterial biofilm formation. In: Lappin-Scott HM, Costerton JW (eds) Microbial biofilms. Cambridge University Press, Cambridge, UK, pp 46–63

Brattka B (1992) A new process for the abatement of slime and odour in the paper machine system. Wochenbl Papierfabr 120(11–12):484–485

Braunsperger F, Oberkofler J, Moser T (1996) Slime control without chemicals. Wochenbl Papierfabr 124(5):192–194

Brewer D (1960) Studies on slime accumulations in pulp and paper mills. IV. Fungal floras of slime accumulations. Tappi J 43:609–611

Brown MRW, Gilbert P (1993) Sensitivity of biofilms to antimicrobial agents. J Appl Bacteriol Symp Suppl 74:87S

Bruce U (2003) Combined halogens: new products to combat an old problem. Tappi J 86(3):22

Bryers JD (1990) Biofilms in biotechnology. In: Characklis WG. Marsall KC (eds) Biofilms. Wiley, New York, NY, p 733

Buchert J, Verhoef R, Schols H, Ratto M, Blanco A, Craperi D, Lenon G, Wilting R, Carreno A, Lamot J, Siika-Aho M (2004) Development of enzymatic slime control approaches for paper machines. In: 9th International conference on biotechnology in the pulp and paper industry, book of abstracts, Durban, South Africa, 10–14 Oct 2004, pp 31–32

Bunnage W, Schenker A (1995) A new biocide for North America. In: Proceedings of the 1995 TAPPI papermakers conference, Atlanta, GA, pp 189–196

Bunnage WJ, Singleton FL, Cross K (2000) Inhibitor treatment program offers option for clearing biofilm buildup. Pulp Pap 74(6):72–81

Camp V (1989) Microbiology in alkaline fine paper machine systems and the control of slime and deposits using pretreated sewage water as a fresh water source. Pap S Af 9(6):12–17

Carvalho DF (1978) Microbiology in paper manufacture. Papel 39:53–62 (in Portuguese)

Chaudhary A (1992) Study and control of biological slimes in a paper mill. Ph.D. Thesis, Punjab University, Chandigarh, India

Chaudhary A, Gupta LK, Gupta JK, Banerjee UC (1996) Study of slime forming organisms of a paper mill-slime production, characterization and control. Folia Microbiol 41:353–356

Chaudhary A, Gupta LK, Gupta JK, Banerjee UC (1997) Study of slime forming organisms of a paper mill-slime production, characterization and control. J Ind Microbiol Biotechnol 18:348–352

Chaudhary A, Gupta LK, Gupta JK, Banerjee UC (1998) Levanase for control of slime in paper manufacture. Biotechnol Adv 16:899–912

Chiari C, Maggian I, Schirch P (1990) Biocidal treatment of white water circuits with hydrogen peroxide. Papel 8:37–43

Chisti Y (1992a) Build better industrial bioreactors. Chem Eng Prog 88(1):55–58

Chisti Y (1992b) Assure bioreactor sterility. Chem Eng Prog 88(9):80–85

Cloete TE, Brozel V (1991) The effect of stress conditions on bacterial species diversity in water systems. Pap S Af 11(1):12–22

Cloete TE, Gray F (1985) Microbiological control in paper mills. Pap S Af 5(4):26–32

Colasurdo AR, Wilton J (1988) Sonoco utilizes enzymes to control problems with slime and deposits. Pulp Pap 62(1):89–93

Corpe WA (1980) Microbial surface components involved in adsorption of microorganisms onto surfaces. In: Biltton G, Marshall KC (eds) Adsorption of microorganisms to surfaces. Wiley, New York, NY, p 105

Costerton JW, Geesey GG, Cheng KJ (1978) How bacteria stick. Sci Am 238:86
Costerton JW, Lewandowski Z, Caldwell DE, Korber DR, Lappin-Scott HM (1995) Microbial biofilms. Annu Rev Microbiol 49:711–745
Costerton JW, Stewart PS, Greenberg EP (1999) Bacterial biofilms: a common cause of persistent infections. Science 284:318–1322
Czechowski MH, Stoodley P (2002) Antimicrobials and biofilms. J Ind Microbiol Biotechnol 29(6):325
Das JR, Bhakoo M, Jones MV, Gilbert P (1998) Changes in the biocide susceptibility of *Staphylococcus* epidermidis and *Escherichia coli* cells associated with rapid attachment to plastic surfaces. J Appl Microbiol 84:852–858
Davis CK, Singleton FL, Schenker AP (1999) Breakthrough biofilm control technology provides superior deposit control on paper machines. In: TAPPI international environmental conference, vol 2, Nashville, TN, pp 553–560
Dedonder R (1966) Levansucrase from *Bacillus subtilus*. Meth Enzymol 8:500
Donlan RM, Costerton JW (2002) Biofilms: survival mechanisms of clinically relevant microorganisms. Clin Microbiol Rev 15:167–193
Dubout G (1979) The effect of recycling of process water on slime formation – solutions. Rev ATIP 33(10):488–492 (in French)
Dudman WF (1977) The role of surface polysaccharides in natural environments. In: Sutherland IW (ed) Surface carbohydrates of the prokaryotic cell. Academic Press, New York, NY, pp 357–414
Dunne WM Jr (2002) Bacterial adhesion: seen any good biofilms lately? Clin Microbiol Rev 15:155–166
Eriksson U, Johnson A, Tornlund M (1995) Risk assessment of slimicides. KEMI Report No. 9/95. The Swedish National Chemicals Inspectorate, Stockholm, Sweden
Eveleigh DE, Brewer D (1963) Studies on slime accumulation in pulp and paper mills. VI. Isolation of thermophilic and themotolerant fungi from paper mills. Can J Bot 41:1377
Eveleigh DE, Brewer D (1964) Nutritional requirements of the microflora of a slime accumulation in a paper mill. Can J Bot 42:35–43
Farkas JP (1990) Alkaline papermaking and biological control. PIMA 72(7):24–26
Farkas JP, Jones EH, Ormerod D (1987) A simple, rapid means for detecting excessive biological activity in pulp and paper mill systems. Tappi J 70(8):165–168
Ferris FG, Fyfe WS, Witten T, Schultz S, Beveridge TJ (1989) Effect of mineral substrate hardness on the population density of epilithic microorganisms in two Ontario rivers. Can J Microbiol 35:744
Fischer K, Baurich C (1999) Environmentally compatible slime control in PM circuits. In: 5th PTS-symposium pulp technology, Dresden, Germany, pp 10-1–10-11
Flemming HC (2002) Biofouling in water systems-cases, causes and counter measures. Appl Microbiol Biotechnol 59:629–640
Freis RE (1984) The effect of a specific enzyme on biocide use. Tappi J 67(10):100–102
Galon E (1997) Use of enzymes as slimicides. Rev Pap Carton 14(54):57–58
Gavelin G (1996) Efficiency can soar as the slime retreats. Pulp Pap Eur 1(2):29
Geller AN (1996) Slime control in closed water systems without hazardous chemicals. In: Proceedings of the European conference on pulp and paper research: the present and the future, Stockholm, Sweden, pp 288–295
Giatti R (1993) Utilisation of dioxides of chlorine as antislime in the productive process of the paper manufacturers. Valchiampo Ind Carta 31(3):131–133
Glazer JA (1991) Overview of deposit control. Tappi J 74(7):72
Goldstein SD (1983) Slime and deposit control in alkaline papermaking systems. In: Proceedings of the TAPPI, 1983 papermakers conference, Portland, OR, pp 55–61
Goldstein SD (1987) Some overlooked fundamentals of slime control. Appita 40(3):213–216
Gould I (1992) Alternative systems for slime control. In: Chemistry of papermaking conference, Manchester, UK, 13 pp
Gould I (1998) Biofilm control through non toxic additives. Papeterie 221:12–15

Gould I (2001) Non-biocidal methods of biofilm control. Pap Technol 42(1):41–45
Grant R (1998) Enzymes come under the microscope. Pulp Pap Int 40(8):35–37
Grobe KJ, Zahller J, Stewart PS (2002) Role of dose concentration in biocide efficacy against *Pseudomonas aeruginosa* biofilms. J Ind Microbiol Biotechnol 29:10–15
Grussenmeyer H, Wollenweber HW (1992) Microbial slime control in paper machine circuit waters using an enzyme preparation – part I. Wochenbl Papierfabr 22:915–917
Grussenmeyer H, Wollenweber HW (1993) Microbial slime control in paper machine circuit waters: part 2. Wochenbl Papierfabr 121(13):541–544
Haack TK, Downward B, Talbot B (1997) Tetrakishydroxymethyl phosphonium sulfate (THPS): a new biocide with environmental benefits for paper mills. 1997 Engineering & Papermakers Conference, Nashville, TN, USA, Book 3, pp 1115–1119
Hagelsieb AM, Turrado J, Perez S (1996) Enzymatic control of slimes in paper industry: part 2. Invest Tecnol Pap 33(127):106–123
Hagelsieb AM, Turrado J, Perez S (1999) Enzymatic control of slimes in paper industry: part 2. Invest Tecnol Pap 36(139):74–88
Han YW, Clarke MA (1990) Production and characterization of microbial levan. J Agric Food Chem 38:393–396
Harju-Jeanty P, Vaatanen P (1984) Detrimental micro-organisms in paper and board mills. Pap Puu 66(3):245–259
Hart BG (2001) Neoteric enzymes for the paper industry. In: Papermakers conference, Cincinnati, OH, 8 pp
Hatcher HJ (1973) US Patent 3 773 623
Hatcher HJ (1983) Enzymatic control of biological deposits in papermaking. Presented at Pira for WAPRI, 'Biotechnology in the pulp and paper industry', London, UK, 12–14 Sept 1983, pp 178–192
Hestrin S, Avineri-Shapiro S, Aschner M (1943) The enzymatic production of levan. Biochem J 37:450
Himpler FJ, Sweeny PG, Ludensky ML (2001) The benefits of Hydantoin-based slimicide in papermaking applications. In: 55th Appita annual conference, Hobart, Australia, pp 107–111
Hootmann U (2002) Experiences made with an in-situ produced biocide. Int Papwirtsch 2:34–38
Hoyle BD, Jass J, Costerton JW (1990) The biofilm glycocalyx as a resistance factor. J Antimicrob Chemother 26:1–6
Hughes-van, Kregten MC (1988) Slime flora of New Zealand paper mills. Appita 41(6):470–474
Humiston CG (1955) Microbiology of pulp and paper VII. Deterioration of coatings, sizes and adhesives. Tappi, Tappi Monograph Series No. 15, p 157
Huster R (1992a) Formation of mucilaginous residues in paper mill water circulation systems. Wochenbl Papierfabr 120(19):779–793
Huster R (1992b) Controlling the use of biocides in practice. Wochenbl Papierfabr 120(20):820–824
Jaquess PA (1994) Two approaches to biofilm dispersion. In: Biological sciences symposium, Minneapolis, MN, pp 233–237
Johnsrud SC (1997) Biotechnology for solving slime problems in the pulp and paper industry. Adv Biochem Eng Biotechnol 57:311–328
Jokinen K (1999) Paper machine microbiology. In: Gullichsen J, Paulapuro H, Neimo L (eds) Papermaking science and technology 4. Papermaking chemistry Fapet Oy, Helsinki, Finland, pp 252–267
Jung WK, Kutzner HJ (1978) Microbiological problems associated with closed process water systems in the paper industry. Eur J Appl Microbiol Biotechnol 5(3):215–224
Kanto C, Brutar J (1996) Environmentally friendly programme for slime control. Stenquist B Svensk Papperstidn 99(11):29–30
Kemira Chemicals Oy (Vaasa, Finland) (2003) Käyttöturvallisuustiedote (Safety data sheet, in Finnish). The newest updates available at 30.1.2003 for all authorized biocide products
King VM (1990) Microbial problems in neutral/alkaline paper machine systems. In: Neutral-alkaline papermaking short course, Orlando, FL, pp 211–216

Knapick EG, Anker LS, Knauer KE, Pidane KW (2003) A brominated methylethylhydantoin slimicide in a tissue mill. Tappi J 2(5):21–24
Kolari M, Nautinen J, Rainey FA, Salkinoja-Salonen MS (2003) Colored moderately thermophilic bacteria in paper machine. Biofilm J Ind Microbial Biotechnol 30:225–238
Koopmans B, de Vreese T (2002) Towards clean closed water loops. In: Mill experiences with novel in-process water treatment techniques, Doorwerth, The Netherlands, 12 Sept 2002, 29 pp
Kuchma SL, O'Toole GA (2000) Surface-induced, biofilm-induced changes in gene expression. Curr Opin Biotechnol 11:429–433
Kulkarni AG, Mathur RM, Jain RK, Gupta A (2003) Microbial slime in papermaking operations: problems, monitoring and control practices. IPPTA Convention Issue, Mumbai, India, pp 121–125
Kupfer P, Baurich C (1999) Enzymatic slime control in white water system. Wochenbl Papierfabr 127(2):103–108
Latorre WC, Canales CYG, Zimmer C, Gallardo VRB (1991) On-line monitoring of biofouling. In: Rossmoore HW (ed) Biodeterioration and biodegradation, vol 8. Elsevier, London, UK, p 370
Lawrence JR, Wolfaardt GM, Korber DR (1994) Monitoring diffusion in biofilm matrices using confocal laser microscopy. Appl Environ Microbiol 60:1166–1173
Lawrence JR, Neu TR, Marshall KC (2002) Colonization – adhesion, bioaggregates and biofilms. In: Hurst CJ, Crawford RL, Mclnerney MJ, Knudsen GR, Stetzenbach LD (eds) Manual of environmental microbiology, 2nd edn. American Society for Microbiology, ASM Press, Washington, DC, pp 466–477
Lindvall O (1998a) Microbe control and changing environmental demands on the paper industry. Pap Puu 80(3):151–153
Lindvall O (1998b) Bacterial control within paper industry. Skogindustri 52(2):20–21
Lindvall O (2000) A clean paper machine has seldom microbiological problems. Invest Tecnol Pap 37(146):689–692
Loosvelt I, Datweiler C (2007) Enzymatic products: uncharted territory for the pulp and paper industry. In: PTS pulp technology symposium, Dresden, Germany, 27–28 Nov 2007, Paper 6, 2 pp
Lustenberger M, Deuber R (1991) On the environmental friendliness of antislime agents in the paper industry. Wochenbl Papierfabr 119(6):204–206
Lutey RW (1972) Microbial deposit control. In: Tappi papermakers conference, Atlanta, GA, p 133
Maillard JY (2002) Bacterial sites for target action. J Appl Microbiol Symp Suppl 92:16S–27S
Marshall KC, Stout R, Mitchell R (1971) Mechanisms of the initial events in the sorption of marine bacteria to surfaces. J Gen Microbiol 68(3):337–348
Martin RB (1955) Microbiology of pulp and paper. III. Microbiology of fresh water. Tappi, Tappi Monograph Series No. 5, p 55
Mattila K, Weber A, Salkinoja-Salonen MS (2002) Structure and on-site formation of biofilms in paper machine water flow. J Ind Microbiol Biotechnol 28:268–279
McBain AJ, Rickard AH, Gilbert P (2002) Possible implications of biocide accumulation in the environment on the prevalence of bacterial antibiotic resistance. J Ind Microbiol Biotechnol 29:326–330
Mobius CH, Demel I, Garhammer J, Lottes K (1986) Use of biocides in paper mill white waters. Papier 40:242–249
Morros J (1995) New bacteriological treatments of paper mill waters. Rev ATIP 49(3):96–98
Mueller RF (1994) Biofilm formation in water systems and their industrial relevance. In: Biological sciences symposium, Minneapolis, MN, pp 195–201
Nagy LA, Olson BH (1986) A comparison of media for the enumeration of filamentous fungi from aqueduct biofilm. Zbl Bact Hyg B182:478–484
Nason HK, Shumard RS, Flemming JD (1940) Tappi J 23:337
Nelson TR (1982) Appleton papers finds chlorine dioxide to be an alternative to conventional biocides in alkaline systems. Tappi J 65(6):69–73

Neu TR, Swerhone GDW, Lawrence JR (2001) Assessment of lectin-binding analysis for in situ detection of glycoconjugates in biofilm systems. Microbiology 147:299–313
Nurmiaho-lassila EL, Lehtinen SA, Marmo SA, Salkinoja-Salinen MS (1990) Institute of physics conference series, No. 98. IOP, London, UK, p 727
Oberkofler J (1987) Biological process for slime and deposit control. Pap Osterreich 11:52–54
Oberkofler J (1989) Biocide-free slime and deposit control on the basis of biological equilibrium. Wochenbl Papierfabr 117(20):920, 922, 923
Oberkofler J (1993) US Patent 5 242 593
Oberkofler J, Braunsperger F (1994) Chemical-free slime control in white-water circuits of paper machines. In: Chemical technology of papermaking, Munich, Germany
Oberkopfler J (1992) Environmentally friendly slime control. Pap Osterreich 1:24
Oppong KVM, Zhou X, Bowen JA (2000) Cultural and biochemical diversity of pink-pigmented bacteria isolated from paper mill slimes. J Ind Microbiol Biotechnol 25:74–80
Orndorff SA (1983) US Patent 4 370 199
Oyaas K (2001) Closed system: more focus on slime problem. Skogindustri 55(2):20
Paice M, Zhang X (2005) Enzymes find their niche. Pulp Pap Can 106(6):17–20
Palcic MM, Teodorescu G (2002) 88th Annual meeting, Montreal, QC, pp C163–C168
Patterson JV (1986) TAPPI seminar notes, pp 23–27
Paulus W (1993) Microbicides for the protection of materials – A handbook. Chapman & Hall, London, UK, 496 pp
Pauly D (2001) Studies into the mechanisms of slime formation in water circuits. In: PTS symposium interface processes in paperboard manufacturing, Munich, Germany, 14 pp
Pellegrin V, Juretschko S, Wagner M, Cottenceau G (1999) Morphological and biochemical properties of a *Sphaerotilus* sp. isolated from paper mill slimes. Appl Environ Microbiol 65:156–162
Piluso AL (1977) Logical approach to wet end problems and deposit control. South Pulp Pap Manuf 40:14
Pirttijärvi TSM (2000) Contaminant aerobic spore forming bacteria in the manufacturing processes of food packaging board and food. Dissertationes Biocentri Viikki Universitatis Helsingiensis 14/2000. Ph.D. thesis, University of Helsinki, Helsinki, Finland
Pirttijärvi TSM, Graeffe TH, Salkinoja-Salonen MS (1996) Bacterial contaminants in liquid packaging boards: assessment of potential for food spoilage. J Appl Bacteriol 81:445–458
Postgate JR (1979) The sulphate-reducing bacteria. Cambridge University Press, Cambridge, UK
Purkiss BE (1973) Control of slime in the paper industry. Papeterie 1:26–39
Purvis MR, Tomlin JL (1991) Microbiological growth and control in the papermaking process. In: TAPPI chemical processing aids short course, Seattle, WA, pp 69–78
Raaska L, Sillanpaa J, Sjöberg AM, Suihko ML (2002) Potential microbiological hazards in the production of refined paper products for food applications. J Ind Microbiol Biotechnol 28:225–231
Rantakokko J, Maunuksela J, Malone J (1994) Paper mill slime control with peracetic acid. Papier 48(11):681, 684–685
Rivera F, Jara A (2007) Enzyme boilout in paper machines. Cellul Pap (Chile) 23(5):14–17
Robertson LR (1994) Prevention of microbial adhesion. In: Biological sciences symposium, Minneapolis, MN, pp 225–232
Robertson LR, Taylor NR (1994) Biofilms and disbursents: a less-toxic approach to deposit control. Tappi J 77:99–103
Robichaud WT (1991) Controlling anaerobic bacteria to improve product quality and mill safety. Tappi J 74(2):149–153
Safade ML (1988) Tackling the slime problem in a paper mill. Paper Technol Ind 29:280–285
Salzburger W (1996) Paper and additives. In: Heitz E, Flemming HC, Sand W (eds) Microbially influenced corrosion of materials. Springer, Berlin, Germany, pp 415–427
Sanborn JR (1965) Pap Trade J (15 Feb):42
Saner M (1998) Biodeposit control by non-toxic procedures and online monitoring of the biofilms. In: 51st Annual meeting, Grenoble, France, 6 pp
Scharpf S (1998) Wet-end chemistry: multiplying demands, developing solutions. Eur Papermaker 6(4):35–39

Scharschmied B (1975) Microbiological growth in the acid, neutral and alkaline range. Wochenbl Papierfabr 103(4):148–150 (in German)
Schenker AP (1996) Biodispersion – microbiological growth control for the future. Svensk Papperstidn 99(1):24–25
Schenker AP, Singleton FL, Davis CK (1998) Non biocidal programmes for biofilm control in paper machine circuits. In: EUCEPA symposium 1998 – chemistry in papermaking, Florence, Italy, pp 331–354
Schirch PFT, Santos CAS, do A, Walsh P (1993) Hydrogen peroxide in white water treatment: analysis of four years of industrial treatment. In: Environmental conference, Boston, MA, Book 1, pp 165–173
Schuetz J, Wollenweber HW (1999) Enzymes in papermaking. Pap Technol 40(8):52–54
Shema BF (1955) Tappi Monograph Series No. 15, pp 28–54
Siika-aho M, Ratto M, Piskonen R, Salo S, Buchert J, Viikari L (2000) Enzymatic control of paper machine slimes. Invest Technol Pap 37(146):667–675
Srinivasan R, Stewart PS, Griebe T, Chen CI, Xu X (1995) Biofilm parameters influencing biocide efficacy. Biotech Bioeng 46:553–560
Stanier RY, Doudoroff M, Adelberg EA (1968) In: Stanier RY, Doudoroff M, Adelberg EA (eds.) General microbiology, 2nd edn, Chap 17:393
Stenqvist B (1992) Nalco chemical ab: new dispersion agent to fight sludge. Svensk Papperstidn 95(7):16–18
Stomps LE (1995) Microbiological control of slime on glass mat machines. In: Nonwovens conference, St Petersburg, FL, pp 9–10
Stoner MT, King VM (1994) Industrial biofilms: an overview. In: Biological sciences symposium, Minneapolis, MN, pp 185–193
Stoodley P, Sauer K, Davies DG, Costerton JW (2002) Biofilms as complex differentiated communities. Annu Rev Microbiol 56:187–209
Sugi T (1999) Slime control agent. Jpn Tappi J 53(2):37–44
Suominen I, Suihko ML, Salkinoja-Salonen MS (1997) Microscopic study of migration of microbes in food-packaging paper and board. J Ind Microbiol Biotechnol 19:104–113
Tanaka SO, Yamamoto T (1979) Synthesis of levan by levansucrase: some factors affecting the rate of synthesis and degree of polymerization of levan. J Biochem 85:287–293
Thomas GS (1999) Microbiological control evolution or revolution? Pap Age 115(4):14–15
Vaatanen P, Harju-Jeanty P (1986) 3rd International conference on biotechnology in the pulp and paper industry, Stockholm, Sweden
Vaatanen P, Niemela SI (1983) Factors regulating the density of bacteria in process waters of a paper mill. J Appl Bacteriol 54:367–371
Väisänen OM, Elo S, Marmo S, Salkinoja-Salonen M (1989) Enzymatic characterization of Bacilli from food packaging paper and board machines. J Ind Microbiol 4:419–428
Väisänen OM, Mentu J, Salkinoja-Salonen MS (1991) Bacteria in food packaging paper and board. J Appl Bacteriol 71:130–133
Väisänen OM, Nurmiaho-Lassila EL, Marmo SA, Salkinoja-Salonen MS (1994) Structure and composition of biological slimes on paper and board machines. Appl Environ Microbiol 60(2):641–653
Väisänen OM, Wiber A, Bennasar A, Fainey FA, Busse H-J, Salkinoja Salonen MS (1998) Microbial communities of printing paper machines. J Appl Microbiol 84:1069–1084
Van Haute E (1997a) Modern deposit control: biodispersants and enzymatic treatments. In: PITA annual conference on chemicals in papermaking, Manchester, UK, pp 112–113
Van Haute E (1997b) Legislation and economics bring on the enzymes. Pulp Pap Eur 2(2):11–13
Van Haute E (1999) Biodispersant and enzyme treatments. A new approach to deposit control on paper machines. In: DITP – 26th international annual symposium, Bled, Slovenia, 17–19 Nov 1999, pp 132–136
Van Haute E (2000) Enzymatic deposit control on tissue machines. In: Hygiene and absorbency products – scientific and technical advances in materials and process technology, Brussels, Belgium, 8 pp

Verhoef R, de Waard P, Schols HA, Ratto M, Siika-aho M, Voragen A (2002) Carbohydr Res 337:1821–1831
Watnick P, Kolter R (2000) Biofilm, city of microbes. J Bacteriol 182:2675–2679
Weir B, Pear JDM, Webb LJ (1981) An evaluation of pathogenic micro-organisms in recycled fibre from garbage, with a view to assessing hazards to health. Pira Rep PB 8(R). Pira, Leatherhead, UK, 104 pp
Weisshuhn A, Riessner H, Schulte J (2000) Use of a biodispersant as a slimicide in an integrated pulp and paper mill. Wochenbl Papierfabr 128(17):1146–1151
Werres J (1998) A successful antislime concept. Pap Technol 39(7):41–44
Whiteley M, Bangera MG, Bumgarner RE, Parsek MR, Teitzel GM, Lory S, Greenberg EP (2001) Nature 413:860–964
Wiatr CL (1990) Application of cellulase to control industrial slime. US Patent 4 936 994
Wiatr CL (1994) Development of biofilms. In: Biological sciences symposium, Minneapolis, MN, pp 203–223
Wolfaardt GM, Lawrence JR, Robarts RD, Caldwell DE (1998) In situ characterization of biofilm exo polymers involved in the accumulation of accumulated chloroorganics. Microbiol Ecol 35:213–223
Wolfanger D (2001) Stabilized enzymes: new options for paper machine boilouts. Pap Age 117(3):34–36
Wright JB (1997) Significantly reduced toxicity approach to paper machine deposit control. In: Proceedings of the 1997 Tappi engineering & papermakers conference. Tappi Press, Atlanta, GA, pp 1083–1088
Xu H (2005) Enzymes: a versatile tool to alter fibre and paper performance. In: Scientific and technical advances in refining and mechanical pulping, Barcelona, Spain, 28 Feb–4 Mar 2005, Impact forum: fibre engineering, Paper 6, 11 pp

Chapter 15
Stickies Control

15.1 Introduction

Recovery and use of secondary fiber for paper production is increasing all over the world. About 40% of the total paper production in the world was based on the secondary fiber in 2006 (Sixta 2006). Paper recycling offers several advantages (Bajpai 2006). Substitution of virgin pulp with recycled fibers saves on wood for making pulp, which reduces the exploitation of old forests, important for their biodiversity. Every ton of recycled fiber saves an average of 17 trees plus related pulping energy. By using wastepaper to produce new paper, disposal problems are reduced.

Use of lower quality recycled paper is increasing each year, as mills strive to save money on their fiber furnish. Lower quality furnish will be heavily contaminated with sticky contaminants, or stickies, and it is essential to find the best way to control them. A variety of stickies are encountered in wastepapers. Stickies are tacky, hydrophobic, pliable organic materials primarily found in recycled paper systems (Doshi and Dyer 2002; Doshi 1999; Delagoutte 2005). They have a broad range of melting points and different degrees of tackiness depending on their composition. Stickies found in recycled fiber can be adhesives, styrene–butadiene rubber (SBR), rubber, vinyl acrylates, polyisoprene, polybutadiene, and hot melts, to name a few. Stickies live up to their name – they stick – and that is why they present a problem to the paper industry. They stick very well to each other and to piping, wires, felts, and dryers. After the particles break down in the pulper, there are only two places where materials can exit the system: with the sludge or with the sheet.

Stickies will become a growing process problem in the months and years to come with the increased use of recycled fibers and the closing of the mill water loops (Patrick 2006). Many significant operational and quality problems are caused by stickies in pulp and papermaking systems. Cleaning up fouled sections of the paper machine causes valuable machine downtime, which diminishes paper quality and reduces output, all costing millions of dollars per year.

P. Bajpai, *Biotechnology for Pulp and Paper Processing*,
DOI 10.1007/978-1-4614-1409-4_15, © Springer Science+Business Media, LLC 2012

Stickies may be classified in different ways. The simplest and most common classification is based on the size of the sticky particles encountered in the recycled pulp. The stickies are classified as macrostickies, microstickies, and colloidal stickies (Hamann and Strauss 2003). Macrostickies are solid and tacky contaminants rejected by a 100-μm slotted laboratory screen. These macrostickies are therefore considered to be larger than 100 μm in diameter. The levels of macrostickies vary with different wastepaper grades, ranging from zero in newsprint to 45,000 mm^2/kg in corrugated board. Microstickies are tacky particles of size 100 μm down to 1–5 μm. Macrostickies are retained on a laboratory screen, but the microstickies pass through the screen and are difficult to isolate. Microstickies are predominant in recycled pulp (Johansson et al. 2003; Delagoutte et al. 2003). Up to 70–80% of the total stickies content comes from microstickies. Colloidal stickies are smaller than 1–5 μm and belong to the dissolved and colloidal fraction. In this category, the stickies may be considered more as potential stickies. Indeed, due to their size, these stickies do not have a real detrimental impact until they remain in colloidal form. Nevertheless, it is reported that destabilization, especially by cationic polymers, may induce their precipitation and sometimes the formation of tacky precipitates called secondary stickies.

Stickies may be divided into primary and secondary stickies. Primary stickies are identical with macro- and microstickies; they are intact tacky particles of adhesive such as hot melts or pressure-sensitive adhesives, inks, binders, waxes, plastics, and wet-strength resins. Shock-type, chemico-physical alterations of a pulp suspension may form secondary stickies. This includes changes in temperature, pH, or charge that destabilize a colloid and cause agglomeration of dissolved or colloidal substances. Secondary stickies can then lead to deposits on the paper machine or its clothing. Primary or secondary stickies may be responsible for stickies problems at a paper mill. And stickies problems are often easier to solve if it can be identified whether the stickies are primary or secondary. If no recovered paper is used, the stickies problems are usually caused by secondary stickies. This can be counteracted by changing the wet-end chemistry or the coating binder system. A survey for the International Association of the Deinking Industry (INGEDE) shows that macrostickies in undeinked pulp have increased by a factor of 2.5 since 1996 (Hamann and Strauss 2003). The stickies problem has also been aggravated by the increase in closed-loop water systems that contain more finely dispersed and colloidal contaminants, as well as by higher machine speeds and the trend to lower grammage papers.

15.2 Problems Caused by Stickies

Stickies can cause problems with paper machine operation and product quality (Doshi and Dyer 2002; Doshi 1999; Delagoutte 2005; Putz 2000; McKinney 1995). They deposit on wires, felts, press rolls, and drying cylinders. They prevent good fiber-to-fiber bonding and increase the risk of web breaks on paper machines, particularly with newsprint and tissue grades and also in the printing press. Even if breaks do not occur, holes or spots in the paper cause loss of quality that means the product has to be used as second-quality material or processed as broke. All problems caused by stickies in the paper machine become more serious with lower basis weight or caliper, lower strength of the paper web, higher paper machine speed, and higher dynamic stress on the paper web. Even more costly are the stickies problems that occur after finishing the paper or board sheet. Due to certain processing conditions of temperature or pressure, stickies bound into the inside of board or thicker paper can reach the outsides of paper or board sheets by melting. This creates an unacceptable appearance after printing, varnishing, or laminating. The same applies to stickies on the surface of a paper or board web that has passed through the papermaking process, including subsequent finishing, without any problem; these stickies cause problems during converting by adhering to paper or board blanks. Layers sometimes stick together when a reel unwinds, causing breaks or surface blemishes. In corrugated board production, a particular problem is wax introduced by recovered paper. If this is insufficiently removed or not neutralized in recovered paper processing, it can lower the coefficient of friction of the liner, then reels or rolls may telescope during transport. The finished corrugated board blanks could also slip. This would cause imperfect stacking and affect further processing. Stickies also cause problems in high-speed printing and converting.

15.3 Control of Stickies

Several methods have been proposed for controlling stickies (Doshi and Dyer 2002; Doshi 1999; Delagoutte 2005; Putz 2000; McKinney 1995; Lamot 2004). Stickies control methods fall into two main categories, mechanical and chemical. Mechanical methods, screens, cleaners, DAFs, and washing stages can all remove stickies. Each equipment type is effective in removing stickies in a given size range. In most cases stickies removal has to be balanced against yield loss. With all of the mechanical methods only a certain percentage of the stickies will be removed. This is true even with the best equipment and process conditions. Chemical control programs take two main approaches. There are stock additions, trying to tie up the stickies and to stop them from depositing and "point of problem" type applications. Traditional chemical programs have been effective but do not really attack the stickie itself. A new approach has been taken to control stickies by looking at enzymes that would break down and change the nature of the stickie.

15.3.1 Enzyme Approach

Use of enzymes is gaining wider acceptance in the pulp and paper industry for a variety of applications such as pulp mill and paper machine boilouts, deposit control by dispersion of accumulated slime, pitch control, and drainage assistance.

A recent area of research involves using enzymes to control stickies (Jones 2005; Fitzhenry et al. 2000a, b; Van Haute 2003; Anon 2003a, b; Toland 2003; Jones and Fitzhenry 2003). Buckman has developed the enzymes that control stickies.

A study of the chemical composition of stickies reveals that most contain a number of ester-type chemical bonds that link the basic building blocks of the stickies together. A number of esterase-type enzyme mixtures have been studied to find the one that had the ability to break down the stickies. Breaking the ester bonds reduces the size of the sticky by breaking it into smaller components. A key advantage of this approach is that once broken down, the chance of the particles re-agglomerating further along the process is greatly reduced. Another important effect on the stickies is the enzymatic modification of the surface of the stickies. This change results in less tacky stickies (Jones 2005). These enzymes have been proven to reduce downtime, decrease cleaning chemical costs, and increase machine-clothing life better than historical stickies control technologies (Eng and Covarrubias 2005; Covarrubias and Eng 2006; Kimura 2006; Sokol and Huszar 2005). More mills now have single-stream recycling. The increased use of single-stream recycling processes has resulted in more stickies. Buckman Laboratories, Canada, manufactures Optimyze, which contains an esterase, which breaks up the ester bonds in PVA stickies. This is particularly effective in pH range 6.5–10 and temperature range of 25–60°C (Covarrubias and Jones 2005). Buckman's Optimyze Plus range also includes an enzyme, which acts on wood pitch. Buckman is developing an enzyme product combined with a dispersant that would disperse and stabilize the stickies. The enzyme is then more likely to penetrate the stickies rather than just acting on the surface (Jones 2008). In all the early work, esterase was added to the stock, but Jones (2005) has shown that applying the enzyme directly onto paper machine clothing also reduces stickies deposition. Positive results were obtained by applying enzyme to forming fabric and press felts. The enzyme is applied through a shower bar and full coverage is important. The enzyme detackifies the stickies, so the cleaning showers can remove them more effectively. In 2004, the US-EPA gave Optimyze a Presidential Green Chemistry Challenge Award. Optimyze is now used in many mills around the world.

Figure 15.1 shows the effects of Optimyze treatment of stickies. On the left, polyvinyl acetate particles have agglomerated to form large globules that can cause not only breaks and downtime on the paper machine but create operating system breaks and problems on converting equipment as well. The globular particles can also cause paper to double feed through printers, etc. On the right in this figure, Optimyze enzyme treatment keeps the particles small, dispersed, and nontacky. Figure 15.2 is a scanning electron microscope photo of the surface of a stickies particle showing the effects before and after enzyme treatment. As can be seen on the right, the surface is very smooth after treatment compared with that before treatment on the left.

The mills currently using Optimyze enzyme to control stickies all report significant gains over traditional chemical approaches they were using. On treatment of old newsprints (ONP) and OMG with esterase-type enzyme in a mill trial, a dramatic reduction in the size of the sticky particles was observed (Jones and Fitzhenry 2003).

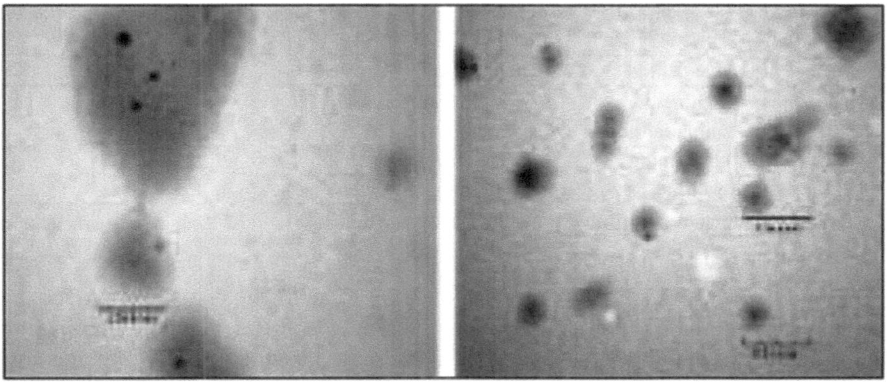

Fig. 15.1 Results of enzyme treatment on stickies (no treatment on the *left* and after Optimyze treatment on the *right*) (Reproduced with permission from Patrick (2004))

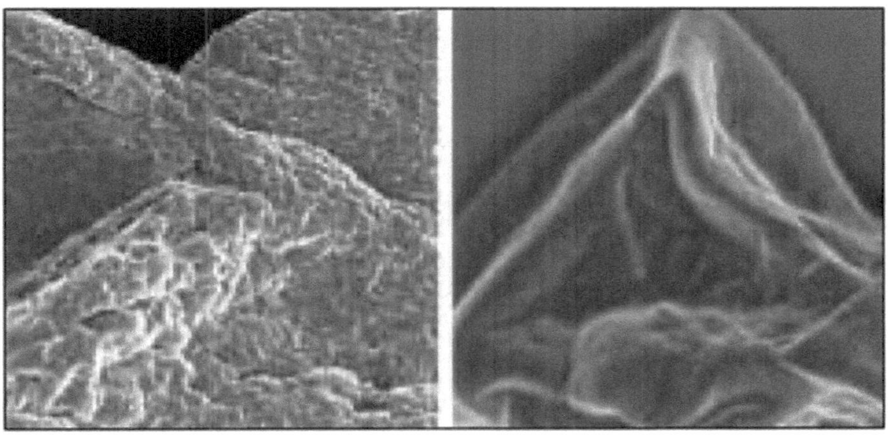

Fig. 15.2 Electron photomicrograph of the surfaces of a stickies particle before enzyme treatment (*left*) and after treatment (*right*) Patrick (2004); Reproduced with permission

The stickies content of all the sizes and the total stickies were much less on enzyme treatment as compared to those without enzyme (pretrial results). The bigger size stickies are totally absent in the recycled fiber treated with enzyme When the recycled fiber from mixed office waste (MOW) was treated with esterase enzyme in a mill trial, the total stickies content was reduced appreciably (Jones and Fitzhenry 2003). Without enzyme treatment of MOW, it was not possible to increase the recycled fiber content in the final furnish beyond 50%. Even then, the total stickies were more than 250 ppm. However, on treating the recycled fiber from MOW with enzyme, total stickies content could be reduced to around 100 ppm, and that was with a higher content (60%) of recycled fiber.

Table 15.1 Savings realized by switching to enzymatic stickies control at a 400 tpd coated paperboard mill (Based on Patrick (2004))

Problems caused by stickies	Benefits of control	Annual savings ($)
Reduced life of forming fabrics	Reduce four fabrics/yr	100,000
Downtime for cleaning fabrics	Increase production	170,000
Chemicals used for cleaning	Replace chemicals	160,000
Paper machine downtime	Eliminate 45 breaks/year at 2 h each	900,000

Total possible annual return from eliminating stickies: $1,330,000

In another MOW mill, the recovered fiber was treated with enzyme to reduce the percentage of high brightness virgin pulp in the final furnish without compromising on brightness of the finished stock. With addition of 35% high-bright pulp, the normal brightness gain from coarse screen to finish stock was only 12–14 points when no enzyme treatment was given to the recovered fiber (Jones and Fitzhenry 2003). However, after esterase enzyme (Optimyze of Buckman) treatment of MOW recovered fiber, it became possible to get a brightness gain of more than 15 points, even when the high bright pulp content was only 15–20%.

In one US mill producing coated paperboard from 100% OCC for conversion to food boxes, stickies problems were causing significant deposition problems on this mill's paper machine and related runnability and off quality problems in the sheet. The high level of paper machine breaks was causing significant operating downtime. Kerosene and other harsh chemical solvents were being used to remove the deposits. After launching an Optimyze enzyme program, paper machine deposition was reduced 75% and machine breaks/downtime were reduced accordingly. Off-quality product tonnage dropped and the mill's use of cleaning chemicals was practically eliminated. With holes and other stickies related quality problems rectified, customer complaints fell to almost none. This coated paperboard mill experienced a $1.33 million annual return by eliminating stickies (Table 15.1). Forming fabric changes were cut by four per year and paper machine breaks were reduced to the tune of 45 annually, together representing a million dollar savings (Patrick 2004).

A mill in Brazil producing 270 tpd of linerboard using 100% OCC was having significant downtime problems due to stickies. Stickies in the sheet were also causing breaks at the mill's customer converting operations. The mill had to use a higher quality, more costly OCC in an attempt to minimize these problems. When the mill switched to Optimyze technology, product quality improved immediately. Production breaks were almost totally eliminated (reduced by 30/month), giving a much more efficient operation. In fact, the mill achieved record production after the switch. In addition, it was able to use a poorer quality OCC, cutting production costs even further. The stickies were eliminated after the Optimyze program was started. Continuing with a few other sample cases, use of Optimyze technology allowed one US mill to produce tissue and toweling products that easily met a fast food chain's stringent quality and fiber content requirements (at least 65% postconsumer fiber). In fact, this mill was able to boost recycle fiber content to 100% and still meet the requirements, while other mills supplying the restaurant chain had to use much high percentages of virgin fiber (Patrick 2004).

Another US mill that produces more than 1,000 tpd of various grades of tissue from 100% recycle fiber was experiencing severe stickies deposits in its press felts. When the deposition approached a critical level, which was happening much too frequently, production had to be stopped and solvents used for cleaning. The mill was using several hundred gallons of kerosene and other solvents per day during these cleaning cycles. Also, press section showers were used continuously in an attempt to keep this machine's press felts clean. The water used in these showers is recycled within the paper mill system, which meant that this source too was building up high levels of stickies and re-depositing them in the cleaned felts and on other paper machine components. Clearly, another solution was needed. The mill now adds Optimyze to the water used in the felt cleaning showers. As a result, downtime for solvent cleaning has been reduced from 1.81 h per week before the treatment to an average of 0.73 h per week. This represents a 60% reduction in downtime and the amount of solvent used for cleaning the machine is reduced by a corresponding amount. Similar problems were occurring on another paper machine in this mill and the same strategy was used to solve them. With this machine, the average downtime for solvent cleaning dropped from an average of 1.6–0.46 h per week, representing a 70% reduction in downtime and solvent usage. This 70% gain would have been closer to 80% if, during 3 weeks of the recent period, a cleaning shower had not been inadvertently turned off (Patrick 2004).

In many other case studies involving almost every grade sector, from multi-ply coated boxboard to packaging grades, stickies have been reduced dramatically with the enzymatic control program, resulting not only in higher quality products and fewer converting customer complaints, but significant reductions in culled production and increased use of lower quality and much less costly recovered fiber. Finally, but certainly not least important, using enzymes to control stickies problems has many safety and health related benefits. Not only are most solvents and mineral oils costly, they also can create potential safety and environmental issues. As shown by extensive paper machine air sampling studies, "ecotox" toxicity evaluations, and skin patch testing programs, enzymes represent absolutely no threat to the health and well-being of humans as well as other terrestrial and aquatic life. Enzymes themselves are not living things. They are just proteins that catalyze specific reactions. One of the nice things about enzymes being focused on just one specific chemical reaction is the fact that they will not interfere with other chemistry in the sheet, such as sizing or strength additives (Patrick 2004).

Several mill-scale studies related to the production of tissues, newsprint, writing, and packaging papers from recovered or recycled paper have been reported (Sokol and Huszar 2005). The use of enzyme has been found to enhance the screening efficiency, reducing the amount of rejects. It resulted in the reduction of stickies by 70–90%, generated clean cellulose fiber and big contaminant particles that were easily removable from the system. In other mill studies with enzymes produced results such as 90% reduction in stickies and increased brightness across the deinking facility, production performance improvements, reduction in the use of cleaning solvents, and a 95% reduction in downtime caused by stickies (Covarrubias and Eng 2006). Buckman is developing an enzyme product combined with a dispersant that would disperse and stabilize the stickies (Jones 2008). The enzyme is then more likely to penetrate the stickies rather than just acting on the surface.

Zeng et al. (2009) used lipase type enzyme and the esterase type enzyme for degradation of stickies in deinked pulp (DIP) and chemithermomechanical pulp (CTMP). These enzymes removed stickies by over 30% in DIP and 75% in CTMP. There was a remarkable change in the molecular weight of the stickies in DIP. Enzyme application had no negative effect on the properties of either pulp.

15.4 Conclusion

The use of recycled fiber continues to grow. As the demand increases, the quality of the recycled fiber will decrease. The result is more stickies. More stickies, more stickie headaches. Mechanical and traditional chemical control approaches have worked but they are not the complete answer. Enzymes are a new approach in the battle against stickies and have proven very effective. Enzymes will continue to play a major role in stickies control. The challenge is to develop new enzymes that work in different temperature and pH ranges and on all types of stickies chemistry. The Optimyze product line from Buckman is successfully helping to minimize or eliminate the problems caused by stickies in many mills around the world.

References

Anon (2003a) Buckman: preventing stickies. Pap Making Distrib 2(4):14

Anon (2003b) Buckman: expansion geared towards better serving customers. Pulp & Paper (Canada) 104(1):32

Bajpai P (2006) Advances in recycling and deinking (Chap. 1). PIRA International, Leatherhead, pp 1–2

Covarrubias RM, Eng GH (2006) Optimyze: enzymatic stickies control developments. Pap Asia 22(8):31–34

Covarrubias RM, Jones DR (2005) Optimyze: enzymatic stickies control developments. 91st Annual Meeting Pulp and Paper Technical Association of Canada, Montreal, QC, Canada, 8–10 Feb 2005, Book A, A107–A116

Delagoutte T, Brun J, Galland G (2003) Drying section deposits: identification of their origin. In: IPE international symposium new technological developments in paper recycling, Valencia, Spain, June, 2003

Delagoutte T (2005) Management and control of stickies. Progr Pap Recycl 15(1):31

Doshi MR, Dyer JM (2002) Overview, recent advances in paper recycling – stickies, paper recycling challenge. In: Doshi MR (ed). Doshi, Appleton, p 1

Doshi MR (1999) Properties and control of stickies. In: Doshi MR, Dyer JM (eds) Paper recycling challenge, vol 4. Doshi, Appleton, p 67

Eng GH, Covarrubias RM (2005) Optimyze: enzymatic stickies control developments. Proceedings of the 7th international conference on pulp, paper and conversion industry. New Delhi, India, 3–5 Dec 2005, pp 135–142

Fitzhenry J, Hoekstra PM, Glover D (2000a) Enzymatic stickies control in MOW, OCC, and ONP furnishes. Tappi pulping, process and product quality conference, Boston, MA, 2000

Fitzhenry J, Hoekstra P, Glover D (2000b) New measurement techniques and new technologies for stickies control. Conference on scientific and technical advances in wet end chemistry, Barcelona, Spain, 19–20 June 2000, p 16

Hamann L, Strauss J (2003) Stickies: definitions, causes and control options. Wochenblatt für Papierfabrikation 131(11/12):65

Johansson H, Wikman B, Lindström E, Österberg F (2003) Detection and evaluation of micro-stickies. Progr Pap Recycl 12(2):4

Jones DR, Fitzhenry JW (2003) Esterase type enzymes offer recycled mills an alternative approach to stickies control. Pulp & Paper (Canada) 28–31

Jones DR (2005) Enzymes: Using Mother Nature's tools to control man-made stickies. Pulp & Paper Canada 106(2):23–25

Jones DR (2008) The next steps in enzymatic stickies control. Pulp Pap 82(6):21

Kimura M (2006) Stickies control agent for recycling paper by OPTMYZE. Jpn TAPPI J 60(7): 35–41

Lamot J (2004) Effective chemistry for improved recycled fibre quality. Eighth Pira international conference on paper recycling technology, Prague 2004

McKinney RWJ (ed) (1995) Wastepaper preparation and contaminant removal, technology of paper recycling. Blackie Academic, Glasgow, p 401

Patrick K (2004) Enzyme technology improves efficiency, cost, safety of stickies removal program. Pap Age 22–26

Patrick K (2006) Stickies still a critical concern for today's recycling plants. Pap Age 122:28–31

Putz H (2000) Stickies in recycled fiber pulp (Chap. 11). In: Gottsching LD, Pakarinen H (eds) Papermaking science and technology, vol 7. Fapet Oy, Helsinki

Sixta, H. (ed.) (2006) Introduction (Chap. 1). In: Handbook of pulp. Wiley, Weinheim, pp 2–19

Sokol A, Huszar L (2005) Optimyze modern enzymatic programme for stickies control in paper-making process, Progress '05: 15th international papermaking conference, Warsaw, Poland, 28–30 Sept 2005, 11pp

Toland J (2003) Developments in deinking: rounding up some of the latest trends in the recovered paper sector. Pulp Paper Int 45(4):25

Van Haute E (2003) Optimyze: Enzymatic stickies control products. Invest Tec Pap 40(149): 47–51

Zeng X, Fu S, Yu J, Li K, Zhan H, Li X (2009) Study on the degradation of stickies in the pulps by complex enzymes. China Pulp Pap 28(2):1–4

Chapter 16
Enzymatic Modification of Starch for Surface Sizing

16.1 Introduction

In the pulp and paper industry, modified starches are used in the preparation of starch pastes for the coating and surface sizing of paper. Surface size and coating colors require a target viscosity at the water content specified for the formulation. Unmodified starches produce an adhesive that are too viscous to meet the water balance requirements for high-solids coating colors. Enzymatic conversion is an important process to modify the native starch to meet the viscosity requirements for surface sizing or paper coating. Various alpha-amylase enzymes offer the paper mill a cost-effective, in-house process for modifying starch, thereby eliminating the need to buy modified starch from third parties. The viscosity of the starch paste can be reduced and controlled with enzyme so that both the water balance and final coating color viscosity requirements are met even for high-solids formulations. Alpha-amylase enzymes are not expensive, and the quantity required is small. The economic advantage is the major reason for the use of enzymatic starch for coating and surface sizing. However, preparation of enzymatic starch requires more careful and closer supervision than does the preparation of adhesives from modified or derivatized starches.

Surface sizing is the application of hydrophobic materials to the surface of paper to improve certain physical properties of sheet and to reduce the porosity. Surface sizing differs from internal sizing in that the sizing agent is applied to the surface of the paper where it cements the fibers to the body of the paper and deposits a more or less continuous film on the paper surface. Surface sizing is primarily concerned with surface films, and hence it is usually desirable to keep the size on the surface of the paper as much as possible. This is particularly true when using expensive sizing agents such as polyvinyl alcohol, carboxymethyl cellulose, and even animal glue. On the other hand, it is desirable on some grades of paper, when starch is the sizing agent, to obtain considerable penetration of starch into the paper in order to build up the burst, internal sizing, and other internal properties of the sheet. The main objective of surface sizing is to improve paper printability. Surface sizing increases

P. Bajpai, *Biotechnology for Pulp and Paper Processing*,
DOI 10.1007/978-1-4614-1409-4_16,

the liquid wetting resistance (water, oil, and grease), the surface strength, the sheet strength, and the dimensional stability. It can also control the coefficient of friction (Tompkins, 1992). Therefore, the use of surface sizing agents has increased due to its advantages (Tsai et al. 1995). Surface sizing has more specific action compared to internal sizing and is less sensitive to changes in wet end operation (Maurer, 2001a; Smook, 1982). There is 100% retention of solids while only the volatile material is lost during the drying. The wet end deposits are reduced which results in increased press clothing life. The size press chemicals generally do not interfere with fiber-to-fiber bonding as takes place in many wet end chemicals since the bonds have already been formed when the size press formulations are applied. Size press application minimizes the contamination of the environment. Almost all the materials get retained at the surface, so wastage of surface sizing chemicals is low. Surface sizing is used to achieve special effects such as those obtained by the polyurethane foam polymers, which make the strong hydrophobes needed for carbonless copy paper.

16.2 Enzymes Used for Starch Conversion

Enzymes use for starch conversion are amylases which are among the most important enzymes and are of great significance for biotechnology, constituting a class of industrial enzymes having approximately 25% of the world enzyme market (Danilienko 1993; Gorinstein, 1993; de Souza and de Oliveira e Magalhães 2010; Kirk et al. 2002; Rajagopalan and Krishnan, 2008; Reddy et al. 2003). They can be obtained from several sources, such as plants, animals, and microorganisms. Today a large number of microbial amylases are available commercially and they have almost completely replaced chemical hydrolysis of starch in starch processing industry. The amylases of microorganisms have a broad spectrum of industrial applications as they are more stable than when prepared with plant and animal alpha-amylases (Tanyildizi et al., 2005). Many amylases require the dispersed state of starch for enzymatic action, while others may already digest starch granules. Pancreas amylase effectively hydrolyzes granular and retrograded starch. Alpha-amylase from *Aspergillus oryzae* is less capable to act likewise. However, *A. oryzae* enzyme is useful to evaluate the extent of starch dispersion and the degree of retrogradation (Tsuge, 1992). The most commonly used alpha-amylase is obtained from *Bacillus subtilis*. Alpha-amylase for use in starch conversion needs to be free of beta amylase and glucoamylase. Commercial amylases are distinguished by their potency, which can be determined by simple laboratory tests (Marciniak and Kuba, 1982). New alpha-amylases with optimized properties, such as enhanced thermal stability, acid tolerance, and ability to function without the addition of calcium, have been developed (Bisgaard-Frantzen et al. 1999; Shaw et al., 1999; Declerck et al., 2000) offering obvious benefits to the industry.

Alpha-amylase hydrolyzes the starch molecule by breaking α-D-(1,4) glucosidic bonds at random, generating α-D-(1,4) (1,6) glucosyl oligosaccharides and, ultimately, maltose as depolymerization products. The enzyme cannot hydrolyze α-D-(1,6) bonds in amylopectin, thus leaving starch fragments (limiting dextrins) in the product. Beta amylase attacks the starch molecule at the nonreducing ends of the outer chains and proceeds by stepwise removal of maltose units. Since it also cannot break α-D-(1,6) bonds, a bimodal product of α-(1,4) (1,6) glucan fragments and reducing sugar is obtained. Glucoamylase hydrolyzes starch to α-D glucose, its ultimate building block. Alpha-amylase is most active at its pH optimum of 6.3–6.8. It is inactive at pH levels below 4 and above 9. As a protein, it is heat sensitive. Enzymatic starch conversion is terminated by raising the temperature until enzyme denaturation occurs, or by the addition of enzyme poisons such as the ions of copper, mercury, zinc, or of oxidizing agents. A quantity of 0.1–0.2% inactivating agent, based on starch, is usually required. Inactivation can also be achieved by moving the pH outside the enzyme's active limits.

Alpha-amylase enzyme for the enzymatic conversion of starch is produced by various manufacturers and sold to the paper mill under a variety of trade names. Each manufacturer produces enzyme at specific levels of potency or strength of enzymatic activity per unit weight or, in the case of liquid enzymes, per unit volume.

16.3 Starches Used for Surface Sizing

Starch is the most frequently used binder in surface sizing (Lipponen et al. 2004). Worldwide, 65% of the total starch volume in the paper industry is applied in surface sizing. Besides surface strength, starch addition reduces dimensional changes, impart stiffness, and improve other properties of the sheet. It is much cheaper than animal glue and other chemicals. Starch contains two major chemical entities, amylose (a straight-chain polymer of glucose) and amylopectin (a branched chain polymer of glucose). Certain waxy starches contain only amylopectin. All starches contain minor amount of fatty acids and other lipids, proteins, and inorganic salts. Starch is a natural polymer with high molecular weight that can be depolymerized with a great degree of control. It is a hydrophilic polymer that disperses in water and attaches to cellulose fibers and pigments through hydrogen bonding. Starch has hydroxyl groups that allow a wide range of substitution or oxidation reactions to adjust the rheological characteristics and to eliminate retrogradation. Cationic, anionic, or amphoteric groups can be added to induce specific charges. Starch may be grafted to produce new materials with properties that combine the advantages of natural and synthetic polymers. Important factors favoring starch are the relatively low cost of the raw material and the fact that it is derived from a renewable resource, and it is biodegradable.

In addition to these advantages, there are several weaknesses that need to be considered. The hydrophilic character of starch makes it sensitive to water and water vapor. Previously used insolubilizers, based on urea- or melamine formaldehyde, are being replaced, and there is a continuing research for effective and environmentally acceptable insolubilizers. Starch is a less efficient binder on a weight basis than polyvinyl alcohol or latex dispersions. Drying conditions for starch binder must be carefully controlled to avoid nonuniform porosity and print mottle. But, most important, starch can retrograde irreversibly, change its rheological characteristics and binding power, and be a ready nutrient for microbiological systems.

Before use, starch is heated in water to a temperature of about 88–98°C and kept at constant temperature for approximately 10–20 min. During heating, the granules take on water and swell, part of the granule dissolves, and the granule starts to disintegrate. The degree of granule disintegration and the viscosity of the starch solution depend on the type of starch, degree of modification, time and temperature of heating, and amount of agitation. Starch solutions are reasonably stable, but there is some retrogradation on aging, thus causing the solution to cloud up, increase in viscosity, and even set back or gel if the concentration is high enough. Retrogradation is increased by low pH, low temperature, presence of certain cations like calcium or aluminum, and slow cooling. The leaching of alum from the paper as it passes through the size is a factor in this connection. Solutions of native or unmodified starch have too high viscosity for ordinary surface sizing, and hence modified starches of reduced viscosity are generally used.

The type of starch used strongly affects the surface sizing. The selection of starch or modified starch is mainly governed by the dispersion viscosity, film formation, and resistant to retrogradation (Tehomaa et al. 1992). Different types of starch – such as unmodified starch, oxidized starch, acid depolymerized starch, substituted starch, cationic and anionic starch, grafted starch, hydrophobic starch – are being used for surface sizing.

For low-cost applications, native cornstarch is depolymerized by thermal/chemical or enzymatic conversion. In oxidative starch conversion at low pH, the product has to be neutralized immediately after discharge. Acid depolymerized cornstarch is also available with wide range of viscosity from thick boiling to thin boiling. This type of starch is suitable for use at both the size press and the water box for improving surface sizing.

For production of high value, surface sized papers, these wide variations in intrinsic quality of oxidized starch have adverse effect on the properties of paper. Moreover, starch pickup depends upon its viscosity and solid concentration and optimum starch pickup is essential for good quality surface sized papers. In case of starch having such a low viscosity, starch pickup is more which not only increases the starch consumption but also adversely affects the printability of paper and reduces the opacity of the paper.

Besides its higher price, there is huge variation in quality of oxidized starch. Its viscosity may vary from 45 to 500 cp (measured at 50°C and 5% solids). In spite of serious quality limitations, most of the papermakers still use the oxidized starch because of nonavailability of a better quality starch. About 15–20% loss in yield takes place during oxidative breakdown of native starch while preparing the chemically oxidized starch. This factor is primarily responsible for the higher price of oxidized starch. Enzymatic modification of starch is cost-effective for surface sizing of paper. Enzymatically modified starch is the major starch used in the developed countries for surface sizing (Bajpai, 2005). It provides better control on quality of starch by replacing the oxidized starch with enzymatically modified starch.

16.4 Process for Enzymatic Modification of Starch

Alpha-amylase enzyme used in the process is measured either by weight or by volume. Equal weights or volumes of different potency enzymes will introduce different amounts of enzymatic activity into the system Therefore, potency must be carefully considered in order to calculate the dose level of enzyme when changing from one source of enzyme to another. If potency is not taken into account, erratic viscosity will result when changing the source of enzyme, even if the same starch is used in the conversion process. The actual enzyme dosage level used is dependent upon individual mill requirements. This dosage may range from 0.01 to 0.1% enzyme, calculated on the dry basis starch content of the system. As a general rule, most paper mills are operating around a 0.05% enzyme-dose level. Another method for calculating enzyme dose is based on the conversion of potency to liquefons per pound of dry basis starch (Johnston and Jozsa 1935). A liquefon is defined as the quantity of enzyme that will dextrinize 0.35 mg of a standard starch substrate in 1 min under a specified experimental condition. If the liquefon value of an enzyme is known, the operator can calculate the weight or volume of enzyme needed in the conversion system to produce the desired starch paste viscosity. In general, the enzyme dose level on a liquefon basis will vary from 2,000 to 20,000 liquefans/kg of dry basis starch. Most paper mills are operating in the area of 9,000 liquefons/kg of dry basis starch. A comparison of the liquefon values of enzymes from two different sources will also allow conversion to potency when changing from one source to another.

Enzyme dose levels can be adversely affected by system variables or components. Equipment size is dependent on individual requirements. Ordinarily, the conversion tank is sized to supply sufficient adhesive for several batches. Stainless steel is the preferred material for tanks, pumps, agitators, and other equipments. Some metals tend to inhibit enzyme activity to some degree; heavy metals such as copper should be avoided. Steam is the usual heat carrier to the system. Heat transfer occurs from a jacket on the conversion tank or by direct injection. In the latter case, condensate will be formed in the paste, and the solids content of the starch dispersion is reduced accordingly. Agitation in the conversion tank must be sufficient to achieve uniformity in adhesive viscosity from conversion to conversion. Viscosity change during high-solids enzyme conversion ranges from less than 100 centipoise (cp) in the slurry to approximately 500,000 cp during its peak, and then to several hundred cp as the enzyme reduces the viscosity to the target level for the adhesive. The driving motors require high horsepower and are gear reduced to control agitation speed. It is better to be slightly "overpowered" rather than "under-powered"; this allows some latitude in changing conversion conditions to accommodate changes in coating formulations. All equipment should be kept in the best possible operating and repair condition to minimize improper enzyme conversion due to equipment malfunction. Time and temperature recorders and controllers are especially sensitive in this regard, and a partial or complete malfunction of them inevitably will result in improper final adhesive viscosity. A high level of sanitation of the equipment is also necessary, and good housekeeping practices should be maintained at all times. Contaminated or dirty equipment will result in nonuniform viscosity. The total starch solids applied to the production of surface size or coating adhesive by enzymatic conversion range from 25 to 38%; the majority of the mills are working with around 35% solids. Lower starch solids concentrations can be used for the preparation of surface size, but preferably a concentrated, converted paste is diluted to application concentration. A time–temperature process cycle is basic to successful enzyme conversion operations; however, this relationship is flexible in either time or temperature. Its sequence governs the enzymatic breakdown of starch molecules and produces the paste viscosity for use as coating adhesive or surface size (Maurer, 2001b). Once a specific sequence is established, it should be followed without deviation from conversion to conversion in order to achieve uniform and consistent results. The importance of careful and precise control of the cycle cannot be overemphasized.

In many cases, storage tanks for converted starch are simply insulated vessels with no temperature control. Such units are generally satisfactory when the starch is consumed within 6–8 h after conversion. However, in instances when it is necessary to hold converted starch for a period of 12–16 h, a jacketed storage tank with a temperature controller to maintain storage temperature at about 71°C should be used. In some instances, an elevated storage temperature of 88°C is used, which aids in inhibiting starch retrogradation (Maurer, 2001b).

Enzymatic reaction of gelatinized starch is dependent on enzyme dose and reaction time when the pH of starch slurry and temperature of reaction are within a specified range. Most of the enzymes are active in hydrolyzing starch in the pH range of 6–7 and at temperature range of 70–80°C for batch cooking. For the application of enzymatic starch for surface sizing of paper, cessation of hydrolysis at a certain stage is essential by deactivating the residual enzyme. At a specific set of pH and temperature, enzyme dose, reaction time, and deactivation of enzyme are the three major operating variables on which the enzymatic hydrolysis of starch depends. Starch amylase reaction starts when the starch gets gelatinized at 70–71°C. Optimum temperature of reaction is 75–78°C. In the lower range of temperature (70–75°C), the reaction is marginally slower and results in a little higher viscosity. Starch is generally cooked at 20% solids with enzyme at the specific dose. The slurry is diluted to desired level of solids say 10% and applied to paper at size press. Enzymatically modified starch is completely miscible with all the ingredients normally used in the size press formulation.

Two types of reactor systems used for enzymatic starch hydrolysis in paper mills are batch and continuous (Figs. 16.1 and 16.2) (Tolan, 2002). Native starch is dispersed in water in a batch starch cooker with agitator on. Enzyme at the specific dose per ton of starch is added and direct (and indirect) heating of the slurry is started with the low pressure steam maintaining constant stirring. At 70–71°C, starch gets gelled. After allowing a predetermined time (10–15 min) for the reaction to occur at the temperature of 70–80°C, more steam is released to raise the temperature to about 95°C to deactivate the enzyme. This thermal condition is maintained for about 5–7 min followed by the addition of residual dilution water to maintain specific solid concentration level. All the operations are carried out under constant and thorough mixing. The slurry is then cooled to 60°C and the volume is made up

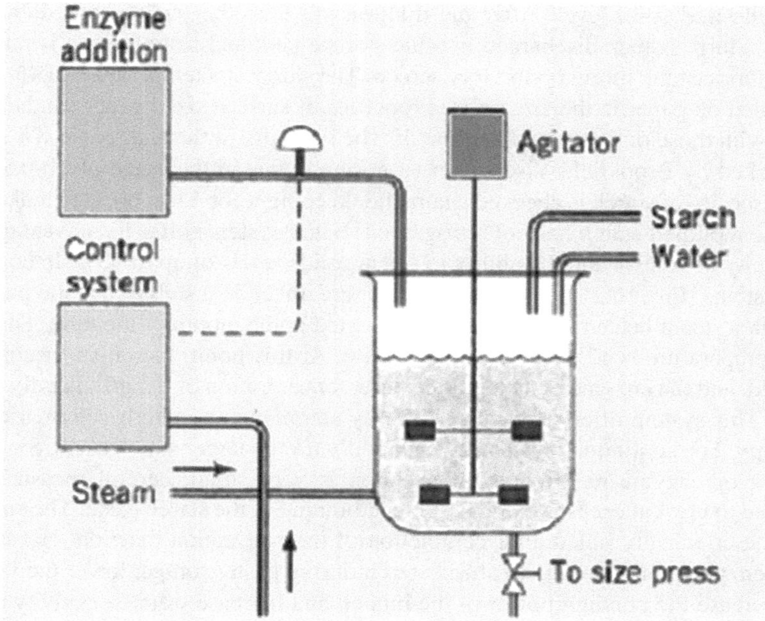

Fig. 16.1 Batch starch conversion system (Based on Tolan (2002))

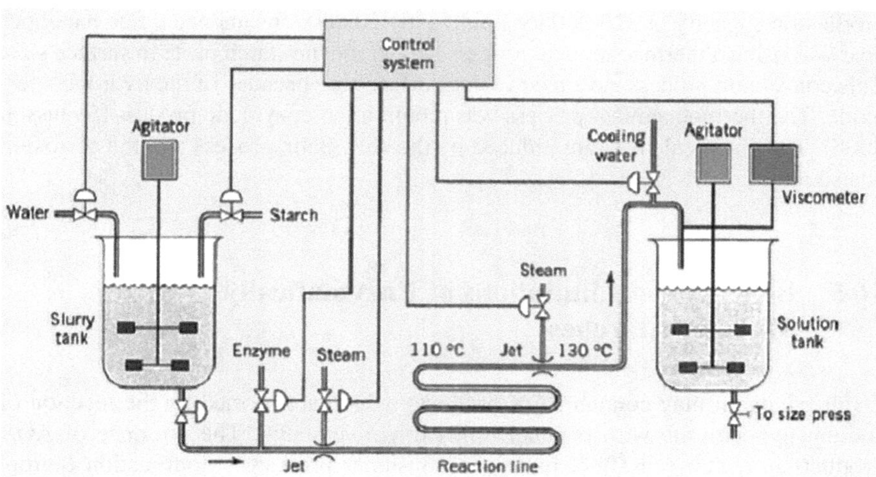

Fig. 16.2 Continuous starch conversion system (Based on Tolan (2C02))

to the desired solid level. After the dilution, all the other sizing ingredients are mixed. Slurry is then discharged into the storage tank and from there it is pumped to the service tank through vibratory screen. The slurry at a temperature of 45–50°C is applied on paper in the size press. Properties of surface sized paper can be compared with those of using oxidized starch. The viscosity of the slurry at 7–8% solids is checked by Brookfield Viscometer (it normally lies in the range of 10–15 (cp). The viscosity of starch is checked again after keeping it for 4 h at 60°C to make sure that the modified starch has not retrograded. Batch systems offer the advantages of simplicity of control and the ability to run at solids levels of up to 40%. In continuous systems (Fig. 16.2), the starch and water are mixed in a slurry tank and pumped through a steam heated tube. Enzyme is injected at the mouth of the tube. The running temperature is 125°C for 2 min at pH 6. At this point, the target viscosity is reached, and the enzyme, which is added to a concentration of 0.5 g/L, is fully denatured. This system offers advantages of easy automation and high volumetric productivity. The continuous systems are generally for the larger starch users. Since the amount of enzyme required for conversion is very small, careful measuring is required to prevent excessive or insufficient thinning of the starch paste. The enzyme must be effectively killed after completion of the conversion reaction. Not inactivated enzyme will continue to attack starch during paste storage, lower the viscosity, decrease the bonding power of the binder, and increase water sensitivity of the coating. For continuous enzymatic conversion, an enzyme with more temperature stability, than sufficient for batch conversion, is required. The operators must be well trained in controlling the time and temperature cycles and able to respond to upsets due to variations in starch slurry concentration and enzyme activity.

Stora Enso Nymölla Mill, an integrated pulp and paper mill in Sweden with a production capacity of 325,000 tons pulp and 450,000 tons uncoated fine paper per year was using a thermochemical process to degrade the starch paste to surface size. This conversion process gave a corrosive surface size because of the hydrogen peroxide. The thermochemical process was rebuilt to an enzymatic process (Svensson 2006). The chemical costs are reduced by the enzymatic process and the corrosion is also reduced.

16.5 Benefits and Limitations of Enzymatically Modified Starches

Oxidized starch may contain AOX products, which are formed by the reaction of sodium hypochlorite with residual lipids in native starch. The presence of AOX products in starch can affect its use in consumer products. Modification (partial hydrolysis) of starch with enzyme does not involve any chemicals; it is totally free from AOX products. Oxidation of native starch with sodium hypochlorite though takes place at relatively lower temperature, it requires longer reaction time. As the reaction is not so selective, it results in appreciable loss of starch (30–40%) in the form of water soluble material, which goes into the wastewater requiring elaborate treatment. It results in increased cost of oxidized starch. On the other hand, enzymatic reaction is highly selective and hydrolysis can be controlled precisely to avoid generation of any soluble material but reduce the viscosity to a desired value. Oxidized starch is produced by chemical modification at the site of starch manufacturers. Therefore, papermaker has no control over its quality in terms of viscosity, etc. The cooking of oxidized starch is done at the papermaker's site for its dispersion and gelation only. The enzymatic modification is done by the papermakers at their site where the final viscosity is controlled by the paper mill. The cost of surface sizing using enzymatically modified starch is much lower as compared to that using oxidized starch. As the native starch contains some residual protein, the brightness of enzymatically modified starch is slightly lower than the oxidized starch. It can be compensated with the use of optical brightening agents. Sometimes, there is a coloration problem due to the presence of metals in the native starch. Therefore, the native starch used for enzymatic modification should have negligible protein and ash contents. Process conditions in terms of pH of the starch slurry, temperature–time profile, enzyme dose, and reaction times are very sensitive to control the quality of enzymatically modified starch. The paper mills can realize substantial saving by switching over from oxidized starch to the enzymatically modified starch as there is an appreciable difference in the cost of oxidized starch and enzymatically modified starch.

References

Bajpai P (2005) Surface sizing (Chap. 8). In: Emerging technologies in sizing. PIRA International, Leatherhead, pp 135

Bisgaard-Frantzen H, Svendsen A, Norman B, Pedersen S, Kjærulff S, Outtrup H, Borchert TV (1999) Development of industrially important α-amylases. J Appl Glycosci 46:199–206

Danilienko AN (1993) Effect of the polymerization degree, moisture content, and temperature on kinetics of hydrolysis of corn starch by alpha-amylase. Starch/Starke 46(2):63

Declerck N, Machius M, Wiegand G, Huber R, Gaillardin C (2000) Probing structural determinants specifying high thermostability in *Bacillus licheniformis alpha*-amylase. J Mol Biol 301:1041–1057

de Souza PM, de Oliveira e Magalhães P (2010) Application of microbial alpha amylase in industry – a review. Braz J Microbiol 41:850–861

Gorinstein S (1993) Kinetic studies during enzyme hydrolysis of potato and cassava starche. Starch/Stiirke 46(3):91

Svensson G (2006). Alternative enzymatic conversion of surface sizing starch at Nymölla Mill. Department of Chemical Engineering, Lund Institute of Technology, Lund. www.chemeng.lth.se/exjobb/E256.pdf. Accessed 9 February 2006

Johnston WR, Jozsa SJ (1935) J Am Chem Soc 57(4):701

Kirk O, Borchert TV, Fuglsang CC (2002) Industrial enzyme applications. Curr Opin Biotechnol 13(4):345–51

Lipponen J, Gron J, Bruun S, Laine T (2004) Surface sizing with starch solutions at high solids contents for up to 18%. J Pulp Paper Sci 30(3):82

Marciniak GP, Kuba MR (1982) A comparative study of the methods for the determination of the activity of bacterial a-amylase. Starch/Starke 34(12):42

Maurer HW (2001a) Surface sizing of paper. In: Maurer HW (ed) Starch and starch products in surface sizing and paper coating. Tappi, Atlanta, GA, p 83
Maurer HW (2001b) Enzyme conversion of starch for paper sizing and coating. In: Maurer HW (ed) Starch and starch products in surface sizing and paper coating. Tappi, Atlanta, GA, p 65
Rajagopalan G, Krishnan C (2008) Alpha-amylase production from catabolite derepressed *Bacillus subtilis* KCC103 utilizing sugarcane bagasse hydrolysate. Bioresour Technol 99:3044–3050
Reddy NS, Nimmagadda A, Sambasiva Rao KRS (2003) An overview of the microbial – alpha amylase family. Afr J Biotechnol 2:645–648
Shaw A, Bott R, Day AG (1999) Protein engineering of alpha amylases for low pH performance. Curr Opin Biotechnol 1999(10):349–352
Smook GA (1982). Surface treatments. In: Handbook for pulp and paper technologists. Joint Textbook Committee of the Paper Industry, p 283
Tanyildizi MS, Ozer D, Elibol M (2005) Optimization of alpha amylase production by *Bacillus* sp. using response surface methodology. Process Biochem 40:2291–2296
Tehomaa J, Palokangas E, Makimattila J, Tuomisto M (1992) A comparison of techniques for surface sizing of fine papers. TAPPI J 75(8):79–84
Tolan J (2002) Enzymes, pulp and paper processing. In: Encyclopedia of bioprocess technology. Wiley, New York
Tsuge H (1992) Screening of a-amylase suitable for evaluating the degree of starch retrogradation. Starch/Starke 44(1):29
Tompkins TW (1992) Surface bonding/sizing agents: keys to improve performance runnability under alkaline conditions. In: Papermakers Conference Proceedings
Tsai YG, Colasurdo AR, Cordoba C (1995) Alternative chemistry for paper surface size with high performance and a large window of operation. In: Papermakers Conference Proceedings

Chapter 17
Biofiltration of Odorous Gases*

17.1 Introduction

Malodorous gas streams from pulping processes are frequent targets of public complaints (Chan 2006; Burgess et al. 2001). Owing to public health concerns and the personal comfort of neighboring residential communities, the industry is put under increasingly stringent regulations. Odors can also seriously lower real estate property values and there are indications that odor causing stress-induced illnesses can result in lower working productivity and lost workdays. People living in or near a kraft pulp mill complain of the bad smell associated with the mill's operations. These complaints are directly related to the production of odorous compounds during the cooking of wood chips with white liquor and subsequent points of gaseous release to the atmosphere. Even when pure sodium hydroxide is used to treat wood and straw, odors are produced. The cause of these odors is to be found in the residual sulfur-containing protoplasm, which reacts with the alkali to form mercaptans and organic sulfides during the digestion phase. It was found that the mercaptans are formed by the saponification of lignin methoxyl groups by sulfide ions.

---

*Excerpted from Bajpai P, Bajpai Pramod K, Kondo R (1999) Biotechnology for environmental protection in the pulp and paper, Chap. 11, Biofiltration of exhaust gases, with kind permission from Springer Science+Business Media.

P. Bajpai, *Biotechnology for Pulp and Paper Processing*,
DOI 10.1007/978-1-4614-1409-4_17,

17.2 Emissions from Pulping

17.2.1 Kraft Pulping

The foul smelling gases released from the kraft process include the following:

- H_2S
- Methyl mercaptan (CH_3SH)
- Organic sulfides (such as dimethyl sulfide (CH_3–S–CH_3) and dimethyl disulfide (CH_3–S–S–CH_3)), collectively referred to as total reduced sulfur (TRS). They are formed during kraft pulping by reaction of sulfides with methoxy groups of lignin via nucleophilic substitution reactions.

The major source of TRS emissions include:

- Digester blow and relief gases
- Multiple effect evaporator vent and condensates
- Recovery furnace with direct-contact evaporators
- Smelt dissolving tank and slacker vents
- Brown-stock washers
- Seal tank vents
- Lime kiln exit vents

Table 17.1 shows the typical characteristics of the gaseous emissions from kraft pulp mill. It is apparent that the source of largest volume of potential emissions is the recovery furnace, followed closely by the digester blow gases and the washer hood vents. However, the most concentrated emissions come from the digester blow and relief gases. Overall, the three most important source of odor production are

Table 17.1 Typical off-gas characteristics of kraft pulp mill

Emission source	Offgas flow rate (m^3/ton pulp)	Concentration (ppm by volume) H_2S	CH_3SH	CH_3SCH_3	CH_3SSCH_3
Batch digester					
Blow gases	3–6,000	0–1,000	0–10,000	100–45,000	10–10,000
Relief gases	0.3–100	0–2,000	10–5,000	100–60,000	100–60,000
Continuous digester	0.6–6	10–300	500–10,000	1,500–7,500	500–3,000
Washer hood vent	1,500–6,000	0–5	0–5	0–15	0–3
Washer seal tank	300–1,000	0–2	10–50	10–700	1–150
Evaporator hotwell	0.3–12	600–9,000	300–3,000	500–5,000	500–6,000
BLO tower exhaust	500–1,500	0–10	0–25	10–500	2–95
Recovery furnace	6,000–12,000	(After direct-contact evaporator)			
		0–1,500	0–200	0–100	2–95
Smelt dissolving tank	500–1,000	0–75	0–2	0–4	0–3
Lime kiln exhaust	1,000–1,600	0–250	0–100	0–50	0–20
Lime slacker vent	12–30	0–20	0–1	0–1	0–1

Based on data from Andersson et al. (1973) and Environmental Pollution Control Pulp and Paper Industry (1976)

Table 17.2 Odor threshold concentration of TRS pollutants

Reduced sulfur compound	Odor threshold concentration (ppb)
Hydrogen sulfide (H_2S)	8–20
Methyl mercaptan (CH_3–SH)	2.4
Dimethyl sulfide (CH_3–S–CH_3)	1.2
Dimethyl disulfide (CH_3–S–S–CH_3)	15.5

Based on data from Springer and Courtney (1993)

Table 17.3 Typical emissions of So*x* and NO*x* from kraft pulp mill combustion sources

	Concentration (ppm by volume)			Emission rate (kg/ton[a])		
Emission source	SO_2	SO_3	NO_x (as NO_2)	SO_2	SO_3	NO_x (as NO_2)
Recovery furnace						
No auxiliary fuel	0–1,200	0–100	10–70	0–40	0–4	0.7–5
Auxiliary fuel added	0–1,500	0–150	50–400	0–50	0–6	1.2–10
Lime kiln exhaust	0–200		100–260	0–1.4		10–25
Smelt-dissolving tank	0–100	–	–	0–0.2		
Power boiler	–	–	161–232	–	–	5–10[b]

Based on data from Environmental Pollution Control Pulp and Paper Industry (1976) and Someshwar (1989)

[a]kg/ton of air dried pulp

[b]kg/ton of oil

black liquor combustion, weak black liquor concentration and the digestion process. About 0.1–0.4 kg of TRS is emitted per ton of pulp at 5 ppm in the recovery boiler flue gases. The principal difficulty with TRS emission is their nauseous odor, which are detected by the human nose at very low concentrations. Table 17.2 presents the odor threshold (odor detectable by 50% of the subjects) concentrations of the principal TRS compounds emitted by kraft mills which are only few parts per billion by volume (Springer and Courtney 1993). At low concentrations, TRS is more of a nuisance than a serious health hazard. Thus, odor control is one of the main air pollution problems in a kraft mill.

Oxides of both sulfur and nitrogen are also emitted in varying quantities from few points in the kraft system. The main source of SO_2 emission is the recovery furnace due to the presence of sulfur in the spent liquor used as a fuel. SO_3 is sometimes emitted when fuel oil is used as an auxiliary fuel. The lime kiln and smelt dissolving tank also emit some SO_2. The emission of nitrogen oxides is more general because nitric oxide is formed whenever oxygen and nitrogen, which are both present in air, are exposed to high temperatures. A small part of the nitric oxide formed may further oxidize to nitrogen dioxide. These two compounds, nitric oxide and nitrogen dioxide, are termed as total oxide of nitrogen. Under normal operating conditions, the temperature in the recovery furnace is not high enough to form large quantities of oxides of nitrogen (NO_x). The main source of NO_x emissions is the lime kiln. Table 17.3 presents SO_x and NO_x emission rates from various kraft mill sources. Large variations in the emission rates are due to the variations in operating conditions at different mills. Large amounts of NO_x are produced if the flame temperature is above 1,300°C and oxygen concentration greater than 2%. Modern recovery boilers should have So_x emissions below 100 ppm when properly operated. Sulfur emissions from power boilers are controlled by using fuels of low sulfur content.

Another type of odorous emissions of nonsulfur compounds is produced by the hydrocarbons associated with the extractive components of wood, such as terpenes and fatty and resin acids, as well as those from materials used in processing and converting operations, such as defoamers, pitch control agents, bleach plant chemicals, etc. These hydrocarbon emissions are small compared to TRS emissions, but they may be odorous, or act as liquid aerosol carriers contaminated with TRS, or undergo photochemical reactions.

17.2.2 Emissions from Neutral Sulfite Semichemical (NSSC) Pulping

In general, the emissions from neutral sulfite semichemical (NSSC) are much less than those from the kraft process. Because no Na_2S is present in the pulping liquor, both methyl mercaptan and dimethyl sulfide (DMS) are absent from the gaseous emissions, a very low amount of reduced sulfur is emitted (Dallons 1979). The sulfur emissions from the Na_2CO_3 (sulfur free) process has been traced to sulfur in the fuel oil and process water streams used. The emissions of SO_2 and NO_x are similar to those of a kraft mill.

17.2.3 Emissions from Sulfite Pulping

The sulfite process mainly operates with acidic SO_2 solutions and as a consequence SO_2 is the principal emission. Organic reduced sulfur (RS) compounds are not produced if proper conditions are maintained in the process. Because the odor threshold is about 1,000 times higher for SO_2 than for RS compounds, sulfite mills generally do not experience the odor problem of a kraft mill. The method of attack on lignin by sulfite liquor is quite different than that by kraft liquor. The sulfite process involves sulfonation, acid hydrolysis, and acid condensation reactions (Rydholm 1965). Volatile compounds such as methyl mercaptan and DMS are not produced in sulfite pulping.

Typical emissions in the sulfite process are SO_2 with special oxides of nitrogen (problems arising in the ammonium-base process). SO_2 is also emitted during sulfite liquor preparation and recovery. Very little SO_2 emission occurs with continuous digesters. However, batch digesters have the potential for releasing large quantities of SO_2, depending on how the digester is emptied. Digester and blow-pit emissions in the sulfite process vary depending on the type of system in operation. These areas have the potential for being a major source of SO_2 emission. Pulp washer and multiple-effect evaporator also emit SO_2.

17.3 Methods for the Elimination of Odorous Compounds

Following methods are available to remove odors components from gaseous emissions (Ottengraf 1986):

- Gas-phase methods
- Liquid-phase methods
- Solid-phase methods
- Combustion
- Biological methods

The most important one is biological methods. These methods generally have the specific advantage that the pollutants are converted to harmless or much less harmful oxidation products (e.g., CO_2, H_2O, etc.). These processes do not generally give rise to new environmental problems, or if they do these problems are minimal. An exhaust air problem should preferably not become a solid waste or waste-water problem. Another advantage of biological treatment is the possibility of carrying out the process at normal temperature and pressure. Moreover, the process is reliable and relatively cheap, while the process equipment is simple and generally easy to operate. The elimination of volatile compounds present in waste gases by microbial activity is due to the fact that these compounds can serve as an energy source and/or a carbon source for microbial metabolism. Hence, a broad range of compounds of organic as well as of inorganic origin can be eliminated by microbiological processes.

As microorganisms need a relatively high water activity, these reactions generally take place in the aqueous phase and as a consequence the compounds to be degraded as well as the oxygen required for their oxidation first have to be transferred from the gas phase to the liquid phase. Therefore, mass transfer processes play an important role in this methodology. The microbial population can either be freely dispersed in the water phase or is immobilized on a packing or carrier material. The first-mentioned operation is carried out in bioscrubbers, the second one in trickling filters and biofilters. Bioscrubbers and tricking filters are more energy intensive than biofilters, as water circulation in these two systems requires relatively much more energy than gas transport through a biofilter. Also, the reliability of operation of bioscrubbers is relatively low due to possible washing away of active microorganisms. On the contrary, the presence of a large amount of packing material with a buffering capacity diminishes the sensitivity of biofilters to different kinds of fluctuations. Therefore, biofiltration technology is receiving a significant attention (Singhal et al. 1996).

17.3.1 Biofiltration Technology

Biofiltration technology is a promising method of odor, volatile organic compounds (VOCs) and air toxic removal from waste-gas streams because of low capital and operating costs, low energy requirements, and an absence of residual products requiring further treatment or disposal (Bajpai et al. 1999; Cáceres et al. 2010; McNevin and Barford 2000; Burgess et al. 2001; Wani et al. 1997; Govind and Bishop 1996; Swanson and Loehr 1997). Biofiltration utilizes microorganisms that are capable of oxidizing many compounds and thus having potential for being used for the abatement of odors, VOCs, and air toxics (Kennes et al. 2007; Ottengraf 1987). The concept of biofiltration is actually not new; it is an adaptation of the process by which the atmosphere is cleaned naturally (Bohn 1992).

Biofiltration is similar to the biological treatment of wastewater or in situ bioremediation of contaminated soils and hazardous sludge (Rozich 1995). It is becoming more popular as stringent emission regulations are implemented. The acceptance of biofiltration has followed from biotechnological advances that provide an increasingly thorough knowledge of the system and how the process can be optimized not only to achieve high removal efficiencies with low energy consumption but importantly, to achieve these elimination efficiencies over long periods of time with minimal operator intervention and/or need for maintenance (Marsh 1994). VOC emissions have become a substantive issue for industrial operators as a result of the implementation of the US 1990 Clean Air Act Amendments and similar regulations in Europe, and thus a major driving force for the exploration of cost effective control options. Biofiltration is a promising control technology for processes that emit large off-gas volumes with relatively low concentrations of contaminants. With respect to the purification of polluted air, biofiltration is a commonly applied technique to odor abatement, where it is an established control method. It has also demonstrated limited success in controlling VOCs.

Biofiltration uses naturally occurring microorganisms immobilized in the form of a biofilm on a porous substrate such as soil, compost, peat, bark, synthetic substances, or their combination. The substrate provides the microorganisms with both a hospitable environment in terms of oxygen, temperature, moisture, nutrients, pH, and a carbon source of energy for their growth and development. As the contaminated air stream passes through the filter bed, contaminants are transferred from the vapor phase to a thin water layer (biofilm) covering the microorganisms held over the surface of the packing particles. The microorganisms utilize these favorable conditions to metabolize carbon-based compounds to their primary components – carbon dioxide and water, plus additional biomass and innocuous metabolic products (Ottengraf 1987; Rozich 1995; Marsh 1994). The absorption and/or adsorption capacity of the filter media is thus continuously renewed by the biological oxidation of the sorbed contaminants (Bohn and Bohn 1988; Hodge et al. 1991).

Biofiltration has the advantage that the pollutants are not transferred to another phase and therefore, new environmental problems are not created or are only minimal (Ottengraf 1986, 1987; Bohn 1992, 1993). Moreover, the process is said to be cheap and reliable and does not usually require complex process facilities (Ottengraf 1987).

Biofilters do extremely well in two main domains; in the removal of odoriferous compounds and in the elimination of volatile organic chemicals (Ottengraf 1986; Hirai et al. 1990; Deshusses and Hammer 1993; Leson and Wikener 1991), primarily solvents, from air. Under optimum conditions, the pollutants are fully biodegraded without the formation of aqueous effluents. As gases pass through a biofilter, odorous compounds are removed by processes thought to include sorption (absorption/adsorption) and biooxidation (Williams and Miller 1992). The odorous gases adsorb onto the surface of the biofilter medium and/or are absorbed into the moisture film on the biofilter particles. Given a sufficient rate of biological activity in the filter, the sorbed compounds are then oxidized (degraded) by microorganisms. End products from the complete biooxidation of the air contaminants are CO_2, water, mineral salts, and microbial biomass. The elimination of a gaseous pollutant in a biofilter is the result of a complex combination of different physicochemical and biological phenomena.

Biofilters are commonly constructed in a vessel packed with loose beds of solid material, soil, or compressed cakes with microbes attached to their surface. Waste gases are passed through these units via induced or forced draft. Biofilters are capable of handling rapid air flow rates and VOC concentrations in excess of 1,000 ppm. These units are gaining importance in bioremediation also and are timely in that they are a cost-effective means by which to deal with the more stringent regulations on VOC emission levels.

There are essentially two types of biofilters. The first and simplest is the soil filter. Contaminated air from a small waste stream or other treatment process is passed through a soil–compost type design, so-called open system (Ottengraf 1986). Sometimes, nutrients are preblended into the compost pile to provide conditions for microbial growth and biodegradation of the waste by indigenous microorganisms. Being usually installed in the open air and partly underground, these systems are exposed to many weather conditions: rain, frost, temperature fluctuations, etc. These filters are usually over designed; they require a very large area. To increase the reliability of these filters, a number of another type (closed type) of systems have been developed which house the treatment beds or disks of different packing materials/media. In the treatment bed, the waste air stream and the filter are humidified as the waste is passed through one, two, or more beds. In this approach, a series of humidified disks or beds are placed inside a reactor shell (Shareefdeen et al. 1993). These layered disks contain packing material/media, nutrients, microbial cultures, and/or compost material. The waste air stream organics undergo biodegradation as they pass through the system. Any collected water condensate from the process is returned to the humidification system for reuse. Biofilters have reportedly been built to handle up to 3,000 m^3/min of air flow using filters up to 6,500 m in wetted area (Anon 1991). The filters can be customized with specific carriers, nutrients blends, or microbial cultures. Some biofilters can endure up to 5 years before replacement is necessary (Holusha 1991). Spent filters can be utilized as fertilizer since they present no hazard.

17.3.2 Microorganisms in Biofilter

Several microorganisms are involved in the degradation of air pollutants in biofilters including bacteria, actinomycetes, and fungi (Ottengraf 1987). The microbial population is generally made up of autotrophic microorganisms (feed directly

Table 17.4 Microbial cultures used for degradation of pollutants

Culture	Pollutant(s)
Soil (indigenous microbes)	Sewage odors-H_2S
Compost (indigenous microbes)	Various VOCs
Aerobically digested sludge of night soil	Sulfur compounds (H_2S, DMS, methanethiol)
Peat (indigenous microbes)	H_2S
Sludge from sewage treatment	H_2S, C_2H_5SH, $(C_2H_5)_2NH$, C_4H_9CHO
Thiobacillus sp. strain MS_1	Methyl sulfides (DMS, DMDS)
Thiobacillus thioparus TK-m	H_2S, DMS, DMDS, methanethiol
Thiobacillus thioparus strain E_6	DMDS
Hyphomicrobium sp. strain S	DMS, DMSO
Hyphomicrobium sp. strain EG	Methylated sulfur compounds
Pseudomonas fluorescens	Methanol, isopropanol, butanol, etc.
Bacterial consortium consisting of *Pseudomonas*, *Methylomonas*, *Aeromonas*, *Achromobacter*, *Flavobacterium*, *Alcaligenes*	Methanol

Based on data from Leson and Wikener (1991), van Lith et al. (1997), Bohn (1975), Pomeroy (1982), Hirai et al. (1990), Lee and Shoda (1989), Furusawa et al. (1984), Ottengraph et al. (1983), Sivela and Sundman (1975), Kanagawa and Kelly (1986), Kanagawa and Mikami (1989), Smith and Kelly (1988a, b), DeBont et al. (1981), Suylen et al. (1986, 1987), Kirchner et al. (1987), and Shareefdeen et al. (1993)

from inorganic compounds) and heterotrophic microorganisms (utilize organic compounds as source of energy and carbon) (Marsh 1994). The composition and survival of microorganisms on the filter bed are the main process parameters. Their growth and activity depends on the physical and chemical conditions in the packing material. The diversity of the active microorganisms is a function of the inlet gas stream composition. Some packing materials of natural origin, such as compost, contain a sufficient number of different microorganisms to initiate the reactions for the elimination of simple contaminants. The efficiency of the purification process is generally increased following the growth of active strains during the adaptation time after the start up of the biofilter. For easily biodegradable organic compounds acclimatization can typically take about 10 days (Ottengraf 1986), and for less biodegradable and those contaminants for which the microorganisms are less likely to be initially present in the biofilter material, the period can be longer (Leson and Wikener 1991).

Different types of representative microorganisms/cultures used by various investigators have been given in Table 17.4.

Soil and compost contain a large variety of indigenous microorganisms which degrade the odorous compounds in air. The common soil bacteria, *Bacillus cereus* var. *mycoides* and strains of *Streptomyces* are most frequently identified in the soil samples. Autotrophic bacteria such as *Thiobacillus*, which grow on thiosulfate medium, are also present in the soil. But the counts of heterotrophic bacteria are much higher. They have been demonstrated to reduce the sewage odors, especially by eliminating the hydrogen sulfide present in the waste air stream (Carlson and Leiser 1966; van Lith et al. 1997; Bohn and Bohn 1988). Following bacteria and microfungi: *Actinomyces globisporus*, *Penicillium* sp., *Cephalosporium* sp., *Mucor* sp., *Micromonospora albus*, *Micrococcus albus*, *Ovularia* sp., etc. are the most frequently occurring microorganisms in the compost cultures. Compost has been a common choice of microbial source in biofiltration (Leson and Wikener 1991; van Lith et al. 1997; Bohn 1975; Pomeroy 1982). In addition to the source of microorganisms, the soil and compost provide a physical support for the microorganisms; these materials also provide water holding capacity and some amount of minor and trace nutrients. Aerobically digested sludge of night soil has also been used as a source of microbial cultures in the biofilters for the removal of H_2S, DMS and methanethiol (Hirai et al. 1990; Lee and Shoda 1989). The digested sludge of night soil is supposed to contain several types of microorganisms, useful in biooxidation of air pollutants. The indigenous microorganisms in the peat have been tried for biooxidation of H_2S in a biofilter (Furusawa et al. 1984). In few cases, sludge from sewage treatment works is used as the source of microorganisms (Ottengraph and Van Denoever 1983). Classical microbiological techniques have revealed the presence of mixed populations of bacteria, yeast, fungi, and higher organisms in the biofilters. Bacterial species of *Thiobacillus* and *Hyphomicrobium* degrade many sulfur compounds such as H_2S, methyl sulfide, DMS, DMDS, DMSO, methanethiol, etc. (Sivela and Sundman 1975; Kanagawa and Kelly 1986; Kanagawa and Mikami 1989; Smith and Kelly 1988a, b; DeBont et al. 1981; Suylen et al. 1986, 1987). For methanol biooxidation, *Pseudomonas fluorescens* (Kirchner et al. 1987) and a bacterial consortium (Shareefdeen et al. 1993) consisting of *Methylomonas*, *Aeromonas*, *Achromobacter*, *Flavobacterium*, *Alcaligenes*, and *Pseudomonas* have been used.

17.3.3 Packing Materials for Biofilters

The conventional packing material for biofilter have been soil, compost, peat moss, bark, or other material that contain a large variety of indigenous microorganisms (Carlson and Leiser 1966; van Lith et al. 1997 ; Bohn and Bohn 1988; Bohn 1975; Furusawa et al. 1984; Ottengraph and Van Denoever 1983; Sivela and Sundman 1975; Van Langenhove et al. 1986; Luo and van Oostrom 1997; Qiao et al. 2008). These materials provide water holding capacity and some amount of minor and trace nutrients in addition to providing a physical support for the microorganisms. Soil, peat, and compost materials exhibit low biodegradation rates, have limited supply of nitrogen and phosphorus, eventually begin to plug due to growth of microorganisms, and have limited capacity to neutralize acidic products of degradation. Hence, compost biofilters are capable of treating low concentration contaminants and are not ideally suited for treating air contaminated with high concentration organics. Sometimes, the bed material is amended with bulking agents such as wood chips, saw dust, bark, sand, bagasse, etc. to improve air flow or with other additives such as limestone for pH control in systems removing sulfur based odors (Ottengraf 1986; Deshusses and Hammer 1993; Sivela and Sundman 1975; Luo and van Oostrom 1997; Chou and Chen 1997; Campbell and Connor 1997; Deshusses et al. 1995; Ottengraf et al. 1986). Peat has the advantages over soil or compost of broadness of the maximum permeable value of the moisture content and a lower pressure drop due to its fibrous structure (Hirai et al. 1990; Furusawa et al. 1984). The peat has been reported to possess a unique combination of chemical and physical properties, such as adsorbency, which could be employed in environmental protection applications (Martin 1992).

Other types of support media used in biofilters are synthetic media, such as ceramic, plastic, etc., with active bacteria immobilized on the surface in the form of biofilms. These synthetic media biofilters are known as biotrickling filters. Synthetic support media are used in trickling filters for wastewater treatment, gas absorption towers, catalytic reactors, etc. However, the design of support media in biotrickling filters is different than in any other application, the major difference being the growth of biomass. In trickling filters, used for waste water treatment, the water flows as a liquid film on the biofilm surface, and sufficient distance between the support media is designed to accommodate biomass growth and air, which provides oxygen for the biodegradation reaction. The contaminants, present in the waste water, diffuse into the biofilm as the water flows over the biofilms and biodegrades. In a biotrickling filter, the contaminants, present in air, diffuse perpendicular to the direction of flow, and biodegrade in the supported biofilms. Since the process is diffusion controlled, designing a large distance between the supported biofilms reduces the overall degradation rate in the filter. Further, unlike the submerged biofilms in the case of the wastewater trickling filter, the biofilms in a biotrickling filter have to be kept moist to maintain bioactivity. Air flowing through the biotrickling filter draws moisture away from the biofilms, and a trickling flow of aqueous nutrients has to be maintained to provide nutrients and water to the active bacteria in the biofilms.

Synthetic support media can be in the form of high surface area pellets, with either a porous or nonporous surface. In some cases, the support media may be coated with activated carbon, to enhance adsorption of contaminant(s). The synthetic support media can be synthesized from plastic, ceramic, metallic, or any other composite material. The desired features of a good support media are as follows: High void fraction, High surface area per unit volume of the biofilter bed, Low gas-phase pressure drop, Hydrophilic surface, to allow good water wettability, and Low cost.

17.3.4 Mechanisms in Biofilter Operation

There are many mechanisms which operate simultaneously or in sequence in a biotrickling filter. These mechanisms include:

- Diffusion of the contaminant(s) from the bulk gas flow to the active biofilm surface
- Sorption of the contaminants directly on the biofilm surface
- Solubilization of the contaminant(s) into the water content of the biofilms
- Direct adsorption of the contaminant(s) on the surface of the support media
- Diffusion and biodegradation of the contaminant(s) in the active biofilm
- Surface diffusion of the contaminant(s) in the support media surface
- Back diffusion of the adsorbed contaminant(s) from the support media surface into the active biofilms. The effect of adsorption of contaminant(s) on support media surface, surface diffusion, and back diffusion of the adsorbed contaminant(s) from the support media surface into the active biofilms, predominantly occurs in activated carbon-coated support media and contaminant(s), which have affinity for the support media surface

In the case of compost biofilters, the contaminant(s) diffuse into the porous compost particles, dissolve into the sorbed water films, adsorb on the organic and inorganic fraction of the compost, and biodegrade by the attached active compost bacteria, entrapped within the compost particles.

17.3.5 Development of Biofiltration Technology

Biofiltration has been used to control odors for several years in many countries (Germany, The Netherlands, UK, Japan, and to a limited extent in the USA) but the use of biofilter to degrade more complex air emissions from chemical plants has occurred within the last 2 decades. This vapor-phase biological treatment is rapidly gaining acceptance as an abatement technology for use in the treatment of VOCs, including odorous chemicals and air toxics because of its technical and economic advantages (Tonga and Skladany 1994). The process was initially applied to odor abatement in composting works, waste water treatment plants and similar situations. It is known that in 1953 a soil biofilter system was used for the treatment of odorous air in Long Beach, California (Pomeroy 1982). In Europe the first attempt with a soil bed was made in Geneva for deodorization at a composting facility (Ottengraf 1986). Around 1959 a soil bed system was used at municipal sewage treatment in Nuremberg, Germany (Leson and Wikener 1991; Shimko et al. 1988). In early 1960s Carlson and Leiser (1966) started systematic research on biofiltration in the USA and used biofilters to treat hydrogen sulfide emissions from sewage. After that, biological gas cleaning has made considerable progress, but is still in its developing stages for application to the control of VOCs and air toxics in industrial use.

During the last 3 decades research activities, especially on the soil bed systems, have intensified in USA with the installation of some full scale operations (Bohn 1975; Prokop and Bohn 1985). Excellent reviews of the historical development of biofiltration have been presented by Ottengraf (1986), Leson and Wikener (1991), and Shimko et al. (1988). Having proven its success in deodorization, current research and application of biofiltration has been focused on the removal of VOCs and air toxics from the chemical and other process industrial exhausts. Current research activities are aiming at understanding the practical behavior of the biofiltration process, optimizing its operational parameters and modeling the system on the basis of reaction kinetics for single as well as multiple contaminant gas streams (Ottengraf 1986; Ottengraph and Van Denoever 1983; Deshusses et al. 1995).

Furusawa et al. (1984) used a packed bed of fibrous peat for the removal of hydrogen sulfide from air. H_2S was almost completely removed irrespective of its inlet concentration when the loading was less than 0.44 g sulfur per day per kg of dry peat. The removal rate of hydrogen sulfide by the acclimatized peat was fairly constant under a constant inlet concentration but the reaction rate constant was proportional to the influent concentration of H_2S. In another study, the elimination of H_2S from odorous air using a wood bark filter to improve the low permeability of soil beds has been reported (Van Langenhove et al. 1986). Lee and Shoda (1989) reported the biological deodorization of methyl mercaptan using an activated carbon fabric as a carrier of microorganisms for the biofilters. The activated carbon fabric seeded with digested night soil was found to be best packing material amongst the five materials evaluated. The critical load of methyl mercaptan, in which the gas can be completely removed, was determined as 0.48 g S/kg activated carbon fabric/day. About 80% of methyl mercaptan removed in the biofilter was converted into the sulfate ion. Effluent gas concentrations of methyl mercaptan and dimethyl disulfide were not detected below 50 ppm inlet concentration at a space velocity of 50/h. Fibrous materials that are flexible, light, and less microbially degradable may become significant as carriers of microorganisms.

The kinetics of removal of three kinds of odorous sulfur compounds – H_2S, methanethiol (MT), and DMS – in acclimatized peat were compared by Hirai et al. (1990) by supplying single or mixed odorous gases. H_2S and MT were found to be degraded on peat irrespective of the acclimatizing gas, and their maximum removal rates were unaffected by the presence of DMS. On the contrary, DMS was degraded only in DMS acclimatized peat. It has been reported that the peat has the advantages over soil or compost of broadness of the maximum permeability of the moisture content and a lower pressure drop due to its fibrous structure. The same laboratory has reported earlier about the characteristics of the peat as a packing material in deodorization device with the following results: zero-order kinetics in complete H_2S removal by peat biofilters (Furusawa et al. 1984), characteristics of isolated H_2S oxidizing bacteria inhabiting a peat biofilter (Wada et al. 1986), and biological removal of organosulfur compounds by peat biofilters (Hirai et al. 1988). Gradual increase of load was better for obtaining a high removal rate than the high load at the start of the experiment. Acclimation periods for H_2S, MT, and DMS were 19, 17, and 24 days, respectively. During this period, the pH of the peat gradually decreased due to accumulation of sulfate ions.

The maximum removal rate of H_2S in its acclimatized peat was one order larger than those in MT and DMS acclimatized peat. The removability of DMS was affected by the mixed gasses. Although the removal of DMS decreased when present with MT, the existence of H_2S will weaken the effect of MT on DMS removal to a certain extent. Thus, it would be better to maintain the space velocity (SV) value lower to guarantee DMS removal (Hirai et al. 1990). At a high SV, two stage columns in series are recommended. In the first column, most of the H_2S and MT can be removed, while the second column will be exclusively for DMS removal. This method is also appropriate for the maintenance of operation including the washing of accumulated ions and the exchange of packing material.

Shareefdeen et al. (1993) used an eight-membered bacterial consortium, obtained from methanol-exposed soil, and a peat–perlite column for the biofiltration of methanol vapors. The biofilter was found to be effective in removing methanol at rates up to 112.8 g/h/m^3 packing. They also derived a mathematical model and validated it. Both experimental data and model predictions suggested that the methanol biofiltration process was limited by oxygen diffusion and methanol degradation kinetics. Bench scale experiments and a numerical model were used by Hodge and Devinny (1994) to test the effectiveness of biofiltration in treating air contaminated with ethanol vapors. Out of the three different packing materials used viz., granulated activated carbon (GAC), compost, and a mixture of compost and diatomaceous earth, the GAC supported the highest elimination rates, ranging from 53 to 219 g/m^2/h for a range of loading rates. Partitioning coefficients for the contaminant on the biofilter packing material had a strong effect on the efficiency of the biofilters. Several studies on removal of volatile solvents such as ketone mixtures, toluene, ethyl acetate by biofiltration have also been reported (Kirchner et al. 1987; Campbell and Connor 1997; Bibeau et al. 1997; Deshusses et al. 1997).

The performance of biofiltration to remove odors (about 40 compounds) from animal rendering plant's gaseous emissions was investigated by Luo and Oostrom (1997) using pilot-scale biofilters containing different media (sand, sawdust, bark, bark–soil mixture). Biofilter odor removal efficiencies of 75–99% were obtained at various air loading rates (0.074–0.057 m^3/m^3 medium/min) and medium moisture contents. Bio-Reaction Industries Inc., Tualatin, OR, the USA has reported to develop a modular vapor-phase biofilter that is capable of treating extremely high concentrations of VOC in low air volumes (Stewart and Thom 1997). These systems are more suitable for point source industrial process air streams, storage tanks and other vent emissions.

Biofiltration of NO_x is reported to be enhanced by the addition of an exogenous carbon and energy source (Apel et al. 1995). pH control is found to be an important operating parameter due to acidic nature of the gas. Addition of calcite to the biofilter bed provided an effective internal buffer and the optimum temperature was found to be 50–60°C. The biofilter using activated carbon or anthracite as the packing material was reported to be most acceptable process for the removal of malodorous compounds containing nitrogen or sulfur (Hwang et al. 1995), since it produced no oxidized organics noticed with ozonation, and it had an equally high removal efficiency of both sulfur and nitrogen containing odorous compounds.

Biofiltration has been successfully applied to remove α-pinene, a very hydrophobic VOC discharged in pulp and paper and wood products emissions, from a contaminated air stream (Mohseni and Grant 1997). Two identical bench scale biofilters were utilized for more than 4 months of experiment. The biofilter medium consisted of a mixture of wood chips and spent mushroom compost that was amended with higher perlite, for the first filter and with GAC, for the second biofilter, the experiment was conducted at loading rates between 5 and 40 g α-pinene/m^3 bed medium/h. Under steady state operating conditions, both biofilters, amended with perlite and GAC, performed similarly and provided removal rates of up to 30–35 g α-pinene/m^3 bed medium/h with gas retention times as low as 30 s. The adsorption characteristics of GAC were significant only during the start-up period where the GAC biofilter had a significantly better performance than perlite biofilter. When the biofilters were subjected to a sudden increase in the loading rate, the performance of the biofilters decreased significantly. The reacclimation period, however, was not long and biofilters reached more than 99% removal within less than 48 h of the spike load.

Studies on the transient behavior of a laboratory-scale compost based biofilters have been reported (Deshusses 1997). This included start-up, carbon balances, and interactions between pollutants in the aerobic biodegradation of VOC mixtures from effluent air streams. The study of transient behavior offers a genuine basis for the development of a conceptual explanation of the complex phenomena that occur in biofilters during pollutant elimination, thereby providing an opportunity for further progress in establishing fundamental understanding of such reactors (Shareefdeen and Baltzis 1994; Tang et al. 1995; Deshusses et al. 1995). During long-term operation of a biofilter, the mandatory absence of net cell growth forces the cells into maintenance metabolism, which is of relatively low rate compared to substrate consumption during the active growth of the acclimation phase. Postacclimation nutrient addition increases activity primarily by allowing a return to the high substrate consumption rate of active growth, and only secondarily helps raise bed activity because of the ultimately higher amount of biomass in the bed (Cherry and Thompson 1997). The biomass content of a biofilter during the acclimation phase can be estimated using two approximate methods. The first follows the cumulative amount of substrate converted and uses the yield of cells from substrate during active growth to estimate the total biomass created. The second method follows a rate constant for conversion of substrate in the bed. This number is proportional to the amount of biomass as long as the conditions in the bed (e.g., temperature, pH, substrate concentration) are relatively constant (Cherry and Thompson 1997).

Generally, the empirical knowledge dictates the design and scale-up of biofiltration plants, even though substantial performance improvement could be expected from a more comprehensive knowledge of the individual processes involved in pollutant elimination. For improved design and performance, an appropriate model for the whole process is required. Deshusses (1997) and Deshusses et al. (1995) have developed a novel diffusion reaction model for the determination of both the steady-state and transient-state behavior of biofilters for waste air treatment, and experimentally evaluated/verified the same. Although this model deals with the aerobic biodegradation of methyl ethyl ketone (MEK) and methyl isobutyl ketone (MIBK) vapors from air, similar mathematical treatment can be given to other biofilters degrading H_2S, organosulfur compounds, and other volatile organics. Most of the mathematical models have been developed mainly to correlate a particular set of experimental data, to explain the influence of selected parameters on the efficiency of the process, and sometimes to seek a better fundamental understanding of the phenomena occurring in a biofilter (Shareefdeen et al. 1993; Hodge and Devinny 1994; Deshusses et al. 1995). More promising quantitative structure–activity relationships for biofiltration have been presented by Choi et al. (1996).

Qiao et al. (2008) studied the removal characteristics of hydrogen sulfide experimentally in the biofilters with fibrous peat and resin as the packed materials. The biofilter with 100% of the peat showed higher removal capacity than the resin biofilter, but the gas flow resistance was lower in the latter. The mixture of the peat and resin as the packed material of the biofilter was proved to be an advisable method to keep the high removal capacity and reduce the gas flow resistance for a long-term operation. The flow resistance can decrease by 50% when 50% of the resin mixed with the peat, but the removal capacity was still considerable high.

Goncalves and Govind (2010) treated H_2S polluted airstreams in two biotrickling filter columns packed with polyurethane (PU) foam cubes, one with cubes coated with a solution of 25 mg/L of polyethyleneimine (PEI, coated reactor) and the other containing just plain PU cubes (uncoated reactor) at empty bed residence times (EBRT) ranging from 6 to 60 s. and inlet H_2S concentrations ranging from 30 to 235 ppm_v (overall loads of up to 44 g H_2S/m^3 bed/h), with overall removal efficiencies (RE) in the range of 90–100% over 125 days. The acclimatization characteristics of the coated reactor outperformed those of the uncoated one, and both the observed elimination capacity (EC) of 77 g H_2S/m^3 bed/h and retention of volatile solids (VS) of 42 mg VS/cube were maxima in the coated reactor. Insights into the controlling removal mechanisms were also provided by means of dimensionless analysis of the experimental data. Denaturing gradient gel electrophoresis (DGGE) showed that the dominant surviving species in both units belonged to the genus *Acidithiobacillus.*

Wani et al. (2001) studied biofiltration using compost and hog and a mixture of two to remove reduced sulfur (RS) gases emitted from pulp mills. The hog fuel showed more resistance to microbially induced bed degradation than compost or mixtures of both and was found to be effective at RS gas removal as compost, with the advantage of costing less.

Biological Filtration Oxygenated Reactor (Biofor) is a new generation of modern apparatus, an aerobic biological reactor from Degremont, with fixed biomass on a support material (Brenna 2000). The principal advantages of biofiltration are a high concentration of biomass that brings the reactor to operation without the problems of bulking with the elimination of pollutants difficult to degrade biologically. Biofor gives these results as a result of an ideal support material, an efficient aeration system, a process of ascending equal currents of air and water, and optimized washing processes. The support material, Biolite, presents optimal qualities of density, hardness friction, and porosity. As well as working without odors and noise, Biofor is adapted for plants to limit environmental impact

Domtar's kraft mill, Cornwall, Ontario, Canada carried out research to find a way of reducing the odors from the plant (Lau et al. 2006). Three types of biofiltration technology were researched: biofilters, bioscrubbers, and biotrickling reactors. This last option seemed the most favorable for treating the gas leaving the brownstock reactor. With a biotrickling reactor conditions such as temperature, pH, and growth of the biomass can be controlled. Four types of packing material were tried. The packing material should have a high void fraction, have a high specific surface area, be made from an acid-resistant material; have a low bulk density, and the microorganisms should stick to the packing. Lantec's HD Q-PAC gave the optimum results

17.3.6 Present Status

Biofiltration is now a well established air pollution control technology. In Europe several chemical process industries are using biofilters for deodorization and treatment of VOCs from the waste gas. In Netherlands and Germany, biofiltration has developed since the early 1960s into a widely used APC technology which is now considered "best available control technology" in a variety of VOC and odor control applications. Successful biofilter applications in Europe include the following: chemical manufacture, chemical storage, adhesive production, coating operations, iron foundries, waste oil recycling, flavors and fragrances, tobacco processing, industrial waste treatment plants, composting facilities, other food processing industries, oil mills, beer yeast drying, etc. (Singhal et al. 1996) with odor control efficiency of 91–99% and organic removal efficiency of 71–95%. Compounds that are typically well degraded include alcohols, ethers, aldehydes, ketones, amines, sulfides, and inorganic compounds such as ammonia and hydrogen sulfide. Higher chlorinated organics show relatively lower ratio of biodegradation. More than 40% of New Zealand animal rendering plants now use biofilters which are usually effective (Luo and Oostrom 1997). Commercial use of biofilters has been less extensive in the United States, although the need for cost-effective air emission technology is clearly acute (Shareefdeen et al. 1993). But lately biofiltration technology has started picking up in the US also. Although very little information is available in the literature about the application of biofilters, in pulp and paper industry for odor removal, substantial information is available for the removal of various compounds similar to those generated in pulp and paper industry. This information could be very useful in installing biofiltration systems in pulp and paper mills.

17.3.7 Parameters Affecting the Performance of Biofilter

In addition to the microbial culture and packing materials, several other parameters are also important which affect the performance of a biofilter. In order to avoid deposits in the filter layer, dusts and aerosols are to be removed to a great extent from the waste gas by means of appropriate separators. Before it enters the filter, the waste gas should be humidified to saturation. The raw gas is humidified in a spray humidifier or by adding steam to it. The dust separation and humidification can be combined in wet scrubbers wherein scrubbing is done by water. Sometimes, the biofilters can be poisoned by the presence of off-gas constituents that are toxic to the microorganisms. Elimination of these substances or changing the vent system can make the off-gas suitable for biofiltration. High particulate loads in the raw gas can adversely affect the operation of a filter in different ways. Clogging of the air distribution system and the filter material itself by grease and resin can also occur. The deposition of dust in the humidifier will generate sludge and can result in the improper humidification. In such cases, the installation of particulate filter is required (Leson and Wikener 1991). Pollutant concentration and pollutant loading rates affect the performance. For example, in cases of H_2S removal by compost biofilter, the efficiency does not change as long as the H_2S loading rate is less than the maximum acceptable value for the compost. The concentration of H_2S as high as 4,000–4,500 ppm can be treated with an efficiency of 99%, but if the concentration increases drastically, say more than 100,000 ppm, then fresh air can be introduced to reduce the H_2S concentration and increase the oxygen concentration (Yang and Allen 1994). The maximum elimination capacity is a function of the biofilter material and the operating conditions. The pollutant loading should be applied accordingly.

To ensure the maximum pollutant elimination capacity of the biofilter system, the gas should stay on the bed for sufficient time. It is 30–40 s for H_2S elimination in compost bed (Yang and Allen 1994). There is no significant increase in the efficiency if the time is greater than 25 s. but when it is decreased to say 10 s, the efficiency decreases by about 80%. The reduction of H_2S removal efficiency at shorter residence time is not necessarily due to the insufficient reaction time between the H_2S molecule and the biomass, but may be due to the slow step involved in the overall process. This slow step comprises of H_2S diffusion from the gas phase into the liquid phase where the microorganisms exist (Yang and Allen 1994).

The moisture content and pH of the packing bed are other important parameters. For the compost, moisture level should be held between 40 and 60%. If the moisture content is reduced below 30%, the H_2S removal efficiency decreases proportionately. Proper moistening equipment such as sprinklers should be installed and operated in such a way that moisture content stays in the prescribed limits. Since the dominant active species present in this biofilter are primarily acidophiles, which prefer an optimum pH value near 3, maximum H_2S removal occurs at a compost pH of 3.2. Sulfur-oxidizing bacteria can live in environments having a wide range of pH (1–8). At the pH below 3, the efficiency decreases drastically. At the higher pH range, chemical reaction between H_2S and the compost material or reaction products can significantly enhance its removal, in addition to biological oxidation (Yang and Allen 1994).

For high pollutant removal efficiency, the temperature of the filter bed should be in the optimum range. The optimum range is 35–50°C for H_2S removal. The efficiency drops rapidly with decreases in temperature. For example, if the temperature reduces to, say 7°C, the H_2S removal efficiency decreases by about 80%. The decrease in H_2S removal at the higher temperature is less significant than that at lower temperature. The removal of H_2S at higher temperatures is probably due to increased chemical oxidation reactions in addition to biological oxidation. Normally, the temperature of biofilter is 10–15°C higher than the ambient temperature. This is due to the biological respiration of the microbes and the exothermic reactions in the filter. Thus, the biofilter can function properly even if the ambient temperature is low.

Since sulfate is the final product of the biofiltration, involving sulfur compounds, it may accumulate in the filter bed if not removed. Accumulation of sulfate can easily reach a level that can significantly reduce the biological function of the biofilter. Therefore, sulfate should be washed off periodically before it reaches the toxic level. A sulfate content of 25 mg/g is a critical level for the microbial environment.

The pressure drop increases approximately linearly with packing height. It increases in significantly larger increments with packing height for smaller particles than that for larger ones. It also depends on the water content of the packing. If the water content is increased, the coagulation of small viscous particles is enhanced and the pressure drop increases sharply. However, the rapid buildup of pressure can be suddenly released by channeling, i.e., a breakdown of filter bulk with much less resistance caused by a separation of packing materials. This situation is undesirable because it allows pollutants to exit the system without treatment. To prevent the high back pressure build up, the surface load of up to 300 m^3 off-gases/h/m^2 of filter should be maintained for proper functioning of the compost filter. Mineralization and compaction of the compost packing during extended operation may eventually increase the bed pressure drop. Practically, the bed needs to be repacked or the compost replaced when the overall pressure drop is greater than 25 kPa (Yang and Allen 1994).

17.3.8 Advantages, Limitations and Future Prospects

Since biofilters compete with incineration and carbon adsorption in many situations, they are attractive in terms of not having to deal with landfilling costs or regeneration headaches. This has already been recognized in Europe, and some biofilter technology has found its way to the US (Anon 1991). Also, the thought of not simply transferring contaminants from one medium to another is particularly appealing. The biofilter creates a truly destructive process.

The use of microbial filter techniques in the treatment of air effluents containing organic pollutants can offer a number of advantages. They are inexpensive, work efficiently at ambient temperature, self-generating, maintenance free with low running cost, long life, environment safety, and oxidize most common VOCs to carbon dioxide and water producing virtually no by-products. The microbial flora survive a fairly long period during which the filter bed is not loaded (periods of a fortnight are easily spanned with hardly any loss of microbial activity). This is important in view of the dynamic behavior of filter bed at discontinuous operation, and means a very short starting time after longer periods of not operating the filter bed (Ottengraph and Van Denoever 1983). Moreover, the presence of a large amount of packing material with a buffering capacity diminishes the sensitivity of biofilters to different kinds of fluctuations.

Although such methods have long been known to be cost-effective, they have not found general acceptance in practice, even when the exhaust gas components to be removed are biodegradable. Long adaptation periods of the biomass (in particular with large exhaust gas flow discontinuities) or low space velocities i.e., low specific purification capacities, are the reasons often cited. Bed compaction problems, specially with soil and compost biofilters, have also been noticed. This results in high pressure drop across the filter. However, with the help of GAC and other synthetic packing materials, individually or in combination with soil–peat–compost materials, have solved these problems to a great extent.

While biooxidizing H_2S and organic sulfur compounds in a filter, accumulation of sulfate can easily reach a level that can significantly reduce the biological activity of the biofilter. Therefore, sulfate should be periodically washed off before it reaches the toxic level. The removal of DMS decreases considerably if methanethiol (MT) is also present in the exhaust gas (Hirai et al. 1990). However, the existence of H_2S weakens the effect of MT on DMS removal rate to a certain extent. In this case, it would be desirable to maintain a low space velocity to ensure DMS removal. At high space velocity, two stage columns in series are recommended. So that, in the first column, most of the H_2S and MT can be removed, while the second column will be exclusively for DMS removal. This method may also be appropriate for the maintenance of operation, including the washing of accumulated ions and the replacement of packing material. Multistage operation of biofilters may also be necessary when the waste gases contain components, which require different conditions for their microbial degradation. This way, optimal growth conditions for the different microbial population can be provided in separate stages. Also, more stages may be necessary when the waste gases include one component in a concentration so high that the capacity of one stage is inadequate for a sufficient degradation. Depending on the nature of the organic compounds present in the waste, the filter sometimes needs inoculation with appropriate microorganisms to start biological activity.

In recent years, there has been significant maturation of biological waste air treatment research. This has resulted in a large number of studies concerning the performance and operation of the biofilters. Biofilter technology has a high potential for exhaust gas clean up, but as with many biological processes, the design requirements have not been fully appreciated. Interestingly, the fundamental processes involved during the elimination of a pollutant in a gas-phase bioreactor are still very poorly understood.

Biofilter technology was utilized in the field well before there was a basic understanding of its fundamental principles. This has resulted in several cases of unsuccessful or suboptimum operation of large-scale bioreactors. Today, with recent advances in the understanding of the fundamental principles underlying biofiltration, promises exist for better reactor design with optimal operating conditions. However, a number of fundamental questions remain unanswered or require further clarification, e.g., the quantification of biomass turnover, biodegradation kinetic relationships and factors influencing these relationships ecology of biofilter microflora, the determination of the availability and cycles of pollutant, oxygen and essential nutrients. The above factors have been found to significantly influence the performance and long-term stability of biofilters, and thus require further investigation in quantitative term. The expanding use of modern tools of biotechnology should be able to make it easier. The largest problem to overcome will be the translation of recent and future basic advances into real process improvements for biofiltration technology to mature from the mysterious black box reactor to a well-engineered process based on solid science rather than on trial and error.

Biofiltration technology for purification of exhaust gases from pulp and paper industry has a great potential. Very little information directly related to the industry is available although reasonably good information is available on the biofiltration of organic compounds similar to those found in the exhaust gases of pulp and paper industry. More studies are needed to obtain a better understanding of the heat transfer, mass transfer and reaction processes occurring within the biofilter beds. Comprehensive long-term studies of full-scale biofilter systems would also be valuable in improving our understanding of biofilters used to remove VOCs from off-gases generated in the paper industry. Extended studies of transient behavior of biofilters are also needed to provide the basic empirical knowledge necessary for plant design, scale-up, and performance evaluation under real conditions.

References

Andersson B, Lovblod R, Grennbelt P (1973) Diffuse emissions of odorous sulfur compounds from kraft pulp mills, 1 VLB145. Swedish Water and Air Pollution Research Laboratory, Gutenberg

Anon (1991) Air pollution control may be reduced with biotechnology. RMT Network 6(1):5–8

Apel WA, Barnes JM, Barrett KB (1995) Biofiltration of nitrogen oxides from fuel combustion gas streams. In: Proceedings of the annual meet-air waste management association, San Antonio, Texas, 88th (vol 4A): 95-TP9C.04

Bajpai P, Bajpai PK, Kondo R (1999) Biotechnology for environmental protection in pulp and paper industry. Springer, Germany, pp 239–261

Bibeau L, Kiared K, Leroux A, Brzezinski VG, Heitz M (1997) Biological purification of exhaust air containing toluene vapor in a filter-bed reactor. Can J Chem Eng 75:921–929

Bohn HL (1975) Soil and compost filters for malodorant gases. J Air Waste Manage Assoc 25:953–955

Bohn HL (1992) Considering biofiltration for decontaminating gases. Chem Eng Prog 88:34–40

Bohn HL (1993) Biofiltration: Design principles and pitfalls. In: Proceedings of the 86th annual meeting of air & waste management association. Denver, Paper #93-TP-52A.01

Bohn H, Bohn R (1988) Soil beds weed out air pollutants. Chem Eng 95(4):73–76

Brenna V (2000) Biofor: biofiltration of paper mill effluents. Ind Carta 38(5):59–63

Burgess JE, Parsons SA, Stuetz RM (2001) Developments in odour control and waste gas treatment biotechnology: a review. Biotechnol Adv 19(1):35–63

Cáceres M, Morales M, San Martín R, Urrutia H, Aroca G (2010) Oxidation of volatile reduced sulphur compounds in biotrickling filter inoculated with *Thiobacillus thioparus*. Electron J Biotechnol 13(5). http://dx.doi.org/10.2225/vol13-issue5-fulltext-9. Accessed Sept 2010

Campbell HJ, Connor MA (1997) Practical experience with an industrial biofilter treating solvent vapor loads of varying magnitude and composition. Pure Appl Chem 69(11):2411–2424

Carlson DA, Leiser CP (1966) Soil bed for control of sewage odors. J Water Pollut Control Fed 38:829–833

Chan AA (2006) Attempted biofiltration of reduced sulphur compounds from a pulp and paper mill in Northern Sweden. Environ Prog 25(2):152–160

Cherry RS, Thompson DN (1997) Shift from growth to nutrient – limited maintenance kinetics during biofilter acclimation. Biotechnol Bioeng 56(3):330–339

Choi DS, Webster TS, Chankg AN, Devinny JS (1996) Quantitative structure – activity relationships for biofiltration of volatile organic compounds. In: Reynolds FE Jr (ed) Proceedings of the 1996 conference on biofiltration. The Reynolds Group, Tustin, pp 231–238

Chou MS, Chen WH (1997) Screening of biofiltering material for VOC treatment. Air Waste Manage Assoc 47(6):674–681

Dallons V (1979) Multimedia assessment of pollution potentials of non-sulfur chemical pulping technology. EPA-600/2-79-026, Jan 1979

DeBont JAM, vanDijken JP, Harder W (1981) Dimethyl sulfoxide and dimethyl sulfide as a carbon, sulfur and energy source for growth of *Hyphomicrobium* S. J Gen Microbiol 127:315–323

Deshusses MA (1997) Transient behavior of biofilters: start-up, carbon balance, and interactions between pollutants. J Environ Eng 123:563–568

Deshusses MA, Hammer G (1993) The removal of volatile ketone mixtures from air in biofilters. Bioprocess Eng 9:141–146

Deshusses MA, Hamer G, Dunn IJ (1995) Behavior of biofilters for waste air biotreatment. I: Dynamic model development. Environ Sci Technol 29(4):1048–1058

Deshusses MA, Johnson CT, Hohenstein GA, Leson G (1997) Treating high loads of ethyl acetate and toluene in a biofilter. In: Air and waste management association 90th annual meeting and exhibition, Toronto, Canada, 8–13 June 1997, p 13

Environmental Pollution Control Pulp and Paper Industry (1976) Part 1, air, U.S. EPA technology transfer series, EPA-625/7-76-001, Oct 1976

Furusawa N, Togashi I, Hirai M, Shoda M, Kubota H (1984) Removal of hydrogen sulfide by a biofilter with fibrous peat. J Ferment Technol 62(6):589–594
Goncalves JJ, Govind R (2010) Enhanced biofiltration using cell attachment promotors. Environ Sci Technol 43(4):1049–1054
Govind R, Bishop DF (1996) Overview of air biofiltration – basic technology, economics and integration with other control technologies for effective treatment of air toxics. In: Emerging solutions VOC air toxics control, Proceedings of the spec. conference, Pittsburgh, PA, pp 324–350
Hirai M, Terasawa M, Inamura I, Fujie K, Shoda M, Kubota H (1988) Biological removal of organosulfur compounds using peat biofilter. J Odor Res Eng 19:305–312
Hirai M, Ohtake M, Soda M (1990) Removal of kinetics of hydrogen sulfide, methanethiol and dimethyl sulfide by peat biofilters. J Ferment Bioeng 70:334–339
Hodge DS, Devinny JS (1994) Biofilter treatment of ethanol vapors. Environ Prog 13(3):167–173
Hodge DS, Medina VF, Islander RL, Devinny JS (1991) Treatment of hydrocarbonfuel vapors in biofilters. Environ Technol 12:655–662
Holusha J (1991) Using bacteria to control pollution. The New York Times, March 13:C6
Hwang Y, Matsuo T, Hanaki K, Suzuki N (1995) Identification and quantification of sulfur and nitrogen containing odorous compounds in wastewater. Water Res 29(2):711–718
Kanagawa T, Kelly DP (1986) Breakdown of dimethyl sulfide by mixed cultures and by *Thiobacillus thioparus*. FEMS Microbiol Lett 34:13–19
Kanagawa T, Mikami E (1989) Removal of methanethiol, dimethyl sulfide, dimethyl disulfide and hydrogen sulfide from contaminated air by *Thiobacillus thioparus* TK-m. Appl Environ Microbiol 55(3):555–558
Kennes C, Veiga MC, Yaomin J (2007) Co-treatment of hydrogen sulfide and methanol in a single-stage biotrickling filter under acidic conditions. Chemosphere 63(6):1186–1193
Kirchner K, Hauk G, Rehm HJ (1987) Exhaust gas purification using immobilized monocultures (biocatalyst). Appl Microbiol Biotechnol 26:579–587
Lau S, Groody K, Chan A (2006) Control of reduced sulphur and VOC emissions via biofiltration. Pulp Pap Canada 107(12):57–63
Lee S-K, Shoda M (1989) Biological deodorization using activated carbon fabric as a carrier of microorganisms. J Ferment Bioeng 68(6):437–442
Leson G, Wikener AM (1991) Biofiltration: an innovative air pollution control technology for VOC emissions. J Air Manage Assoc 41:1045–1054
Luo J, van Oostrom A (1997) Biofilters for controlling animal rendering odor – a pilot-scale study. Pure Appl Chem 69(11):2403–2410
Marsh A (1994) Biofiltration for emission abatement. Eur Coating J 7/8:528–536
Martin AM (1992) Peat as an agent in biological degradation: peat biofilters. In: Martin AM (ed) Biological degradation of waste. Elsevier, New York, pp 341–362
McNevin D, Barford J (2000) Biofiltration as an odour abatement strategy. Biochem Eng J 5(3):231–242
Mohseni M, Grant AD (1997) Biofiltration of α-pinene and its application to the treatment of pulp and paper air emissions. Tappi Environ Conf Exhib 2:587–592
Ottengraf SPP (1986) Exhaust gas purification. In: Schonborn W (vol ed) Biotechnology, vol 8, Rehm H-J, Reed G (series eds) Microbial degradations. VCH, Weinheim, Germany, Chap. 12, pp 425–452
Ottengraf SPP (1987) Biological system for waste gas elimination. Trends Biotechnol 5:132–136
Ottengraf SPP, Meesters JPP, van den Oever AHC, Rozema HR (1986) Biological elimination of volatile xenobiotic compounds in biofilters. Bioprocess Eng 1:61–69
Ottengraph SPP, Van Denoever AHC (1983) Kinetics of organic compound removal from waste gases with a biological filter. Biotechnol Bioeng 25:3089–3102
Pomeroy D (1982) Biological treatment of odorous air. J Water Pollut Control Fed 54:1541–1545
Prokop WH, Bohn HL (1985) Soil bed systems for control of rendering plant odors. J Air Waste Manage Assoc 35:1332–1338
Qiao S, Fu L, Chi Y, Yan N (2008) Removal characteristics of hydrogen sulfide in biofilters with fibrous peat and resin. Bioinform Biomed Eng 2008:3996–3999

Rozich A (1995) Tackle airborne organic vapors with biofiltration. Environ Eng World 1:32–34
Rydholm SA (1965) Pulping process. Wiley, New York, p 452
Shareefdeen Z, Baltzis BC (1994) Biofiltration of toluene vapor under steady-state and transient conditions-theory and experimental results. Chem Eng Sci 49:4347–4360
Shareefdeen Z, Baltzis BC, Oh YS, Bartha R (1993) Biofiltration of methanol vapor. Biotechnol Bioeng 41:512–524
Shimko IG, Spasov VA, Chinennaya SK, Zakirova RI, Tananina IN, Perchugor GY, Pavlova OI (1988) Biochemical methods of freeing gas-air mixture from sulfur containing compounds. Fiber Chem 19:373–378
Singhal V, Singla R, Walia AS, Jain SC (1996) Biofiltration – an innovative air pollution control technology for H_2S emissions. Chem Eng World 31(9):117–124
Sivela S, Sundman V (1975) Demonstration of *Thiobacillus* type bacteria which utilize methyl sulfides. Arch Microbiol 103:303–304
Smith NA, Kelly DP (1988a) Isolation and physiological characterization of autotrophic sulfur bacteria oxidizing dimethyl disulfide as sole source of energy. J Gen Microbiol 134:1407–1417
Smith NA, Kelly DP (1988b) Mechanism of oxidation of dimethyl disulfide by *Thiobacillus thioparus* strain E6. J Gen Microbiol 134:3031–3039
Someshwar AV (1989) Impact of burning oil as auxiliary fuel in kraft recovery furnaces upon SO_2 emissions. NCASI Technical Bulletin No. 578, Dec 1989
Springer AM, Courtney FE (1993) Air pollution: a problem without boundaries. In: Springer AM (ed) Industrial environmental control pulp and paper industry, 2nd edn. Tappi Press, Atlanta, pp 525–533
Stewart WC, Thom RC (1997) High VOC loading in biofilters industrial applications. In: Emerging solutions VOC air toxics control, Proceedings of the spec. conference, Pittsburgh, PA, pp 38–65
Suylen GM, Stefess GC, Kuenen JG (1986) Chemolithotropic potential of a *Hyphomicrobium* species capable of growth on methylated sulfur compounds. Arch Microbiol 146:192–198
Suylen GM, Large PJ, vanDijken JP, Kuenen JG (1987) Methyl mercaptan oxidase, a key enzyme in the metabolism of methylated sulfur compounds by *Hyphomicrobium* EG. J Gen Microbiol 133:2989–2997
Swanson W, Loehr R (1997) Biofiltration: fundamentals, design and operation principles, and applications. J Environ Eng 123:538–546
Tang HM, Hwang SJ, Hwang SC (1995) Dynamics of toluene degradation in biofilters. Hazard Waste Hazard Mat 2(3):207–219
Tonga AP, Skladany GJ (1994) Field pilot-scale vapor-phase treatment of styrene using biofiltration. In: Flathman PE, Jerger DE, Exner JH (eds) Bioremediation: field experience. Lewis Publishers, Ann Arbor, pp 507–521
Van Langenhove H, Wuyts E, Schamp N (1986) Elimination of hydrogen sulfide from odorous air by a wood bark biofilter. Water Res 20:1471–1476
Van Lith C, Leson G, Michelson R (1997) Evaluating design options for biofilters. J Air Waste Manage Assoc 47:37–48
Wada A, Shoda M, Kubota H, Kobayashi T, Katayama FY, Kuraishi H (1986) Characteristics of H_2S oxidizing bacteria inhabiting a peat biofilter. J Ferment Technol 64:161–167
Wani A, Branion R, Lau AK (1997) Biofiltration: a promising and cost-effective control technology for odors, VOCs and air toxics. J Environ Sci Health 32(7):2027–2055
Wani AH, Branion RMR, Lau AK (2001) Biofiltration using compost and hog fuel as a means of removing reduced sulphur gases from air emissions. Pulp Paper Can 102(5):27–32
Williams TQ, Miller FC (1992) Odor control using biofilters. BioCycle 33(10):72–77
Yang Y, Allen ER (1994) Biofiltration control of hydrogen sulfide design and operational parameters. J Air Waste Manage Assoc 44:863–868

Chapter 18
Management/Utilization of Wastewater Treatment Sludges*

18.1 Introduction

Sludge is the largest by-product of the pulp and paper industry and disposal of sludge is a major solid waste problem for the industry (Battaglia et al. 2003; Geng et al. 2006; Suriyanarayanan et al. 2010; Krigstin and Sain 2005; Mladenov and Pelovski 2010). The nature of sludge generated from paper industries mainly depends on the raw materials used in different unit processes. The quantity of sludge generation varies with the type of pulping and papermaking or both (Springer 1993). Sludge generated from the industrial sludge contains a large number of ingredients, some of which are toxic. Solid waste is generated from the both large and small categories of paper mills. Solid waste from paper industries is generated usually in various stages of paper production. The production of chemical pulp generates various fractions of solid waste:

- Inorganic sludge (i.e. dregs and lime mud) from the chemical recovery
- Bark and wood residues from woodhandling
- Sludge from the biological wastewater treatment plant (i.e. inorganic material, fibers, and biological sludge)
- Dust from boilers and furnaces
- Rejects containing mainly sand from woodhandling
- Fly and bottom ashes from the fluidized bed boiler

Dregs and lime mud are separated from the chemical recovery cycle in order to keep the amount of inert material and nonprocess chemicals in the cycle at an acceptable level and thus secure high reaction rates in the chemical recovery system.

---

*Excerpted from Bajpai P, Bajpai Pramod K, Kondo R (1999) Biotechnology for environmental protection in the pulp and paper, Chap. 10, Management of wastewater treatment sludges, with kind permission from Springer Science+Business Media.

P. Bajpai, *Biotechnology for Pulp and Paper Processing*,
DOI 10.1007/978-1-4614-1409-4_18, © Springer Science+Business Media, LLC 2012

Solid waste disposal is usually to landfill, although incineration is becoming increasingly widespread. Prior to any land application of solid residues, the levels of chemicals of concern need to be routinely demonstrated to fall below realistic regulatory levels. Kenny et al. (1997) have reported that Canadian pulp and paper mills with activated sludge wastewater treatment system produce primary sludge of 31 kg(od)/tons of pulp while the secondary (biological) sludge generation is 16 kg/tons. A typical floatation deinking plant produces 80–150 kg of dry sludge per tons of recycled pulp (Latva-Somppi et al. 1994). The amount of sludge generated by a recycled fiber mill depends very much on the type of furnish being used and the end product being manufactured. For example, a recycled tissue mill and a recycled newsprint mill may use the same old newspaper as furnish, but the higher brightness and lower dirt requirements of the tissue will result in a lower yield and higher sludge generation.

The composition of sludge depends on the raw material, manufacturing process, chemicals used, final products, and the wastewater treatment technique. In case of recycled papers, it also depends on the type of paper used and the number and types of cleaning stages used in the recycling operation. For example, sludge from mixed office wastepaper (MOW) may contain high levels of clay and other types of fillers, printing inks, stickies from envelope adhesives, as well as fibers and paper fines. In fact, sludges from MOW recycling operations may contain as much as 2% ash from fillers in the wastepaper. Sludge solids produced by pulp and paper mills typically include a majority fraction of fiber. Depending on the mill ink, sand, rock, biological solids, clay/fillers, boiler ash, grits from recausticizing, etc. may make up the other fractions. Because of the constituents that may exist, along with the water fraction, typical sludge analysis can vary widely. So, it is very important to characterize the sludge carefully for determining the best method for sludge disposal.

18.2 Dewatering of Sludge

Sludge is usually disposed of through landfilling, incineration, land spreading or through alternate uses. All these approaches have one feature in common: the sludge must be as dry as possible. Hence, dewatering the sludge to as high a solids level as possible is important for both economical and environmental reasons.

The primary sludges can be dewatered easily as these are high in fiber and low in ash. The most difficult are the solids from the high-rate biological treatment systems. The primary sludge most difficult to dewater is that containing groundwood fines. Primary sludges are normally tertiary or quaternary pulp and paper mill rejects, but often consist of quality fibers having a high monetary value. As the percentage of secondary sludge increases, the dewatering characteristics deteriorate, resulting in decreased cake solid contents. Tissue mills, NSSC plants, and recycle paperboard plants have problems with dewaterability of combined sludges.

Sometimes, it may be desirable to dewater the primary sludge separately from the secondary sludge. One example is a situation in which the secondary sludge can be disposed of through land application. Blended sludges are not usually suitable for such disposal. Another example is a situation in which the primary sludge can be used to produce a by-product or can be reused within the production process, but the blended sludge cannot be used. If the combined consistency is less than 4–5%, sludges must be predewatered. It actually helps the dewatering process by reducing solution volume while increasing solid content for further dewatering, absorbing fluctuations of inlet solids consistency while stabilizing the output consistency, increasing outlet solids content and solids capture efficiency, and reducing overall polymer consumption.

Predewatering technologies which are commonly used include rotary sludge thickeners (RSTs) and gravity thickeners. Other technologies in use include gravity table thickeners, dissolved air floatation (DAF) clarifiers, and belt presses. Centrifuges, V-presses, coil vacuum filters, and fabric vacuum filters are also used but these are not very common. The floatation thickener used on secondary sludge can achieve approximately 4% solids, giving it an advantage over the gravity thickener, which can achieve only about 2%. The advantages of gravity thickeners include: simplicity, low operating costs, low operator attention and a degree of sludge storage. Conditioning chemicals are not normally required and there is minimal power consumption. However, these advantages are often offset by potential septicity/odor, less dewatering capability (as compared to other technologies) and large space requirements. These disadvantages have limited the use of gravity thickeners in recent installations. An RST is a rotary screen where water is removed by gravity and tumbling action. RTSs have been installed in many mills as predewatering units before the screw presses. This type of predewatering device is capable of increasing consistency to between 4 and 10% depending upon the proportion of secondary sludge and the percentage of solids from the secondary and primary clarifiers. In a gravity table filter, sludge is drained on a rotary wire. Drainage is assisted by moving paddles. The paddles are required to prevent wire pluggage. Gravity tables and RSTs produce sludges of similar consistency. Gravity tables are normally placed over screw presses to allow feeding by gravity. As with RSTs, polymers are applied before the table filter.

With DAF clarifiers, secondary sludge is floated with dissolved air, usually with the aid of some dewatering chemicals. Sludge is skimmed from the surface of the clarifier and the underflow re-treated in the aeration pond or the primary clarifier. In the DAF process, solids can be increased to 3–6% for secondary sludges. The actual performance is frequently dependent upon the type of chemical applied and the dosage rate. DAF units also have the potential to eliminate odor problems. Few mills rely solely on DAF units for sludge pre-dewatering (Kenny et al. 1997). Few activated sludge treatment plants use DAF units in combination with RSTs. One mill in Canada uses coil filters and V-presses to dewater primary sludge (Kenny et al. 1997). After pre-dewatering of the primary sludge, the secondary clarifier sludge and pre-dewatered primary sludge are mixed in a paddle mixer and then discharged for final dewatering on screw presses. No dewatering chemicals are required. Vacuum filter dewatering of biological sludges has been phased out of service in North America. Problems with poor capture rates, blinding and landfilling difficulties have eliminated this option.

In order to obtain good dewatering efficiency, sludge from wastewater treatment plants is frequently conditioned, using chemical or physical means to alter the floc structures of the sludge, i.e. imparting sufficient stiffness and incompressibility to the structures so that water entrained in the sludge can rapidly be drained through filtering or other means (Benitez et al. 1993; Wu et al. 1998). The functions of conditioners are to improve the sludge dewatering properties, to reduce specific filtration resistance, and to enhance the dewatering efficiency. These increase the solids content after dewatering. There are four mechanisms through which chemical conditioners added to sludge at wastewater treatment plants act: compression of the electrical double layer, neutralization of charges, retention of precipitates, and the bridging effect. These actions destroy the stability of the existing flocs, causing them to re-aggregate and precipitate into a tighter sludge filtration cake, thus enhancing dewatering (Huang and Chang 1997). In general, following chemicals are used for conditioning, regardless of the type of equipment: lime, ferric chloride, and polymers. The three can be used separately or in combination. Ferric chloride has the disadvantage of being highly corrosive but is a very effective conditioning agent. The sludges that are difficult to dewater require high polymer addition rates. Wet air oxidation has also been used as a conditioning process to aid sludge dewatering and has been commercially applied in the paper industry (Mertz and Jayne 1984) where filler recovery was a side benefit. However, the brightness of the filler recovered was lower than that of the filler grade clay and the installation experienced considerable down time and high maintenance costs.

Polymers are typically added to flocculate sludge during wastewater treatment. Banerjee (2009) has reported that cyclodextrins (CDs) boost the performance of these polymers by increasing the cake solids and drainage rates of belt- or screw-pressed biological or primary sludge. These benefits are obtained at very low CD dosage. CDs also decrease the specific resistance to filtration (SRF) and increase the capture rate of solids during belt pressing. In three different full-scale trials, a combination of higher cake solids, better drainage, better filtrate clarity and lower polymer use was obtained. The CD application for sludge dewatering has been implemented at the Mississippi mill and has provided stable benefits of a 30% reduction in polymer costs for several months. Several successful trials at other facilities in the United States have been run successfully and additional implementations are anticipated. From a cost standpoint, the CD is approximately twice the cost of the polymer. It displaces a much higher proportion of the polymer, so the cost:benefits are attractive. Finally, CDs are biologically derived products in that they are prepared from starch. Sludge-conditioning polymers are derived from hydrocarbons, so that the displacement of polymers by CD carries both economic and socio-political benefit. The cost of α-CD is about three times higher than that of a typical sludge-conditioning polymer, but it is applied at very low doses so the increase in overall chemical cost is relatively small. This cost is more than offset by the savings realized from the reduced polymer dosage. The benefits of the CD are incremental; the CD essentially boosts the performance of the polymer(s) applied with regard to cake solids, drying rate, and capture efficiency. The cost:benefits are site-specific, but they are especially attractive at locations where sludge disposal costs are high. Finally, the CDs are biologically derived products. Sludge-conditioning polymers are derived from hydrocarbons, so their partial replacement with CDs carries both economic and environmental benefit.

Taiwan researchers (Perng et al. 2006) have studied the application of nanosilica for paper mill dewatering. The study was conducted in a paper mill in Taiwan which produces cultural and industrial paper products and which applies sedimentation and a single-stage activated sludge (AC) process to treat its mill effluent. The primary sludge from sedimentation and the waste biosludge from the AC stage were collected for the experiment. A conventional cationic polymer and a nano-silica preparation were respectively used as a dewatering agent and co-agent to see whether the dewatering efficiency could be enhanced. Sludge dewatering efficiencies were quantified using the SRF and capillary suction time (CST). A 23-factorial experimental design was used to delineate the effects and interactions of the sequence of polymer addition and the dosages. Analyses of the factorial design on the CST and SRF tests showed that both the primary sludge and biosludge had similar treatment behaviors. All three variables under investigation were significant, but none showed interactions with each other. The biosludge had a poorer dewatering efficiency than did the primary sludge on the CST and SRF tests. They found that the cationic polymer should be added first, followed by the anionic nano-silica. The reverse sequence of addition was largely deleterious to the dewatering of the primary sludge. Both the cationic polymer and nano-silica showed close weighting factors on the dewatering efficiency.

The commonly used dewatering devices used in the paper industry are rotary vacuum filters, centrifuges, V-presses, twin-wire presses, and screw presses. In some situations, the primary dewatering device is followed by a press to further increase the solids content. The vacuum filter had been the most popular device. Solids capture in vacuum filter is 90–95%, and the cake produced is about 20% solids. In order for the filter cake to discharge properly from the filter, 10–20% long fiber (>100 mesh) must be present in the sludge (Miner and Marshall 1976). Vacuum filter cakes containing combined sludge solids can be further dewatered on V-presses to approximately 35–40% consistency. A V-press is just two disks providing a converging nip that applies pressure to the sludge to squeeze out the water. Vacuum filters can be equipped with either fabric media or steel coils. Fabric media are often used in situations when fiber content is low, the ash content is high, or the solids are otherwise difficult to dewater on a coil filter. The power costs for operating the large vacuum pump required by a vacuum filter are quite high. Vacuum filters are being replaced by belt presses, which seem to perform as well if not better, at lower operating cost.

Voith Paper has developed Thune, a new design of screw press for dewatering pulp and paper mill sludge (Norli and Smedsrud 2006). The trial was taken at the new Adolf Jass Schwarza mill at Rudolstadt Germany in 2005. The new screw press achieves high torque distributed evenly along the axis by integrating the inlet and discharge housings and the screen supports into the machine frame. The centerline of the press is kept low in order to minimize deflection at high loadings, the height above mountings being only 270 mm. The operating cost is kept low and the machine has been designed to facilitate maintenance and servicing. Above all the new press achieves a higher dewatering per screen area than comparable sludge presses. The Thune SPS70 screw press at Schwarza handles all fine and sludge for dewatering, fed by a Meri BlueDrain gravity table. A Meri Sediphant is used to predewater cleaner reject and prescreened sewer matter. Dynamic torque control ensures a uniform consistency of the discharge. Voith Paper dewatering center at Tranby Norway have also installed a smaller system at Orbro Kartong in Sweden, with a Meri Elephant filter and a Thune screw press.

Disk centrifuges have found little application in the paper industry. They have been tried as thickening devices but experience has been unsatisfactory. Basket centrifuges have been used to a limited extent for sludges that are very difficult to dewater, but they operate in a batch mode rather than continuously. Usually, it is desirable to use the continuous decanter scroll centrifuge. Special scroll units have been developed for secondary sludge, and they are usually preferred over the basket centrifuge. Scroll centrifuges dewatering combined paper industry sludges generally produce cakes of 20–40% consistency at solids capture efficiencies of 85–98% from sludges conditioned with polymer. As the centrifuges operate on the basis of density difference separation, the sludges which are much denser than water, such as high-ash sludges, provide the best application of centrifuges. Specially designed scroll centrifuges can dewater secondary sludge from 2 to 11% solids with 99.9% capture efficiency (Reilly and Krepps 1982). However, it required 6–8 kg polymer per tons of sludge for conditioning. Centrifuges have a relatively low capital cost but can be expensive to operate due to requirement of chemical conditioning agents, their high power requirements, and their maintenance costs. Dissatisfaction with centrifugation has been attributed to the following: (a) generation of poor quality supernatant that could cause a buildup of fines in the treatment system, (b) susceptibility of centrifuges to plugging with pieces of bark, and (c) the severe screw conveyor abrasion experienced at many mills.

V-presses have been applied successfully to the dewatering centrifuge and vacuum filter cakes containing as much as 30% biological solids. However, the combined sludges normally encountered require sufficient conditioning for vacuum filtration or centrifugation to render them amenable to V-pressing (Miner and Marshall 1976). V-pressing can be performed to raise the solids content of the sludge high enough for incineration (Stovall and Berry 1969). V-presses generally produce primary sludge cake consistencies of 30–45%. Either a V-press or a screw press would precede most bark boilers burning bark and sludge. The sludge would enter the press at 15–25% solids and be subjected to a pressure of 690 kPa to raise the solids to the 30–45% suitable for incineration (McKeown 1979).

Pressure filters are the most powerful dewatering devices available. For combined sludge, cake of 30–35% consistency can be produced with solids capture efficiency of 95–100% (Miner and Marshall 1976). However, it is necessary to precoat the filter cloth to facilitate cake discharge and minimize the frequency of media cleaning. Diatomaceous earth, flyash, cement dust, etc. can be used for precoating. Media cleanliness has been indicated as a crucial parameter in determining the pressure filter cycle time. Pressure filtration also requires conditioning of the sludge before filtration. On pure secondary sludge, 35–40% cake solids can be achieved with a conditioning agent and a pressure of 200–250 psi. The main drawback of the pressure filter is that it is a batch operation and requires a lot of operator attention. Continuously operating automatic units have also been developed, but they are mechanically complex and therefore subject to many maintenance problems.

Moving belt press (Twin-wire press) has received intensive industry interest in the past. Many paper mills have installed moving-belt presses. On primary or combined sludges, moving-belt presses have generated cakes of a consistency comparable to that of two-stage dewatering with V-presses, and with similar or somewhat higher conditioning costs and generally lower power consumption. Polymers are commonly used for the sludge conditioning, and some processes use dual-polymer systems. The cake solids are 20–50% for the primary sludge whereas they are 10–20% for the secondary sludge. Capture efficiency is very high for belt presses, about 95–99% of the solids fed. Requirement of operator attention is low. These presses require power only to drive the belt, thus they are energy-efficient. Another advantage is their ability to operate on secondary biological sludge. However, the major operating problem is belt life, which is only few months. The usual cause of failure is puncture of the belt by incompressible objects in the sludge. The press is also subjected to corrosion due to hydrogen sulfide gas that is sometimes generated if there is any sulfur content in the sludge.

The latest development in sludge-dewatering is screw press of new design. These presses produce cake solids of 50–55% when operated as the only sludge dewatering device, solids capture ranges from 70 to 88% with no polymer addition on primary sludge (Toole and Kirkland 1984). Biological solids adversely affect solids recovery. Polymer can be used to improve efficiency but it has little or no effect on final sludge consistency; therefore is often not used on primary sludge. With secondary sludge, polymer is used. These presses appear to be energy-efficient. Screw presses are replacing twin-wire presses as the dewatering technology of choice for the pulp and paper industry.

18.3 Methods of Disposal

The pulp and paper industry disposes of its dewatered solids by landfilling, incineration, land spreading, or through alternate uses (Monte et al. 2009).

18.3.1 Landfill Application

Landfill has been the most common method till recent past for disposal of sludge, etc. (Gavrilescu 2004; Monte et al. 2009). However, the major factors to be considered when planning for landfill site include:

- Environmental suitability of area for landfill
- Geology of the area
- Environmental impact of run off water from the site
- Impact on ground water
- Composition and volume of the sludge
- Transportation cost

Mills favor landfilling whenever disposal sites are readily available and handling costs are low (Russel and Odendahl 1996). Landfilling is preferred because of the relatively low capital investment and the availability of mill owned land. In recent years, however, regulatory agencies have recognized the potential for far-reaching adverse environmental effects from landfilling activities. This has resulted in the tightening of regulations and requirements for more monitoring, environmental impact assessments, closure plans, and public consideration.

Normal sanitary landfill practices should be observed in constructing an industrial landfill. Some of the requirements that must be met are as follows (McKeown 1979):

- The disposal site should be a minimum distance above groundwater
- All subsurface conduits – such as culverts, gas and water lines – should be removed
- The site should be above the flood plain and be protected from flooding
- The site should be a minimum distance from a public well, highways and watercourse, and
- The nearest property line should be a certain distance away. After a site is chosen, according to the listed criteria, it should be used in accordance with good operating procedures for sanitary landfills.

Studies of the specific requirements for the design of papermill landfills are described by several researchers (Wardwell et al. 1978; Holt 1983; Ledbetter 1976). Modern landfill will require a liner design. A leachate collection system is required plus FML liners and a clay liner. In daily use, intermediate cover is usually not required, but a final cover will be, and it must be impermeable, properly sloped, vented, and have the ability to support vegetation.

Most of the environmental effects from landfills arise from the runoff of liquid leached from the waste, that is, the leachate. Leachate is generated at solid waste landfills as a result of physical, chemical, and biological activity within the landfill. Leachate characteristics are effected by

1. Precipitation
2. Run-off from and run-on into the landfill
3. Groundwater flow into the landfill
4. Evapotranspiration
5. Consolidation and water generated during the decomposition of the waste

These factors depend on local conditions such as climate, topography, soils, hydrogeology, the type of cover on the filled sections, and the type of waste. Leachates from pulp and paper industry landfills are known to contain conventional pollutants as well as metals, volatile organic compounds, phenolic compounds, volatile fatty acids, and some base neutral compounds (NCASI 1992). A NCASI study (1992) found that metals were usually present at fairly low concentrations. Volatile organic compounds were detected; toluene being the most common with a median concentration of 35 μg/L which is well below the Canadian Council of Resource and Environment Minister's goal of 300 μg/L for protection of aquatic life. The only base/neutral compounds found in detectable quantities, more than once were bis-(2 ethyl-hexyl)-pthalate and di-*n*-octyl pthalate. Pthalates are used in plasticizers, defoamers, and lubricating oils. Several kinds of phenolic compounds may be found in pulp mill landfill leachates including cresol isomers, phenols, and chlorinated phenols. Volatile fatty acids are produced from the decomposition of organic matter under anaerobic conditions and are common to leachates from much type of landfills. Acetic acid and propionic acid were found in the highest concentrations in pulp and paper mill landfills. A comparison of the average TOC and COD concentrations and the total UFA concentrations showed that UFAs contributed from 7 to 100% of the organic material in kraft mill landfill leachates (NCASI 1992). These leachates if not properly collected and treated may contaminate groundwater or surface water bodies. When landfills are on relatively permeable soils such as sand or gravel, leachate migration may cause contamination over areas many times longer than the area of the landfill. This can also occur over impermeable surfaces such as bed rock where the leachate can flow quickly toward a receptor. Groundwater contamination is a concern if the groundwater is a drinking water source or if it flows to a surface water body. If groundwater contamination directly affects the drinking water supply, the liability implications for the landfill owner/operator may be enormous. In addition to impairment of drinking water quality, leachate contamination of ground or surface water may result in the impairment of biological communities, aesthetics and recreational uses. Recognition of these potential effects, together with public awareness of landfilling issues dictates the necessity for a thorough EIA of new landfill sites.

In Canada, while the regulatory framework does not typically require an EIA for pulp and paper landfill proposals, many of the components of an EIA are fundamental to a successful permitting process. The key components include establishing a site development and approval plan, conducting effective public consultation throughout the process and undertaking solid technical studies and impact assessment analysis in support of the project (Russel and Odendahl 1996). The mill will need to decide on the specific scope of work based on the environmental conditions of the site, the community needs and the input from local regulatory agencies. Regardless of scope or approach, the mill as a proponent of a new landfill development must recognize the long-term commitment associated with landfill effects and adopt a management approach which incorporates public involvement with solid technical design and assessment.

A cost-effective approach has been developed and applied to a landfill in Ontario (Russel and Odendahl 1996). Essentially, a control chart method is used where warning and control limits are established for selected leachate indicators. Leachate indicators are selected based on the ratios between background and leachate concentrations, with the highest ratios indicating the most appropriate indicators. The leachate indicators selected should also represent different chemical groups such as metals, nutrients, ions, and organic compounds. Before landfill operation, the selected leachate indicators (three to five chemical constituents) are monitored monthly and the concentration differential is used to establish the warning and control limits. The landfill is monitored monthly during the operation and the concentration differential is plotted on a graph for each leachate indicator with the warning and control limits. If the value is within the warning limit, no action is required, however if the value is above warning or control limits, an established response is implemented to determine the cause and if necessary, initiate mitigative measures. The use of control charts for tracking water quality is beneficial as it is easily interpretable by the public and the mill's environmental managers.

The main disadvantages linked with the landfill is the possible risk of contamination of land and ground water due to which most of the developed countries are banning landfill in near future.

18.3.2 Incineration

The solid wastes rich in organics are incinerated mainly to reduce its volume and ultimate disposal in a feasible way which is easier and cheaper to landfill. Sludge is mainly burnt in fluidized bed and grate boilers. Burning of sludge is also associated with several limitations such as high capital investment, need of auxiliary fuel due to high moisture content, emissions of dioxin, NO_x, heavy metals, etc. in addition to other problems like:

- Storage
- Handling
- Low combustion efficiency
- Opaque stack gas
- Sticky ash formation

The following three types of incineration are in practice in the industry:

(a) Burning in an incinerator specifically designed for the sludge
(b) Burning in the bark boiler
(c) Burning in a power boiler that also burns fossil fuel

Burning the sludge in the bark boiler, which is a hogged fuel (combination fuel) boiler, seems to cause few problems except for reduced steam generation and reduced boiler efficiency (Miner 1981). Incineration in the bark boiler appears to be acceptable for sludge incineration if such a boiler is available on the mill site and if it can take the increased water load. Dewatering to higher levels, 45–50% solids, will make bark boiler incineration an even more attractive and will minimize the effect on boiler operation.

Combustion properties of a sludge are generally related to the amount of fiber present. Energy available is usually inversely related to the ash content. High ash values (up to 50% on dry basis) correlate with relatively low heating values. Sulfur values are important as related to emissions. Dewatering of the sludge stream will be required to increase solids up to some minimum level before combustion will be beneficial or even breakeven. Self-sustained combustion is available with some sludges generated depending on the moisture and organic levels. Cost and benefit evaluations can be made that will indicate the moisture level for optimum performance. Removal of additional water to increase solids above 50% requires a different method similar to paper passing from the press section to the dryer section on a paper machine (Busbin 1995). Thermal drying with hot gases or air can be done in a conveyor dryer, cascade system, or a stand alone drying unit. Reduced water content obviously helps improve efficiency and also can improve long-term storage options through reduced microbial growth.

The sludge product may be available in several forms depending on the method of combustion and the boiler used. Dewatered sludge straight off a screw press will be lumpy and after moving through several conveying operations begin to break up into a fuel that is fine, uniform and fibrous in nature. Sludge may also be processed further into briquettes or pellets (David 1995; Nichols and Flanders 1995; Sell and McIntosh 1988) to improve handling, storage or combustion characteristics. Blending dewatered sludge with other fuel (chip fines or saw dust) can help improve conveying characteristics. Pelletizing has come to the forefront as a method to convert combustible solid waste into a usable fuel. Waste to energy via pellet fuels needs to be examined more closely and regarded more highly as a successful solution to landfill crisis. They are quickly becoming a very viable and profitable alternative (Bezigian 1995).

Various types of combustion methods are available which include traveling grate boilers, vibrating grate boilers, other hog fuel boilers, bubbling bed combustors, circulating fluidized boilers, stage combustors, rotary kilns, and pyrolysis/pulse combustors (Kraft and Orender 1993; King et al. 1994; Fitzpatrick and Seiler 1995). The practicality of the above would be based on the sludge characteristics (contaminant contents, fuel size, volatility, ash characteristics, heat content, etc.) and to a great degree the volume to be fired (Busbin 1995). Operating experiences with stoker firing of TMP clarifier sludge with wood waste and combustion of the wastewater clarifier underflow solids in a hog fuel boiler with a new high energy air system have also been reported (King et al. 1994; La Fond et al. 1995). Combined cycle fluidized bed combustion of sludges and other pulp and paper mill wastes to useful energy has been suggested (Davis et al. 1995). Pulp and paper companies can improve the cost of operation by using proven, readily available power plant and combustion equipment and systems to efficiently convert the energy available in mill wastes to useful thermal energy and electrical power. By using the combined cycle concept, either as the combustion turbine combined cycle or the diesel combined cycle, the firing of wood waste and sludge provides net energy gain for the operation of facility rather than merely a means of disposal.

Other alternatives of recovering energy from the sludge have also been tried. A sludge gasification plant has been tested to generate the clean fuel (AghaMohammadi et al. 1995). Steam reforming as an alternative method for disposal of waste sludge has been suggested (AghaMohammadi and Durai-Swamy 1995). A novel method of thermal treatment of contaminated de-inking sludge has been proposed which is based on the application of the low-high-low temperature (LHL) regions during the combustion (Kozinski et al. 1997). The LHL approach allows for the simultaneous encapsulation of heavy metals within solid particles, removal of submicron particulate, and destruction of polycyclic aromatic hydrocarbons before they are emitted into the atmosphere. The encapsulation of the heavy metal layers surrounding the heavy metal-rich cores of the ash particles may prevent the metals from leaching under acidic conditions.

Sludge can be easy to burn with the right combustion technology. Knowing that the right technology is very fuel-specific and having the technology characterization customized for site-specific conditions is essential to make proper combustion technology choices. Incineration is not practical for high-ash sludges. Stringent air pollution emission requirements for combination boilers have diminished the amount of incineration practiced. One of the Finnish mills incinerate sludge if the solids content is over 32%, and landfills the sludge if it is less than 32% (Kenny et al. 1995). Operation of the boiler must also be considered when the sludge is not available as a fuel. Several points of consideration include the combustion temperature, fuel feed systems, and boiler rating. Older boilers burning sludge as an alternative fuel should be able to simply return to earlier operating states.

Some of the chlorinated organics not eliminated through process modifications could be trapped on the sludge from the external treatment process(es). The disposal of pulp and paper mill sludges, which may contain chlorinated organic compounds, represents an increasing problem. However, if those sludges could be dried to 90% dry content, in an energy-efficient manner, they could provide high enough flame temperature upon combustion in order to destroy the organic chlorides entrapped in the sludges. In addition, this approach could improve mills' fuel self-sufficiency.

18.3.3 Land Application (Composting)

Two factors viz., continued decrease in availability of landfill space and increase in energy cost in incineration, have forced the pulp and paper mills internationally to look for the land application of the same as a low cost disposal method. In composting process microorganism break down the organic matter of the sludge under aerobic conditions. It is suitable both for biosludge and sludge from primary clarifier.

Much work has been done with land application of pulp and paper mill sludge in the last 2 decades. In Canada, several mills are routinely doing land application and several have conducted serious field trials. QUNO Inc. Thorold, Ontario, Canada has experience with land application of primary, secondary, and deinking sludges (Pridham and Cline 1988). Primary and deinking sludges have been found to have similar characteristics – low nitrogen and high fiber content. Conversely, secondary sludges (biosolids) have relatively high nitrogen and phosphorus content and low fiber content making them more suitable for land application. Tests at QUNO found that the heavy metal content of the combined paper mill sludge was equivalent to that of the cattle manure, and about one-tenth that of municipal sludge. The sludge has been successfully used as a replacement for manure in agricultural applications, as well as for land reclamation projects of old sand pits, coal/clinker sites and a former foundry site. Work has been completed with Alberta pulp and paper mills in conjunction with the Alberta Research Council on land application (Macyk 1993). Land spreading trials have been completed on both agricultural and forest cut-block sites. Research is also being conducted by the Alberta Newsprint Corporation and Alberta Research Council to evaluate the environmental effect of land spreading conventional and deinking sludge (Pickell and Wunderlich 1995). Preliminary research indicated that the procedure should not present any problems in regard to soil quality or plant growth. Trials with land spreading around the mill site have been successfully completed by applying the sludge on top of a gravel base. Alberta Research Council has also completed research on ash and sludge land spreading in conjunction with the Slave Lake Pulp Corporation (SLPC) (Pickell and Wunderlich 1995). Grass yield on the test plot site at SLPC indicated as much as five times the yield of control plots. SLPC has had favorable results with sludge application on the surrounding agricultural area. Previously, landfilled sludge has been reclaimed and distributed to the farming community and applied using manure spreaders.

There has been considerable interest in use of paper mill biosolids and ink waste in agricultural land for many years (Pridham and Cline 1988). Sludges function only as amendments and not as fertilizers because they do not contain the elemental analysis required of a fertilizer (Atwell 1981). For a soil amendment, the carbon nitrogen ratio should be 20:1–30:1. An average composition of seven different paper mill combined sludges from ten different mill types was 26:1, so this criterion is being met. The calcium/magnesium ratio should be above 6:1; many combined sludges fail to meet this criterion but the addition of lime to the sludge fulfills it. Sludges are good soil amendments for sandy soils. Detailed analysis of the seven combined sludges did not indicate a heavy metal problem (McGovern et al. 1983). Trials have been conducted in which fly ash and either primary sludge or secondary sludge were applied to crop land. The fly ash–sludge blends were as effective as commercial fertilizer. In these same trials, lime mud applied to agricultural land performed better than dolomite limestone used for the same purpose (Simpson et al. 1983). Australian Newsprint mills Ltd. (ANM) used small quantities of biosolids on vegetable and horticultural gardens with good results and no observed detrimental effects (Hoffman et al. 1995). Several farmers have also used the material on small pasture areas and on orchards, but no objective evaluations have been carried out. Because of the high level of farmer interest, ANM carried land spreading trials on crops and pastures (Hoffman et al. 1995). The biosolids were utilized on a farm land close to mill. For this, a desk study and a survey of local farmers were conducted. It was found that biosolids would be readily used by farmers, if it could be demonstrated that it was a viable fertilizer, that it was safe to apply to the environment and the cost was competitive with existing practices. This study also confirmed that about 2,000 ha/year of land would be required to dispose of the material. It identified the area of interest of land for economic disposal as areas of crops and pasture land within 20 km of the mill and luceme flats where disposal could take place in winter. In 1992, a field experimental program started with a large area experiment on oats at a location known as Waitara. Biosolids were found to be slow to release their nutrients and produced an effect similar to fertilizer without producing adverse environmental effects. Rates of 16–64 tons/ha were required to substitute for normal rates of conventional fertilizer. ANM also conducted trials to spread the biosolids on forest land (Hoffman et al. 1995). Trials started in the Carabost and Green hills State Forests, near Tumbarumba. The major disadvantage with forest spreading over agricultural land spreading is the higher cost of transport to the disposal site. So, the cost of transport would normally make forest spreading unattractive. However, if the solid could be back loaded on log trucks then the economic disadvantage decreases. In Canada, Greater Vancouver Regional District (GVRD) and the University of British Columbia's Forest Sciences department embarked on a 3-year research program at UBC's Malcolm Knapp Research Forest in Maple Ridge to determine the environmental and silvicultural application of recycling pulp and paper sludge and treated sewage sludge as an organic forest fertilizer called Nutrifor (Pickell and Wunderlich 1995). The second phase of the program introduced Nutrifor as a viable fertilizer for forestry and other users.

Scott Paper Ltd. in New Westminster conducted a full-scale land application project with GVRD in which paper mill sludge is combined with municipal sludge and then applied to a tree farm in the Fraser Valley (Pickell and Wunderlich 1995). In 1990, the GVRD, Western Forest Products Ltd. and the IBEC Aquaculture participated in a fertilization project in which various mixtures of pulp mill wastes, sewage sludge and fish mort silage were applied to forest sites in Southern British Columbia near Port McNeil on Vancouver Island (Taylor et al. 1992). Initial results indicate a rapid response by young conifers to organic fertilization. In 1992, a project cosponsored by Nutrifor was completed at Malaspina College where 600 dry tons (2,500 wet tons) of sludge were applied over an area of 26 ha in the Malaspina College Research Forest on Central Vancouver Island (Braman 1993). Full-scale projections were made using data obtained from the trials to determine cost per tons of sludge for each of three application methods (Braman 1993). The lowest cost method of spreading the sludge was found to be dry application. Projected cost could be reduced to $56/wet tons to apply approximately 36,000 wet tons onto 400 ha.

Seattle, Washington has a sludge management plan which calls for the development of a number of alternative methods (Pridham and Cline 1988). Since halting ocean disposal in 1972, the system has made compost, undertaken strip mine reclamation and is said to have been one of the first to use biosolids in forestry. An innovative application is the growing of hops for the beer industry. Seattle is making use of about 101,000 dry tons/year at 20% moisture. The effects of lands spreading wastewater sludges from pulp and paper mills were investigated by examining (a) the fate of chlorinated organic materials in land spread sludge and (b) the impact of sludge on plant growth and wild life (Sherman 1995). The results indicated that high-molecular-weight chlorolignins were rapidly absorbed by soil or humic matter and organic chlorine was slowly released as inorganic chloride. There was no detectable release of new monomeric chlorolignin-related chloro-compounds. Even under severe extraction conditions, the extractability of low-molecular-weight chloroaromatic compounds decreased rapidly (half lives of 6–70 days), apparently the result of biodegradation and biologically mediated chemical binding into the soil humic structure. No persistent biotransformation products were detected. Sludge applications produced an increase in plant growth (grass, hay, corn, trees). Studies of wildlife on sludge-amended soils did not detect any adverse effects on the health of individuals or on reproductive parameters. Criteria have also been proposed for the land spreading of solid waste (Springer 1993). Briefly, the proposed criteria are:

1. The soil sludge mixture must not have a high content of heavy metal that can be taken up by growing plants
2. The soil-waste pH should be 6.5 or higher
3. Excess nitrogen should not be applied beyond that normally taken up by the crop in one season
4. The sludge applied should be free of living pathogenic organisms
5. Solids must be applied in such a manner that they are not available for direct ingestion by domestic animals or humans

Land application is not a trouble-free technology however (Springer 1993). The most commonly noted problems are odors, groundwater contamination, heavy metals, and specific organic toxics. Other problems are noise, surface water contamination, pathogens, and excessive nitrogen application. The process of applying sludge is dirty and noisy, so if there are houses in the vicinity, potential difficulties will arise. Actually, public and user acceptance has been very good because sludge is applied mostly to rural areas close to the mill and in some cases on mill-owned land.

Pulp and paper mill sludges are usually amenable to well-controlled composting techniques. Markets for compost include land application for agriculture, horticulture, land reclamation, landscaping, and individual consumer use. One mill has had considerable success with marketing its composted sludge. This mill presently composts about 50% of its sludge. The mill sells the compost to a limited number of distributors who market the material in an area within a 250-mill radius from the mill. Initiation of new composting operations within the industry has slowed considerably since the mid-1980s. Lack of sufficiently large, locally available markets for compost and regulatory concerns about the possible presence of chlorinated dioxins and furans in industry sludges are two common reasons for the limited utilization of this management alternative. Recent industry initiatives to reduce the presence of dioxin in sludges are likely to relieve some regulatory concerns about land application of sludges.

A mill in the northeastern United States began working with a third party company to produce synthetic topsoil using sludge (Weigand and Unwin 1994). The process involves the homogenization of sludge with varying proportions of sand, gravel, and fertilizer to produce a synthetic soil. More than a dozen landfills have used the soil as part of the final cover. It also has use in other applications requiring vegetative cover. The pulp fiber content of the synthetic soil probably allows for an increased resistance to erosion before the establishment of vegetative cover.

18.3.4 Recovery of Raw Materials

Paper industry sludges usually contain significant percentages of both cellulose fiber and paper making fillers such as clay and titanium dioxide. Attempts have been made to reduce sludge volume by reclaiming the fiber or filler or both for reuse (Weigand and Unwin 1994). There are several methods to recover raw materials from sludge. One of the most common is to recycle primary sludge back into the mills' fiber processing system. Recycled paperboard mills commonly use this technique. Some manufacturers of unbleached and bleached pulp and paper have also practiced recycling primary sludge back to the mill with limited success (Rosenqvist 1978). Segregated effluents from paper machines, bleach plants, and various cleaning and screening operations can be good targets for fiber reclamation because they usually lack contaminants such as bark or causticizing waste solids. Using some fractionation scheme for the sludge may also provide recovery of fiber alone. The complexity of fiber recovery systems varies widely and depends on the nature of the constituents in the sludge. Mills producing bleached pulp sometimes add recovered fiber to the unbleached pulp entering the bleach plant. This strategy allows for both the reclamation of unbleached fiber and the brightening of previously bleached fiber which may have dirtied by exposure to contaminants in the wastewater. Some mills have associated the reuse of fiber recovered from sludge with increased deposits of pitch on equipment. Use of fractionation system helps to recover filler. Most systems for which pilot- or full-scale data are available have employed a thermal oxidation technique for destroying the organic fraction of the sludge to yield filler in the form of ash (Weigand and Unwin 1994). Experiments with calcination systems have revealed that controlling the kiln temperature 816 and 843°C helps to avoid formation of fused agglomerates which can cause the recovered filler to be excessively abrasive. Wet air oxidation can be also used to recover filler materials from sludge. One U.S. mill is practicing this process on a full scale (Weigand and Unwin 1994). Wet air oxidation is an oxidation reaction carried out in a liquid environment under high temperature and pressure. This process is capable of reducing sludge volume through oxidation of the organic fraction to yield an ash composed of inert materials, e.g., filler clay, titanium dioxide and calcium carbonate for reuse in the papermaking process. Initial experience with the operation of WAO unit for filler recovery revealed problems with Ca-sulfate and Ca-oxalate scale deposition. Both pilot- and full-scale systems have demonstrated some problems with low brightness of the recovered filler. In Turkey, primary sludge has been successfully used in the manufacture of hardboard (Ozturk et al. 1992). Full-scale studies using sludge at a 1:4 ratio indicate that the use of 28 bdt/day of waste primary sludge mill save $455,000/year on wood costs and $130,000/year on electricity costs.

18.3.5 Production of Ethanol and Animal Feed

Ethanol is a common additive in automobile gasoline. Traditionally, it is produced by fermentation of starches and syrups. Interest has been shown to produce ethanol from agricultural waste, municipal solid waste, and pulp and paper mill sludge in order to reduce production cost and to make ethanol more widely available. Laboratory and pilot scale studies to produce ethanol from wood-based feedstocks have used acid and enzymatic hydrolysis followed by fermentation of the resulting sugars into ethanol (Goldstein and Easter 1992; Alterthum and Ingram 1989; Lee and McCaskey 1983). Primary sludges can be used as feedstock for ethanol production because they are widely available in sufficient quantity and that they have little economic value. In University of Florida, Dr. Ingram's group has conducted research on conversion of cellulose and hemicellulose fractions of wood-based feedstocks into hexose and pentose sugars followed by fermentation to ethanol using a genetically engineered strain of *Escherichia coli* (Ingram and Conway 1988). The advantage of this process is that it can ferment both the pentose and hexose sugars into ethanol thereby increasing the overall yield.

Sludge has been also used for production of animal feed. There are two methods for using sludge in animal feed. One method involves production of single cell protein. Cell protein is present in secondary sludge and derives from the fermentation of fibrous sludge. It is possible to dry these proteins and incorporate them into feed mixtures. In the United States, one mill used a process to convert secondary sludge into saleable protein product for use in animal feed. Mechanically, dewatering secondary sludge to 12% solids with further dewatering by feeding a mixture of sludge and oil to specially designed multiple effect falling film evaporators produced a 45% protein material. Centrifugation of the evaporator discharge gave 83% dry solids, 1% water, and 16% oil. Targeted markets for the finished product included feed for cattle and poultry and use in agricultural composting. However, acceptance of this product in the market was not sufficient to support continued production.

The second method incorporates sludge directly into animal feed mixtures (Weigand and Unwin 1994). This method exploits the presence of carbohydrates which are primarily in the form of cellulose and other nutrients present in primary or combined sludges. Research in the early 1970s included experiments on the palatability and digestibility of sludge-augmented feed mixture on goats, sheep, and cattle. It was found that the digestibility of sludge relates directly to the carbohydrate content and inversely to the ash and lignin content. Hardwood pulp residues were found to be more digestible than softwood residues (Millet et al. 1973).

18.3.6 Pelletization of Sludge

The reasons for producing sludge pellets are:

1. Volume reduction
2. Odor control
3. Recovery of fuel value
4. By-product applications

The most common reason for production of pellets is for use as an alternative fuel. One mill transports dewatered sludge to an off-site pellet mill for drying and formation into pellets. The mill purchases the finished pellets as a fuel supplement. The finished pellets contain 15–20% moisture and 10% ash. They have a heating value of 14.7×10^6 J/kg (Weigand and Unwin 1994).

Two companies are now manufacturing pellets by using mixtures of sludge and nonrecyclable paper (Bajpai et al. 1999). These pellets are being marketed as an alternative fuel compatible for use in most stoker and some pulverized coal boilers. The amount of sludge in these pellets can range between 10 and 66%. It is possible to control the fuel value of the pellets by manipulating both the sludge content and the grade of nonrecyclable paper used. The fuel values of the finished pellets are in the range of $14–23 \times 10^6$ J/kg. The regulatory agencies require evaluation of alternative fuels for by-products of combustion before widespread use of the fuel. Companies involved in both production and use of sludge and NRP fuel pellets have indicated that regulatory reaction to trial run data has generally been positive. NCASI has developed a proprietary process to convert combined sludge from a recovered paper deinking mill into a granular product. The product has been used as a carrier material for agricultural as well as home and garden pesticides and can compete with other common pesticide carrier materials composed of clay, vermiculite, diatomaceous and cob products. Claims for the product indicate that it is superior to some of these conventional carriers because it is dust-free and attrition-resistant (Weigand and Unwin 1994). The company's production facility has a capacity of 180 tons/day of the granular product.

Kitty littre, poultry littre, and large animal bedding have all used pelletized sludge. One U.S. mill processes all of its primary sludge into several varieties of animal litter sold to a distributor for marketing. The litter production process is proprietary. It involves sanitizing and deodorizing primary sludge followed by drying and pelletization. Kitty litter is the primary product manufactured, but other products include large animal bedding, pet bedding and bedding for laboratory animals. Grocery stores market kitty litter and feed stores market bedding products. Bedding sells in 25- and 50-lb bags and 1,000 lb totebins (Weigand and Unwin 1994). Several other companies have studied the feasibility of using sludge to produce kitty or poultry littre. In these cases, they have usually demonstrated production of a quality litter product from primary sludge. Initial capital costs, distribution and marketing issues and incompatibility with company business strategies have inhibited some companies from persuing this byproduct alternative.

18.3.7 Manufacture of Building and Ceramic Materials and Lightweight Aggregate

Sludge use in building products has followed three general techniques. One method is the use of sludge as a feedstock to a cement kiln. Raw materials used to produce cement can include calcium carbonate, clay, silica, and smaller amounts of aluminum and iron. Some sludges contain significant quantities of these materials. Two companies have extensively investigated this alternative and one mill currently practices this on full scale (Bajpai et al. 1999). The mill sends all its primary sludge and all its coal boiler ash to the cement manufacturer. This is a combined total of approximately

100 tons/day. For the kiln involved, this amount of material represents only about 2% of the total feed stock.

Another alternative is the use of sludge in cementitious products. Lot of work has been done on the use of organic fibers including wood pulp in cementitious composites. The advantages include increased durability and pumpability as well as reduced shrinkage-related cracking (Thomas et al. 1987). Two studies undertaken to assess the performance characteristics of composites which included paper industry sludge concluded that a composite material potentially useful in building blocks, wallboards, panels, shingles, fire retardants, and filler materials for fireproof doors could result from combining Portland cement with sludge from deinking mill (Thomas et al. 1987). It was found that mixtures including Portland cement, ash, sand, and sludge yielded a compressive strength comparable to conventional concrete and superior flexural strength (Thomas et al. 1987).

Sludge has been also used in the production of LWAs (Weigand and Unwin 1994). Aggregate is a term describing a collection of materials used as a filler in construction materials. Aggregates find use in cementitious products such as concrete, masonary, building blocks, and asphalt. Sand and gravel or both are typical aggregate materials mixed with cement to produce concrete. LWA refers to a select group of materials which allow for reductions in final density while maintaining acceptable strength properties. Products which sometimes incorporate LWA include concrete block, architectural panels, and decorative stone.

18.3.8 Landfill Cover Barrier

Paper industry sludges have been found to show low hydraulic conductivity (permeability). This finding has led to research by many groups on the potential utilization of sludge as hydraulic barrier layer in landfill cover systems. In 1987, NCASI completed construction of four pilot-scale field test cells designed to allow investigation and comparison of the performance of hydraulic barrier layers made from sludge and made from clay (Weigand and Unwin 1994). Results obtained from these cells during the first 5 years of operation indicate that the sludge barriers perform as well or better than the clay barriers. Experience with the use of paper industry sludge as daily, interim and final cover for paper industry and municipal landfills is available. Worthy of special mention is the experience of one organization. To demonstrate the utility of paper mill sludge as landfill-capping material, this recovered fiber processing mill constructed six test cells to compare the performance of primary sludge combined sludge and clay as hydraulic barriers (Weigand and Unwin 1994). Data from these test cells sufficiently supported a petition to the Massachusetts Department of Environmental Protection for a full-scale demonstration project. The project involved capping a 2 ha municipal landfill with combined mill sludge. To date, monitoring of cap performance indicates that the demonstration has been successful.

18.3.9 Other Uses

Pyrolysis, gasification, and supercritical water oxidation have been studied as a way of reducing sludge volume. During pyrolysis, oil like liquids and gases are formed which have fuel value. Study has been conducted on pyrolysis of cellulose-based waste materials but there is not much published information on experience with pyrolysis of pulp and paper industry wastes (Weigand and Unwin 1994). Pilot studies have been conducted on the application of this technology to wood chips, recycle mill sludge, and bleached kraft mill sludge. There is no report on a full-scale experience with the pyrolysis of sludge. Supercritical water oxidation has undergone research as a waste management technology for approximately 10 years. The process involves the decomposition of organic and some inorganic materials in the aqueous phase above the critical point of water (374°C and a pressure of 22×10^3 kPa). In this state, organic materials become much more soluble in water and oxidize readily. The principal of supercritical water oxidation except that wet air oxidation maintains subcritical conditions. No full-scale supercritical water oxidation units are currently in operation. Laboratory scale research has been conducted on supercritical water oxidation of pulp and paper mill sludge. This work used an 80 cm^3/min bench top system. Operating limits for the reactor were 600°C and 25.5×10^8 kPa. Residence time in the reactor ranged from 10 s to 10 min. In the experiments, a 99% reduction of total organic carbon was possible. The problems anticipated with large-scale and or full-scale systems involve (1) corrosion of equipment particularly for low pH and high chloride concentration wastes and (2) deposition of salts or pyrolytic chars which could lead to plugging or increased cleaning requirements.

In Canada, Ensyn Technologies has developed a rapid thermal processing (RTP) reactor which heats biomass to an extremely high temperature (400–900°C) for 0.5 s at atmospheric pressure with no oxygen (Rodden 1993). RTP is also called fast cracking and is similar to the catalytic cracking process used by the oil industry. The rapid heating of the biomass cracks the chemical bonds and produces a liquid bio-oil. Rapid cooling prevents the completion of chemical reactions. The feed stock can vary: pulp sludge, wood waste, rice husks, and agricultural residue. The bio-oil created from the process has been used as a fuel oil substitute. Destructive distillation as a resource recovery process for solid waste was evaluated during 1982–1984 at Marcel Paper Mills, Elmwood Park, New Jersey (USA) (FioRito 1995). The results indicate that the process is environmentally friendly and has the edibility to provide substantial energy savings utilizing organic solid waste as its sole source of fuel. The technology is able to fractionate the biomass content of municipal and industrial wastewater sludge to a combustible gas and inert char in an environmentally safe manner. Full-scale operation of the process was carried out on sewage and deinked paper mill sludge at installations in California and New Jersey.

The expense of solids disposal could be eliminated by destroying the microorganisms in the excess secondary sludge and recycling the material through the treatment process. Springer et al. (1996) used a simple mechanical device – a kady mill to breakdown the microorganisms in the excess sludge allowing all of the material to be recycled to the treatment process. The kady mill combines the effects of high shear and temperatures, both of which are required for efficient cell destruction. Based on 60 days of operating data, it was found feasible to operate an activated sludge plant in extended aeration mode by recycling sludge that has been lysed in a kady mill. This process could be an alternative wastewater treatment system for use in the pulp and paper industry. The system would be most suitable for use in mills operating well within EPA permit discharge limits for BOD. This system operated with an average COD-removal efficiency of 80%, compared with 87% removal for the conventional system. Both systems operated with an influent COD of 260 mg/L. The sludge-lysis-and-recycle process operated free of bulking problems. This process appears to be an economically attractive alternative to conventional treatment if higher BOD values can be accommodated.

The biosolids generated by activated sludge process can also be anaerobically digested to reduce its volatile solids and generate energy in the form of methane gas (Krogmann et al. 1997).

Hammond and Empie (2007) have reported that secondary wastewater sludge can be added to the black liquor gasification process at a paper mill to produce a combustible fuel gas. The gas is fed to a combined cycle boiler plant and turbo-generator system to generate electricity.

Anaerobic digestion is found to be an effective alternative for sludge management in pulp and paper treatment plants (Guiot and Frigon 2006). Waste characteristics, organic loading rate, hydraulic retention time (HRT), temperature, pH, mixing, and the presence of inhibitory matters are shown to affect the rate of anaerobic digestion. An increase in temperature increases the digestion rate and lowers the HRT and digester volume, resulting in higher amounts of treated waste loads. Biogas recovery in anaerobic digestion avoids odor release and lowers greenhouse gas (GHG) emissions from landfill diffusion and from burning fossil fuels. The solution is believed to provide 100% digestion of the sludge generated, thus offering an improved means of disposal, green energy and lowered GHG discharges. Methane and carbon dioxide are also generated during the process. Purified methane from sludge digestion can be used as natural gas, which can replace fossil fuel and reduce GHG emission. Anaerobic digestion is believed to be a cost-effective approach in the valorization of waste sludge, especially when the cost of natural gas is high.

A method of treating paper mill sludge treatment as raw material for the manufacture of animal bedding won a National Recycling Award for EnviroSystems, Cheshire, UK (Anon 2005). The sludge is dried down to 90% dry material and broken in small pieces, and then is heat-treated. The finished product is called EnviroBed and is being used as bedding for 50,000 dairy cows in the UK. Sludge from Bridgewater Paper and Shotton Paper is being processed at EnviroSystems plant in Cheshire. A second plant at Brent Pelham, Hertfordshire, is being supplied by material from Aylesford. EnviroSystems is looking for additional supplies of suitable paper crumble, with 40–45% organic matter or above and without a high moisture content.

The wastewater sludge of Neenah Paper, Neenah, WI, USA, is recycled into useful forms, including electrical power and glass aggregate (Anon 2004). 5,000 tpy of paper sludge are recycled using a system installed by Minergy Corp, also located in Neenah. Solids are melted in a glass furnace, destroying organic compounds. The inorganic mineral waste exits the furnace as liquid glass which is used in the manufacture of floor tiles, abrasives, roofing shingles, asphalt and decorative landscaping materials. Via a steam generator furnace heat produces electricity which dries the wastewater solids. The recycling process provides many environmental benefits, in Neenah Paper's case preserving green space and reducing landfill use. The company has developed an online tool for individuals and businesses to calculate the environmental benefits of using recycled paper.

Oxycair is an innovative treatment technology developed to treat various types of wastewaters, which has been shown to generate substantial savings over conventional treatment costs (Gagnon and Haney 2005). The technology uses patented processes, is based on concurrent physical mechanisms taking place within multiple reactor vessels and uses no chemicals. The destructive mechanisms include physical destruction, thermic stabilization, air supersaturation, oxidation, explosive decompression, cavitation, and microbubble oxidation. The technology has been tested at both laboratory and industrial scale, transforming excess sludge stream into a nearly sterile stream rich in dissolved oxygen and the nutrients and micronutrients contained in bacterial cells. This stream can be returned directly to the bioreactor as a nutrient supplement. Oxycair is a service provided by WR3 Technologies Inc., Canada.

References

AghaMohammadi B, Durai-Swamy K (1995) A disposal alternative for sludge waste from recycled paper and cardboard. In: Joyce TW (ed) Environmental issues and technology in the pulp and paper industry – a Tappi Press anthology of published papers 1991–1994. Tappi Press, Atlanta, pp 445–458

AghaMohammadi B, Shekarchi S, Durai-Swamy K, Steedman W, Dauber R (1995) Testing of a sludge gasification plant at Inland Containers Ontario (California) Mill. In: Joyce TW (ed) Environmental issues and technology in the pulp and paper industry – a Tappi Press anthology of published papers 1991–1994. Tappi Press, Atlanta, pp 431–443

Alterthum F, Ingram LO (1989) Ethanol production from glucose, lactose, and xylose by recombinant *E. coil*. Appl Environ Microbiol 55(8):1943–1948

Anon (2004) Paper mill sludge converted to glass aggregate. Recy Pap News 14(7):2–3

Anon (2005) Turning sludge into animal bedding. Pap Technol 46(9):7

Atwell JS (1981) Disposal of boiler ash. Tappi J 64(8):67–70

Bajpai P, Bajpai PK, Kondo R (1999) Biotechnology for environmental protection in pulp and paper industry. Springer-Verlag, Germany, pp 209–238

Banerjee S (2009) Sludge dewatering with cyclodextrins: a new cost-effective approach. In: Thirteenth international water technology conference, IWTC, Hurghada, Egypt, 13, 2009

Battaglia A, Calace N, Nardi E, Petronio BM, Pietroletti M (2003) Paper mill sludge-soil mixture: kinetic and thermodynamic tests of cadmium and lead sorption capability. Microchem J 75: 97–102

Benitez J, Rodriguez A, Suarez A (1993) Optimization technique for sewage sludge conditioning with polymer and skeleton builders. Water Sci Technol 28(10):2067–2073

Bezigian T (1995) Alternative solutions to landfilling paper mill packaging waste. In: Joyce TW (ed) Environmental issues and technology in the pulp and paper industry – a Tappi Press anthology of published papers 1991–1994. Tappi Press, Atlanta, pp 459–473

Braman JR (1993) Forest fertilization with sludge in Malaspina College research forest. Operations report on Malaspina project 1992, Feb 1993, pp 1–46
Busbin SJ (1995) Fuel specifications – sludge. In: Joyce TW (ed) Environmental issues and technology in the pulp and paper industry – a Tappi Press anthology of published Papers 1991–1994. Tappi Press, Atlanta, pp 349–355
David PK (1995) Converting paper, paper mill sludge and other industrial wastes into pellet fuel. In: Joyce TW (ed) Environmental issues and technology in the pulp and paper industry – a Tappi Press anthology of published papers 1991–1994. Tappi Press, Atlanta, pp 365–367
Davis DA, Gounder PK, Shelor FM (1995) Combined cycle-fluidized bed combustion of sludges and other pulp and paper mill wastes to useful energy. In: Joyce TW (ed) Environmental issues and technology in the pulp and paper industry – a Tappi Press anthology of published papers 1991–1994. Tappi Press, Atlanta, pp 379–384
FioRito WA (1995) Destructive distillation – paper mill sludge management alternative. In: Joyce TW (ed) Environmental issues and technology in the pulp and paper industry – a Tappi Press anthology of published papers 1991–1994. Tappi Press, Atlanta, pp 425–429
Fitzpatrick J, Seiler GS (1995) Fluid bed incineration of paper mill sludge. In: Joyce TW (ed) Environmental issues and technology in the pulp and paper industry – a Tappi Press anthology of published papers 1991–1994. Tappi Press, Atlanta, pp 369–378
Gagnon D, Haney HE (2005) Oxycair solution: new and unique technology for pulp and paper secondary sludge management. In: 91st annual meeting pulp and paper technical association of Canada, Montreal, Canada, 8–10 Feb 2005, Book A, pp A123–A126
Gavrilescu D (2004) Solid waste generation in kraft pulp mills. Environ Eng Manage J 3: 399–404
Geng X, Deng J, Zhang SY (2006) Effects of hot-pressing parameters and wax content on the properties of fiberboard made from paper mill sludge. Wood Fiber Sci 38(4):736–741
Goldstein IS, Easter JM (1992) An improved process for converting cellulose to ethanol. Tappi J 75(8):135–140
Guiot SR, Frigon J-C (2006) Anaerobic digestion as a sustainable solution for biosolids management in the pulp and paper sector. In: 92nd annual meeting of the pulp and paper technical association of Canada, Montreal, QC, Canada, 7–9 Feb 2006, Book A, pp A261–A264
Hammond D, Empie HJ (2007) Gasification of mixtures of black liquor and secondary sludge. Tappi J 6(3):9–15
Hoffman R, Coghill R, Sykes J (1995) Solid waste management at ANM, Albury – from waste problems to resource opportunity. Appita 48(1):12–14
Holt WH (1983) Solid waste landfills at paper mills. Tappi J 66(9):51–54
Huang CP, Chang MC (1997) Conditioning of sludge and selection of polymers for the purpose. Ind Pollut Abatement 64:88–111
Ingram LO, Conway T (1988) Expression of different levels of ethanologenic enzymes from *Zymomonas mobilis* in recombinant strains of *E. coli*. Appl Environ Microbiol 54(2):397–404
Kenny R, Coghill R, Almost S, Easton C (1995) CPPA international sludge dewatering survey. In: Proceedings of the 1995 Tappi environmental conference, Atlanta, GA, April 1995.
Kenny R, Almost S, Coghill R, Easton C, Osterberg F (1997) CPPA/international review of pulp and paper activated sludge dewatering practices. Pulp Pap Canada 98(8):T277–T281
King B, McBurney B, Barnes TW, Cantrell M (1994) Operating experience with stoker firing TMP clarifier sludge with wood waste. In: Joyce TW (ed) Environmental issues and technology in the pulp and paper industry – a Tappi Press anthology of published papers 1991–1994. Tappi Press, Atlanta, pp 393–403
Kozinski JA, Zheng G, Saade R, DiLalla S (1997) On the clean and efficient thermal treatment of deinking solid residues. Can J Chem Eng 75(1):113–120
Kraft DL, Orender HC (1993) Considerations for using sludge as a fuel. Tappi J 76(3):175–183
Krigstin S, Sain M (2005) Characterization and potential utilization of recycled paper mill sludge. In: Paper presented at the PAPTAC 91st annual meeting 2005, Montreal Quebec
Krogmann U, Boyles LS, Martel CJ, McComas KA (1997) Biosolids and sludge management. Water Environ Res 69(4):534–550

La Fond JF, Lantz D, Ritter LG (1995) Combustion of clarifier underflow solids in a hog fuel boiler with a new high energy air system. In: Joyce TW (ed) Environmental issues and technology in the pulp and paper industry – a Tappi Press anthology of published papers 1991–1994. Tappi Press, Atlanta, pp 385–392

Latva-Somppi J, Tran HM, Barham D (1994) Characterization of deinking sludge and its ashed residue. Pulp Pap Canada 95(10):31–35

Ledbetter RH (1976) Design considerations for pulp and paper mill sludge landfills. U.S. Environmental Protection Agency EPA-600/3-76-11, December 1976

Lee YY, McCaskey TA (1983) Hemicellulose hydrolysis and fermentation of resulting pentoses to ethanol. Tappi J 66(5):102–107

Macyk T (1993) Research relative to land application of BCTMP mill waste in Alberta. Preprints 1993 Pacific Paper Expo 1993, pp 91–95

McGovern JN, Berbee JG, Bockheim TG, Baker AJ (1983) Characteristics of combined effluent treatment sludges from several types of pulp and paper mills. Tappi J 66(3):115–118

McKeown JJ (1979) Sludge dewatering and disposal. A review of practices in the U.S. paper industry. Tappi J 62(8):97–100

Mertz HA, Jayne TG (1984) Start up and operating experience with Zimpro high pressure wet oxidation system for sludge treatment and clay reclamation. In: Proceedings of the Tappi environmental conference, Savannah, GA, 9–11 April 1984

Millet MA, Baker AJ, Satter LD, McGovern JN, Dinius DA (1973) Pulp and papermaking residues as feedstuffs for ruminants. J Anim Sci 37(2):599–607

Miner RA (1981) A review of sludge burning practices in combination fuel-fired boilers. National Council of the Paper Industry for Air and Stream Improvement, New York, Nov 1981. Technical Bulletin No. 360

Miner RA, Marshall DW (1976). Sludge dewatering practice in the pulp and paper industry. Technical Bulletin no. 286, National Council of the Paper Industry for Air and Steam Improvement, New York

Mladenov M, Pelovski Y (2010) Utilization of wastes from pulp and paper industry. J Univ Chem Technol Met 45(1):33–38

Monte MC, Fuente E, Blanco A, Negro C (2009) Waste management from pulp and paper production in the European Union. Waste Manag 29(1):293–308

National Council of the Paper Industry for Air and Stream Improvement (NCASI) (1992) Chemical composition of pulp and paper industry landfill leachates. Technical Bulletin No. 643, Sept 1992

Nichols WE, Flanders LN (1995) An evaluation of pelletizing technology. In: Joyce TW (ed) Environmental issues and technology in the pulp and paper industry – a Tappi Press anthology of published papers 1991–1994. Tappi Press, Atlanta, pp 357–363

Norli L, Smedsrud L (2006) Thune screw presses for sludge: the innovative screw press design for high dry contents. Twogether 22:24–27

Ozturk I, Eroglu V, Basturk A (1992) Sludge utilization and reduction experiences in the pulp and paper industry. Water Sci Technol 26(9–11):2105–2108

Perng YS, Wang IC, Yu ST, Gong HY, Dinh L, Kuo LS (2006) Application of nano-silica to paper mill sludge dewatering. Taiwan J For Sci 21(3):353–362

Pickell J, Wunderlich R (1995) Sludge disposal: current practices and future options. Pulp Pap Canada 96(9):T300–T306

Pridham NF, Cline RA (1988) Paper mill sludge disposal: completing the ecological cycle. Pulp Pap Canada 89(2):T73–T75

Reilly MT, Krepps WE (1982) A case study-trials with a mobile unit demonstrate centrifugation of secondary sludge. Tappi J 65(3):83–85

Rodden G (1993) The new Alchemy: turning waste into oils and chemicals. Can Chem News 45(8):45–48

Rosenqvist GV (1978) The use of primary waste water treatment sludge in the manufacture of printing paper at Kymi Kymmene. Paperi Ja Puu 60(4a):205–217

Russel C, Odendahl S (1996) Environmental considerations for landfill development in the pulp and paper industry. Pulp Pap Canada 97(1):T17–T22

Sell NJ, McIntosh TH (1988) Technical and economic feasibility of briquetting mill sludge for boiler fuel. Tappi J 71(3):135–139
Sherman WR (1995) A review of the Maine "Appendix A" sludge research program. Tappi J 78(6):135–150
Simpson GG, King LD, Corlile BL, Blickensderfer PS (1983) Paper mill sludges, coal flyash, and surplus lime mud as soil amendments in crop production. Tappi J 66(7):71–74
Springer AM (1993) Solid waste management and disposal. In: Springer AM (ed) Industrial environmental control-pulp and paper industry, 2nd edn. Tappi Press, Atlanta, pp 458–493
Springer AM, Dietrich-Velazquez HCM, Digiacomo D (1996) Feasibility study of sludge lysis and recycle in the activated-sludge process. Tappi J 79(5):162–170
Stovall JH, Berry DA (1969) Pressing and incineration of kraft mill primary clarifier sludge. Tappi J 52(11):2093–2097
Suriyanarayanan S, Mailappa AS, Jayakumar D, Nanthakumar K, Karthikeyan K, Balasubramanian S (2010) Studies on the characterization and possibilities of reutilization of solid wastes from a waste paper based paper industry. Global J Environ Res 4(1):18–22
Taylor BR, Mcdonald MA, Kimmins JP, Hawkins BJ (1992) Combining pulp mill sludges with municipal sewage to produce slow-release forest fertilizers. Pacific Paper Expo, pp 63–65
Thomas CO, Thomas RC, Hover KC (1987) Wastepaper fibers in cementitious composites. J Environ Eng 113(1):16–31
Toole NK, Kirkland JH (1984) Pilot studies of screw presses for dewatering primary sludges. In: Proceedings of the Tappi environmental conference, Savannah, Georgia, 9–11 April 1984
Wardwell RE, Cooper SR, Charlie WA (1978) Disposal of paper mill sludge in landfills. Tappi J 61(12):72–76
Weigand PS, Unwin JP (1994) Alternative management of pulp and paper industry solid wastes. Tappi J 77(4):91–97
Wu CC, Chien SH, Chuang HH, Wen PC, Kang YW (1998) Investigation on the mechanisms of polymer conditioners for sludge. In: Proceedings of the 13th waste disposal technology symposium, National Sun Yat-Sen University, Kaohsiung, Taiwan, 21–22 Nov 1998, pp 107–112

Chapter 19
Integrated Forest Biorefinery

19.1 Introduction

The pulp and paper industry is facing an economic stalemate due to new market constraints, which include lower selling prices as well as increased competition and fuel costs. This new environment weakens the economic health of all paper mills. The pulp and paper industry must therefore identify new opportunities in order to define a business plan that is better adapted to current market conditions while preserving its main activity – production of pulp and paper products.

Forest biorefinery (FBR) has been defined as the "full integration of the incoming biomass and other raw materials, including energy, for simultaneous production of fibers for paper products, chemicals and energy" (Axegård 2005; Axegård et al. 2007; Chambost and Stuart 2007). By integrating FBR activities at an existing plant, pulp and paper mills have the opportunity to produce significant amounts of bioenergy and bioproducts and to drastically increase their revenues while continuing to produce wood, pulp, and paper products. Manufacturing new value-added byproducts (e.g. biofuels, bulk and specialty chemicals, pharmaceuticals, etc.) from biomass represents for some forestry companies an unprecedented opportunity for revenue diversification.

Biorefining is an exciting concept for the pulp and paper industry, however in many ways, the industry has been considering its implementation for decades (Wising and Stuart 2006). There have been many examples where mills have produced organic chemicals in addition to pulp and paper. The biorefinery builds on the same principles as the petrochemical refinery. In a petrochemical refinery the raw material is normally crude oil, whereas in the FBR the raw material is wood/biomass. The raw material stream is fractionated into several product streams. The products can be a final product or a raw material for another process. New technology is being developed that could be integrated into an existing pulp and paper mill,

P. Bajpai, *Biotechnology for Pulp and Paper Processing*,

DOI 10.1007/978-1-4614-1409-4_19,

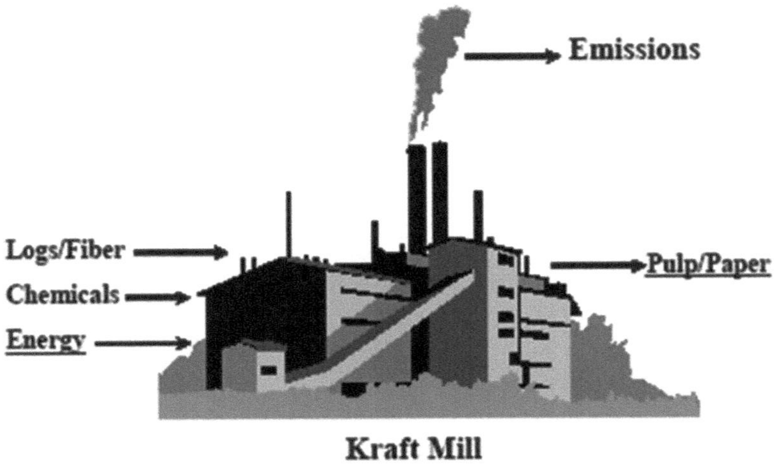

Fig. 19.1 Current pulp mill; reproduced from Thorp et al. (2008) with permission

transforming it into a FBR. There are still significant challenges associated with these new technologies, but several of them look promising. Research is initiating focusing on biorefinery technology development in North America and around the world (Closset 2004; Mabee et al. 2005). However, these process technology development activities alone do not address most of the significant risks associated with implementing the FBR.

Biorefinery technology development will typically be implemented in retrofit, and must be accompanied by a careful process systems analysis in order to understand the impact on existing processes, e.g., pulp yield reductions since carbon is used to make alternative products, and the potential for changed black liquor scaling characteristics in evaporators. The objective of this process systems analysis would be to preserve the value of the existing pulp and paper producing asset.

In addition to process technology development, product development will be essential for identifying successful new markets for biorefinery products, and their supply chain management strategies. These are again systems-oriented issues whose evaluation will be critical for the success of the FBR.

The current pulp and paper mill (Fig. 19.1) uses logs and fiber, chemicals and energy to produce commodity pulp and paper products (Connor 2007; Thorp et al. 2008). Future mills (Fig. 19.2), Integrated Forest Biorefineries, will import regional biomass instead of purchased energy. They will expand the industry's mission from simply manufacturing low margin paper products to creating new revenue streams by producing "green" power and creating new, high-value products such as biofuels and biochemicals, all while improving the efficiency and profitability of their core paper-making operations. Figure 19.3 shows possible products from a pulp mill biorefinery.

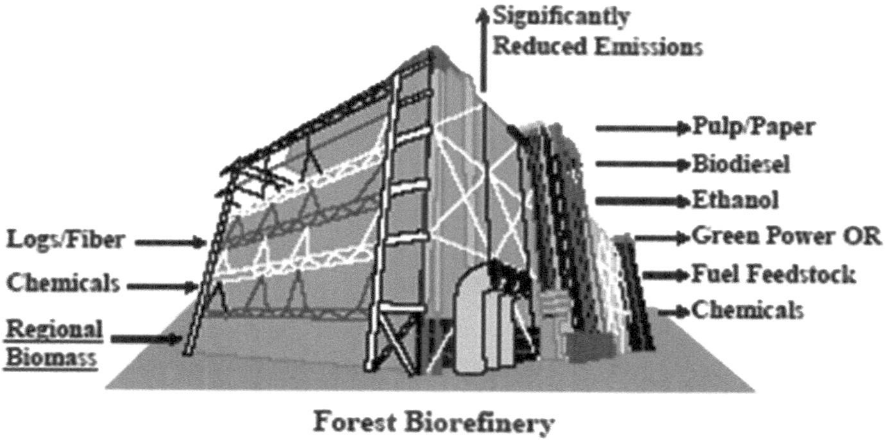

Fig. 19.2 Future mill; reproduced from Thorp et al. (2008) with permission

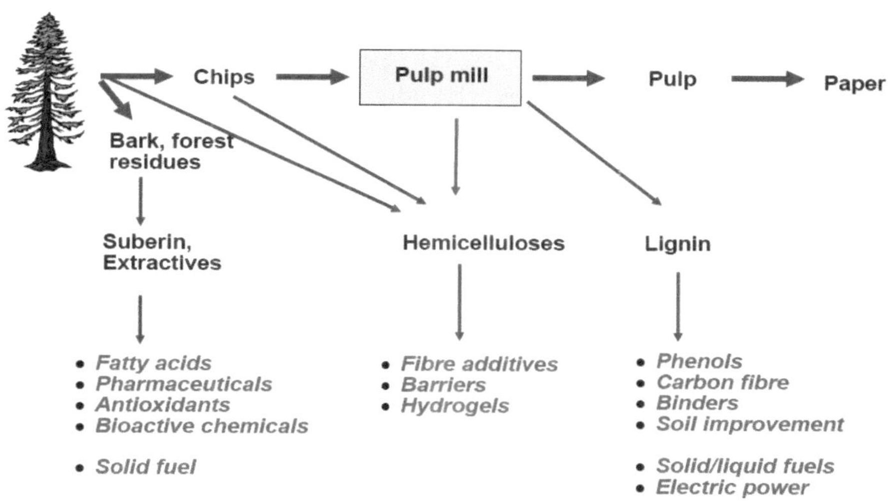

Fig. 19.3 Possible products from a pulp mill biorefinery; reproduced from Axegård (2005) with permission

19.2 Forest Biorefinery Options

Several process alternatives should be considered for implementation of biorefinery in a pulp and paper mill. These are recovering more of the biomass left in the forest, removing lignin from the black liquor in the digester, pyrolysis of bark, etc. In one of the biorefinery workshop (Montréal Workshop 2005), one important consensus

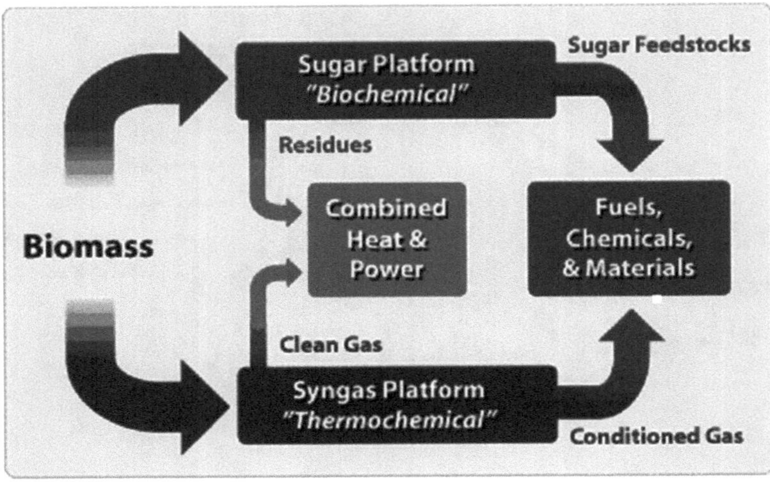

Fig. 19.4 Biorefinery concept; reproduced from the National Renewable Energy Laboratory Biomass Research website: http://www.nrel.gov/biomass/biorefinery.html with permission. Accessed April 20, 2011

Table 19.1 Emerging biorefining technologies

Technology	Yield	Capital cost
Hemicellulose preextraction	Low	Medium
Black liquor gasification	High	High
Removal of lignin from black liquor	Low/high	Low/high
Tall oil extraction	Low	Low

Based on Wising and Stuart (2006)

reached was that before mills can implement the FBR, they need to increase its energy efficiency, eliminate fossil fuels from their operations, and maximize carbon availability for the FBR. This appears to be a valid point since many of the activities today regarding the FBR are motivated by the Kyoto Protocol.

The biorefinery technologies currently under development are typically characterized as biochemical and thermochemical processes (Fig. 19.4). Biochemical processes use steam, dilute acid, concentrated acid, and/or enzyme hydrolysis to convert the hemicellulose and cellulose of biomass into simpler pentoses and glucose. The thermochemical processes use slow or medium temperature gasification or higher temperature pyrolysis to create a high hydrogen content synthetic gas (syngas) that can be used for electricity generation or catalytically converted into liquid biofuels. In Table 19.1, the different technologies discussed here are presented.

The technologies – Hemicellulose preextraction, lignin precipitation, Tall oil extraction – are biochemical, and black liquor gasification (BLG) is thermochemical. The choice of biorefinery technology will depend firstly on the choice of appropriate products as they relate to markets and the supply chain. Depending on the choice of technologies implemented, the yield, the impact on the pulp and paper process and the capital cost will vary. Since the processes in a pulp and paper mill are strongly linked, it is difficult to foresee the impact implementing these different technologies might have on the entire mill. Plus, adding two or more technologies to one mill bring process issues that are even complex to anticipate.

One of the key criteria for FBR options is that the processes are adaptable (Farmer 2005). For many of the products that could be produced in a FBR follows different value cycles. If these products could be changed the most profitable product could be produced at a time where the value of said product is the highest. By developing a concept of adaptable FBR, the mill would be less economically vulnerable, since the product produced could change over time.

19.2.1 Hemicellulose Extraction Prior to Pulping

This is the most extensively investigated concept of the biorefinery platform. During kraft pulping, hemicelluloses are degraded into low molecular weight isosaccharinic acids and end up in the black liquor, with degraded lignin. To prevent an environmental impact and recover energy, black liquors are concentrated and burned. As the heating value of hemicelluloses is considerably lower than that of lignin, extracting the hemicelluloses before the pulping stage for generation of high value products has the potential to improve overall economics. Hemicelluloses can be used directly in polymeric form for novel industrial applications such as:

- Biopolymers (Ebringerova et al. 1994)
- Hydrogels (Gabrielii et al. 2000) or
- Thermoplastic xylan derivatives (Jain et al. 2000) or
- Source of sugars for fermentation to fuels, such as ethanol, or chemicals, such as 1,2,4-butanetriol, a less hazardous alternative to nitroglycerine (Niu et al. 2003)

The cosmetics industry uses hemicelluloses as emulsifiers to prepare water and oil emulsions. Research has also been carried out into hemicelluloses as immunomodulators or those properties that fight infections. The building blocks of hemicelluloses also include sugars with interesting physiological effects. One example of such a sugar is mannose, which has been shown to help combat certain stomach infections. These monosaccharides are currently being studied for example converting xylose into xylitol and mannose into mannitol. These sugars are packed with potential. If hemicelluloses are broken down into smaller pieces or so-called oligomers, there is evidence that these pieces are highly bioactive. There are also data that they promote tree growth or function as growth hormones. Hemicellulose can also be used as a dietary fiber. Their sugars are so-called slow carbohydrates, which help balance blood sugar levels and promote weight loss.

Little work has been done on extracting and utilizing hemicelluloses prior to the pulping process. Removal of hemicelluloses from wood as a pretreatment step is presently being practiced commercially in the production of dissolving pulps.

Table 19.2 Benefits of hemicellulose preextraction

Reduction in kraft cooking times
Enhancing kraft cooking liquor impregnation
Yielding improved pulp properties
Improving pulp production capacity for kraft pulp mills that are recovery-furnace limited

The hemicelluloses are removed to allow the production of pure cellulose. Dissolving pulps are processed into products such as cellulose nitrate, cellulose xanthate (rayon fibers), and cellulose acetate. Preextraction of hemicellulose can provide a totally new feedstock for biofuel/bioethanol production, thus increasing the total revenue stream for the pulp and paper industry (Ragauskas et al. 2006; van Heiningen 2006). It is therefore desirable to develop a pretreatment process that can solubilize hemicellulose sugars with minimal formation of fermentation inhibitors, while preserving the fiber integrity.

It is expected that preextraction of these "waste" hemicelluloses prior to kraft pulping could substantially improve pulp mill operations (Thorp and Raymond 2005; Ragauskas et al. 2006) (Table 19.2).

These process benefits and biofuel possibilities are strong drivers for the development of wood hemicellulose preextraction technologies for kraft pulp mills. An important consideration that must be taken into account with any preextraction of wood chips prior to kraft pulping is the need to develop a system that is readily integrated with modern pulping operations and will not deteriorate the quality of kraft pulps. A key physical parameter in the production of many grades of paper is the strength of the final paper sheet. It has been well documented that if the DP of cellulose is decreased beyond its normal ~1,600 postpulping to ~700 after bleaching (Yanagisawa et al. 2005), the strength properties of the sheet are degraded. This relationship is due to the fact that cellulose is the primary load-bearing element in a lignocellulosic fiber and has a direct relationship with the fiber strength, which contributes to paper strength. Hence, any hemicellulose preextraction technology employed prior to kraft pulping needs to minimize the hydrolysis of cellulose. Furthermore, it has been reported that hemicellulose content is related to paper bond strength, which has been attributed to the adhesive properties of hemicellulose. Studies suggest that for kraft pulps with an α-cellulose content higher than ~80%, a decrease in paper sheet strength properties occurs (Page and Seth 1985; Molin and Teder 2002; Schönberg et al. 2001). This product specification defines a limit for hemicellulose preextraction technologies.

In an ideal scenario, if one could extract 15–20% hemicellulose before pulping and get the same pulp yield as obtained before – it will be possible to keep the same pulp mill production level without increasing the wood demand and would also reduce black liquor solids (BLS) going to the recovery boiler. Removing the recovery boiler bottleneck may allow the manufacturing of more tonnage, which will further improve the profitability of the Kraft mill.

The most common commercial procedures for extracting hemicellulose are presteaming to release natural wood acids (autohydrolysis) followed by water extraction or acid hydrolysis with small amounts of mineral acids (sulfuric acid or hydrochloric acid). The use of water as prehydrolysis stage relies on the in situ hydrolysis of acetate groups on the hemicellulose chains yielding acetic acid. The liberated acid lowers the solution pH to a range of 3–4. This results in the hydrolysis and solubilization of hemicelluloses. Control of the prehydrolysis parameters is an important consideration, as more vigorous conditions will degrade the fiber resource. Pretreatments of lignocellulosic materials by water or steam are referred to in literature as autohydrolysis (Lora and Wayman 1978), hydrothermolysis or hydrothermal pretreatment (Kubikova et al. 1996), and aqueous liquefaction or extraction (Heitz et al. 1986). Microwave heat-fractionation of wood has been recently used to extract hemicelluloses (Lundqvist et al. 2002; Palm and Zacchi 2003). This method requires a treatment temperature of 180–200°C for 2–5 min. Other methods for hemicellulose extraction include mild alkaline solutions with and without addition of cations such as Na, K, Li, and borate at low temperatures, organosolv fractionation, supercritical carbon dioxide, ionic liquids (new class of solvents with nonmolecular, ionic character that are liquids at room temperature) (Hashimoto and Hashimoto 1975; Cunningham et al. 1986; Scott 1989; Bozell et al. 1995; Lu et al. 2004; Wai et al. 2003; Eckert et al. 2000, 2004; Lazzaroni et al. 2005; Wyatt et al. 2005; Lesutis et al. 2001; Nolen et al. 2003; Fitzpatrick 1997; Moens and Khan 2003; Swatloski et al. 2002; Li et al. 2004). Organosolv fractionation technology developed by National Renewable Energy Laboratory utilizes a ternary mixture of methyl isobutyl ketone, ethanol, and water in the presence of low concentrations of sulfuric acid to effect a separation of cellulose, hemicellulose, and lignin. The method typically requires a treatment temperature of 140°C for 1 h. This approach has worked well to fractionate hardwoods, yielding high purity cellulose and selectively dissolving lignin and hemicellulose (Bozell et al. 1995). However, the method proves difficult with softwoods, requiring more acid, higher temperatures, and longer retention times, resulting in poor cellulose pulps. For integration into a kraft biorefinery, the organosolv extraction method would need to be studied further.

Water prehydrolysis is found to be more effective at removing hemicelluloses than steam prehydrolysis, especially for softwoods. All prehydrolysis treatments also extract low levels of lignin and extractives. A key consideration for extracting hemicelluloses prior to kraft pulping for nondissolving grades of paper is the need to yield a wood furnish that still yields excellent physical strength pulp properties. This will undoubtedly require an optimization of hemicellulose preextraction technologies providing optimal removal of hemicelluloses for biofuel production and sufficient retention of select hemicelluloses for the production of high quality kraft pulps.

19.2.1.1 Production of Ethanol from Preextracted Hemicelluloses

After extraction, the hemicelluloses must be converted to monomeric sugars. Two techniques are available for the conversion of wood hemicelluloses into a fermentable sugar solution. These are acid hydrolysis and enzymatic hydrolysis processes. In these processes, monosaccharides are produced which are converted to ethanol via fermentation (Nguyen et al. 2000; Kim 2005; Wright and Power 1987; Wyman and

Goodman 1993). Depending on what technologies are optimized for the preextraction of hemicelluloses from wood chips, an acid hydrolysis of polysaccharides to hexoses and pentoses may be preferred.

The enzymatic hydrolysis of pretreated cellulosic biomass has been commercialized for the processing of wheat straw to bioethanol and is being actively pursued for other agricultural waste resources (Tolan 2003). An important consideration for hemicellulose preextraction and depolymerization treatment protocol is to reduce byproducts that are inhibitors of the fermentation of sugars to ethanol, such as furans, carboxylic acids, and phenolic compounds (Palmqvist and Hahn-Hägerdal 2000). Some inhibitors are present in the raw material, but others can be formed during the hydrolysis process (Klinke et al. 2004). The nature, composition, and concentration of these compounds are dependent on the hydrolysis conditions and may have a profound influence on the fermentation production rate of biofuels from the hydrolysate (Taherzadeh et al. 2000a, b). There are several strategies for dealing with the inhibitors in hydrolysates. First, the hydrolysis conditions may be optimized not only with respect to maximal sugar yields but also to generating reduced amounts of inhibitor compounds (Larsson et al. 1999). Detoxification prior to fermentation is another option, including alkali, sulfite, evaporation, anion exchange, or enzymatic treatments (Alriksson et al. 2005; Horváth et al. 2005 ; Persson et al. 2002). The hydrolyzed hemicellulose sugar solution will finally need to undergo fermentation for the production of ethanol. The microorganisms that are able to ferment sugars to ethanol can be either yeasts or bacteria (Kuyper et al. 2005a, b; Senthilkumar and Gunasekaran 2005). Recent advances in genetic engineering, forced evolution, and mutation and selection strategies have enhanced the biological utilization of hexoses and pentoses for the biological production of ethanol. The well-documented fermentation of wood hydrolysates to ethanol provides a strong technical basis from which practical fermentation technologies can be designed for the conversion of preextracted wood hemicelluloses to ethanol. The fermentation of dilute acid hydrolysates from aspen, birch, willow, pine and spruce using *Saccharomyces cerevisiae* has been reported (Taherzadeh et al. 1997). These wood hydrolysates contained varying amounts of xylose, glucose, and mannose, and the efficiency of fermentation varied substantially, depending upon wood species employed. The use of other yeast and fungi for the production of ethanol from wood hydrolysates has also been reported, and their efficiencies and cost-performance properties continue to be enhanced (Sreenath and Jeffries 1999; Millati et al. 2005; Zaldivar et al. 2001; Brandberg et al. 2004).

The concept of hemicellulose preextraction prior to pulping has been funded by a consortium of large pulp and paper manufacturers and is being operated under the auspices of Agenda 2020. In the United States, wood chip preextraction technologies could make available to the biofuels industry about 14 million tons of hemicelluloses annually while at the same time enhancing the production of kraft pulps (Ragauskas et al. 2006). These extractable hemicelluloses could provide a valuable, high-volume resource of sugars for bioethanol production generating ~20–40 million gallons ethanol/year/mill (Amidon et al. 2007). Thorp (2005a, b) has reported that the potential annual production of ethanol from preextraction of hemicellulose could approach two billion gallons of ethanol/year. Extracting the hemicellulose from the wood chips prior to pulping and depositing the oligomer portion onto the pulp stream after the digester could increase pulp yield by 2%, resulting in approximately $600 million a year in extra pulp production (http://www1.eere.energy.gov/industry/forest/pdfs/hemicellulose_extraction.pdf).

Research studies have already established the viability of extracting hemicelluloses from wood chips prior to kraft pulping for dissolving pulps. The challenge for the biofuel and forest product industries is to develop optimized preextraction technologies that provide a hemicellulose stream for biofuels production and a lignocellulosics stream for pulp production. This vision will, undoubtedly, require a cooperative research program with multipartner stakeholders. These efforts have already begun and will accelerate in the near future, given the significant benefits to all interested parties.

19.2.1.2 Production of Chemicals, Materials, and Polymers

The number of chemicals, materials, and polymers which may potentially be produced in an integrated forest product biorefinery is very large similar to a petrochemical refinery. However, this number may be reduced significantly when guided by a DOE study (Werpy and Petersen 2004) which identifies the top 12 building blocks that may be produced from sugars. Itaconic acid is one of the 12 building block chemicals identified by DOE. Itaconic acid can be produced by fermentation from C5 and C6 monomers. Subsequently, itaconic acid can be converted into polymers through two major routes:

- First route involves the radical homopolymerization of itaconic acid to polyitaconic acid (Yang and Lu 2000). Polyitaconic acid is a highly water soluble and highly hydroscopic material and may be used in paper coating to allow optimal dispersion of the pigment for paper leveling.
- Second route involves the formation by step polymerization of an unsaturated polyester from itaconic acid and a sugar-derived polyol such as propane diol, butane diol, or methyl butane diol (Werpy and Petersen 2004). Such polymers are essentially hydrophobic and can react with vinylic monomers such as styrene and methylmethacrylate to produce tough thermosets for usage in structural material such as wood composites and sheet molding compounds.

Conversion of hemicelluloses into polymers of itaconic acid presents a great economic opportunity for an IFBR.

Another example is the production of carbon fibers using lignin precipitated from alkaline hardwood black liquor. Carbon fibers can be made from hardwood kraft lignin when mixed with commercial polymers such as polyesters, polyolefins, and polyethylene oxide (PEO) (Kadla et al. 2002). A main requirement for processing the lignin is that it contains a minimum of volatile compounds, sugars, and ash. Since the actual spinning of the fibers occurs at a temperature of about 220°C, a minimal amount of gaseous components should release at this temperature to avoid

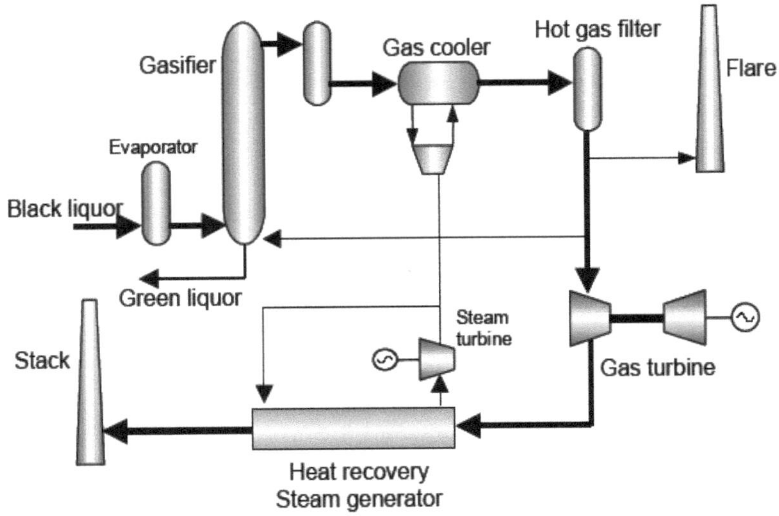

Fig. 19.5 Integrated gasification and combined cycle (IGCC); based on Sricharoenchaikul (2001)

bubbles in the fibers and thus lower physical properties and avoid spinning problems. Thus filtration to remove particulates, carbohydrate stripping, and washing of (almost) sulfur free lignin will be needed to obtain a suitable feed stock for carbon fiber production (Griffith et al. 2003).

19.2.2 Black Liquor Gasification

BLG has excited particular interest in recent years (Bajpai 2008). It offers a way to generate electricity and to reclaim pulping chemicals from black liquor. This is accomplished by converting the fixed carbon to a combustible gas mixture using oxygen-containing gases such as oxygen, carbon dioxide, and water vapor. The combustible gas is then burned to generate electrical power. BLG has been a popular topic in several conferences on biorefining, engineering, pulping, and environmental matters.

This technology has been under development for many years now, and today there are a small number of installations and some additional ones being planned. BLG would replace the Tomlinson recovery boiler for the recovery of spent chemicals and energy. Gasification may become part of integrated gasification and combined cycle (IGCC) operation, or lead to pulp mills becoming biorefineries (Larsen et al. 2003). Figure 19.5 shows a simplified schematic for the black liquor IGCC.

The organic matter in black liquor is partially oxidized with an oxidizing agent to form syngas in the gasifier, while leaving behind a condensed phase. The syngas is cleaned to remove particulates and tars and to absorb inorganic species (i.e., alkali vapor species, SO_2, and H_2S), and this is done to prevent damage to the gas turbine and to reduce pollutant emissions. The clean syngas is burned in gas turbines coupled with generators to produce electricity, and gas turbines are inherently more efficient than the steam turbines of recovery boilers due to their high overall air fuel ratios (Nilsson et al. 1995). The hot exhaust gas is then passed through a heat exchanger (typically a waste-heat boiler) to produce high-pressure steam for a steam turbine and/or process steam. The condensed phase (smelt) continuously leaves the bottom of the gasifier and must be processed further in the lime cycle to recover pulping chemicals.

Essentially all of the alkali species and sulfur species leave in the smelt (mostly as Na_2S and Na_2CO_3) in the recovery boilers, but in gasifiers, there is a natural partitioning of sulfur to the gas phase (primarily H_2S) and alkali species to the condensed phase after the black liquor is gasified. Because of this inherent separation, it is possible to implement alternative pulping chemistries that would yield higher amounts of pulp per unit of wood consumed (Larsen et al. 1998, 2003). Gasification at low temperatures thermodynamically favors a higher sodium/sulfur split than gasification at high temperatures, which results in higher amounts of sulfur gases at low temperatures. Because a large amount of the black liquor sulfur species leaves the low-temperature process as H_2S, H_2S may be recovered via absorption to facilitate alternative pulping chemistries. Industry has numerous patented processes for accomplishing the absorption, including using green or white liquor as an absorbing solvent (Larsen et al. 1998, 2003; Martin et al. 2000).

The partitioning of sodium and sulfur in BLG requires a higher capacity for the lime cycle compared to the current technology. The sodium/sulfur split results in a higher amount of Na_2CO_3 in the green liquor because less sulfur is available in the smelt to form Na_2S. For each mole of sulfur that goes into the gas phase, one more mole of Na_2CO_3 is formed in the condensed phase (Larsen et al. 2003). The increase in Na_2CO_3 results in higher causticization loads, increases in lime kiln capacity, and increases in fossil fuel consumption to run the lime kiln. This leads to higher raw material and operating costs, which must be reduced in order to make the gasification process economically favorable.

BLG may be performed either at low temperatures or at high temperatures, based on whether the process is conducted above or below the melting temperature range (650–800°C) of the spent pulping chemicals (Sricharoenchaikul 2001). In low temperature gasification, the alkali salts in the condensed phase remain as solid products while molten salts are produced in high-temperature gasification. Low-temperature gasification is advantageous over high-temperature gasification because gasification at low temperatures yields improved sodium and sulfur separation. Additionally, low-temperature gasification requires fewer constraints for materials of construction because of the solid product. However, the syngas of low-temperature gasification may contain larger amounts of tars, which can contaminate gas clean-up operations in addition to contaminating gas turbines upstream of the gasifier. These contamination problems can result in a loss of fuel product from the gasifier (Sricharoenchaikul 2001).

19.2.2.1 Gasification Processes

The different gasification processes can roughly be categorized into:

- Low-temperature processes – work below 715°C and the inorganic salts are removed as dry solids.
- High-temperature processes – operate above 900°C and an inorganic salt smelt is obtained.

Several companies have performed trials to develop a commercially feasible process for BLG. History of BLG development is well described by Whitty and Baxter (2001) and Whitty and Verrill (2004). Only two technologies are currently being commercially pursued: the MTCI (low temperature) and Chemrec (high temperature) technologies. Weyerhaeuser, New Bern, uses a Chemrec booster for BLG but it operates at atmospheric pressure, which does not give maximum energy efficiency. Energy efficiency is enhanced by going to higher pressures. Trials are currently under way at Kappa Kraftliner, Sweden, in which the black liquor is gasified at high temperature and pressure in a reactor then the gas is cooled and separated from droplets of smelt. The condensate is dissolved to form low-sulfidity green liquor. The raw gas containing carbon monoxide and carbon dioxide is saturated with steam at high pressure then cooled and stripped of particles. The gas can be used as a feedstock in a combined-cycle (CC) technology or for chemical synthesis (Larson et al. 2000).

MTCI Gasification

MTCI technology – also known as TRI (ThermoChem Recovery International, Inc.) – uses a low-temperature gasification with a bubbling fluidized bed steam reformer (Durai-Swamy et al. 1991; Mansour et al. 1992, 1993, 1997; Rockvam 2001; Whitty and Verrill 2004) operating at 580–620°C. The bed is indirectly heated by several bundles of pulsed combustion tubes, which burn some of the produced gas. Black liquor is sprayed into the fluidized bed and coats the solids, where it is quickly dried and pyrolyzed. The remaining char reacts with steam to produce a hydrogen-rich fuel gas (Rockvam 2001). Part of the bed material is continuously removed, dissolved in water, and cleaned from unburned carbon to obtain green liquor. The produced gas is passed through a cyclone to separate solids and then to a heat recovery steam generator. Part of the generated steam is used in the gasifier as both reactant and fluidizing medium. The gas continues through a Venturi, a gas cooler and is finally cleaned from H_2S in a scrubber with some of the green liquor. The cleaned gas contains about 73% H_2, 14% CO_2, 5% CH_4, and 5% CO (Rockvam 2001). The heating value of the gas is high (~13 MJ/Nm^3). It can be burned in an auxiliary boiler, used in a fuel cell to generate electricity and pressurized it can be fired in a gas turbine.

MTCI has two projects running today, both in mills with a Na_2CO_3 semi-chemical cooking process. The first project is for Georgia Pacific Corporation's Big Island mill in Virginia. This system is a full-scale gasifier, designed to process 200 ton dry solids per day and is fully integrated with the mill (DeCarrera 2006). The second project is for the Norampac Trenton mill, Ontario, Canada (Middleton 2006; Newport et al. 2004; Vakkilainen et al. 2008). Prior to the start-up of the gasifier, the mill had no chemical recovery system. For over 40 years, the mill's spent liquor was sold to local counties for use as a binder and dust suppressant on gravel roads. The discontinuance of the spreading of spent liquor required Norampac to select, purchase, and install a technology to process spent liquor. The TRI BLG system was selected. The capacity of the TRI spent liquor gasification system is 125 tons/day of BLS. TRI's scope of supply included the steam reformer, pulse combustors and fuel train, detailed engineering and start-up support, materials handling equipment, and instrumentation. The project, which started operations in 2003, is operating day in and day out meeting all of the needs of the mill's chemical recovery requirements. Process optimization is continuing in the area of energy recovery. TRI's gasification process is ideal for use in a forest products biorefinery as it is uniquely configured for high-performance integration with pulp and paper facilities and is capable of handling a wide variety of cellulosic feedstocks, including woodchips, forest residuals, agricultural wastes and energy crops, as well as mill byproducts (spent liquor). Compared to other biomass gasification technologies that are based on partial oxidation, TRI's steam reformer converts biomass to syngas more efficiently, producing more syngas per ton of biomass with a higher Btu content. This medium-Btu syngas can be used as a substitute for natural gas and fuel oil, and as a feedstock for the production of value-added products such as biodiesel, ethanol, methanol, acetic acid, and other biochemicals. TRI's technology can be integrated with a wide variety of catalytic and fermentation technologies to convert the syngas to high-value bio-based fuels and chemicals. For example, syngas generated by TRI's technology can be conditioned and sent to a commercially proven gas-to-liquids ("GTL") facility (i.e., Fischer-Tropsch or other catalytic technologies) inside the biorefinery. The GTL process produces a range of products (naphtha, gasoline, diesel/kerosene, wax, methanol, dimethyl ether [DME], etc.) that are stabilized for storage and transported offsite to a downstream refinery for conversion to marketable products. The unreacted syngas and light noncondensable gases (tail gas) are utilized in the process to replace fossil fuels. Additionally, the GTL conversion, which is exothermic, provides another source of process heat that is recovered and used. A fully integrated forest products biorefinery utilizing TRI's technology will achieve thermal efficiencies from 70 to 80% depending upon process configuration and biomass feedstock. Figure 19.6 shows MTCI steam reformer.

A TRI system is also in the trial stage at Georgia Pacific, Big Island. Technical issues have included excessive tar formation (over 30% of the organic content of the processed liquor was lost to the sewer as tar), lower than expected carbon conversion (approximately 80% vs. the expected 99%) and concerns about the design of the fluidization system.

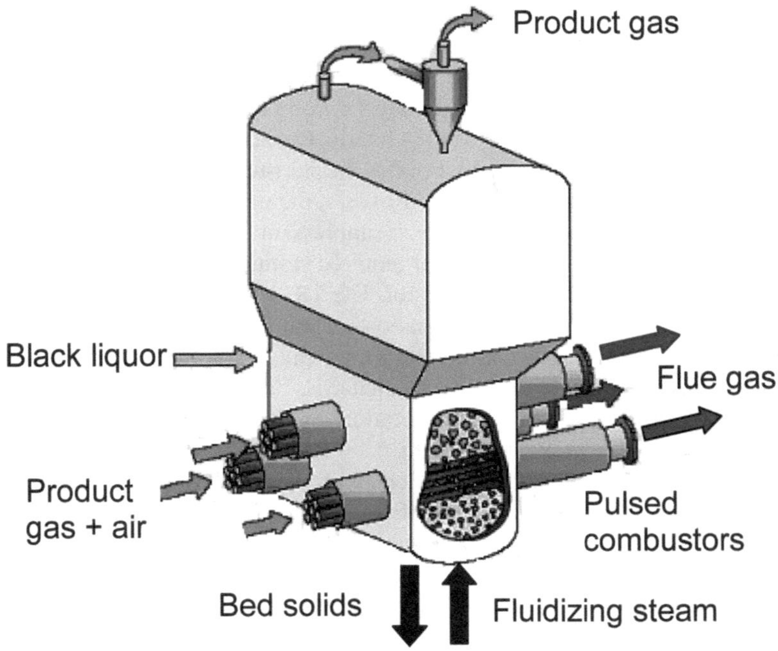

Fig. 19.6 MTCI steam reformer; based on Whitty and Baxter (2001)

Chemrec Gasification

Chemrec is working on both an atmospheric version and a pressurized version of a high-temperature downflow entrained flow reactor (Brown and Landälv 2001; Kignell 1989; Stigsson 1998; Whitty and Nilsson 2001; Whitty and Verrill 2004). The atmospheric versions are mainly considered as a booster to give additional black liquor processing capacity. The pressurized version is more advanced and would replace a recovery boiler or function as a booster.

In the atmospheric system, black liquor is fed as droplets through a burner at the top of the reactor. The droplets are partially combusted with air or oxygen at 950–1,000°C and atmospheric pressure. The heat generated sustains the gasification reactions. The salt smelt is separated from the gas, falls into a sump, and dissolves to form green liquor. The produced gas passes a cooling and scrubbing system to condense water vapor and remove H_2S. The gas has low heating value (~2.8 MJ/ Nm^3) and is suitable for firing in an auxiliary boiler. It consists of 15–17% CO_2, 10–15% H_2, 8–12% CO, 0.2–1% CH_4, and 55–65% N_2 (Lindblom 2003). The thermal efficiency is quite low. An atmospheric Chemrec Booster system with a firing rate of 270 ton DS/day is in use at Weyerhaeuser's New Bern mill since 1997. However, it was shut down in 2001 due to extensive cracking in the reactor shell and it was started again in 2003. The gasifier had then been rebuilt with a new reactor vessel as well as a modified refractory lining design and it has operated well since then (Brown et al. 2004).

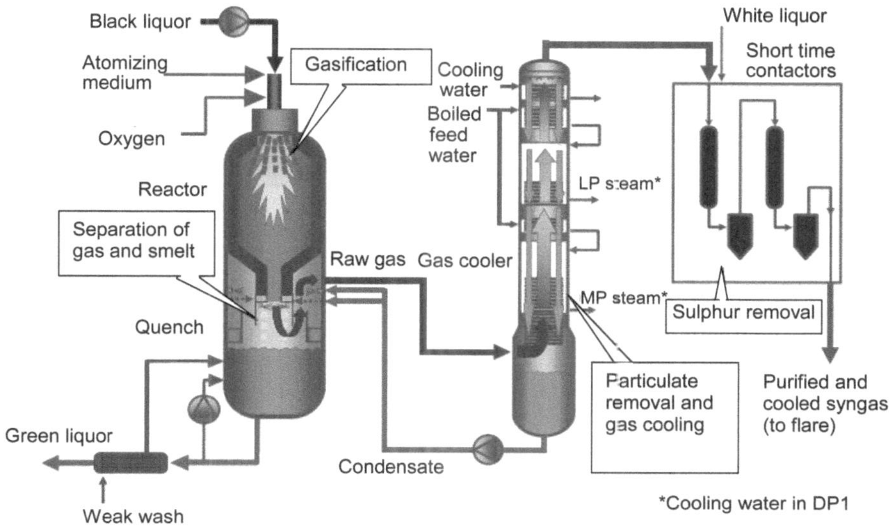

Fig. 19.7 The CHEMREC DP-1 plant. Source: www.chemrec.se/admin/UploadFile.aspx?path=/UserUploadFiles/2005%20DP-1%20brochure.pdf (reproduced with permission)

The pressurized system is similar but operates at a pressure of 30 atm. The salt smelt is separated from the gas in a quench device. The gas cleanup system is more advanced, cleaning the gas of fine particles and condensed hydrocarbons. The sulfur-rich gas stream separated in an absorber/stripper system can be used to prepare advanced pulping solutions. The gas produced has a higher heating value (~7.5 MJ/Nm^3) and can be, e.g., fired in a gas turbine to produce electricity or used to produce biofuels such as methanol or DME. The exhaust from the turbine is passed through a heat recovery steam generator. The thermal efficiency is above 80%.

A pressurized system has been built within the Swedish national BLG program (2004–2006) in Piteå, Sweden. It is a development plant built for 20 ton DS/day. The system includes the processes of gasification and quenching, gas cooling, and gas cleaning. The produced gas has been determined to contain about 41% H_2, 31% CO_2, 25% CO, 2% CH_4, and 1.4% H_2S (Lindblom 2006). The aim of the program is a verified process that will be ready for scale up (15 times) as well as an optimized integration of the process with the pulping cycle. Figure 19.7 shows the CHEMREC DP-1 plant.

The CHEMREC BLGCC system has several advantages over recovery boilers; the most significant being dramatically improved electricity yield. The CHEMREC BLGMF system combines BLG with a chemical synthesis plant for production of green automotive fuels such as methanol or DME. The new combined pulp and chemicals production facility requires additional energy to compensate the pulp mill for the withdrawal of the new green automotive fuels. The efficiency of the CHEMREC BLGMF system for generating the new green automotive fuels is very high and the cost of these fuels from a full scale unit is competitive with

Table 19.3 Possible products from syngas

Hydrogen
Methanol
DME
Fischer-Tropsch fuels
Ethanol
MTBE

Based on Tampier et al. (2004)

petroleum-based alternatives. The CHEMREC $BLGH_2$ system utilizes the syngas from the black liquor gasifier as feedstock for novel green hydrogen production. Table 19.3 shows the list of possible chemicals that can be produced from the syngas.

The investment cost for a full-scaled PBLG unit is estimated to be slightly higher than for a new conventional recovery boiler (Warnqvist et al. 2000). However, pressurized BLG with an integrated combined cycle (BLGCC) has the potential to double the amount of net electrical energy for a kraft pulp mill compared to a modern recovery boiler with a steam turbine (Axegård 1999). For more closed systems with less need of steam, this increase in electrical energy will be even higher. Another advantage with the PBLG process is the increased control of the fate of sulfur and sodium in the process that can be used to improve the pulp yield and the quality for the mill. This control is very important for the green liquor quality and is quite limited with a conventional recovery boiler. A disadvantage with gasification is that it will increase the causticizing load. However, BLG has a lower requirement for make-up salt cake compared to the recovery boiler. Even though the PBLG process might have a lot of advantages compared to the recovery boiler there are still a number of uncertainties for this technology.

BLG is still a developing technology. Only small (100–350 tds/days) commercial atmospheric units have been built. Similar size pressurized demonstration units do not yet exist. It will take some time before reliable large units are available. BLG can produce more electricity (Vakkilainen et al. 2008). Current commercial atmospheric processes are not as energy efficient as the kraft recovery boiler process (Grace and Timmer 1995; Mckeough 2003). The black liquor gasifier needs to operate under pressure to have an electricity advantage.

Even though there are significant gains to be made, there still remain many unresolved issues (Tucker 2002; Katofsky et al. 2003): finding materials that survive in a gasifier, mitigating increased causticizing load, how to startup and shutdown, tar destruction, alkali removal, and achieving high reliability. The full impact of the BLG on recovery cycle chemistry needs to be carefully studied with commercial units. The first large demonstration units will cost two to three times more than a conventional recovery boiler. Although this will improve with time, price will hinder the progress of BLG. A small BLG with a commercial gas turbine size of 70 MWe requires a mill size of over 500,000 ADt/a. Commercial gasifiers probably need to be over 250 MWe in size. It is therefore expected that full size black liquor gasifiers will be built in new greenfield mills, and not as replacement units of old recovery boilers.

BLG whether conducted at high or low temperatures is still superior to the current recovery boiler combustion technology. The thermal efficiency of gasifiers is estimated to be 74% compared to 64% in modern recovery boilers, and the IGCC power plant could potentially generate twice the electricity output of recovery boiler power plants given the same amount of fuel (Farmer and Sinquefield 2003). While the electrical production ratio of conventional recovery boiler power plants is 0.025–0.10 MWe/MWt, the IGCC power plant can produce an estimated 0.20–0.22 MWe/MWt (Farmer and Sinquefield 2003; Sricharoenchaikul 2001). This increase in electrical efficiency is significant enough to make pulp and paper mills potential exporters of renewable electric power. Alternatively, pulp mills could become manufacturers of biobased products by becoming biorefineries. Additionally, the new technology could potentially save more than 100 trillion BTUs of energy consumption annually, and within 25 years of implementation, it could save up to 360 trillion BTU/year of fossil fuel energy (Larsen et al. 2003). The new technology also offers the benefits of improved pulp yields if alternative pulping chemistries are included, and reductions in solid waste discharges. Also, the process is inherently safer because the gasifier does not contain a bed of char smelt unlike in recovery boilers, which reduces the risk of deadly smelt-water explosions (Sricharoenchaikul 2001).

IGCC power plants will reduce wastewater discharges at pulp and paper mills, even though they most likely will not significantly impact water quality (Larsen et al. 2003). Also, IGCC power plants will reduce cooling water and make-up water discharges locally at the mill, and because the efficient gasifiers will cause grid power reductions, substantial reductions in cooling water requirements at central station power plants will also occur (Larsen et al. 2003). Central station power plants have large water requirements for cooling towers in order to provide grid power to customers. Overall, the implementation of IGCC power plants will cause net savings in cooling water requirements and net reductions in wastewater discharges.

The most significant environmental impact caused by BLG will occur in air emissions. Compared to the current recovery technology, the IGCC system could cause low emissions of many pollutants, such as SO_2, nitrogen oxides (NO_x), CO, VOCs, particulates and TRS gases, and overall reductions in CO_2 emissions. Even with improved add-on pollution control features, the recovery boiler system still causes higher overall emissions than the IGCC system (Larsen et al. 1998, 2003). Table 19.4 shows a list of different emissions and their qualitative environmental impact, along with relative emissions rates for both recovery boilers and gasifiers.

Because the biomass sources at pulp and paper mills are sustainably grown, a BLG-based IGCC plant or biorefinery would transfer smaller amounts of CO_2 to the atmosphere as compared to using fossil fuels. The vast majority of the CO_2 emitted would be captured from the atmosphere for photosynthesis and used for replacement biomass growth, producing O_2 (Larsen et al. 2003). According to Larsen, if the pulp and paper industry converts the 1.6 quads of total biomass energy to electricity, 130 billion kWh/year of electricity could be generated. This electricity generation in a BLG-based IGCC plant could displace net CO_2 emissions by 35 million tons of carbon per year within 25 years of implementation. Within 25 years of implementation, the IGCC could displace 160,000 net tons of SO_2, since most of the SO_2 produced in the process would be absorbed during H_2S recovery. Moreover, the overall reduction

Table 19.4 Relative emissions rates of different emissions

Pollutant	Relative environmental Impact	Relative emissions rates with controls on recovery boilers	Relative emissions rates with gasification technology
SO_2	High	Low	Very low
NO_x	High	Medium	Very low
CO	Low	Medium	Very low
VOC's	High	Low	Very low
Particulates	High	Low–medium	Very low
CH_4	Low–medium	Low	Very low
HAP's	Medium–high	Low	Very low
TRS	Low	Low	Very low
Wastewater	Medium–high	Low	Very low–low
Solids	Very low	Low	Low

Based on Larsen et al. (2003)

of TRS gases (i.e., H_2S) using gasification technology will also reduce odor, which will improve public acceptance of pulp and paper mills, particularly in populated areas.

Clearly, BLG technology offers tremendous potential to make an impact on society. However, before it can totally replace the current recovery boiler technology, some work must be done to make it more economically attractive. One major area that requires attention is the causticization process. Gasification technology can cause significant increases in capacity for the lime cycle, requiring significant increases in fossil fuel consumption, and to improve economic viability, alternative causticization technologies must be considered.

Gasification is a well-established technique, but its application to black liquor is new and creates specific research needs. Perhaps, the highest priority is to deal with the materials for constructing the gasifier. The process can operate at very high temperatures (up to 1,000°C) and involves very aggressive molten salts (Na_2S, Na_2CO_3, NaCl) that tend to react strongly with ceramics and other materials. There is a very aggressive gas atmosphere (HCl, CO). This was an issue with the gasification system at Weyerhaeuser, New Bern. The problem has now been solved by using new materials and making some design changes (Brown et al. 2004). There are issues concerning the formation of tar and condensable organic matter. Approximately 1–5% of the carbon in black liquor is converted to methanol, ethanol, cresol, xylene, and a variety of other tar and condensable organic components. Several other questions need to be addressed. For example, can sodium and sulfur separation be controlled by process design or operation? How much H_2S is produced, rather than other sulfur-containing gases? And can H_2S be recovered efficiently from the product gases? Researchers around the world are trying to find answers.

19.2.3 *Removal of Lignin from Black Liquor*

STFI-Packforsk has developed a new and cost-effective process for extracting high-quality lignin from kraft black liquor. This process is named LignoBoost (Axegård

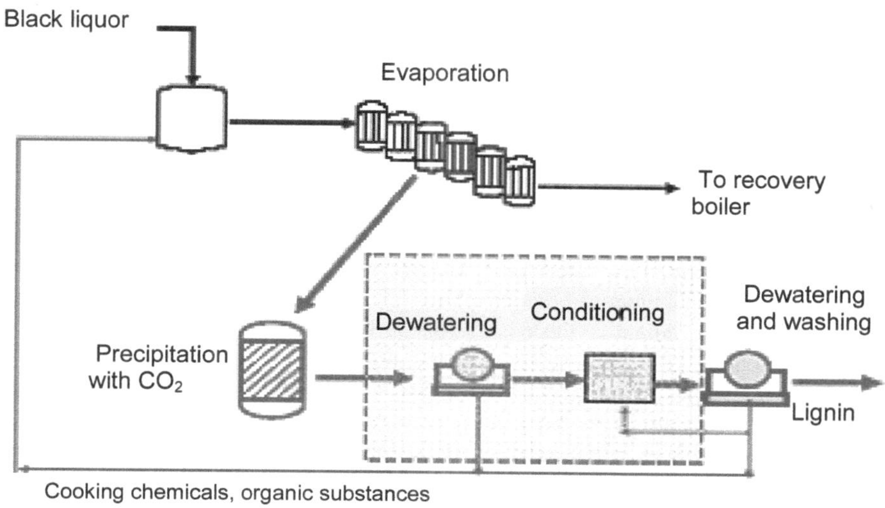

Fig. 19.8 The two-stage washing/dewatering process, LignoBoost, for washing lignin precipitated from black liquor; reproduced from Axegård (2007b) with permission

2006a, 2007a, b; Frisell 2008; Wallmo and Theliander 2007). Carbon dioxide is used to precipitate lignin. It is then dewatered in the first stage and dewatered/washed in a second washing stage (Fig. 19.8). Washing is done counter-currently. This reduces the risk for lignin dissolution which is a main disadvantage in the conventional one-stage process. Compared to the one-stage process the water use is lower, lignin is cleaner with respect to ash and sodium, and the capacity is significantly higher. The lignin has very good properties including 65–70% dry solids content, ash content of 0.1–0.5%, sodium 0.01–0.4% and heating value of 26 GJ/tons. It can be used as biofuel, replacing coal and oil, i.e. in pulp mill's power generation or in lime kilns. LignoBoost gives customers the possibility to increase the capacity of a pulp mill and turn pulp mills into significant energy suppliers. At the same time the extracted lignin is also of interest for other process industries as a raw material for plastics, coal fibers, and chemicals (Axegård 2007a, b; Neumann 2008). There are four key operations in the LignoBoost process. These are:

- Precipitation – In precipitation, the black liquor is acidified by absorption of black liquor and solid lignin precipitates.
- Dewatering – During the dewatering operation, the solid lignin is filtered off and dewatered by gas displacement.
- Re-suspension – The re-suspending phase involves the re-suspension of the solid lignin and the reduction of the pH.
- Final washing – In final washing, the solid lignin is filtered off, washed by means of displacement washing and finally dewatered by gas displacement.

The LignoBoost process enables the fast production of high-quality lignin at a low cost. Low-filtration resistances can be maintained throughout the process and an even lignin filter cake that is easy to wash and finally dewater is formed in the second filtration/washing stage. Using the novel process, the specific filtration resistance is one to two orders of magnitude lower compared with the separation and washing made in a single filtration step. The separation of the pH, and the ion strength reduction in two different steps results in the lignin becoming much more stable in all process stages with only a small amount of lignin dissolved during the final displacement washing.

The LignoBoost technology has proven its technical maturity over several years of research and laboratory testing as well as during operation in an industrial-size demonstration plant integrated into the pulping process of Nordic Paper – Bäckhammar, Kristinehamn, Sweden (Anon 2007; Lennholm 2007). The demonstration plant will remain in the possession of STFI-Packforsk. For production of lignin, acid precipitation was selected as the most potentially promising route. For production of xylan, membrane fractionation was selected as the most promising route. These two methods can be successfully combined.

Lignins are used as binders, dispersants, emulsifiers, and sequestrants. It has been proposed to isolate phenols from lignin and to produce carbon fibers. The LignoBoost-technology is currently being tested in a demonstration plant which is located at the Bäckhammar unbleached kraft pulp mill and is operated as a subsidiary to STFI-Packforsk, LignoBoost Demo AB. The demonstration plant is expected to achieve an annual lignin production of about 4,000 tons. Nearly all lignin produced will be used in different incinerators such as lime kilns, bark boilers, and Fortum's heat & power plant in Stockholm. The process also offers new opportunities for further use of a kraft pulp mill as a biorefinery such as in xylan removal from black liquor, biomass gasification, and ethanol fermentation (Axegård 2007a, b; Rodden 2007).

On May 27, 2008, Metso and STFI-Packforsk AB have signed a purchase agreement regarding the shares of Lignoboost AB, a Swedish research company. The transaction includes all the intellectual property rights as well as the LignoBoost brand and its related know-how. In addition, Metso and STFI-Packforsk have signed a research and development agreement related to LignoBoost technology. Both agreements come into force with immediate effect. The acquired company will become part of Metso Power, a part of Metso Paper business area. The acquisition supports Metso's profitable growth strategy and opens an interesting biofuel business opportunity within pulping processes. Metso Power sees great value in getting a process with such high future expectations. Recently, Södra and STFI-Packforsk have demonstrated the use of lignin for up to 100% replacement of fossil fuel in the lime kiln of the pulp mill.

A mill trial was carried out in FRAM (the Future Resource-adapted Pulp Mill) using a ceramic membrane in the black liquor in continuous two-vessel digester system (Öhman 2006). The lignin separation was performed between 145 and 155°C at full digester pressure without adjustment of the pH. Ceramic membranes with cut-off between 5 and 15 kDa were used. The retentate is mixture of lignin and xylan and further fractionation is needed. Another option is to apply membrane separation immediately before or after the LignoBoost process. In the former case,

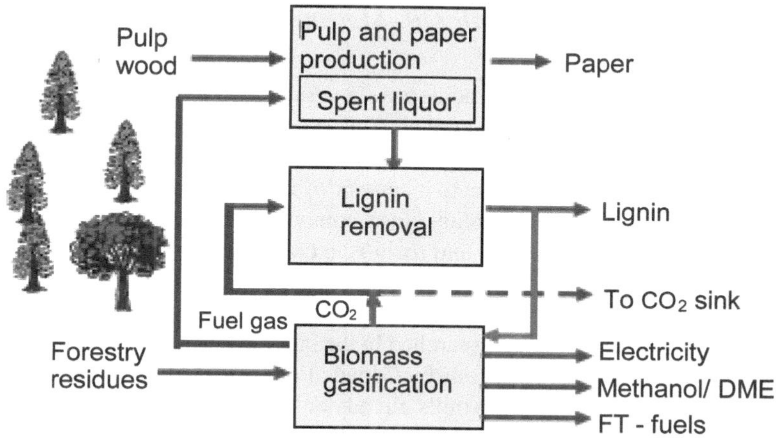

Fig. 19.9 Integration opportunities between LignoBoost and gasification of forestry residues proposed by STFI-Packforsk and VTT; reproduced from Axegård (2007b) with permission

the performance of LingoBoost will be improved and the lignin will be purer. In the latter case, the retentate will be relative pure xylan as the high molecular weight lignin is precipitated in LignoBoost.

Precipitation of lignin requires carbon dioxide. The bulk of the variable cost is due to carbon dioxide if commercial product is used. It may be possible to use carbon dioxide from the lime kiln, but gas cleaning is a challenge. Carbon dioxide from ethanol fermentation yields about 1 tonne of pure carbon dioxide per tonne of ethanol produced. Currently sized ethanol plants are too small to justify recovery of the produced carbon dioxide. By combining lignin production with ethanol production, the carbon dioxide can efficiently be utilized and the economical performance significantly improved.

The amount of lignin (and xylan) that can be removed from black liquor is depending on mainly on the status of the recovery boiler. At a certain amount of heat value in the fired black liquor, the performance is deteriorated. In many mills, this critical level is between 10 and 30% of lignin removed. One interesting way to handle this is to add fuel gas from gasified biomass and thus compensate for lost heat value (Fig. 19.9). Produced carbon dioxide can also be used for lignin precipitation (Axegård 2006b). The ultimate development would be removal of all valuable organic components from the black liquor such as lignin, xylan and sugar acids and instead obtain all the fuel need from gasified biomass such as forestry residuals. Such an approach would make a complete removal of organic components in black liquor possible. The traditional recovery boiler may also be replaced with less capital demanding and less complicated techniques. STFI Packforsk and VTT are currently, together with selected partners, applying for a large collaborative EU-financed project based on these ideas in the framework of EU seventh research program.

19.2.4 Other Products (Tall Oil, Methanol, etc.)

Extractives such as rosin and fatty acids are sometimes removed from the spent pulping liquor and processed into crude tall oil (CTO). In Canada, most CTO is currently incinerated as fuel in the lime kilns of pulp mills to displace fossil fuel. In the south eastern United States, where extractive content of the wood is much higher, tall oil plants fractionate the CTO into value-added components. Processes have also been proposed to convert both the fatty and rosin acid components of the CTO into green diesel fuel. Thorp (2005a, b) has reported production-rate potential of 530 million liters diesel per year in the United States. The processing of tall oil into a high-quality diesel additive has been researched in the laboratory and pilot scale. The later studies included promising road tests by Canada Post Corporation (Ragauskas et al. 2006). Given that many kraft pulp mills already collect these extractives, their future utilization for fuels will be based on competing economic considerations. Fatty acids can be directly esterified by alcohols into diesel fuel, while the rosin acids can be converted by the "Super Cetane" hydrogenation process developed in Canada. Turpentine recovered from process condensates in Canadian mills is generally incinerated as fuel in one of the onsite boilers. Processing it into consumer grade products is possible but, in many cases, it is more valuable as a fuel.

The average 1,000 tonnes/day softwood kraft mill has approximately 7 tonnes/day of methanol in its foul condensate streams. Most mills use steam strippers to concentrate the methanol to about half its volume before incineration. Some mills use air strippers, which do not remove methanol effectively or simply send foul condensates to effluent treatment where the methanol is consumed by biological activity. It is possible to purify this methanol for alternative uses, either onsite or for sale. One pilot project has used the catalytic conversion process for converting the methanol to formaldehyde. Waste organics sent to effluent treatment at pulp and paper mills are unique compared with municipal organic wastes, which have a very high carbon-to-nitrogen ratio. Certain bacteria in activated sludge treatment systems under such conditions accumulate 3-hydroxybutyric acid (PHB), a potential building block for biopolymers. Extraction of PHB remains the significant hurdle to this process. Pulp and paper waste treatment sludge is typically buried in landfills, incinerated, or spread on land as a nutrient enhancer. Research is under way to improve the performance of microbes in the conversion of nutrients in effluents to PHB and other fermentation products.

U.S. pulp and paper industry processes 108 million tons pulpwood per annum. At least 14 million tons of hemicellulose (two billion gallons ethanol; 600 million gallons acetic acid; $3.3 billion net cash flow), five million tons of paper mill sludge (feedstock for ethanol; no pretreatment), and 700 million liters of turpentine and tall oil (feedstock for biodiesel) per annum are available.

In an optimized FBR, part of the hemicellulose that is now burned would be used to create new, more valuable products. A portion of hemicellulose can be extracted from wood chips prior to pulping using hot water extraction in low-pressure digesters. Some acetic acid is formed during the extraction process and this must be separated from the sugar solution. The sugars can then be fermented to ethanol or other high value chemicals, creating an additional product stream. Removing part of the hemicellulose prior to the digester will increase the throughput potential of the pulping process. However, utilizing some of the hemicellulose as a sugar feedstock reduces the energy content of the pulping byproduct black liquor, which is an important renewable energy source for kraft pulp mills. In the future, to fully optimize the FBR, the economic and energy implications of diverting a portion of hemicellulose to other products will need to be balanced. The loss of this energy source can be offset by improved energy efficiency in the pulp and paper manufacturing process. Ultimately, forest biorefineries would potentially use a combination of new technologies that result in more complete, energy efficient and cost effective use of the wood feedstock.

19.3 Environmental Impacts of Forest Biorefineries

Forest biorefineries could produce fewer emissions and support sustainable forestry. The overall environmental implications and life cycle of the FBR are still being studied. However, there could be a number of positive environmental impacts. For example, a FBR utilizing gasification (in a BLG combined cycle configuration) rather than a Tomlinson boiler is predicted to produce significantly fewer pollutant emissions due to the intrinsic characteristics of the BLGCC technology. Syngas clean-up conditioning removes a considerable amount of contaminants and gas turbine combustion is more efficient and complete than boiler combustion. There could also be reductions in pollutant emissions and hazardous wastes resulting from cleaner production of chemicals and fuels that are now manufactured using fossil energy resources. In addition, it is generally accepted that production of power, fuels, chemicals and other products from biomass resources creates a net zero generation of carbon dioxide (a greenhouse gas), as plants are renewable carbon sinks. A key component of the FBR concept is sustainable forestry. The FBR concept utilizes advanced technologies to convert sustainable woody biomass to electricity and other valuable products, and would support the sustainable management of forest lands. In addition, the FBR offers a productive value-added use for renewable resources such as wood thinnings and forestry residues as well as urban wood waste.

References

Alriksson B, Horváth IS, Sjöede A, Nilvebrant NO, Jönsson LJ (2005) Ammonium hydroxide detoxification of spruce acid hydrolysates. Appl Biochem Biotechnol 121–124:911–922

Amidon TE, Francis R, Scott GM, Bartholomew J, Ramarao BV, Wood CD (2007) Pulp and pulping processes from an integrated forest biorefinery. Appl. No. PCT/US2005/013216

Anon XX (2007) LignoBoost does business with lignin fuel. Beyond 2:4–5

Axegård P (1999) Kretsloppsanpassad massafabrik-Slutrapport, KAM 1 1996–1999, KAMrapport A31, Stiftelsen för Miljöstrategisk forskning

Axegård P (2005) The future pulp mill – a biorefinery. In: First international biorefinery workshop, Washington
Axegård P (2006a) Lignin removal from black liquor for increased energy efficiency and pulp capacity increase. In: Energy management for pulp and papermakers, Budapest, Hungary, 16–18 Oct 2006, Paper 12, 31pp
Axegård P (2006b) Presentation "utilization of black liquor and forestry residues in a pulp mill biorefinery" at the forest based sector technology platform conference, Lahti, Finland, 22–23 Nov 2006
Axegård P (2007a) Lignin from black liquor: a valuable fuel and chemical feedstock. In: Biorefining for the pulp and paper industry, Stockholm, Sweden, 10–11 Dec 2007, 34pp
Axegård P (2007b) The kraft pulp mill as a biorefinery. In: Third ICEP international colloquium on eucalyptus pulp, Belo Horizonte, Brazil, 4–7 March 2007, 6pp
Axegård P, Backlund B, Tomani P (2007) The pulp mill based biorefinery. In: Pulp paper 2007 conference. Biomass conversions, Helsinki, Finland, 5–7 June 2007, pp 19–26
Bajpai P (2008) Chemical recovery in pulp and paper making. In: PIRA international, UK, 166pp
Bozell JJ, Black SK, Myers M (1995) Clean fractionation of lignocellulosics – a new process for preparation of alternative feedstocks for the chemical industry. In: 8th international symposium on wood and pulping chemistry, Helsinki, Finland, pp 697–704
Brandberg T, Franzén CJ, Gustafsson L (2004) The fermentation performance of nine strains of *Saccharomyces cerevisiae* in batch and fed-batch cultures in dilute acid wood hydrolysate. J Biosci Bioeng 98(2):122–125
Brown C, Landälv I (2001) The Chemrec Black liquor recovery technology – a status report. In: International chemical recovery conference, Whistler, Canada, 11–14 June 2001
Brown CA, Gorog JP, Leary R, Abdullah Z (2004) The Chemrec black liquor gasifier at New Bern – a status report. In: International chemical recovery conference, Charleston, 6–10 June 2004
Chambost V, Stuart PR (2007) Selecting the most appropriate products for the forest biorefinery. Ind Biotechnol 3(2):112–119
Closset G (2004) Advancing the forest biorefinery. In: Presentation at forest products techno-business forum, Atlanta, GA, 26–27 Oct 2004
Connor E (2007) The integrated forest biorefinery: the pathway to our bio-future. In: International chemical recovery conference: efficiency and energy management, Quebec City, QC, 29 May to 1 June 2007, pp 323–327
Cunningham RL, Carr ME, Bagby MO (1986) Hemicellulose isolation of annual plants. In: Biotechnology bioengineering symposium, no. 17, 8th symposium biotechnology for fuels and chemicals, Gatlinburg, 13–16 May 1986, pp 159–168
DeCarrera R (2006) Quarterly technical progress report 20 demonstration of black liquor gasification at Big Island. Report 40850R20 http://www.gp.com/containerboard/mills/big/pdf/rpt40850R20.pdf (06-04-28). Accessed on Dec. 2010
Durai-Swamy K, Mansour MN, Warren DW (1991) Pulsed combustion process for black liquor gasification. U.S. DOE Report DOE/CE/40893-T1 (DE92003672)
Ebringerova A, Hromadova Z, Kaucurakova M, Antal M (1994) Quaternized xylans: synthesis and structural characterization. Carbohyd Polym 24:301–308
Eckert CA, Bush D, Brown JS, Liotta CL (2000) Tuning solvents for sustainable technology. Ind Eng Chem Res 39(12):4615–4621
Eckert CA, Liotta CL, Bush D, Brown J, Hallett J (2004) Sustainable reactions in tunable solvents. J Phys Chem B 108:18108–18118
Farmer, MC (2005) The adaptable integrated biorefinery for existing pulp mills. In: Presentation at TAPPI engineering, pulping, and environmental conference, Philadelphia, PA, 28–31 Aug 2005
Farmer M, Sinquefield S (2003) An external benefits study of black liquor gasification. Final report, Georgia Institute of Technology, 15 June 2003
Fitzpatrick SW (1997) US Patent 5,608,105
Frisell H (2008) Breakthrough for new Swedish environmental technology. Dagens Ind 33(69):26

Gabrielii I, Gatenholm P, Glasser WG, Jain RK, Kenne L (2000) Separation, characterization and hydrogel-formation of hemicellulose from aspen wood. Carbohyd Polym 43:367–374
Grace TM, Timmer WM (1995) A comparison of alternative black liquor recovery technologies. In: Proceedings of the international chemical recovery conference, Toronto, pp B269–B275
Griffith WL, Compere AL, Leitten CF, Shaffer JT (2003) Low-cost, lignin-based carbon fiber for transportation applications. In: International SAMPE technical conference, vol 35, pp 142–149
Hashimoto T, Hashimoto K (1975) Studies on the utilization of xylan and glucomannan in woods. I. Purification and separation. Yakugaku Zasshi 95(10):1239–1244
Heitz M, Carrasco F, Rubio M, Chauvette G, Chornet E, Julian L, Overend RP (1986) Generalised correlations for the aqueous liquefaction of lignocellulosics. Canad J Chem Eng 64:647–650
Horváth IS, Sjoede A, Alriksson B, Jönsson LJ, Nilvebrant NO (2005) Critical conditions for improved fermentability during overliming of acid hydrolysates from spruce. Appl Biochem Biotechnol 121–124:1031–1044
Jain RK, Sjostedt M, Glasser WG (2000) Thermoplastic xylan derivatives with propylene oxide. Cellulose 7(4):319–336
Kadla JF, Kubo S, Venditti RA, Gilbert RD, Compere AL, Griffith W (2002) Lignin-based carbon fibers for composite fiber applications. Carbon 40:2913–2920
Katofsky R, Consonni S, Larson ED (2003) A cost-benefit analysis of black liquor gasification combined cycle systems. In: Proceedings of the TAPPI fall technical conference: engineering, pulping & PCE&I, Chicago, p 22
Kignell JE (1989) Process for chemicals and energy recovery from waste liquors. US Patent 4,808,264
Kim KH (2005) Two-stage dilute acid-catalyzed hydrolytic conversion of softwood sawdust into sugars fermentable by ethanologenic microorganisms. J Sci Food Agric 85(14):2461–2467
Klinke HB, Thomsen AB, Ahring BK (2004) Inhibition of ethanol-producing yeast and bacteria by degradation products produced during pre-treatment of biomass. Appl Microbiol Biotechnol 66(1):10–26
Kubikova J, Zemann A, Krkoska P, Bobleter O (1996) Hydrothermal pretreatment of wheat straw for the production of pulp and paper. Tappi J 79:163–169
Kuyper M, Hartog MMP, Toirkens MJ, Almering MJH, Winkler AA, van Dijken JP, Pronk JT (2005a) Metabolic engineering of a xylose-isomerase-expressing *Saccharomyces cerevisiae* strain for rapid anaerobic xylose fermentation. FEMS Yeast Res 5(4–5):399–409
Kuyper M, Toirkens MJ, Diderich JA, Winkler AA, van Dijken JP, Pronk JT (2005b) Evolutionary engineering of mixed-sugar utilization by a xylose-fermenting *Saccharomyces cerevisiae* strain. FEMS Yeast Res 5(10):925–934
Larsen E, Kreutz T, Consonni S (1998) Performance and preliminary economics of black liquor gasification combined cycles for a range of Kraft pulp mill sizes. In: International chemical recovery conference, Tampa, FL, 1–4 June 1998, vol 2, pp 675–692
Larsen E, Consonni S, Katofsky R (2003) A cost-benefit assessment of biomass gasification power generation in the pulp and paper industry. Final report, Princeton Environmental Institute, 8 Oct 2003
Larson GW, McDonald ED, Yang W, Frederick WJ, Iisa K, Kreutz TG, Malcolm EW, Brown CA (2000) A cost-benefit assessment of BLGCC technology. Tappi J 83(6):1–15
Larsson S, Palmqvist E, Hahn-Hägerdal B, Tengborg C, Stenberg K, Zacchi G, Nilvebrant NO (1999) The generation of fermentation inhibitors during dilute acid hydrolysis of softwood. Enzyme Microbiol Technol 24(3/4):151–159
Lazzaroni MJ, Bush D, Brown JS, Eckert CA (2005) High pressure vapor and liquid equilibria of some carbon dioxide and organic binary systems. J Chem Eng Data 50(1):60–65
Lennholm B (2007) Lignin from the pulp mills' black liquor: new biofuel with promising potential, Nord. Papperstidn. no. 6, June, p 16
Lesutis HP, Gläser R, Griffith K, Liotta CL, Eckert CA (2001) Near critical water: a benign medium for catalytic reactions. Ind Eng Chem Res 40:6063–6067

Li X, Simonsen J, Li K (2004) Wood dissolution and the regeneration of its components using ionic liquids. In: 227th American chemical society national meeting abstracts, Anaheim, CA

Lindblom M (2003) An overview of Chemrec process concepts. In: 6th international colloquium on black liquor combustion and gasification, Park City, Utah, 13–16 May 2003

Lindblom M (2006) Chemrec pressurized black liquor gasification – status and future plans. In: 7th international colloquium on black liquor combustion and gasification, Jyväskylä, Finland, 31 July to 2 Aug 2006

Lora JH, Wayman M (1978) Delignification of hardwoods by autohydrolysis and extraction. Tappi J 61:47–50

Lu J, Lazzaroni MJ, Hallett JP, Bommarius AS, Liotta CL, Eckert CA (2004) Tunable solvents for homogeneous catalyst recycle. Ind Eng Chem Res 43(7):1586–1590

Lundqvist J, Jacobs A, Palm M, Zacchi G, Dahlman O, Stålbrand H (2002) Characterization of galactoglucomannan extracted from spruce (picea abies) by heat-fractionation at different conditions. Carbohyd Polym 51(2):203–211

Mabee WE, Gregg DJ, Saddler JN (2005) Assessing the emerging biorefinery sector in Canada. Appl Biochem Biotechnol 121–124:765–777

Mansour MN, Steedman WG, Durai-Swamy K, Kazares RE, Raman TV (1992) Chemical and energy recovery from black liquor by steam reforming. In: International chemical recovery conference, Seattle, WA, 7–11 June 1992

Mansour MN, Durai-Swamy K, Aghamohammadi B (1993) Pulsed combustion process for black liquor gasification. Second Annual Report U.S. DOE Report DOE/CE/40893-T2 (DE94002668)

Mansour MN, Durai-Swamy K, Warren DW (1997) Endothermic spent liquor recovery process. US Patent 5,637,192

Martin N, Anglani N, Einstein D, Khrushch M, Worrell E, Price, LK (2000) Opportunities to improve energy efficiency and reduce greenhouse gas emissions in the U.S. pulp and paper industry. Report, Ernest O. Lawrence Berkeley National Laboratory, July 2000

Mckeough P (2003) Evaluation of potential improvements to BLG technology. In: Colloquium of black liquor combustion and gasification, Park City, Utah, p 12

Middleton T (2006) Steam reforming technology at the Norampac Trenton mil. In: Presentation at IEA meeting, Annex XV black liquor gasification, Washington, NC, 20–22 Feb 2006

Millati R, Edebo L, Taherzadeh MJ (2005) Performance of Rhizopus, Rhizomucor, and Mucor in ethanol production from glucose, xylose, and wood hydrolyzates. Enzyme Microbiol Technol 36(2–3):294–300

Moens L, Khan N (2003) Application of room-temperature ionic liquids to the chemical processing of biomass-derived feedstocks. NATO Science Series, II. Math Phys Chem 92:157–171

Molin U, Teder A (2002) Importance of cellulose/hemicellulose-ratio for pulp strength. Nord Pulp Pap Res 17(1):14–19, 28

Montréal Workshop on Bio-refineries (2005) Capturing Canada's natural advantage, Montréal, QC, 21 Nov 2005

Neumann M (2008) New uses for lignin in the biorefinery of the future. Nord Papp Mass 1:42–43

Newport DG, Rockvam L, Rowbotton R (2004) Black liquor steam reformer start-up at Norainpac. In: Proceedings of TAPPI international chemical recovery conference, South Carolina

Nguyen QA, Tucker MP, Keller FA, Eddy FP (2000) Two-stage dilute-acid pretreatment of softwoods. Appl Biochem Biotechnol 84–86:561–576

Nilsson LJ, Larson ED, Gilbreath KR, Gupta A (1995) Energy efficiency and the pulp and paper industry. ACEEE, Washington

Niu W, Molefe MN, Frost JW (2003) Microbial synthesis of the energetic material precursor 1,2,4-butanetriol. J Am Chem Soc 125:12998

Nolen SA, Liotta CL, Eckert CA, Gläser R (2003) The catalytic opportunities of near-critical water: a benign medium for conventionally acid and base catalyzed organic synthesis. Green Chem 5:663–669

Öhman F (2006) Precipitation and separation of lignin from kraft black liquor. PhD thesis. Chalmers Technical University, Gothenburg, Sweden
Page DH, Seth RS (1985) Strength and chemical composition of wood pulp fibres. In: The 8th fundamental research symposium, Oxford, UK, pp 77–91
Palm M, Zacchi G (2003) Extraction of hemicellulosic oligosaccharides from spruce using microwave oven or steam treatment. Biomacromolecules 4(3):617–623
Palmqvist E, Hahn-Hägerdal B (2000) Fermentation of lignocellulosic hydrolysates. I: Inhibition and detoxification. Bioresour Technol 74(1):17–24
Persson P, Larsson S, Jönsson LJ, Nilvebrant NO, Sivik B, Munteanu F, Thörneby L, Gorton L (2002) Supercritical fluid extraction of a lignocellulosic hydrolysate of spruce for detoxification and to facilitate analysis of inhibitors. Biotechnol Bioeng 79(6):694–700
Ragauskas AJ, Nagy M, Kim DH, Eckert CA, Hallett JP, Liotta CL (2006) From wood to fuels: integrating biofuels and pulp production. Ind Biotechnol 2(1):55–65
Rockvam LN (2001) Black liquor steam reforming and recovery commercialization. In: International chemical recovery conference, Whistler, Canada, 11–14 June 2001
Rodden G (2007) Lignoboost is proving its worth: Wermland paper is in the forefront of biofuel development thanks to an agreement with STFI-Packforsk. Pulp Pap Int 49(8):26–28
Schönberg C, Oksanen T, Suurnäkki A, Kettunen H, Buchert J (2001) The importance of xylan for the strength properties of spruce kraft pulp fibres. Holzforschung 55(6):639–644
Scott RW (1989) Influence of cations and borate on the alkali extraction of xylan and glucomannan from pine pulps. J Appl Polym Sci 38(5):907–914
Senthilkumar V, Gunasekaran P (2005) Bioethanol production from cellulosic substrates: engineered bacteria and process integration challenges. J Sci Ind Res 64(11):845–853
Sreenath HK, Jeffries TW (1999) Production of ethanol from wood hydrolyzate by yeasts. Bioresour Technol 72(3):253–260
Sricharoenchaikul V (2001) Fate of carbon-containing compounds from gasification of kraft black liquor with subsequent catalytic conditioning of condensable organics. PhD Dissertation, Georgia Institute of Technology, 2001
Stigsson L (1998) Chemrec black liquor gasification. In: International chemical recovery conference, Tampa, FL, 1–4 June 1998
Swatloski RP, Spear SK, Holbrey JD, Rogers RD (2002) Dissolution of cellulose with ionic liquids. J Am Chem Soc 124(18):4974–4975
Taherzadeh MJ, Eklund R, Gustafsson L, Niklasson C, Lidén G (1997) Characterization and fermentation of dilute-acid hydrolyzates from wood. Ind Eng Chem Res 36(11):4659–4665
Taherzadeh MJ, Gustafsson L, Niklasson C, Lidén G (2000a) Physiological effects of 5-hydroxymethylfurfural on *Saccharomyces cerevisiae*. Appl Microbiol Biotechnol 53(6):701–708
Taherzadeh MJ, Gustafsson L, Niklasson C, Lidén G (2000b) Inhibition effects of furfural on aerobic batch cultivation of *Saccharomyces cerevisiae* growing on ethanol and/or acetic acid. J Biosci Bioeng 90(4):374–380
Tampier M, Smith D, Bibeau E, Beauchemin PA (2004) Identifying environmentally preferable uses for biomass resources – stage 1 report: identification of feedstock-to-product threads. Report, Envirochem Services Inc., North Vancouver
Thorp B (2005a) Transition of mills to biorefinery model creates new profit streams. Pulp Paper 79(11):35–39
Thorp B (2005b) Biorefinery offers industry leaders business model for major change. Pulp Pap 79(11):35–39
Thorp B, Raymond D (2005) Forest biorefinery could open door to bright future for P&P industry. PaperAge 120(7):16–18
Thorp BA, Thorp BA, Murdock-Thorp LD (2008) A compelling case for integrated biorefineries. http://www.epoverviews.com/oca/Compellingcaseforbiorefineries.pdf. Accessed on Dec. 2010
Tolan JS (2003) Conversion of cellulosic biomass to ethanol using enzymatic hydrolysis. In: 226th American chemical society national meeting abstracts, New York
Tucker P (2002) Changing the balance of power. Solutions 85(2):34–38

Vakkilainen EK, Kankkonen S, Suutela J (2008) Advanced efficiency options: increasing electricity generating potential from pulp mills. Pulp Pap Canada 109(4):14–18

van Heiningen A (2006) Converting a kraft pulp mill into an integrated biorefinery. Pulp Pap Canada 107(6):T141–T146

Wai CM, Gopalan AS, Jacobs HK (2003) An introduction to separations and processes using supercritical carbon dioxide. In: ACS symposium series, 860 (supercritical carbon dioxide), American Chemical Society, pp 2–8

Wallmo H, Theliander H (2007) The Lignoboost process: comments on key-operations. In: International chemical recovery conference: efficiency and energy management, Quebec City, QC, 29 May to 1 June, pp 333–335

Warnqvist B, Delin L, Theliander H, Nohlgren I (2000) Teknisk ekonomisk utvärdering avsvartlutförgasningsprocesser. Värmeforsk service AB, Stockholm

Werpy T, Petersen G (2004) Top value-added chemicals from biomass, volume I: results of screening for potential candidates from sugars and synthesis gas. Pacific NorthProduct west National Laboratory, Aug 2004 http://www.eere.energy.gov/biomass/pdfs/35523.pdf

Whitty K, Baxter L (2001) State of the art in black liquor gasification technology. In: Joint international combustion symposium, Kauai, Hawaii, 9–12 Sep 2001

Whitty K, Nilsson A (2001) Experience from a high temperature, pressurized black liquor gasification pilot plant. In: International chemical recovery conference, Whistler, Canada, 11–14 June 2001

Whitty K, Verrill CL (2004) A historical look at the development of alternative black liquor recovery technologies and the evolution of black liquor gasifier designs. In: International chemical recovery conference, Charleston, SC, 6–10 June 2004

Wising U, Stuart PR (2006) Identifying the Canadian forest biorefinery. Pulp Pap Canada 107(6):25–30

Wright JD, Power AJ (1987) Comparative technical evaluation of acid hydrolysis processes for conversion of cellulose to alcohol. Energy Biomass Wastes 10:949–971

Wyatt VT, Bush D, Lu J, Hallett JP, Liotta CL, Eckert CA (2005) Determination of solvatochromic solubility parameters for the characterization of gas-expanded liquids. J Supercrit Fluids 36(1):16–22

Wyman CE, Goodman BJ (1993) Biotechnology for production of fuels, chemicals, and materials from biomass. Appl Biochem Biotechnol 39–40:41–59

Yanagisawa M, Shibata I, Isogai A (2005) SEC-MALLS analysis of softwood kraft pulp using LiCl/1,3-dimethyl-2-imidazolidinone as an eluent. Cellulose 12(2):151–158

Yang CQ, Lu Y (2000) Text Res J 70(4):359–362

Zaldivar J, Nielsen J, Olsson L (2001) Fuel ethanol production from lignocellulose: a challenge for metabolic engineering and process integration. Appl Microbiol Biotechnol 56(1–2):17–34

Index

P. Bajpai, *Biotechnology for Pulp and Paper Processing*,
DOI 10.1007/978-1-4614-1409-4, © Springer Science+Business Media, LLC 2012

F

G

T

U

V

W

X

FSC
www.fsc.org
MIX
Papier | Fördert
gute Waldnutzung
FSC® C083411

Kolibri 360 GmbH
Ferdinand-Jühlke-Straße 7
99095 Erfurt, Deutschland
produktsicherheit@kolibri360.de